Lecture Notes in Computer Scien

T0237997

Commenced Publication in 1973
Founding and Former Series Editors:
Gerhard Goos, Juris Hartmanis, and Jan van Leeuwen

Roberto Avanzi   Liam Keliher
Francesco Sica (Eds.)

Selected Areas in Cryptography

15th International Workshop, SAC 2008
Sackville, New Brunswick, Canada, August 14-15
Revised Selected Papers

 Springer

Volume Editors

Roberto Avanzi
Faculty of Mathematics, Ruhr University Bochum, Germany
E-mail: Roberto.Avanzi@ruhr-uni-bochum.de

Liam Keliher
Francesco Sica
Department of Mathematics and Computer Science, Mount Allison University
Sackville, New Brunswick, Canada
E-mail:{lkeliher, fsica}@mta.ca

Library of Congress Control Number: 2009933273

CR Subject Classification (1998): E.3, D.4.6, K.6.5, G.1.8, I.1

LNCS Sublibrary: SL 4 – Security and Cryptology

ISSN   0302-9743
ISBN-10   3-642-04158-2 Springer Berlin Heidelberg New York
ISBN-13   978-3-642-04158-7 Springer Berlin Heidelberg New York

springer.com

Typesetting: Camera-ready by author, data conversion by Scientific Publishing Services, Chennai, India
Printed on acid-free paper   SPIN: 12738137   06/3180   5 4 3 2 1 0

Preface

The book in front of you contains the proceedings of SAC 2008, the 15th annual Workshop on Selected Areas in Cryptography. SAC 2008 took place during August 14–15 at Mount Allison University, Sackville, New Brunswick, Canada. This was the first time that SAC was hosted in New Brunswick, and the second time in an Atlantic Canadian province. Previous SAC workshops were held at Queen's University in Kingston (1994, 1996, 1998, 1999, and 2005), at Carleton University in Ottawa (1995, 1997, 2003), at the University of Waterloo (2000, 2004), at the Fields Institute in Toronto (2001), at Memorial University of Newfoundland at St. John's (2002), at Concordia University in Montreal (2006) and at the University of Ottawa (2007).

The intent of the workshop series is to provide a relaxed atmosphere in which researchers in cryptography can present and discuss new work on selected areas of current interest. The SAC workshop series has firmly established itself as an international forum for intellectual exchange in cryptological research.

The responsibility for choosing the venue of each SAC workshop and appointing the Co-chairs lies with the SAC Organizing Board. The Co-chairs then choose the Program Committee in consultation with the Board. Hence, we would like to express our gratitude to the SAC Organizing Board for giving us the mandate to organize SAC 2008, and for their invaluable feedback while assembling the Program Committee.

Starting with 2008, SAC is organized in cooperation with the International Association for Cryptologic Research (IACR). SAC 2008 witnessed two further significant events in the history of SAC. The first one was a revision of the wording of the fixed themes of the workshop. This revision takes into account trends that emerge from the papers presented at the last SAC workshops, while remaining true to the original spirit of the series. The three fixed themes are:

- *Design and analysis of symmetric key primitives and cryptosystems*
- *Efficient implementations of symmetric and public key algorithms*
- *Mathematical and algorithmic aspects of applied cryptology*

Each SAC workshop has a fourth theme which is changed every year. For SAC 2008 this was:

- *Elliptic and hyperelliptic curve cryptography, including theory and applications of pairings*

The second event was a significant increase in the number of submissions. A total of 99 technical papers were submitted to the conference from an international authorship. Of these, 27 were accepted for presentation at the workshop, a slight increase with respect to previous years, while the rate of accepted papers has been reduced. In addition to these 27 papers, two speakers were invited to give presentations at the conference.

- Jacques Patarin gave the Stafford Tavares Lecture on *The "coefficients H" Technique.*
- Joseph Silverman gave a lecture dealing with our fourth theme, on the subject of *Lifting and the Elliptic Curve Discrete Logarithm Problem.*

The Program Committee (PC) for SAC 2008 was also the largest to date, comprising 21 members in addition to the Co-chairs. The reviewing process was a challenging task. Every paper was refereed by at least three reviewers, with papers (partially) co-authored by members of the Program Committee refereed by at least five reviewers. A total of about 310 reviews were written and uploaded by the PC members, who were helped by 107 subreviewers. The reviews were then followed by through discussions on the papers, which contributed in a decisive way to the quality of the final selection. About 300 additional discussion comments were written by the PC members and the Co-chairs, with up to 30 discussion comments per PC member, with some papers receiving up to 20 discussion comments. The reviews were rewritten, taking these discussions into account, before being sent to the authors. In most cases, extensive comments were sent, with one set of comments totalling 3,444 words on 477 lines – the average being about 200 lines of text per submission. Despite the huge amount of work, the atmosphere in the PC was always serene and friendly, even with some lighter moments. For us it was a honor to work with this PC.

We would like to thank the authors of all the submitted papers, both those whose work is included in these proceedings, and those whose work could not be accommodated.

The submission and review process was done using a Web-based software system developed by Shai Halevi. Changes to the system were made to accommodate our needs, and Shai replied to all our questions very quickly. We thank Shai for making his package available and for his help.

All the contributions are given in this volume in the same order as they appeared in the final program. These include revised versions of all 27 accepted submissions and the two papers related to the invited talks.

We had 71 registered participants from the following countries: Austria, Belgium, Canada, Chile, China, France, Germany, Korea, Luxembourg, Japan, The Netherlands, Singapore, Spain, Switzerland, Turkey, UK, and USA.

We also wish to express our gratitude to Mount Allison University and IEEE New Brunswick Section for financial support.

Finally we would like to thank Cindy Allan, Judith Van Rooyen, Stuart MacDonald, and Amy Adsett for helping with the organization, and all the participants of SAC 2008.

April 2009

Roberto Avanzi
Liam Keliher
Francesco Sica

Organization

SAC 2008 was organized by the SAC Organizing Board and by AceCrypt (Atlantic Centre of Excellence for Cryptographic Research), Sackville, New Brunswick, Canada in cooperation with the International Association for Cryptologic Research (IACR).

SAC Organizing Board

Carlisle Adams (Chair)	University of Ottawa, Canada
Roberto Avanzi	HGI – Ruhr Universität Bochum, Germany
Orr Dunkelman	École Normale Supérieure, France
Phil Eisen	Cloakware Corporation, Canada
Helena Handschuh	Spansion, France
Doug Stinson	University of Waterloo, Canada
Mike Wiener	Cryptographic Clarity, Canada
Adam Young	MITRE Corp, USA
Francesco Sica	AceCrypt – Mount Allison University, Canada

Program Committee

Roberto Avanzi (Co-chair)	HGI – Ruhr Universität Bochum, Germany
Paulo Barreto	Universidade de São Paulo, Brazil
Claude Carlet	Université Paris 8, France
Christophe Doche	Macquarie University, Australia
Orr Dunkelman	École Normale Supérieure, France
Joachim von zur Gathen	BIT Bonn, Germany
Elisa Gorla	Universität Zürich, Switzerland
Laurent Imbert	CNRS, PIMS–Europe – University of Calgary, Canada
Marc Joye	Thomson, France
Liam Keliher (Co-chair)	AceCrypt – Mount Allison University, Canada
Kristin Lauter	Microsoft, USA
Gregor Leander	Danmarks Tekniske Universitet, Denmark
Arjen Lenstra	EPFL Lausanne, Switzerland
Stefan Lucks	Bauhaus Universität Weimar, Germany
Tatsuaki Okamoto	NTT Japan, Japan
Roger Oyono	Université de la Polynesie Française
Bart Preneel	Katholieke Universiteit Leuven, Belgium
Vincent Rijmen	Technische Universität Graz, Austria
Francesco Sica (Co-chair)	AceCrypt – Mount Allison University, Canada
Doug Stinson	University of Waterloo, Canada

Nicolas Thériault   Universidad de Talca, Chile
Michael Wiener   Cryptographic Clarity, Canada
Adam Young   MITRE Corp, USA
Amr Youssef   Concordia University, Canada

External Reviewers

Tolga Acar
Laila El Aimani
Davide Alessio
Elena Andreeva
Kazumaro Aoki
Frederik Armknecht
Gilles Van Assche
Jean-Claude Bajard
Côme Berbain
Thierry Berger
Daniel J. Bernstein
    and Tanja Lange
Andrey Bogdanov
Guilhem Castagnos
Li Chao
Pierre-Louis Cayrel
Jean-Sébastien Coron
Nicolas Courtois
Joan Daemen
Jérémie Detrey
Vassil Dimitrov
Yevgeniy Dodis
Thomas Dullien
Matthieu Finiasz
Ewan Fleischmann
Pierre-Alain Fouque
Julien Francq
Benedikt Gierlichs
Guang Gong
Michael Gorski
Louis Granboulan
Johann Großschädl
Sylvain Guilley
Tim Güneysu
Helena Handschuh
Darrel Hankerson
Martin Hell

Peter Hellekalek
Kevin Henry
Miia Hermelin
Mathias Herrmann
Sebastiaan Indesteege
Tetsu Iwata
Thomas Johansson
Pascal Junod
Timo Kasper
Ulrich Kühn
Alexandre Karlov
Shahram Khazaei
Aleksandar Kircanski
Patrick Lacharme
Mario Lamberger
Kerstin Lemke-Rust
Gaëtan Leurent
Benoît Libert
Daniel Loebenberger
Subhamoy Maitra
Stéphane Manuel
Kerry McKay
Willi Meier
Florian Mendel
Alfred Menezes
Tomislav Nad
Mridul Nandi
Christophe Negre
Michael Nüsken
Katsuyuki Okeya
Francis Olivier
Sıddıka Berna Örs
Onur Özen
Dan Page
Jacques Patarin
Maura Paterson
Josef Pieprzyk

Axel Poschmann
Emmanuel Prouff
Michael Quisquater
Håvard Raddum
Christian Rechberger
Tom Ristenpart
Christophe Ritzenthaler
Matt Robshaw
Robert Rolland
Neyire Deniz Sarier
Juraj Sarinay
Palash Sarkar
Martin Schlaeffer
Peter Schwabe
Taizo Shirai
Igor Shparlinski
Hervé Sibert
Andrey Sidorenko
Martijn Stam
François-Xavier
    Standaert
Dirk Stegemann
Ron Steinfeld
Marc Stevens
Christine Swart
Stefano Tessaro
Arnaud Tisserand
Yukiyasu Tsunoo
Berkant Ustoaglu
Frederik Vercauteren
Damien Vergnaud
Jiang Wu
Brecht Wyseur
Jens Zumbrägel

Table of Contents

Second Invited Talk – Stafford Tavares Lecture

Mathematical Aspects of Applied Cryptography II

Curve-Based Primitives in Hardware

Block Ciphers II

Faster Halvings in Genus 2

Peter Birkner[1] and Nicolas Thériault[2,*]

[1] Department of Mathematics and Computer Science, Coding Theory and
Cryptology, Eindhoven University of Technology,
P.O. Box 513, 5600 MB Eindhoven, The Netherlands
`p.birkner@tue.nl`

[2] Instituto de Matemática y Física, Universidad de Talca, Casilla 747, Talca, Chile
`ntheriau@inst-mat.utalca.cl`

Abstract. We study divisor class halving for hyperelliptic curves of
genus 2 over binary fields. We present explicit halving formulas for the
most interesting curves (from a cryptographic perspective), as well as all
other curves whose group order is not divisible by 4. Each type of curve
is characterized by the degree and factorization form of the polynomial
$h(x)$ in the curve equation. For each of these curves, we provide explicit
halving formulæ for all possible divisor classes, and not only the most
frequent case where the degree of the first polynomial in the Mumford
representation is 2. In the optimal performance case, where $h(x) = x$,
we also improve on the state-of-the-art and when $h(x)$ is irreducible of
degree 2, we achieve significant savings over both the doubling as well as
the previously fastest halving formulas.

Keywords: hyperelliptic curve, genus 2, halving, binary field.

1 Introduction

The double-and-add algorithm is essential to the efficiency of cryptosystems
based on hyperelliptic curves. This algorithm (and many of its variations) is
based on two basic group operations: the addition of two distinct group elements
and the computation of the double of an element. An alternative that proved
very successful in elliptic curves over binary fields is the halve-and-add algorithm,
which relies on the computation of the "half" of a group element (of odd order),
i.e. the computation of a pre-image of the doubling operation [9,14]. Given the
important savings produced by this approach for elliptic curves, it is natural to
ask if similar results can be obtained for hyperelliptic curves over binary fields.

In a double-and-add algorithm, we can use explicit formulæ for the most common cases of the doubling and the addition, going back to Cantor's polynomial-
based algorithm if any special cases are encountered. In practice, using explicit
formulæ for the special cases has no measurable impact on the average performance of a scalar multiplication, so it is not essential to develop them. The same

* The authors would like to thank the Fields Institute in Toronto for supporting this
work. Research was partially supported by FONDECYT (Chile) grants #1070242
and by the Programa Reticulados y Ecuaciones of the Universidad de Talca.

R. Avanzi, L. Keliher, and F. Sica (Eds.): SAC 2008, LNCS 5381, pp. 1–17, 2009.
© Springer-Verlag Berlin Heidelberg 2009

is not true for the halve-and-add algorithm however, since we cannot easily describe the halving operation in terms of polynomial arithmetic. A halve-and-add algorithm should therefore contain explicit formulæ for all possible cases of the halving operation.

In this paper we investigate halving of divisor classes of hyperelliptic curves of genus 2 over finite fields of characteristic 2. Doubling formulæ for the different types of curves can be found in [10,11,1] and halving for some types of curves and/or cases have also been investigated in [7,8,3]. We present halving formulæ for all cases (most frequent and special cases) for all curves having at most one divisor of order 2 and no divisor of order 4 (see the second half of Section 2.2 for the reasons behind this condition). Rather than inverting the best doubling formulæ, we invert Cantor's algorithm, using the divisibility condition on semi-reduced divisors to lower the cost. In the optimal performance case, where $h(x) = x$, we also improve on the state-of-the-art [3] and when $h(x)$ is irreducible of degree 2, we achieve significant savings over both the doubling [1] as well as the currently fastest halving formulas [7].

The remainder of this paper is structured as follows: Section 2 contains some important terminology and background. In Section 3, we develop a complete case study of the halving formulæ in the most efficient curves for cryptographic application, and in Section 4 we do the same for the most general type of curves where halving is of interest. For completeness, full sets of formulæ for the remaining types of curves where halving can be efficient are presented in the appendix, as well as the addition formula for the form of curve equation used in Section 4.

2   Basic Notations and Preliminaries

In this section we briefly recall the definitions of hyperelliptic curves, divisor class groups and the Mumford representation since we will use these notions throughout the whole paper.

A comprehensive resource for the mathematics of finite fields is [12]. For background on hyperelliptic curves we refer the interested reader to [1], from which the following definitions and notations are taken.

Definition 1 (Hyperelliptic curve of genus 2 in characteristic 2). *Let* $\mathbb{F}_q$ *be a field of characteristic 2 and* $\overline{\mathbb{F}}_q$ *its algebraic closure. A curve* C*, given by an equation of the form*

$$C : y^2 + h(x)y = f(x), \tag{1}$$

where $f \in \mathbb{F}_q[x]$ *is a polynomial of degree 5 and* $h \in \mathbb{F}_q[x]$ *is a non-zero polynomial of degree at most 2, is called an* imaginary *hyperelliptic curve of genus 2 over* $\mathbb{F}_q$ *if there is no point* (x, y) *on the curve over* $\overline{\mathbb{F}}_q$ *for which both partial derivatives are 0, i.e. such that* $h(x) = 0$ *and* $f'(x) - h'(x)y = 0$ *(This last condition ensures that the affine curve is non-singular).*

Definition 2 (Divisor class group). *Given a hyperelliptic curve* C *of genus 2 over a binary field* $\mathbb{F}_q$*, the group of degree 0 divisors of* C *is denoted by* Div_C^0*.*

The quotient group of Div_C^0 *by the group of principal divisors of* C *is called the* divisor class group of C *and is denoted by* Pic_C^0. *It is also called the* Picard group of C.

Theorem 1 (Mumford). *Let* C *be a hyperelliptic curve of genus* 2 *over a binary field* $\mathbb{F}_q$. *Each nontrivial divisor class of* C *over* $\mathbb{F}_q$ *can be represented by a unique pair of polynomials* $u, v \in \mathbb{F}_q[x]$, *where*

1. u *is monic,*
2. $\deg v < \deg u \leq 2$,
3. $u \mid v^2 + vh - f$.

A divisor satisfying Theorem 1 is called reduced (i.e. it is the shortest representative of its class), and if the condition $\deg(u) \leq 2$ is removed, the divisor is called semi-reduced. The divisibility condition will be essential in establishing some of the halving formulæ. Our halving formulæ expect the input divisor class to be in Mumford representation, work directly on the coefficients of the polynomials u and v and return an output in Mumford form. Since our goal is to compute pre-images of the group doubling, we refer to Algorithm 1 for a description of how this operation is performed using Cantor's algorithm.

Algorithm 1. Cantor's/Koblitz's doubling algorithm for genus 2 HEC in characteristic 2

INPUT: Reduced divisor $D = [u_a(x), v_a(x)]$
OUTPUT: Reduced divisor $D' = [u_c(x), v_c(x)]$, $D' = [2]D$

1: $d \leftarrow \gcd(u_a, h)$, $u_0 \leftarrow u_a/d$, $v_0 \leftarrow v_a \bmod u_0$
2: $c \leftarrow h^{-1} \bmod u_0$, $u_1 \leftarrow u_0^2$, $v_1 \leftarrow v_0 + c(v_0^2 + v_0 h + f) \bmod u_1$
3: **if** $\deg(u_1) \leq 2$ **then**
4:     $u_c \leftarrow u_1$, $v_c \leftarrow v_1$
5: **else**
6:     $u_c \leftarrow \text{monic}\left(\frac{f+v_1 h+v_1^2}{u_1}\right)$, $v_c \leftarrow v_1 + h \bmod u_c$
7: **end if**
8: **return** $[u_c, v_c]$

We observe that all the pairs of polynomials computed in Algorithm 1 satisfy the divisibility condition of Theorem 1, i.e. $u_0 | v_0^2 + v_0 h + f$, $u_1 | v_1^2 + v_1 h + f$, and $u_c | v_c^2 + v_c h + f$. To obtain our halving formulæ, we sometime use the identities coming from these divisibility conditions rather than those which are more obvious in the polynomial equalities of Algorithm 1. In exchange, any identity which is not used to perform the halving becomes a divisibility condition that must be satisfied for $[u_i(x), v_i(x)]$ to be a semi-reduced divisor.

2.1   Field Arithmetic and Divisor Halving

Throughout this paper, we will assume that the field $\mathbb{F}_q$ has order 2^n where n is not divisible by 2 or 3. This is mainly due to security concerns, since various versions of the Weil descent attack could be applied when n admits a factor of 2 or 3 (for example, see [6,15,5]). In fact, for cryptographic applications it is often assumed that n is a prime. As an added bonus, having n coprime to 2 means that we can take cube and 5-th roots in the field (since α^3 and α^5 are both isomorphism as 3 and 5 are coprime to $2^n - 1$), which allows us to simplify the curve equations a little more. Furthermore, since n will be odd we will have $\mathrm{Tr}(1) = 1$. In various places, we implicitly take advantage of the identity $\mathrm{Tr}(\alpha) = \mathrm{Tr}(\alpha^2)$ to simplify some trace computations.

In finite fields of characteristic 2, some operations which are very expensive in fields of odd characteristic become very efficient, in particular computing the roots of a quadratic equation (when they are available in the field of definition). This observation led to the development of *halve-and-add* algorithms, a variation of the double-and-add scalar multiplication where the doubling operation is replaced with a *halving* (the representation of the scalar is adjusted accordingly). Such an approach was first used for elliptic curves [9,14], and was recently extended to hyperelliptic curves of genus 2 [7,8,3]. In some fields, computing of square roots can be faster than the computation of squares [4,2]. Two other operations, the trace (from $\mathbb{F}_{2^n}$ to $\mathbb{F}_2$) and the half-trace (HT, to solve quadratic equations, see [4,2]), can also be implemented to have similar costs to the squaring operation. For curves over those fields, it can be a good strategy to "replace" squares with square roots in the group arithmetic, which is often what halving does.

To count the number of operations, we denote inverses by I, multiplications by M, squares by S, square roots by SR, traces by TR, and half-traces by HT.

2.2   Choices of Curves

An imaginary hyperelliptic curve of genus 2 over $\mathbb{F}_{2^d}$ is of the form

$$y^2 + (h_2x^2 + h_1x + h_0)y = f_5x^5 + f_4x^4 + f_3x^3 + f_2x^2 + f_1x + f_0, \qquad (2)$$

where $h(x) = h_2x^2 + h_1x + h_0 \neq 0$. It is also customary to use isomorphisms to impose that $f(x)$ is monic, i.e. that $f_5 = 1$, but we will relax this condition for some curves as the halving formulæ are more efficient if we use the isomorphisms to force $h_1 = 1$. We also note that when $h(x)$ is constant (i.e. $h_2 = h_1 = 0$), the curve is known to be supersingular, and therefore of limited interest for cryptography, but we will still cover these curves for completeness.

For the curve (2), the possible isomorphisms are given by $x = \alpha\tilde{x} + \beta$ and $y = \gamma\tilde{y} + \delta\tilde{x}^2 + \epsilon\tilde{x} + \zeta$, where both α and γ are nonzero, after which the equation is divided by γ^2 to get the coefficient of y^2 back to 1. We distinguish five types of curves depending on the degree and factorization type of $h(x)$:

Ia $h_2 \neq 0$ and $h(x)$ irreducible: Using α and γ we can force $h_2 = h_1 = 1$. We can then use β to restrict $h(x)$ to $x^2 + x + 1$. The remaining freedom on β allows us to impose $\mathrm{Tr}(f_5) \cdot \mathrm{Tr}(f_4) = 0$. Taking advantage of δ, we can restrict f_4 to $\{0,1\}$ and then ϵ and ζ allow us to remove f_3 and f_2. We are left with $f(x) = f_5 x^5 + f_4 x^4 + f_1 x + f_0$ where $f_4 \in \mathbb{F}_2$ and $f_4 \cdot \mathrm{Tr}(f_5) = 0$.

Ib $h_2 \neq 0$ and $h(x)$ is the product of two distinct linear factors: Using β and one of the roots of $h(x)$, we can force $h_0 = 0$. We can then use α and γ to restrict $h(x)$ to $x^2 + x$. The remaining freedom on β allows us to impose $\mathrm{Tr}(f_5) \cdot \mathrm{Tr}(f_4) = 0$. Taking advantage of δ, we can restrict f_4 to $\{0,1\}$ and then ϵ and ζ allow us to remove f_3 and f_2. We are left with $f(x) = f_5 x^5 + f_4 x^4 + f_1 x + f_0$ where $f_4 \in \mathbb{F}_2$ and $f_4 \cdot \mathrm{Tr}(f_5) = 0$.

Ic $h_2 \neq 0$ and $h(x)$ a square: Using α and β and γ, we can force $h(x) = x^2$ and make $f(x)$ monic. ϵ and ζ allow us to remove f_3 and f_2. Finally, δ can be used to limit f_4 to $\{0,1\}$, leaving us with $f(x) = x^5 + f_4 x^4 + f_1 x + f_0$ with $f_4 \in \mathbb{F}_2$.

II $h_2 = 0$, $h_1 \neq 0$: Using α and β and γ, we can force $h(x) = x$ and make $f(x)$ monic. δ and ζ allow us to remove f_4 and f_1. Finally, we restrict f_2 using ϵ, giving us $f(x) = x^5 + f_3 x^3 + f_2 x^2 + f_0$ with $f_2 \in \mathbb{F}_2$.

III $h_2 = h_1 = 0$: Using α and γ, we can force $h(x) = 1$ and make $f(x)$ monic. By selecting δ, ϵ and β wisely, we can remove f_4 and f_2 and reduce the number of possible values of f_1 (in general to a set of at most 16 values). Finally, ζ can be used to limit f_0 to $\{0,1\}$, leaving us with $f(x) = x^5 + f_3 x^3 + f_1 x + f_0$ with $f_0 \in \mathbb{F}_2$.

Note that we did not include the non-singularity condition, nor conditions on the group order in the descriptions of the different types. In terms of isomorphism classes, types Ia and Ib are the most common (each with $\frac{3}{2}q^3 + O(q^2)$ classes), followed by types II and Ic (each with $2q^2 + O(q)$ classes) and with type III (supersingular) the less common ($O(q)$ classes). From the point of view of the 2-torsion group, type Ic is closer to type II than type Ia and Ib.

We limit ourselves to curves for which the order of the Jacobian is either odd ($h(x)$ constant) or 2 times an odd number (which eliminates all type Ib curves). This restriction is needed to get a better performance out of the halving. Given any hyperelliptic curve, the halve-and-add algorithm allows us to compute the scalar multiple of a divisor class, given that it is in a (sub)group of odd order. In this way, the pre-image of the doubling can always be computed and "becomes" unique (all other pre-images of the doubling have even order). The group order conditions are due to the following reasons:

1. To verify that the pre-image is in the subgroup of odd order, we make sure that it can be halved again as many times as we want. If the group contains divisors of order 2^r, we must make sure that we can halve the pre-image

(at least) r times, which obviously affects the cost of our halving formulæ. When $r \geq 2$ (i.e. when there are divisors of order 4), the increased work required for this check becomes too expensive for the halving to be interesting.

2. If C is of type Ib, there are four possible pre-images of the doubling. The halving formula must then distinguish which of the four is in the subgroup of odd order, which significantly increases the cost of the halving. We also computed formulæ in this case, and the halving does indeed become much more expensive than the doubling.

When we consider all the isomorphisms classes for a given type of curves (other than type III), between a half and two thirds of them have divisors of order 4, so rejecting these curves has an acceptably small impact on the number of possible curves. Furthermore, because of the attack of Pohlig and Hellman [13], curves with divisors of order 4 are in general (slightly) weaker than those without, so the restriction can be seen as advantageous for the security of the curves. From a cryptographic perspective, the two most interesting types of curves for halving formulæ are type II (most efficient halving) and type Ia (largest number of isomorphism classes). In terms of the benefits of halving over doubling, type Ia gives the best savings.

3   Type II: $h(x) = x$

In this section, the curve C is of the form

$$y^2 + xy = x^5 + f_3 x^3 + f_2 x^2 + f_0, \qquad f_2 \in \mathbb{F}_2.$$

Theorem 2. *Let $D_a = [u_a, v_a]$ be a divisor in Div_C^0. If $\deg(u_a) = 2$, then D_a can be halved if and only if $\mathrm{Tr}(u_{a1}(u_{a0} + f_3 + u_{a1}^2)) = 0$. If $\deg(u_a) = 1$, then D_a can be halved if and only if $\mathrm{Tr}(f_2 + u_{a0}(u_{a0}^2 + f_3)) = 0$.*

Proof. To obtain these trace conditions, we try to invert the explicit formulæ for each of the possible cases of doubling (matching the outputs to the different forms of D_a). We do this by solving a sequence of equations to obtain a divisor D_c such that $[2]D_c = D_a$. We first observe that linear or square equations ($w^2 = \alpha$) pose no problem, whereas quadratic equations of the form $w^2 + w + \alpha = 0$ can be solved if and only if $\mathrm{Tr}(\alpha) = 0$. If $\deg(u_a) = 1$, then we only encounter one quadratic (non-square) equation, with $\alpha = f_2 + u_{a0}(u_{a0}^2 + f_3)$. If $\deg(u_a) = 2$ and $u_{a1} = 0$ (which obviously satisfies $\mathrm{Tr}(u_{a1}(u_{a0} + f_3 + u_{a1}^2)) = 0$), all equations are either linear or square. Finally, if $\deg(u_a) = 2$ and $u_{a1} \neq 0$, then we encounter one quadratic (non-square) equation, with $\alpha = u_{a1}(u_{a0} + f_3 + u_{a1}^2)$.

To show that the trace conditions are sufficient for the existence of a pre-image of the doubling, we show that D_c is indeed a valid divisor (i.e. the pair of polynomials computed correspond to a real divisor). In the computation of $D_c = [u_c(x), v_c(x)]$, we ignored a number of identities (essentially divisibility conditions) which are easily shown to be direct consequences of the divisibility conditions on D_a.

From Theorem 2, we obtain a simple condition for the group order:

Corollary 1. *The Jacobian of the curve C has order $2 \cdot$odd if and only if $f_2 = 1$.*

Proof. The curve C has exactly one divisor of order 2, $[x, \sqrt{f_0}]$. The group order is divisible by 4 if and only if $[x, \sqrt{f_0}]$ can be halved. From Theorem 2, this is possible if and only if $\mathrm{Tr}(f_2) = 0$ and since $f_2 \in \mathbb{F}_2$ we find that C has a divisor of order 4 if and only if $f_2 = 0$.

Let D_{c_1} and D_{c_2} be the two pre-images of D_a, then $D_{c_1} - D_{c_2} = [x, \sqrt{f_0}] = D_{c_2} - D_{c_1}$. From this observation, we get the following corollary which allow us to distinguish the different special cases.

Corollary 2. *Let $D_a = [u_a(x), v_a(x)]$ be a divisor in Div_C^0 that can be halved and $D_c = [u_c(x), v_c(x)] = [\frac{1}{2}]D_a$ its pre-image (under the doubling) of odd order. If $\deg(u_a) = 2$ and $u_{a1} \neq 0$, then $\deg(u_c) = 2$ ("HLV22"). If $\deg(u_a) = 2$ and $u_{a1} = 0$, then $\deg(u_c) = 1$ or 2 (with $u_{c0} = 0$ in the second case) ("HLV21/22"). If $\deg(u_a) = 1$, then $\deg(u_c) = 2$ ("HLV12").*

Finally, to verify that we are computing the pre-image of odd order, we use the conditions of Theorem 2 on $u_c(x)$ to ensure that it could be halved again (i.e. that it has odd order), and we correct the computations if necessary. We obtain the following formulæ:

Algorithm 2. (HLV22, $h(x) = x$, $f(x) = x^5 + f_3 x^3 + x^2 + f_0$)

INPUT: The divisor class $\overline{D}_a = [x^2 + u_{a1}x + u_{a0}, v_{a1}x + v_{a0}]$

OUTPUT: The divisor class $[x^2 + u_{c1}x + u_{c0}, v_{c1}x + v_{c0}] = [1/2]\overline{D}_a$

1:   $s_0 \leftarrow \sqrt{u_{a1}}$, $s_1 \leftarrow 1/s_0$, $s_2 \leftarrow s_1^2$ ▷ 1I+1S+1SR

2:   $s_3 \leftarrow \mathrm{HT}(u_{a1}(u_{a0} + f_3 + s_0))$, $s_4 \leftarrow s_3 s_2$, $s_5 \leftarrow s_4 + f_3$ ▷ 2M+1HT

3:   $s_6 \leftarrow v_{a0} + u_{a0}(s_4 + s_0)$, $s_7 \leftarrow s_6$, $u_{c1} \leftarrow \sqrt{s_5}$ ▷ 1M+1SR

4:   **if** $\mathrm{Tr}(s_5(s_7 + f_3^2 + u_{c1})) = 1$ **then**   ▷ 1M+1TR

5:       $s_3 \leftarrow s_3 + 1$, $s_4 \leftarrow s_4 + s_2$, $s_5 \leftarrow s_5 + s_2$

6:       $s_6 \leftarrow s_6 + s_2 u_{a0}$, $s_7 \leftarrow s_6$, $u_{c1} \leftarrow u_{c1} + s_1$ ▷ 1M

7: **end if**

8: $u_{c0} \leftarrow \sqrt{s_7}$, $v_{c0} \leftarrow \sqrt{s_7 s_1 + f_0}$ ▷ 1M+2SR

9: $v_{c1} \leftarrow v_{a1} + 1 + s_3 + s_1(u_{a0} + u_{a1}^2 + u_{c0} + s_5) + s_4 u_{c1}$ ▷ 2M+1S

10: **return** $[x^2 + u_{c1}x + u_{c0}, v_{c1}x + v_{c0}]$ ▷ 1I+8M+2S+4SR+1HT+1TR

11:                   ▷ Average: 1I+7.5M+2S+4SR+1HT+1TR

Algorithm 3. (HLV12, $h(x) = x$, $f(x) = x^5 + f_3x^3 + x^2 + f_0$)

INPUT: The divisor class $\overline{D}_a = [x + u_{a0}, v_{a0}]$

OUTPUT: The divisor class $[x^2 + u_{c1}x + u_{c0}, v_{c1}x + v_{c0}] = [1/2]\overline{D}_a$

1: $s_0 \leftarrow \sqrt{u_{a0}}$, $s_1 \leftarrow f_3 + s_0$, $s_2 \leftarrow 1 + s_1u_{a0}$	▷ 1M+1SR
2: $s_3 \leftarrow \mathrm{HT}(s_2)$, $s_4 \leftarrow v_{a0} + u_{a0}(s_3 + 1 + s_0u_{a0})$	▷ 2M+1HT
3: $u_{c1} \leftarrow \sqrt{s_1}$, $u_{c0} \leftarrow \sqrt{s_4}$	▷ 2SR
4: **if** $\mathrm{Tr}(u_{c1}(u_{c0} + s1 + f_3)) = 1$ **then**	▷ 1M+1TR
5: $s_3 \leftarrow s_3 + 1$, $s_4 \leftarrow s_4 + u_{a0}$, $u_{c0} \leftarrow \sqrt{s_4}$	▷ 1SR
6: **end if**	
7: $v_{c1} \leftarrow s_3 + s_0u_{c1}$, $v_{c0} \leftarrow s_4 + s_0u_{c0}$	▷ 2M
8: **return** $[x^2 + u_{c1}x + u_{c0}, v_{c1}x + v_{c0}]$	▷ 6M+4SR+1TR+1HT
9:	▷ Average: 6M+3.5SR+1TR+1HT

Algorithm 4. (HLV21/22, $h(x) = x$, $f(x) = x^5 + f_3x^3 + x^2 + f_0$)

INPUT: The divisor class $\overline{D}_a = [x^2 + u_{a0}, v_{a1}x + v_{a0}]$

OUTPUT: The divisor class $[x + u_{c0}, v_{c0}] = [1/2]\overline{D}_a$
           or $[x^2 + u_{c1}x, v_{c1}x + v_{c0}] = [1/2]\overline{D}_a$

1: $s_0 \leftarrow \sqrt{u_{a0}}$, $s_1 \leftarrow v_{a0} + s_0v_{a1}$	▷ 1M+1SR
2: **if** $\mathrm{Tr}(u_{a0}s_0) = 1$ **then**	▷ 1M+1TR
3: $\overline{E} \leftarrow [x + s_0, s_1]$	
4: **else**	
5: $s_2 \leftarrow 1/s_0$, $v_{c1} \leftarrow (s_1 + \sqrt{f_0})s_2$	▷ 1I+1M
6: $v_{c0} \leftarrow s_1 + v_{c1}s_0$, $\overline{E} \leftarrow [x^2 + s_0x, v_{c1}x + v_{c0}]$	▷ 1M
7: **end if**	
8: **return** $\overline{E}$	▷ 1I+4M+1SR+1TR
9:	▷ Average: 0.5I+3M+1SR+1TR

4   Type Ia: $h(x) = x^2 + x + 1$

In this section, the curve C is of the form $y^2 + (x^2 + x + 1)y = f_5x^5 + f_4x^4 + f_1x + f_0$ with $f_4 \in \mathbb{F}_2$ and $f_4 \cdot \mathrm{Tr}(f_5) = 0$. To improve the efficiency of the formulæ, we will assume that f_5^{-1} and f_5^{-2} are precomputed (only once per curve).

Theorem 3. *Let $D_a = [u_a(x), v_a(x)]$ be a divisor in Div_C^0. If $\deg(u_a) = 2$, then D_a can be halved if and only if $\mathrm{Tr}(u_{a1}f_5) = 0$. If $\deg(u_a) = 1$, then D_a can be halved if and only if $\mathrm{Tr}(f_4 + f_5 u_{a0}) = 0$.*

Proof. To prove this, we can follow the same ideas as in Theorem 2, with one difference: some of the halving formulæ require to solve two quadratic (non-square) equations rather than one. In those formulæ, it is easy to verify that switching between the roots of the first quadratic equation (adding one to the half-trace) changes the trace of the constant term of the second quadratic equation by 1. The choice of root for the first equation (when roots exist in $\mathbb{F}_q$) and the trace condition from the second equation are then purely internal to the halving formula.

From this theorem, we obtain a simple condition for the group order:

Corollary 3. *The Jacobian of the curve C has order $2 \cdot odd$ if and only if $\mathrm{Tr}(f_5) = 1$ and $f_4 = 0$.*

Proof. The curve C has exactly one divisor of order 2, which is of the form $[x^2 + x + 1, v_h]$. The group order is divisible by 4 if and only if $[x^2 + x + 1, v_h]$ can be halved. From Theorem 3, this is possible if and only if $\mathrm{Tr}(f_5) = 0$. The condition $f_4 = 0$ is then direct from the restriction $f_4 \cdot \mathrm{Tr}(f_5) = 0$.

Let D_{c_1} and D_{c_2} the two pre-images of D_a, then their difference is $[x^2 + x + 1, v_h]$. From this observation, we get the following corollary which allow us to distinguish the different special cases:

Corollary 4. *Let $D_a = [u_a(x), v_a(x)]$ be a divisor in Div_C^0 that can be halved and $D_c = [u_c(x), v_c(x)] = [\frac{1}{2}]D_a$ its pre-image (under the doubling) of odd order. If $\deg(u_a) = 2$ and $u_{a1} \neq 0$, then $\deg(u_c) = 2$ ("HLV22"). If $\deg(u_a) = 2$ and $u_{a1} = 0$, then $\deg(u_c) = 1$ or 2 ("HLV21/22"). If $\deg(u_a) = 1$, then $\deg(u_c) = 2$ ("HLV12").*

Finally, to verify that we are computing the pre-image of odd order, we use the conditions of Theorem 3 on $u_c(x)$ to ensure it could be halved again (i.e. that it has odd order), and we correct the computations if necessary. We obtain the following formulæ:

Algorithm 5. (HLV22, $h(x) = x^2 + x + 1$, $f(x) = f_5 x^5 + f_1 x + f_0$)

INPUT: The divisor class $\overline{D} = [x^2 + u_{a1}x + u_{a0}, v_{a1}x + v_{a0}]$,
           and f_5^{-2} precomputed

OUTPUT: The divisor class $[x^2 + u_{c1}x + u_{c0}, v_{c1}x + v_{c0}] = [1/2]\overline{D}$

1: $s_0 \leftarrow f_5 u_{a1}$, $s_1 \leftarrow u_{a1}^{-1}$, $s_2 \leftarrow \mathrm{HT}(s_0)$, $s_3 \leftarrow s_2 s_1$ ▷ 1I+2M+1HT

2: $s_4 \leftarrow s_2 u_{a1}$, $s_5 \leftarrow f_5 u_{a0} + s_2 + s_3 + v_{a1} + 1 + u_{a1} + s_4$ ▷ 2M

3: $s_6 \leftarrow s_5 u_{a1}$ ▷ 1M

4: **if** $\text{Tr}(s_6) = 1$ **then**   ▷ 1TR

5:    $s_2 \leftarrow s_2 + 1,\ s_3 \leftarrow s_3 + s_1,\ s_4 \leftarrow s_4 + u_{a1}$

6:    $s_5 \leftarrow s_5 + 1 + s_1 + u_{a1},\ s_6 \leftarrow s_5 u_{a1}$ ▷ 1M

7: **end if**

8: $s_8 \leftarrow \text{HT}(s_6),\ s_9 \leftarrow s_8 s_1$ ▷ 1M+1HT

9: $s_{10} \leftarrow s_3 u_{a0} + s_4 + s_8 + v_{a1} + 1 + u_{a1}$ ▷ 1M

10: **if** $\text{Tr}((s_{10} + s_3 + s_9)(s_3 + f_5 + s_1)) = 1$ **then**   ▷ 1M+1TR

11:    $s_8 \leftarrow s_8 + 1,\ s_9 \leftarrow s_9 + s_1,\ s_{10} \leftarrow s_{10} + 1$

12: **end if**

13: $s_{11} \leftarrow (s_3 + f_5 + s_1) f_5^{-2},\ s_{12} \leftarrow (s_{10} + s_3 + s_9) s_{11}$ ▷ 2M

14: $s_{13} \leftarrow (s_2 + s_9) u_{a0} + v_{a0} + 1 + u_{a0}$ ▷ 1M

15: $s_{14} \leftarrow (s_{13} + s_{10} + f_1) s_{11},\ u_{c0} \leftarrow \sqrt{s_{14}},\ u_{c1} \leftarrow \sqrt{s_{12}}$ ▷ 1M+2SR

16: $s_{15} \leftarrow s_3 u_{c1},\ s_{16} \leftarrow s_{15} + s_9,\ s_{17} \leftarrow s_{16} u_{c0}$ ▷ 2M

17: $v_{c1} \leftarrow s_{10} + (s_3 + s_{16})(u_{c0} + u_{c1}) + s_{15} + s_{17}$ ▷ 1M

18: $v_{c0} \leftarrow s_{13} + s_{17}$

19: $\overline{E} \leftarrow [x^2 + u_{c1} x + u_{c0},\ v_{c1} x + v_{c0}]$

20: **return** $\overline{E}$ ▷ 1I+16M+2SR+2TR+2HT

21:   ▷ Average: 1I+15.5M+2SR+2TR+2HT

Note that for curves with $f_5 = 1$, the worst-case complexity decreases to 1I+13M+2SR+2HT+2TR. If we move the computation of s_{11} and s_{12} before the second trace computation (which becomes $\text{Tr}(s_{12})$), then the average complexity drops to 1I+12M+2SR+2HT+2TR (with that approach, a multiplication is required to correct s_{12}).

Algorithm 6. (HLV12, $h(x) = x^2 + x + 1$, $f(x) = f_5 x^5 + f_1 x + f_0$)

INPUT: The divisor class $\overline{D}_a = [x + u_{a0}, v_{a0}]$ and f_5^{-1} precomputed

OUTPUT: The divisor class $[x^2 + u_{c1} x + u_{c0}, v_{c1} x + v_{c0}] = [1/2]\overline{D}_a$

1: $s_0 \leftarrow \text{HT}(f_5 u_{a0}),\ s_1 \leftarrow u_{a0}^2 + u_{a0}$ ▷ 1M+1S+1HT

2: $s_2 \leftarrow v_{a0} + 1 + s_1(s_0 + 1) + s_0^2$ ▷ 1M+1S

3: **if** $\text{Tr}(s_2) = 1$ **then**   ▷ 1TR

4:    $s_0 \leftarrow s_0 + 1,\ s_2 \leftarrow s_2 + s_1 + 1$

5: **end if**

6:   $s_3 \leftarrow \mathrm{HT}(s_2)$ ▷ 1HT

7:   **if** $\mathrm{Tr}(s_3 f_5) = 1$ **then** $s_3 \leftarrow s_3 + 1$ **end if**   ▷ 1M+1TR

8:   $s_4 \leftarrow s_3 f_5^{-1}$, $s_5 \leftarrow s_0 + s_3$, $s_6 \leftarrow s_0^2 + s_3(1 + u_{a0} + s_3)$ ▷ 2M+1S

9:   $s_7 \leftarrow (f_1 + s_5 + s_6)f_5^{-1}$, $u_{c1} \leftarrow \sqrt{s_4}$, $u_{c0} \leftarrow \sqrt{s_7}$ ▷ 1M+2SR

10: $v_{c1} \leftarrow s_5 + s_0 u_{c1}$, $v_{c0} \leftarrow s_6 + s_0 u_{c0}$ ▷ 2M

11: **return** $[x^2 + u_{c1}x + u_{c0}, v_{c1}x + v_{c0}]$ ▷ 8M+3S+2SR+2TR+2HT

Algorithm 7. (HLV21/22, $h(x) = x^2 + x + 1$, $f(x) = f_5 x^5 + f_1 x + f_0$)

INPUT: The divisor class $\overline{D}_a = [x^2 + u_{a0}, v_{a1}x + v_{a0}]$ and f_5^{-2}

OUTPUT: The divisor class $[x + u_{c0}, v_{a0}] = [1/2]\overline{D}_a$
             or $[x^2 + u_{c1}x + v_{c0}, v_{c1}x + v_{c0}] = [1/2]\overline{D}_a$

1:   $u_{c0} \leftarrow \sqrt{u_{a0}}$ ▷ 1SR

2:   **if** $\mathrm{Tr}(f_5 u_{c0}) = 0$ **then**   ▷ 1M+1TR

3:      $v_{c0} \leftarrow v_{a0} + u_{c0}v_{a1}$, $\overline{E} \leftarrow [x + u_{c0}, v_{c0}]$ ▷ 1M

4:   **else**

5:      $s_0 \leftarrow v_{a1} + 1 + u_{a0}f_5$, $s_1 \leftarrow s_0 + f_1$, $s_2 \leftarrow s_0 + f_5$ ▷ 1M

6:      $s_3 \leftarrow u_{a0} + (s_0 + s_2^2)f_5^{-2}$, $u_{c1} \leftarrow \sqrt{s_3}$, $s_4 \leftarrow f_5 u_{c1}$,   ▷ 2M+1SR+1S

7:      $s_5 \leftarrow s_3 u_{a0} + (s_2 + s_0^2 + f_1)f_5^{-2}$, $u_{c0} \leftarrow \sqrt{s_5}$ ▷ 2M+1SR+1S

8:      $s_6 \leftarrow (s_2 + s_4)u_{c0}$, $v_{c1} \leftarrow (s_0 + s_4)(u_{c1} + u_{c0} + 1) + s_6$ ▷ 2M

9:      $v_{c0} \leftarrow s_1 + s_6$, $\overline{E} \leftarrow [x^2 + u_{c1}x + u_{c0}, v_{c1}x + v_{c0}]$

10: **end if**

11: **return** $\overline{E}$ ▷ 9M+3SR+2S+1TR

12:          ▷ Average: 5M+2SR+1S+1TR

---

5   Conclusion

We investigated the halving of divisor classes of hyperelliptic curves of genus 2 over binary fields. We provided new and improved formulæ for all cases of the halving for all types of curves whose divisor class group has at most one divisor of order 2 and does not contain any divisor of order 4. We summarize our results for the most common cases in the table below, where we also compare with doubling formulæ [10,11,1] and with the fast halving formulæ available in previous papers [7,8,3].

Curve type	Doubling	Halving (previous best)	Halving (this work)
Ia (general)	1I+20M+6S	1I+19.5M+2S +2SR+2TR+2HT	1I+15.5M +2SR+2TR+2HT
Ia, with $h_1 = f_5 = 1$	1I+15M+7S	1I+13.5M+3S +2.5SR+2TR+2HT	1I+12M +2SR+2TR+2HT
Ic	1I+10M+6S	—	1I+10.5M +4SR+1TR+1HT
II	1I+5M+6S	1I+8M +5SR+1TR+1HT	1I+7.5M+2S +4SR+1TR+1HT
III	1I+4M+6S	—	1I+4M+6SR

Note that Kitamura, Katagi and Takagi [7,8] also provide a brief description of halving for type Ic curves, but the best approximation for the cost would be from their formulæ for type Ia curves with $h_1 = f_5 = 1$.

For curves of type Ic and III, our halving formulæ are as efficient as the doubling. For type II curves, our halving formulæ improve on the state-of-the-art, although not sufficiently to match the efficiency of the doubling. Our most important gain comes from type Ia curves, where our halving cost is significantly lower than both the doubling and the previous best halving formulæ.

References

1. Avanzi, R.M., Cohen, H., Doche, C., Frey, G., Lange, T., Nguyen, K., Vercauteren, F.: Handbook of Elliptic and Hyperelliptic Curve Cryptography. Chapman & Hall/CRC, Boca Raton (2006)
2. Avanzi, R.M.: A Note on Square Roots in Binary Fields (preprint)
3. Birkner, P.: Efficient Divisor Class Halving on Genus Two Curves. In: Biham, E., Youssef, A.M. (eds.) SAC 2006. LNCS, vol. 4356, pp. 317–326. Springer, Heidelberg (2007)
4. Fong, K., Hankerson, D., López, J., Menezes, A.: Field Inversion and Point Halving Revisited. IEE Trans. Computers 53(8), 1047–1059 (2004)
5. Gaudry, P.: Index calculus for abelian varieties and the elliptic curve discrete logarithm problem (2004) (preprint), http://eprint.iacr.org/2004/073/
6. Gaudry, P., Hess, F., Smart, N.P.: Constructive and destructive facets of Weil descent on elliptic curves. Journal of Cryptology 15(1), 19–46 (2002)
7. Kitamura, I., Katagi, M., Takagi, T.: A Complete Divisor Class Halving Algorithm for Hyperelliptic Curve Cryptosystems of Genus Two (preprint) (2005), http://eprint.iacr.org/2005/255/
8. Kitamura, I., Katagi, M., Takagi, T.: A complete divisor class halving algorithm for hyperelliptic curve cryptosystems of genus two. In: Boyd, C., González Nieto, J.M. (eds.) ACISP 2005. LNCS, vol. 3574, pp. 146–157. Springer, Heidelberg (2005)
9. Knudsen, E.W.: Elliptic scalar multiplication using point halving. In: Lam, K.-Y., Okamoto, E., Xing, C. (eds.) ASIACRYPT 1999. LNCS, vol. 1716, pp. 135–149. Springer, Heidelberg (1999)

10. Lange, T.: Formulae for Arithmetic on Genus 2 Hyperelliptic Curves. Applicable Algebra in Engineering. Communication and Computing 15(5), 295–328 (2005)
11. Lange, T., Stevens, M.: Efficient doubling on genus two curves over binary fields. In: Handschuh, H., Hasan, M.A. (eds.) SAC 2004. LNCS, vol. 3357, pp. 170–181. Springer, Heidelberg (2004)
12. Lidl, R., Niederreiter, H.: Finite Fields, 2nd edn. Cambridge University Press, Cambridge (1997)
13. Pohlig, S., Hellman, M.: An improved algorithm for computing logarithms over GF(p) and its cryptographic significance. IEEE Trans. Inform. Theory IT-24, 106–110 (1978)
14. Schroeppel, R.: Elliptic curves: Twice as fast! In: Crypto 2000 Rump Session (2000)
15. Thériault, N.: Weil Descent for Artin-Schreier Curves (preprint) (2003), http://homepage.mac.com/ntheriau

A Type III: $h(x) = 1$

In this section, the curve C is of the form $y^2 + y = x^5 + f_3 x^3 + f_1 x + f_0$ with $f_0 \in \mathbb{F}_2$. Doing a case-by-case study of the doubling algorithm gives us the following theorem, and the halving formulæ are obtained by inverting the doubling ones:

Theorem 4. Let $D_a = [u_a(x), v_a(x)]$ be a reduced divisor in Div^0_C and $D_c = [u_c(x), v_c(x)] = \frac{1}{2}D_a$ its pre-image under the doubling. If $\deg(u_a) = 2$, then $\deg(u_c) = 2$ if and only if $u_{a1} \neq 0$ (HLV22), otherwise $\deg(u_c) = 1$ (HLV21). If $\deg(u_a) = 1$, then $\deg(u_c) = 2$ (HLV12).

Algorithm 8. (HLV22, $h(x) = 1$, $f(x) = x^5 + f_3 x^3 + f_1 x + f_0$)

INPUT: The divisor class $\overline{D} = [x^2 + u_{a1}x + u_{a0}, v_{a1}x + v_{a0}]$
OUTPUT: The divisor class $[x^2 + u_{c1}x + u_{c0}, v_{c1}x + v_{c0}] = [1/2]\overline{D}$

1: $s_0 \leftarrow \sqrt{u_{a1}}$, $s_1 \leftarrow 1/s_0$, $s_2 \leftarrow s_1 + f_3$, $u_{c1} \leftarrow \sqrt{s_2}$ ▷ 1I+2SR
2: $s_3 \leftarrow s_1\sqrt{u_{a0} + s_2}$, $v_{c1} \leftarrow \sqrt{s_3}$, $s_4 \leftarrow u_{a1}s_1$, $s_5 \leftarrow s_3 + s_4$ ▷ 2M+2SR
3: $s_6 \leftarrow u_{a0}s_5$, $s_7 \leftarrow 1 + v_{a0} + f_0 + s_6$, $v_{c0} \leftarrow \sqrt{s_7}$ ▷ 1M+1SR
4: $s_8 \leftarrow v_{a1} + f_1 + (u_{a0} + u_{a1})(s_1 + s_5) + s_4 + s_6$, $u_{c0} \leftarrow \sqrt{s_8}$ ▷ 1M+1SR
5: **return** $[x^2 + u_{c1}x + u_{c0}, v_{c1}x + v_{c0}]$ ▷ 1I+4M+6SR

Algorithm 9. (HLV12, $h(x) = 1$, $f(x) = x^5 + f_3 x^3 + f_1 x + f_0$)

INPUT: The divisor class $\overline{D} = [x + u_{a0}, v_{a0}]$
OUTPUT: The divisor class $[x^2 + u_{c1}x + u_{c0}, v_{c1}x + v_{c0}] = [1/2]\overline{D}$

1: $s_0 \leftarrow \sqrt{u_{a0}}$, $s_1 \leftarrow f_3$, $s_2 \leftarrow s_1 u_{a0}$, $s_3 \leftarrow \sqrt{s_0 + s_2}$ ▷ 1M+2SR

2: $s_4 \leftarrow s_3 + f_1$, $s_5 \leftarrow v_{a0} + 1 + u_{a0}(s_3 + s_0 u_{a0})$ ▷ 2M

3: $u_{c1} \leftarrow \sqrt{s_1}$, $u_{c0} \leftarrow \sqrt{s_4}$, $v_{c1} \leftarrow s_3 + \sqrt{s_2}$, $v_{c0} \leftarrow s_5 + s_0 u_{c0}$ ▷ 1M+3SR

4: **return** $[x^2 + u_{c1}x + u_{c0}, v_{c1}x + v_{c0}]$ ▷ 4M+5SR

Algorithm 10. (HLV21, $h(x) = 1$, $f(x) = x^5 + f_3 x^3 + f_1 x + f_0$)

INPUT: The divisor class $\overline{D}_a = [x^2 + u_{a0}, v_{a1}x + v_{a0}]$

OUTPUT: The divisor class $[x + u_{c0}, v_{c0}] = [1/2]\overline{D}_a$

1: $u_{c0} \leftarrow \sqrt{u_{a0}}$, $v_{c0} \leftarrow v_{a0} + u_{c0}v_{a1}$ ▷ 1M+SR

2: **return** $[x + u_{c0}, v_{c0}]$ ▷ 1M+1SR

B   Type Ic: $h(x) = x^2$

In this section, the curve C is of the form $y^2 + x^2 y = x^5 + f_4 x^4 + f_1 x + f_0$ with $f_4 \in \mathbb{F}_2$.

Theorem 5. *Let $D_a = [u_a(x), v_a(x)]$ be a divisor in Div_C^0. If $\deg(u_a) = 2$, then D_a can be halved if and only if $\mathrm{Tr}(u_{a1}) = 0$. If $\deg(u_a) = 1$, then D_a can be halved if and only if $\mathrm{Tr}(f_4 + u_{a0}) = 0$.*

Proof. As in Theorem 2.

From this theorem, we obtain a simple condition for the group order:

Corollary 5. *The Jacobian of the curve C has order $2 \cdot$ odd if and only if $f_4 = 1$.*

Proof. The curve C has exactly one divisor of order 2, which is of the form $[x, \sqrt{f_0}]$. The group order is divisible by 4 if and only if $[x, \sqrt{f_0}]$ can be halved. From Theorem 5, this is possible if and only if $\mathrm{Tr}(f_4) = 0$ and since $f_4 \in \mathbb{F}_2$, C has a divisor of order 4 if and only if $f_4 = 0$.

Let D_{c_1} and D_{c_2} the two pre-images of D_a, then $D_{c_1} - D_{c_2} = D_{c_2} - D_{c_1} = [x, \sqrt{f_0}]$. From this observation, we get the following corollary which allow us to distinguish the different special cases.

Corollary 6. *Let $D_a = [u_a(x), v_a(x)]$ be a divisor in Div_C^0 that can be halved and $D_c = [u_c(x), v_c(x)] = \frac{1}{2}D_a$ its pre-image (under the doubling) of odd order. If $\deg(u_a) = 2$ and $u_{a1} \neq 0$, then $\deg(u_c) = 2$ ("HLV22"). If $\deg(u_a) = 2$ and $u_{a1} = 0$, then $\deg(u_c) = 1$ or 2 (with $u_{c0} = 0$ in the second case) ("HLV21/22"). If $\deg(u_a) = 1$, then $\deg(u_c) = 2$ ("HLV12").*

Finally, to verify that we are computing the pre-image of odd order, we use the conditions of Theorem 5 on $u_c(x)$ to ensure that it could be halved again (i.e. that it has odd order), and we correct the computations if necessary. We obtain the following formulæ:

Algorithm 11. (HLV22, $h(x) = x^2$, $f(x) = x^5 + x^4 + f_1 x + f_0$)

INPUT: The divisor class $\overline{D} = [x^2 + u_{a1}x + u_{a0}, v_{a1}x + v_{a0}]$
OUTPUT: The divisor class $[x^2 + u_{c1}x + u_{c0}, v_{c1}x + v_{c0}] = [1/2]\overline{D}$

1: $s_0 \leftarrow 1/u_{a1}$, $s_1 \leftarrow HT(u_{a1})$, $s_2 \leftarrow s_1 s_0$ ▷ 1I+1M+1HT

2: $s_3 \leftarrow f_1(s_2 + 1 + s_0)$, $s_4 \leftarrow \sqrt{s_1 + (v_{a1} + u_{a0})s_0}$ ▷ 2M+1SR

3: $s_5 \leftarrow v_{a1} + u_{a1}(1 + s_4 + s_1) + s_2 u_{a0}$, $s_6 \leftarrow s_5(s_2 + 1 + s_0)$ ▷ 3M

4: **if** $Tr(s_6) = 1$ **then**   ▷ 1TR

5:      $s_2 \leftarrow s_2 + s_0$, $s_3 \leftarrow s_3 + f_1 s_0$, $s_4 \leftarrow s_4 + 1$ ▷ 1M

6:      $s_5 \leftarrow s_5 + u_{a0} s_0$, $s_6 \leftarrow s_5(s_2 + 1 + s_0)$ ▷ 2M

7: **end if**

8: $u_{c1} \leftarrow \sqrt{s_6}$, $u_{c0} \leftarrow \sqrt{s_3}$, $v_{c0} \leftarrow \sqrt{f_0 + s_3(1 + s_4)}$ ▷ 1M+3SR

9: $v_{c1} \leftarrow s_6 + s_2 u_{c0} + s_4 u_{c1}$ ▷ 2M

10: **return** $[x^2 + u_{c1}x + u_{c0}, v_{c1}x + v_{c0}]$ ▷ 1I+12M+4SR+1HT+1TR

11:        ▷ Average: 1I+10.5M+4SR+1HT+1TR

Algorithm 12. (HLV12, $h(x) = x^2$, $f(x) = x^5 + f_4 x^4 + f_1 x + f_0$)

INPUT: The divisor class $\overline{D}_a = [x + u_{a0}, v_{a0}]$
OUTPUT: The divisor class $[x^2 + u_{c1}x + u_{c0}, v_{c1}x + v_{c0}] = [1/2]\overline{D}_a$

1: $s_1 \leftarrow HT(u_{a0} + 1)$, $s_2 \leftarrow \sqrt{v_{a0} + (s_1 + 1)u_{a0}^2}$ ▷ 1M+1S+1SR+1HT

2: $s_3 \leftarrow s_2(u_{a0} + s_2)$, $s_4 \leftarrow s_2$ ▷ 1M

3: **if** $Tr(s_4) = 1$ **then**   ▷ 1TR

4:      $s_1 \leftarrow s_1 + 1$, $s_2 \leftarrow s_2 + u_{a0}$, $s_4 \leftarrow s_2$

5: **end if**

6: $u_{c1} \leftarrow \sqrt{s_4}$, $u_{c0} \leftarrow \sqrt{f_1}$, $v_{c1} \leftarrow s_2 + s_1 u_{c1}$ ▷ 1M+1SR

7: $v_{c0} \leftarrow s_3 + s_1 u_{c0}$, $\overline{E} \leftarrow [x^2 + u_{c1}x + u_{c0}, v_{c1}x + v_{c0}]$ ▷ 1M

8: **return** $\overline{E}$ ▷ 4M+1S+2SR+1TR+1HT

Algorithm 13. (HLV21/22, $h(x) = x^2$, $f(x) = x^5 + f_4 x^4 + f_1 x + f_0$)

INPUT: The divisor class $\overline{D}_a = [x^2 + u_{a0}, v_{a1}x + v_{a0}]$

OUTPUT: The divisor class $[x + u_{c0}, v_{c0}] = [1/2]\overline{D}_a$
          or $[x^2 + u_{c1}x, v_{c1}x + v_{c0}] = [1/2]\overline{D}_a$

1: $s_0 \leftarrow \sqrt{u_{a0}}$, $s_1 \leftarrow v_{a0} + s_0 v_{a1}$ ▷ 1M+1SR

2: **if** $\mathrm{Tr}(s_0) = 1$ **then**   ▷ 1TR

3:      $\overline{E} \leftarrow [x + s_0, s_1]$

4: **else**

5:      $s_2 \leftarrow 1/s_0$, $v_{c1} \leftarrow (s_1 + \sqrt{f_0})s_2$ ▷ 1I+1M

6:      $v_{c0} \leftarrow s_1 + v_{c1}s_0$, $\overline{E} \leftarrow [x^2 + s_0 x, v_{c1}x + v_{c0}]$ ▷ 1M

7: **end if**

8: **return** $\overline{E}$ ▷ 1I+3M+1SR+1TR

9:        ▷ Average: 0.5I+2M+1SR+1TR

C   Genus 2 Addition for Type Ia, $h(x) = x^2 + x + 1$

The form of the equation we used for curves of type Ia differs from the one more commonly used in explicit formulæ paper (which prefer to force $f_5 = 1$ rather than $h_1 = 1$). It is therefore natural to ask what impact the form of the equation has on the group addition, which is also necessary for the halving-and-add algorithm.

As the following formula shows, allowing $f_5 \neq 1$ allows us to perform the group addition in 1I+21M+4S, which is in fact slightly more efficient than the 1I+22M+3S required with the more common form of the equation.

Algorithm 14. (ADD22, $h(x) = x^2 + x + 1$, $f(x) = f_5 x^5 + f_1 x + f_0$)

INPUT: The divisor classes $\overline{D}_a = [x^2 + u_{a1}x + u_{a0}, v_{c1}x + v_{a0}]$,
          $\overline{D}_b = [x^2 + u_{b1}x + u_{b0}, v_{b1}x + v_{b0}]$

OUTPUT: The divisor class $[x^2 + u_{c1}x + u_{c0}, v_{c1}x + v_{c0}] = \overline{D}_a + \overline{D}_b$

1: $s_0 \leftarrow u_{a1} + u_{b1}$, $s_1 \leftarrow u_{a1}s_0$, $s_2 \leftarrow u_{a0} + u_{b0}$, $s_3 \leftarrow s_1 + s_2$ ▷ 1M

2: $s_4 \leftarrow s_3 s_2 + u_{a0}s_0^2$, $s_5 \leftarrow v_{a0} + v_{b0}$, $s_6 \leftarrow v_{a1} + v_{b1}$ ▷ 2M+1S

3: $s_7 \leftarrow s_3 s_5$, $s_8 \leftarrow s_0 s_6$ ▷ 2M

4: $s_9 \leftarrow (s_3 + s_0)(s_5 + s_6) + s_7 + s_8(1 + u_{a1})$, $s_{10} \leftarrow s_4 s_9$ ▷ 3M

5: **if** $s_{10} = 0$ **then**

6:    Use Cantor's algorithm

7: **else**

8:    $s_{11} \leftarrow s_7 + u_{a0}s_8$, $s_{12} \leftarrow 1/s_{10}$, $s_{13} \leftarrow s_4s_{12}$ ▷ 1I+2M

9:    $s_{14} \leftarrow s_9^2 s_{12}$, $s_{15} \leftarrow s_4s_{13}$, $s_{16} \leftarrow s_{15}^2$, $s_{17} \leftarrow s_{11}s_{13}$ ▷ 3M+2S

10:    $s_{18} \leftarrow f_5s_{16}$, $u_{c1} \leftarrow s_0 + s_{15} + s_{18}$ ▷ 1M

11:    $u_{c0} \leftarrow s_2 + s_{17}^2 + s_1 + (1 + s_{17} + u_{a1})s_{15} + s_0s_{18}$ ▷ 2M+1S

12:    $s_{19} \leftarrow u_{b1} + u_{c1}$, $s_{20} \leftarrow (s_{17} + u_{b1})s_{19}$ ▷ 1M

13:    $s_{21} \leftarrow (u_{c0} + u_{b0})s_{17} + u_{b0}s_{19}$, $v_{c1} \leftarrow u_{c1} + 1 + v_{b1} + s_{14}s_{20}$ ▷ 3M

14:    $v_{c0} \leftarrow u_{c0} + 1 + v_{b0} + s_{14}s_{21}$ ▷ 1M

15: **end if**

16: **return** $[x^2 + u_{c1}x + u_{c0}, v_{c1}x + v_{c0}]$ ▷ 1I+21M+4S

Efficient Pairing Computation on Genus 2 Curves in Projective Coordinates

Xinxin Fan[1,*], Guang Gong[1,*], and David Jao[2,**]

[1] Department of Electrical and Computer Engineering
[2] Department of Combinatorics and Optimization,
University of Waterloo,
Waterloo, Ontario, N2L 3G1, Canada
{x5fan@engmail,ggong@calliope,djao@math}.uwaterloo.ca

Abstract. In recent years there has been much interest in the development and the fast computation of bilinear pairings due to their practical and myriad applications in cryptography. Well known efficient examples are the Weil and Tate pairings and their variants such as the Eta and Ate pairings on the Jacobians of (hyper-)elliptic curves. In this paper, we consider the use of projective coordinates for pairing computations on genus 2 hyperelliptic curves over prime fields. We generalize Chatterjee *et. al.*'s idea of encapsulating the computation of the line function with the group operations to genus 2 hyperelliptic curves, and derive new explicit formulae for the group operations in projective and new coordinates in the context of pairing computations. When applying the encapsulated explicit formulae to pairing computations on supersingular genus 2 curves over prime fields, theoretical analysis shows that our algorithm is faster than previously best known algorithms whenever a field inversion is more expensive than about fifteen field multiplications. We also investigate pairing computations on non-supersingular genus 2 curves over prime fields based on the new formulae, and detail the various techniques required for efficient implementation.

Keywords: Genus 2 hyperelliptic curves, Tate pairing, Miller's algorithm, Projective coordinates, Efficient Implementation.

1   Introduction

Bilinear pairings were first introduced to cryptography by Menezes *et. al.* [24] and Frey and Rück [13] as a tool to attack instances of the discrete logarithm problem (DLP) on (hyper-)elliptic curves. Subsequently, Sakai *et. al.*'s non-interactive key distribution scheme [29] and Joux's tripartite Diffie-Hellman key agreement protocol [19] provided examples of positive usages of pairings. This use of pairings has inspired much research devoted to the design of cryptographic protocols with novel properties and the improvement of existing ones, with some classical

---

* Supported by NSERC Strategic Project Grants (SPG).
** Partially supported by NSERC.

R. Avanzi, L. Keliher, and F. Sica (Eds.): SAC 2008, LNCS 5381, pp. 18–34, 2009.
© Springer-Verlag Berlin Heidelberg 2009

examples being Identity Based Encryption [4] and short signatures [5]. Since pairing computations are generally the most important and expensive operation in any pairing-based cryptosystem, improving the speed of pairing computations has become an important issue in pairing-based cryptography.

Miller proposed the first algorithm [26] for computing the Weil pairing on elliptic curves. In practice, the Tate pairing shows better performance than that of the Weil pairing and therefore is widely used. While many important techniques have been proposed to accelerate the computation of the Tate pairing and its variants on elliptic curves [2,3,17], the subject of pairing computations on hyperelliptic curves is also receiving an increasing amount of attention. Choie and Lee [7] investigated the implementation of the Tate pairing on supersingular genus 2 hyperelliptic curves over prime fields. Later on, Ó hÉigeartaigh and Scott [16] improved the implementation of [7] significantly by using a new variant of Miller's algorithm combined with various optimization techniques. Duursma and Lee [9] presented a closed formula for the Tate pairing computation on a very special family of supersingular hyperelliptic curves. Barreto $et.$ $al.$ [2] generalized the results of [9] and proposed the Eta pairing approach for efficiently computing the Tate pairing on supersingular genus 2 curves over binary fields. In particular, their algorithm leads to the fastest pairing implementation in the literature. In [23], Lee $et.$ $al.$ considered the Eta pairing computation on general divisors on supersingular genus 3 hyperelliptic curves with the form of $y^2 = x^7 - x \pm 1$. Recently, the Ate pairing, which is an extension of the Eta pairing to the setting of ordinary curves, has been generalized to hyperelliptic curves [14] as well. Although the Eta and Ate pairings hold the record for speed at the present time, we will focus our attention on the Tate pairing in this paper. The main reason is that the Tate pairing is uniformly available across a wide range of hyperelliptic curves and subgroups, whereas the Eta pairing is only defined for supersingular curves and the Ate pairing incurs a huge performance penalty in the context of ordinary genus 2 curves [14, Table 6].

Previous work for computing pairings on hyperelliptic curves only considered using affine coordinates. Motivated by Chatterjee $et.$ $al.$'s work [6], we address the efficient implementation of the Tate pairing on genus 2 hyperelliptic curves over large prime fields in projective coordinates in this contribution. We first derive new explicit formulae for the group operations for genus 2 hyperelliptic curves in projective and new (weighted projective) coordinates, respectively. Letting I denote a field inversion, M a field multiplication, and S a field squaring, we find in the context of pairing computations that compared to Lange's formulae [22], our mixed-addition formulae can save $5M$ and $3M$ in projective and new coordinates, respectively, whereas our doubling formulae can save $2M$ in both projective and new coordinates. We then show how to encapsulate the computation of the line function with the mixed addition and doubling formulae in new coordinates, and how to omit some operations which are cancelled by the final exponentiation in the encapsulated method. Our encapsulated explicit formulae can be applied to pairing computations on both supersingular and non-supersingular genus 2 hyperelliptic curves over prime fields. Finally, we describe an efficient implementation of the Tate pairing on a non-supersingular genus 2 hyperelliptic curve with an embedding degree of 2 over prime fields as a case study. To our knowledge, this is the first concrete implementation of pairing computations on non-supersingular genus 2 curves.

This paper is organized as follows. Section 2 gives an overview of the Tate pairing on hyperelliptic curves and Miller's algorithm for computing the pairing. In Section 3 we describe new explicit formulae which encapsulate group operations and line computations for genus 2 curves over prime fields. Section 4 shows how to apply various techniques from the literature to accelerate the pairing computation on a specific non-supersingular genus 2 curve over prime fields, analyzes the computational complexity of computing the Tate pairings and gives implementation results. Finally, Section 5 concludes this contribution.

2   Mathematical Background

2.1   Tate Pairing on Hyperelliptic Curves

Let $\mathbb{F}_q$ be a finite field with q elements, and $\overline{\mathbb{F}}_q$ be its algebraic closure. Let C be a hyperelliptic curve of genus g over $\mathbb{F}_q$, and let $\mathcal{J}_C$ denote the degree zero divisor class group of C. We say that a subgroup of the divisor class group $\mathcal{J}_C(\mathbb{F}_q)$ has *embedding degree* k if the order n of the subgroup divides $q^k - 1$, but does not divide $q^i - 1$ for any $0 < i < k$. For our purpose, n should be a (large) prime with $n \mid \#\mathcal{J}_C(\mathbb{F}_q)$ and $\gcd(n, q) = 1$. Let $\mathcal{J}_C(\mathbb{F}_{q^k})[n]$ be the n-torsion group and $\mathcal{J}_C(\mathbb{F}_{q^k})/n\mathcal{J}_C(\mathbb{F}_{q^k})$ be the quotient group. Then the Tate pairing is a well defined, non-degenerate, bilinear map [13]:

$$\langle \cdot, \cdot \rangle_n : \mathcal{J}_C(\mathbb{F}_{q^k})[n] \times \mathcal{J}_C(\mathbb{F}_{q^k})/n\mathcal{J}_C(\mathbb{F}_{q^k}) \rightarrow \mathbb{F}_{q^k}^*/(\mathbb{F}_{q^k}^*)^n,$$

defined as follows: let $D_1 \in \mathcal{J}_C(\mathbb{F}_{q^k})[n]$, with $\mathrm{div}(f_{n,D_1}) = nD_1$ for some rational function $f_{n,D_1} \in \mathbb{F}_{q^k}(C)^*$. Let $D_2 \in \mathcal{J}_C(\mathbb{F}_{q^k})/n\mathcal{J}_C(\mathbb{F}_{q^k})$ with $\mathrm{supp}(D_1) \cap \mathrm{supp}(D_2) = \emptyset$ (to ensure a non-trivial pairing value). The Tate pairing of two divisor classes $\overline{D}_1$ and $\overline{D}_2$ is then defined as

$$\langle \overline{D}_1, \overline{D}_2 \rangle_n = f_{n,D_1}(D_2) = \prod_{P \in C(\overline{\mathbb{F}}_q)} f_{n,D_1}(P)^{\mathrm{ord}_P(D_2)}.$$

Note that the Tate pairing as detailed above is only defined up to n-th powers. One can show that if the function f_{n,D_1} is properly normalized, we only need to evaluate the rational function f_{n,D_1} at the effective part of the reduced divisor D_2 in order to compute the Tate pairing [3,14].

In practice, the fact that the Tate pairing is only defined up to n-th power is usually undesirable, and many pairing-based protocols require a unique pairing value. Hence one defines the *reduced* pairing as

$$\langle \overline{D}_1, \overline{D}_2 \rangle_n^{(q^k-1)/n} = f_{n,D_1}(D_2)^{(q^k-1)/n} \in \mu_n \subset \mathbb{F}_{q^k}^*,$$

where $\mu_n = \{u \in \mathbb{F}_{q^k}^* \mid u^n = 1\}$ is the group of n-th roots of unity. In the rest of this paper we will refer to the extra powering required to compute the reduced pairing as the *final exponentiation*. Furthermore, we also assume the embedding degree k is greater than 1.

2.2 Miller's Algorithm

The main task involved in the computation of the Tate pairing $\langle \overline{D}_1, \overline{D}_2 \rangle_n$ is to construct a rational function f_{n,D_1} such that $\mathrm{div}(f_{n,D_1}) = nD_1$. In [26], Miller described a polynomial time algorithm, known universally as Miller's algorithm, to construct the function f_{n,D_1} and compute the Weil pairing on elliptic curves. However, the algorithm can be easily adapted to compute the Tate pairing on hyperelliptic curves.

Let $G_{iD_1,jD_1} \in \mathbb{F}_{q^k}(C)^*$ be a rational function with $\mathrm{div}(G_{iD_1,jD_1}) = iD_1 + jD_1 - (iD_1 \oplus jD_1)$ where $\oplus$ is the group law on $\mathcal{J}_C$ and $(iD_1 \oplus jD_1)$ is reduced. Miller's algorithm constructs the rational function f_{n,D_1} based on the following iterative formula:

$$f_{i+j,D_1} = f_{i,D_1} f_{j,D_1} G_{iD_1,jD_1}.$$

Algorithm 1 shows the basic version of Miller's algorithm for computing the reduced Tate pairing on hyperelliptic curves according to the above iterative relation. Essentially, computing the Tate pairing with Miller's algorithm amounts to performing a scalar multiplication of a reduced divisor and evaluating certain intermediate rational functions which appear in the process of the divisor class addition. A more detailed version of Miller's algorithm for hyperelliptic curves can be found in [14].

Algorithm 1. Miller's Algorithm for Hyperelliptic Curves (basic version)

IN: $\overline{D}_1 \in \mathcal{J}_C(\mathbb{F}_{q^k})[n], \overline{D}_2 \in \mathcal{J}_C(\mathbb{F}_{q^k})$, represented by D_1 and D_2
    with $\mathrm{supp}(D_1) \cap \mathrm{supp}(D_2) = \emptyset$
OUT: $\langle D_1, D_2 \rangle_n^{(q^k-1)/n}$
1. $f \leftarrow 1, T \leftarrow D_1$
2. **for** $i \leftarrow \lfloor \log_2(n) \rfloor - 1$ **downto** 0 **do**
3.   ▷ Compute T' and $G_{T,T}(x,y)$ such that $T' = 2T - \mathrm{div}(G_{T,T})$
4. $f \leftarrow f^2 \cdot G_{T,T}(D_2), \overline{T} \leftarrow [2]\overline{T}$
5. **if** $n_i = 1$ **then**
6.   ▷ Compute T' and $G_{T,D_1}(x,y)$ such that $T' = T + D_1 - \mathrm{div}(G_{T,D_1})$
7. $f \leftarrow f \cdot G_{T,D_1}(D_2), \overline{T} \leftarrow \overline{T} \oplus \overline{D}_1$
8. **Return** $f^{(q^k-1)/n}$

---

3 Encapsulated Computation on Genus 2 Curves

In this section we generalize the idea of encapsulated add-and-line and encapsulated double-and-line proposed in [6] to genus 2 hyperelliptic curves over large prime fields. Note that, in the process of computing Tate pairings, one inversion is required for each divisor class addition and doubling, and the calculation of the inversion of an element in large characteristic is usually quite expensive.

Therefore, to avoid inversions, we need to derive efficient inversion-free explicit formulae for genus 2 hyperelliptic curves in the context of pairing computations.

Lange [22] presented efficient explicit formulae for the group operations on genus 2 curves using various systems of coordinates. In the projective coordinate system, the quintuple $[U_1, U_0, V_1, V_0, Z]$ corresponds to the affine class $[x^2 + U_1/Zx + U_0/Z, V_1/Zx + V_0/Z]$ in Mumford representation [28], whereas the sextuple $[U_1, U_0, V_1, V_0, Z_1, Z_2]$ stands for the affine class $[x^2 + U_1/Z_1^2x + U_0/Z_1^2, V_1/(Z_1^3Z_2)x + V_0/(Z_1^3Z_2)]$ in the new coordinate system. Lange's formulae are designed to be used in the context of computing scalar multiplications, and do not explicitly calculate all of the rational functions required in Miller's algorithm. However, one can extract the rational functions required from the formulae in [22] at the cost of 3 extra field multiplications.

Choie and Lee [7] modified Lange's explicit formulae in affine coordinates to reduce the cost of extracting the rational functions required in Miller's algorithm. The formulae presented in [7] require $1I + 23M + 3S$ and $1I + 23M + 5S$ in $\mathbb{F}_p$ for divisor class addition[1] and doubling, respectively, thereby saving 2 field multiplications over the previous method. Ó hÉigeartaigh and Scott [16] further optimized the doubling formula proposed in [7] for supersingular genus 2 curves over $\mathbb{F}_p$ of the form $y^2 = x^5 + a$ by saving 1 multiplication and 1 squaring.

Based on the above explicit formulae in affine coordinates, we derive new explicit mixed-addition and doubling formulae in the projective and new coordinate systems in the context of pairing computations, respectively. Since the explicit formulae in new coordinates are more efficient than those in projective coordinates, we use new coordinates to represent divisor classes in the main presentation. The mixed-addition and doubling formulae in projective coordinates can be found in the appendix. We will explain how to encapsulate the group operations and the line computations in the following subsections. To increase performance, we also enlarge the set of coordinates to $[U_1, U_0, V_1, V_0, Z_1, Z_2, z_1, z_2]$ as in [22], where $z_1 = Z_1^2$ and $z_2 = Z_2^2$.

3.1   Encapsulated Divisor Addition and Line Computation

In this subsection, we show how to encapsulate the computation of the line function with the divisor class addition in new coordinates. Given two divisor classes $\overline{E}_1 = [U_{11}, U_{10}, V_{11}, V_{10}, 1, 1, 1, 1]$ and $\overline{E}_2 = [U_{21}, U_{20}, V_{21}, V_{20}, Z_{21}, Z_{22}, z_{21}, z_{22}]$ in new coordinates as inputs, Table 1 describes an explicit mixed-addition formula which calculates a divisor class $\overline{E}_3 = [u_3(x), v_3(x)]$ and the rational function $l(x)$ such that $E_1 + E_2 = E_3 + \operatorname{div}\left(\frac{y - l(x)}{u_3(x)}\right)$ in the most common case. Our new explicit formula requires $36M + 5S$ for computing the divisor class addition in new coordinates. Table 2 summarizes the computational cost of calculating the divisor class addition and extracting the line function in various coordinate systems. From Table 2 we note that in the context of pairing computations our mixed-addition formulae can save $5M$ in the projective coordinate system and

---

[1] We note that the addition formula in [7] requires $3S$ instead of $2S$ as claimed. Indeed, each of Steps 1, 4, and 6 in [7, Table 5] requires a separate squaring.

Table 1. Mixed-Addition Formula on a Genus 2 Curve over $\mathbb{F}_p$ (New Coordinates)

Input	Genus 2 HEC $C : y^2 = x^5 + f_3x^3 + f_2x^2 + f_1x + f_0$	
	$\overline{E}_1 = [U_{11}, U_{10}, V_{11}, V_{10}, 1, 1, 1, 1]$ and	
	$\overline{E}_2 = [U_{21}, U_{20}, V_{21}, V_{20}, Z_{21}, Z_{22}, z_{21}, z_{22}]$	
Output	$\overline{E}_3 = [U_{31}, U_{30}, V_{31}, V_{30}, Z_{31}, Z_{32}, z_{31}, z_{32}] = \overline{E}_1 \oplus \overline{E}_2$	
	$l(x)$ such that $E_1 + E_2 = E_3 + \mathrm{div}\left(\frac{y - l(x)}{u_3(x)}\right)$	
Step	Expression	Cost
1	**Compute resultant and precomputations:**	$7M, 1S$
	$z_{23} = Z_{21}Z_{22}, z_{24} = z_{21}z_{23}, \tilde{U}_{11} = U_{11}z_{21}, y_1 = \tilde{U}_{11} - U_{21}$	
	$y_2 = U_{20} - U_{10}z_{21}, y_3 = U_{11}y_1, y_4 = y_2 + y_3, r = y_2y_4 + y_1^2U_{10}$	
2	**Compute almost inverse of u_2 mod u_1:**	$-$
	$inv_1 = y_1, inv_0 = y_4$	
3	**Compute s':**	$7M$
	$w_0 = V_{10}z_{24} - V_{20}, w_1 = V_{11}z_{24} - V_{21}, w_2 = inv_0w_0$	
	$w_3 = inv_1w_1, s_1' = y_1w_0 + y_2w_1, s_0' = w_2 - U_{10}w_3$	
4	**Precomputations:**	$4M, 3S$
	$\tilde{r} = rz_{23}, R = \tilde{r}^2, Z_{31} = s_1'Z_{21}, Z_{32} = \tilde{r}Z_{21}$	
	$z_{31} = Z_{31}^2, z_{32} = Z_{32}^2, \tilde{s}_0' = s_0'z_{21}$	
5	**Compute l:**	$5M$
	$l_2 = s_1'U_{21} + \tilde{s}_0', l_0 = s_0'U_{20} + rV_{20}$	
	$l_1 = (s_1' + s_0')(U_{21} + U_{20}) - s_1'U_{21} - s_0'U_{20} + rV_{21}$	
6	**Compute U_3:**	$7M, 1S$
	$w_1 = \tilde{U}_{11} + U_{21}, U_{31} = s_1'(2\tilde{s}_0' - s_1'y_1) - z_{32}, l_1' = l_1s_1'$	
	$U_{30} = \tilde{s}_0'(s_0' - 2s_1'U_{11}) + s_1'^2(y_3 - \tilde{U}_{10} - U_{20}) + 2l_1' + Rw_1$	
7	**Compute V_3:**	$6M$
	$w_1 = l_2s_1' - U_{31}, V_{30} = U_{30}w_1 - z_{31}(l_0s_1')$	
	$V_{31} = U_{31}w_1 + z_{31}(U_{30} - l_1')$	
Sum		$36M, 5S$

Table 2. Divisor Class Addition in Different Systems and in Odd Characteristic

Reference	Coordinate Type	Addition	Mixed Addition	Extracting Line Function $l(x)$
Miyamoto *et al.* [27]	Affine	$1I, 24M, 2S$	$-$	no cost
	Projective	$54M$	$-$	no cost
Lange [22]	Affine	$1I, 22M, 3S$	$-$	$3M$
	Projective	$47M, 4S$	$40M, 3S$	$3M$
	New	$47M, 7S$	$36M, 5S$	$3M$
Choie and Lee [7]	Affine	$1I, 23M, 3S$	$-$	no cost
Our work	Projective	$-$	**$38M, 3S$** Table 9	**no cost**
	New	$-$	**$36M, 5S$** Table 1	**no cost**

Table 3. Doubling Formula on a Genus 2 Curve over $\mathbb{F}_p$ (New Coordinates)

Input	Genus 2 HEC $C : y^2 = x^5 + f_3 x^3 + f_2 x^2 + f_1 x + f_0$	
	$\overline{E}_1 = [U_{11}, U_{10}, V_{11}, V_{10}, Z_{11}, Z_{12}, z_{11}, z_{12}]$	
Output	$E_3 = [U_{31}, U_{30}, V_{31}, V_{30}, Z_{31}, Z_{32}, z_{31}, z_{32}] = [2]\overline{E}_1$	
	$l(x)$ such that $2E_1 = E_3 + \mathrm{div}\left(\frac{y - l(x)}{u_3(x)}\right)$	
Step	Expression	Cost
1	**Compute resultant:**	$4M, 2S$
	$w_0 = V_{11}^2, w_1 = U_{11}^2, w_2 = V_{10} z_{11}$	
	$w_3 = w_2 - U_{11} V_{11}, r = U_{10} w_0 + V_{10} w_3$	
2	**Compute almost inverse:**	–
	$inv_1' = -V_{11}, inv_0' = w_3$	
3	**Compute k':**	$7M, 1S$
	$z_{11}' = z_{11}^2, w_3 = f_3 z_{11}' + w_1, \tilde{U}_{10} = U_{10} z_{11}$	
	$k_1' = z_{12}(2(w_1 - \tilde{U}_{10}) + w_3), z_{11}'' = z_{11} z_{11}'$	
	$k_0' = z_{12}(U_{11}(4\tilde{U}_{10} - w_3) + f_2 z_{11}'') - w_0$	
4	**Compute s':**	$5M$
	$w_0 = k_0' inv_0', w_1 = k_1' inv_1', s_1' = w_2 k_1' - V_{11} k_0', s_0' = w_0 - \tilde{U}_{10} w_1$	
5	**Precomputations:**	$8M, 4S$
	$Z_{31} = s_1' z_{11}, z_{31} = Z_{31}^2, w_0 = r z_{11}, w_1 = w_0 Z_{12}$	
	$Z_{32} = 2w_1 Z_{11}, z_{32} = Z_{32}^2, w_2 = w_1^2, R = r Z_{31}$	
	$S_0 = s_0'^2, S = s_0' Z_{31}, s_0 = s_0' s_1', s_1 = s_1' Z_{31}$	
6	**Compute l:**	$6M$
	$l_2 = s_1 U_{11} + s_0 z_{11}, V_{10}' = R V_{10}, l_0 = s_0 U_{10} + 2V_{10}'$	
	$V_{11}' = R V_{11}, l_1 = (s_1 + s_0)(U_{11} + U_{10}) - s_1 U_{11} - s_0 U_{10} + 2V_{11}'$	
7	**Compute U_3:**	$1M$
	$U_{30} = S_0 + 4(V_{11}' + 2w_2 U_{11}), U_{31} = 2S - z_{32}$	
8	**Compute V_3:**	$4M$
	$w_0 = l_2 - U_{31}, w_1 = w_0 U_{30}, w_2 = w_0 U_{31}$	
	$V_{31} = w_2 + z_{31}(U_{30} - l_1), V_{30} = w_1 - z_{31} l_0$	
Sum		$35M, 7S$

$3M$ in the new coordinate system, respectively, when compared to the formulae given by Lange [22].

In the new coordinate system, the rational function $c(x, y) = y - l(x)$ that is required in Miller's algorithm has the following form:

$$c(x, y) = y - \left(\frac{s_1'}{r z_{23}} x^3 + \frac{l_2}{r z_{24}} x^2 + \frac{l_1}{r z_{24}} x + \frac{l_0}{r z_{24}}\right),$$

where $s_1', l_2, l_1, l_0, r, z_{23}$ and $z_{24} = z_{21} z_{23}$ are computed in Table 1. By defining the auxiliary rational function $c'(x, y) = (r z_{24}) c(x, y)$, we obtain

$$c'(x, y) = (r z_{24}) y - ((s_1' z_{21}) x^3 + l_2 x^2 + l_1 x + l_0).$$

Note that the result of evaluating the function $c(x, y)$ at an image divisor D_2 will be raised to the power $(q^k - 1)/n$ $(k > 1)$ in the last step of Miller's algorithm.

Table 4. Divisor Class Doubling in Different Systems and in Odd Characteristic

Reference	Coordinate Type	Doubling	Extracting Line Function $l(x)$
Miyamoto *et al.* [27]	Affine	$1I, 23M, 4S$	no cost
	Projective	$53M$	no cost
Lange [22]	Affine	$1I, 22M, 5S$	$3M$
	Projective	$38M, 6S$	$3M$
	New	$34M, 7S$	$3M$
Choie and Lee [7]	Affine	$1I, 23M, 5S$	no cost
Ó hÉigeartaigh and Scott [16]	Affine	$1I, 22M, 4S$	no cost
Our work	Projective	**$39M, 6S$** **Table 10**	**no cost**
	New	**$35M, 7S$** **Table 3**	**no cost**

For efficiency reasons, the first input to the Tate pairing is usually restricted to the 1-eigenspace of the Frobenius endomorphism on $\mathcal{J}_C[n]$. Therefore, we have the following relation

$$c(D_2)^{(q^k-1)/n} = ((c'(D_2)/(rz_{24}))^{q-1})^{(q^{k-1}+q^{k-2}+...+1)/n} = c'(D_2)^{(q^k-1)/n}.$$

The above relation means that in new coordinates we can work with the rational function $c'(x, y)$ instead of $c(x, y)$ without altering the value of the resulting Tate pairing. For the same reason we also work with the rational function $u_3'(x) = z_{31}x^2 + U_{31}x + U_{30}$ instead of $u_3(x) = x^2 + \frac{U_{31}}{z_{31}}x + \frac{U_{30}}{z_{31}}$ for both divisor addition and divisor doubling.

3.2   Encapsulated Divisor Doubling and Line Computation

In this subsection, we describe how to encapsulate the computation of the line function with the divisor class doubling in new coordinates. Given a divisor class $\overline{E}_1 = [U_{11}, U_{10}, V_{11}, V_{10}, Z_{11}, Z_{12}, z_{11}, z_{12}]$ in new coordinates as an input, Table 3 describes an explicit doubling formula which calculates a divisor class $\overline{E}_3 = [u_3(x), v_3(x)]$ and the rational function $l(x)$ such that $2E_1 = E_3 + \text{div}\left(\frac{y-l(x)}{u_3(x)}\right)$ in the most common case. Our new explicit formula needs $35M + 7S$ to double a divisor class in new coordinates. Table 4 summarizes the computational cost of doubling a divisor class and extracting the line function in various coordinate systems. From Table 4 we note that in the context of pairing computations our doubling formulae can save $2M$ in both projective and new coordinates, when compared to the formulae given by Lange [22].

In the new coordinate system, the rational function $c(x, y) = y - l(x)$ that is required in Miller's algorithm has the following form:

$$c(x, y) = y - \left(\frac{s_1}{s_1' Z_{32}}x^3 + \frac{l_2}{Z_{31}Z_{32}}x^2 + \frac{l_1}{Z_{31}Z_{32}}x + \frac{l_0}{Z_{31}Z_{32}}\right),$$

where $s_1, s_1', l_2, l_1, l_0, Z_{31}$ and Z_{32} are available in Table 3. By defining the auxiliary rational function $c'(x, y) = (Z_{31}Z_{32})c(x, y)$, we obtain

$$c'(x, y) = (Z_{31}Z_{32})y - ((s_1 z_{11})x^3 + l_2 x^2 + l_1 x + l_0),$$

where z_{11} is also available in Table 3. With the same argument as the case of the mixed-addition, we have the relation $c(D_2)^{(q^k-1)/n} = c'(D_2)^{(q^k-1)/n}$ for an image divisor D_2. Therefore, we can simply work with the rational function $c'(x, y)$ instead of $c(x, y)$ without altering the value of the resulting Tate pairing in the new coordinate system.

4   Implementing the Tate Pairing

4.1   The Non-supersingular Pairing-Friendly Genus 2 Curve

There are only a few techniques that have been proposed for constructing non-supersingular curves of genus $g \geq 2$ and low embedding degree for pairing-based cryptography — see [10,11,18,21] for example. By modeling on the Cocks-Pinch method for constructing pairing-friendly elliptic curves [8], Freeman generated the first examples of non-supersingular pairing-friendly genus 2 curves [11]. In our implementations, we will use an example from [11], which gives a genus 2 curve whose Jacobian has embedding degree 2 with respect to the prime $n = 2^{160} + 7$. The curve is given by the equation

$$C : y^2 = x^5 + f_3 x^3 + f_2 x^2 + f_1 x + f_0$$

over $\mathbb{F}_p$. The curve coefficients f_3, f_2, f_1 and f_0, the subgroup order n, and the characteristic p of the prime field can be found in Appendix A of [11]. Although Freeman [11] also provides examples of non-supersingular genus 2 curves with larger embedding degree, those curves are defined over prime fields with very large characteristics and therefore are not suitable for efficient implementations. Generating non-supersingular pairing-friendly genus 2 curves defined over small prime fields ($|p| \sim 80$) with large embedding degree ($k \geq 12$) remains an open problem.

Let $c \in \mathbb{F}_p$ be a quadratic non-residue over $\mathbb{F}_p$. A quadratic twist of C, denoted by C_t, over $\mathbb{F}_p$ is defined by the following equation

$$C_t : y^2 = x^5 + c^2 f_3 x^3 + c^3 f_2 x^2 + c^4 f_1 x + c^5 f_0.$$

Let $\mathcal{J}_{C_t}(\mathbb{F}_{p^2})$ be the Jacobian of C_t when considering C_t as a curve defined over $\mathbb{F}_{p^2}$, and $\overline{D}_t = [u_t, v_t]$ be an element of $\mathcal{J}_{C_t}(\mathbb{F}_{p^2})$ in Mumford representation. It is known [1] that $C_t(\mathbb{F}_{p^2})$ is isomorphic to $C(\mathbb{F}_{p^2})$. Therefore, we can construct the isomorphism ψ of Jacobians $\mathcal{J}_{C_t}(\mathbb{F}_{p^2})$ and $\mathcal{J}_C(\mathbb{F}_{p^2})$ by applying the isomorphism ϕ to each point $P = (x_t, y_t)$ in the support of the divisor D_t as shown in the following figure.

$$\psi: \mathcal{J}_{C_t}(\mathbb{F}_{p^2}) \longrightarrow \mathcal{J}_C(\mathbb{F}_{p^2}) \qquad \phi: C_t(\mathbb{F}_{p^2}) \longrightarrow C(\mathbb{F}_{p^2})$$
$$\overline{D}_t \longmapsto \overline{D} \qquad (x_t, y_t) \longmapsto (x, y)$$
$$[u_t, v_t] \longmapsto [u, v] \qquad (x_t, y_t) \longmapsto (c^{-1}x_t, c^{-5/2}y_t)$$

The Isomorphism of $\mathcal{J}_{C_t}(\mathbb{F}_{p^2})$ and $\mathcal{J}_C(\mathbb{F}_{p^2})$ The Isomorphism of $C_t(\mathbb{F}_{p^2})$ and $C(\mathbb{F}_{p^2})$

4.2   Finite Field Arithmetic

In our implementation, the curve C has embedding degree $k = 2$. Therefore we first need to construct the quadratic extension field $\mathbb{F}_{p^2}$. Since the prime p in this paper is congruent to 5 modulo 12, the quadratic extension field $\mathbb{F}_{p^2}$ can be constructed by the irreducible binomial $x^2 + 3$. Letting β denote -3, the elements of the field $\mathbb{F}_{p^2}$ can be represented as $a + b\sqrt{\beta}$, where $a, b \in \mathbb{F}_p$. By using the Karatsuba multiplication technique [20], a multiplication of two elements in $\mathbb{F}_{p^2}$ costs 3 multiplications in $\mathbb{F}_p$.

4.3   Using Degenerate Divisors and Denominator Elimination

Degenerate divisors have been widely used in the literature to speed up pairing computations on supersingular hyperelliptic curves [2,9,16]. Frey and Lange [12] have shown that the value of the Tate pairing is non-trivial if one restricts the second input to the embedding of $C(\mathbb{F}_{q^k})$ into $\mathcal{J}_C(\mathbb{F}_{q^k})$. In particular, when the embedding degree k is even, we can use a degenerate divisor class $\overline{P - P_\infty} \in \mathcal{J}_C(\mathbb{F}_{q^k})$ as the second argument of the Tate pairing where the coordinates of $P = (x, y) \in C(\mathbb{F}_{q^k})$ satisfy $x \in \mathbb{F}_{q^{k/2}}$ but $y \notin \mathbb{F}_{q^{k/2}}$. Therefore, in our implementation we first generate a degenerate divisor class $\overline{D}_t = [x - x_t, y_t] \in \mathcal{J}_{C_t}(\mathbb{F}_p)$ on the twisted curve $C_t/\mathbb{F}_p$. We then use the isomorphism ψ given above to obtain the degenerate divisor class $\overline{D} = \psi(\overline{D}_t) = [x - c^{-1}x_t, c^{-5/2}y_t] \in \mathcal{J}_C(\mathbb{F}_{p^2})$ on the curve C defined over $\mathbb{F}_{p^2}$. Hence, $\overline{D}$ can be used as the second argument to Miller's algorithm. Note that the first part of $\overline{D}$, namely $x - c^{-1}x_t$, is defined over $\mathbb{F}_p$, and thus the denominator elimination technique [3] applies in this case.

4.4   Evaluating Line Functions

At each iteration of the loop, we extract the rational functions $y - l(x)$ and $u_3(x)$ from the group operations and evaluate these functions at the second argument D_2. When new coordinates are used, we can work with $c'(x, y) = (rz_{24})y - ((s'_1 z_{21})x^3 + l_2 x^2 + l_1 x + l_0)$ and $u'_3(x) = z_{31}x^2 + U_{31}x + U_{30}$ for group addition, and $c'(x, y) = (Z_{31} Z_{32})y - ((s_1 z_{11})x^3 + l_2 x^2 + l_1 x + l_0)$ and $u'_3(x) = z_{31}x^2 + U_{31}x + U_{30}$ for group doubling as described in Section 3, where $\tilde{r}, z_{11}, z_{21}, Z_{31}, Z_{32}, s_1, s'_1, l_2, l_1$ and l_0 are from Table 1 and Table 3. We consider the following two cases for evaluating line functions at the divisor D_2:

1. D_2 is a degenerate divisor generated by the method in Section 4.3. In this case, D_2 can be represented by $[x - x_2, y_2] \in \mathcal{J}_C(\mathbb{F}_{p^2})$ for which $x_2 \in \mathbb{F}_p$ and $y_2 \notin \mathbb{F}_p$. Since x_2^2 and x_2^3 can be precomputed, this leaves $6M$ over $\mathbb{F}_p$ to be computed each time the function $c'(x, y)$ is evaluated. In particular, denominator elimination is applicable in this case and we do not need to

Table 5. Evaluating $c'(x, y)$ and $u'_3(x)$ at a General Divisor D_2 in New Coordinates

Input	$c'(x, y) = (\tilde{r} z_{21}) y - ((s'_1 z_{21}) x^3 + l_2 x^2 + l_1 x + l_0) \in \mathbb{F}_p[x, y]$	
	$u'_3(x) = z_{31} x^2 + U_{31} x + U_{30} \in \mathbb{F}_p[x]$	
	$D_2 = [x^2 + u_{21} x + u_{20}, v_{21} x + v_{20}] \in \mathcal{J}_C(\mathbb{F}_{p^2})$	
Output	$c'(D_2), u'_3(D_2) \in \mathbb{F}_{p^2}$	

Precomputations	Cost
$t_1 = u_{20} v_{21}, t_2 = u_{21} v_{20}, t_3 = t_1 - t_2, t_4 = v_{21} t_3, t_5 = v_{20}^2$	$13M, 3S$
$t_6 = t_4 + t_5, t_7 = u_{21} v_{21}, t_8 = 2 v_{20} - t_7, t_9 = t_1 + t_3, t_{10} = u_{21} t_3$	in $\mathbb{F}_{p^2}$
$t_{11} = u_{20} v_{20}, t_{12} = t_{10} + 2 t_{11}, t_{13} = u_{21}^2, t_{14} = t_3 t_{13}, t_{15} = 2 t_3 - t_2$	
$t_{16} = u_{20} t_{15}, t_{17} = t_{14} - t_{16}, t_{18} = u_{20} u_{21}, t_{19} = u_{20}^2, t_{20} = t_{19} u_{20}$	
$t_{21} = t_{19} u_{21}, t_{22} = t_{13} - 2 u_{20}, t_{23} = u_{20} t_{22}, t_{24} = t_{22} - u_{20}, t_{25} = u_{21} t_{24}$	

Computing $c'(D_2)$	Cost
$w_1 = \tilde{r} z_{21}, w_2 = s'_1 z_{21}, w_3 = w_1 t_6, w_4 = w_2 t_{17}, w_5 = l_2 t_{12}, w_6 = l_1 t_9$	$38M, 1S$
$w_7 = l_0 t_8, w_8 = w_3 - w_4 + w_5 - w_6 - w_7, w_9 = w_1 w_8, w_{10} = w_2 t_{20}$	in $\mathbb{F}_p$
$w_{11} = l_2 t_{21}, w_{12} = l_1 t_{23}, w_{13} = l_0 t_{25}, w_{14} = w_{10} - w_{11} + w_{12} - w_{13}$	
$w_{15} = w_2 w_{14}, w_{16} = l_2 t_{19}, w_{17} = l_1 t_{18}, w_{18} = l_0 t_{22}, w_{19} = w_{16} - w_{17} + w_{18}$	
$w_{20} = l_2 w_{19}, w_{21} = l_1 u_{20}, w_{22} = l_0 u_{21}, w_{23} = w_{21} - w_{22}, w_{24} = l_1 w_{23}$	
$w_{25} = l_0^2, c'(D_2) = w_9 + w_{15} + w_{20} + w_{24} + w_{25}$	

Computing $u'_3(D_2)$	Cost
$i_1 = z_{31}^2, i_2 = z_{31} U_{30}, i_3 = U_{30}^2, i_4 = i_1 t_{19}, i_5 = U_{31} u_{20}, i_6 = z_{31} t_{18}, i_7 = U_{30} u_{21}$	$13M, 2S$
$i_8 = i_5 - i_6 - i_7, i_9 = U_{31} i_8, i_{10} = i_2 t_{22}, u'_3(D_2) = i_3 + i_4 + i_9 + i_{10}$	in $\mathbb{F}_p$

evaluate the function $u'_3(x)$ at D_2. Therefore, the total cost of evaluating the rational functions at a degenerate divisor D_2 is given as $6M$ in $\mathbb{F}_p$ per iteration of the loop, with a precomputation of $1M + 1S$ in $\mathbb{F}_p$.

2. D_2 is a general divisor in Mumford representation, namely $D_2 = [u_2(x), v_2(x)] = [x^2 + u_{21} x + u_{20}, v_{21} x + v_{20}] \in \mathcal{J}_C(\mathbb{F}_{p^2})$. Note that the Mumford representation of divisors essentially gives the symmetric functions of the coordinates of the points in the support of the divisor. Hence, we can use these symmetric functions to obtain an explicit formula for evaluating the rational functions $c'(x, y)$ and $u'_3(x)$ at D_2, which only uses the coefficients of $u_2(x)$ and $v_2(x)$. Table 5 describes the efficient explicit formula for computing $c'(D_2)$ and $u'_3(D_2)$.

    Note that, in many of the multiplications in Table 5, one of the operands is in $\mathbb{F}_p$. Hence, a multiplication in $\mathbb{F}_{p^2}$ only needs $2M$ in $\mathbb{F}_p$ in this case. The total cost of evaluating the rational functions at a general divisor D_2 is given as $51M + 3S$ in $\mathbb{F}_p$ per iteration of the loop, with a precomputation of $13M + 3S$ in $\mathbb{F}_{p^2}$.

4.5 Final Exponentiation

For a genus 2 curve with an embedding degree of $k = 2$, the output of Miller's algorithm must be exponentiated to the power of $(p^2 - 1)/n$. The final exponentiation can be expressed in terms of operations in the base field $\mathbb{F}_p$. Letting

$f = a + b\sqrt{\beta} \in \mathbb{F}_{p^2}$ denote the output of Miller's algorithm, we can compute the final exponentiation as follows:

$$f^{\frac{p^2-1}{n}} = \left(\frac{a - b\sqrt{\beta}}{a + b\sqrt{\beta}}\right)^{\frac{p+1}{n}} = \left(\frac{a^2 - 3b^2 + \sqrt{\beta}\left((a-b)^2 - (a^2 + b^2)\right)}{a^2 + 3b^2}\right)^{\frac{p+1}{n}}.$$

We first calculate the expression in the parenthesis with $1I + 2M + 3S$ in $\mathbb{F}_p$, followed by an expensive exponentiation by $(p+1)/n$ which is executed in the arithmetic in $\mathbb{F}_{p^2}$.

4.6   Efficiency Comparison and Analysis

Since our encapsulated explicit formulae are applicable to pairing computations on both supersingular and non-supersingular genus 2 curves, we first show how our method can be used to improve previous implementations on supersingular genus 2 curves. We then analyze the case of non-supersingular genus 2 curves.

In [7] and [16], the authors considered the pairing computation on a family of supersingular genus 2 hyperelliptic curves with embedding degree 4 in affine coordinates. The curves are defined by the equation $y^2 = x^5 + a$, where $a \in \mathbb{F}_p^*$ and $p \equiv 2, 3 \bmod 5$. Note that our explicit doubling formulae only need $37M + 6S$ and $32M + 6S$ in projective and new coordinates, respectively, for this family of curves since the curve coefficients f_2 and f_3 are zero. Assume that the order n of the subgroup is about 160 bits. Following the same analysis as in [7] and [16], we compare the cost of computing the Tate pairing on this family of curves in different coordinate systems (without including the cost of the final exponentiation) in Table 6.

Table 6. Theoretical Complexity of Miller's Algorithm in Different Systems

Reference	Coordinate Type	Subgroup Order	Cost
Choie and Lee [7]	Affine	Random	$240I, 17688M, 2163S$
Ó hÉigeartaigh and Scott [16]	Affine	Solinas Prime	$162I, 10375M, 645S$
Our work	Projective	Random	$20017M, 1201S$
		Solinas Prime	$13129M, 967S$
	New	Random	$19297M, 1361S$
		Solinas Prime	$12487M, 971S$

We assume that field squarings have cost $S = 0.8M$. Then our encapsulated method is faster than that of [7] whenever $I/M > 4.03$. Moreover, our algorithm can achieve better performance than that of [16] whenever $I/M > 14.65$. These conditions usually hold for large prime field arithmetic on modern processors [25]. Therefore, for sufficiently large I/M, our method based on the encapsulated explicit formulae will be superior.

Next, we analyze the computational complexity of computing the Tate pairing using the non-supersingular genus 2 curve with embedding degree 2 (see Section

4.1). Note that the group order $n = 2^{160} + 7 = 2^{160} + 2^3 - 1$ is a Solinas prime [32]. The encapsulated explicit formulae for performing the group operations and extracting the rational functions in new coordinates are used (See Section 3). Furthermore, the degenerate divisor generated by using the technique in Section 4.3 is used as the second argument to Miller's algorithm. Based on these optimizations, the theoretical cost for computing the Tate pairing is given as (again without including the cost of the final exponentiation)

$$(\log_2 n)(T_D + T_c + T_{sk} + T_{mk}) + 2(T_A + T_c + T_{mk}),$$

where $T_A = 36M + 5S$ is the cost of adding two general divisors and extracting the rational functions with the formula in Table 1, $T_D = 35M + 7S$ is the cost of doubling a general divisor and extracting the rational functions with the formula in Table 3, $T_c = 6M$ (with a precomputation of $1M + 1S$) is cost of evaluating the rational function $c'(x,y)$ at the degenerate divisor D_2, and $T_{sk} = 3S$ and $T_{mk} = 3M$ are respectively the cost of squaring and multiplication in $\mathbb{F}_{p^2}$. Hence, the total cost of computing the Tate pairing with our optimizations is given as $7175M + 1621S$ in $\mathbb{F}_p$, whereas $163I + 4243M + 975S$ are required when using affine coordinates.

4.7   Experimental Results

In this section, experimental results are given for computing the Tate pairing using the techniques detailed in this paper for the non-supersingular genus 2 curve defined over $\mathbb{F}_p$ with embedding degree 2. All experiments were conducted on a Core 2 Duo^{TM}processor with a clock frequency of 2.67 GHz. The code was written in C and complied and debugged using *Microsoft Developer Studio 6*. The implementation of $\mathbb{F}_p$-arithmetic is based on various efficient algorithms in [15], where p is a 651-bit prime. Table 7 shows the timings of our finite field library and the corresponding IM-ratio. From Table 7, we note that the IM-ratio is 45.8 and $S = 0.89M$ in the target processor. Therefore, using our encapsulated explicit formulae, we can obtain a 31.5% performance improvement over working with affine coordinates.

Table 7. Timings of Prime Field $\mathbb{F}_p$ Library

# of bits of p	Multiplication (M)	Squaring (S)	Inversion (I)	IM-ratio
651	$4.59\mu s$	$4.08\mu s$	$210\mu s$	45.8

Table 8 gives experimental results for the implementation of the Tate pairing for the (160/1024) security level. All of the timings are given in milliseconds and three cases are included in the Table. The first case is the time taken to compute the Tate pairing when a degenerate divisor is used as the second argument to Miller's algorithm and the denominator technique is applied. The second case

Table 8. Experimental Results – (160/1024) Security Level

Case	Description	Running Time (ms)
1	Evaluating at a degenerate divisor	46.5
2	Evaluating using Mumford Representation	85.2
3	Elliptic curve pairing ($k = 2$ and $\log_2 p \approx 512$) [31]	8.9

gives the time when a general divisor with Mumford representation is used as the second input to the algorithm. The third case is the timing for computing the Tate pairing on non-supersingular elliptic curves with embedding degree 2 over $\mathbb{F}_p$ given by Scott [31].

Note that the implementations in [16] and [31] use special assembly language routines in MIRACL [30] for field operations. Therefore, it is hard to compare our implementation with those in [16,31]. The timings in Table 8 indicate that the Tate pairing on the non-supersingular genus 2 curve over $\mathbb{F}_p$ is a valid candidate for practical applications. However, elliptic curve pairings are faster than those in the genus 2 case for a (160/1024) bit security level, due to the more complicated group operations and larger Jacobian sizes of genus 2 curves.

5   Conclusion

In this paper, we have described how to efficiently implement pairing computations on genus 2 hyperelliptic curves over prime fields in projective coordinates. We generalize Chatterjee *et. al.*'s idea of encapsulated double-and-line computation and add-and-line computation to genus 2 curves in projective and new coordinates, respectively. We also show that some of the operations in the encapsulated method do not need to be computed since they are eliminated by the final exponentiation. Our new explicit formulae are applicable to pairing computations on both supersingular and non-supersingular genus 2 curves. Theoretical analysis shows that for pairing computations on supersingular genus 2 curves with embedding degree 4 over prime fields, our encapsulated method is faster than previously best known algorithms whenever $I/M > 14.65$. Furthermore, we also report the first efficient implementation of pairing computations on a non-supersingular genus 2 curve with embedding degree 2 over prime fields using the encapsulated explicit formulae and various known optimization techniques.

References

1. Avanzi, R.M., Cohen, H., Doche, C., Frey, G., Lange, T., Nguyen, K., Vercauteren, F.: Handbook of Elliptic and Hyperelliptic Curve Cryptography. Chapman & Hall/CRC, Boca Raton (2006)
2. Barreto, P.L.S.M., Galbraith, S., Ó hÉigeartaigh, C., Scott, M.: Efficient Pairing Computation on Supersingular Abelian Varieties. Design, Codes and Cryptography 42, 239–271 (2007)

3. Barreto, P.L.S.M., Kim, H.Y., Lynn, B., Scott, M.: Efficient Algorithm for Pairing-Based Cryptosystems. In: Yung, M. (ed.) CRYPTO 2002. LNCS, vol. 2442, p. 354. Springer, Heidelberg (2002)

4. Boneh, D., Franklin, M.: Identity-Based Encryption from the Weil Pairing. SIAM Journal of Computing 32(3), 586–615 (2003)

5. Boneh, D., Lynn, B., Shacham, H.: Short Signatures from the Weil Pairing. In: Boyd, C. (ed.) ASIACRYPT 2001. LNCS, vol. 2248, pp. 514–532. Springer, Heidelberg (2001)

6. Chatterjee, S., Sarkar, P., Barua, R.: Efficient Computation of Tate Pairing in Projective Coordinate over General Characteristic Fields. In: Park, C.-s., Chee, S. (eds.) ICISC 2004. LNCS, vol. 3506, pp. 168–181. Springer, Heidelberg (2005)

7. Choie, Y., Lee, E.: Implementation of Tate Pairing on Hyperelliptic Curve of Genus 2. In: Lim, J.-I., Lee, D.-H. (eds.) ICISC 2003. LNCS, vol. 2971, pp. 97–111. Springer, Heidelberg (2004)

8. Cocks, C., Pinch, R.G.E.: Identity-based Cryptosystems Based on the Weil Pairing (Unpublished manuscript) (2001)

9. Duursma, I.M., Lee, H.-S.: Tate pairing implementation for hyperelliptic curves $y^2 = x^p - x + d$. In: Laih, C.-S. (ed.) ASIACRYPT 2003. LNCS, vol. 2894, pp. 111–123. Springer, Heidelberg (2003)

10. Galbraith, S.D., McKee, J.F., Valença, P.C.: Ordinary Abelian Varieties Having Small Embedding Degree. Finite Fields and Their Applications 13(4), 800–814 (2007)

11. Freeman, D.: Constructing Pairing-Friendly Genus 2 Curves over Prime Fields with Ordinary Jacobians. In: Takagi, T., Okamoto, T., Okamoto, E., Okamoto, T. (eds.) Pairing 2007. LNCS, vol. 4575, pp. 152–176. Springer, Heidelberg (2007)

12. Frey, G., Lange, T.: Fast Bilinear Maps from The Tate-Lichtenbaum Pairing on Hyperelliptic Curves. In: Hess, F., Pauli, S., Pohst, M. (eds.) ANTS 2006. LNCS, vol. 4076, pp. 466–479. Springer, Heidelberg (2006)

13. Frey, G., Rück, H.-G.: A Remark Concerning m-Divisibility and the Discrete Logarithm Problem in the Divisor Class Group of Curves. Mathematics of Computation 62(206), 865–874 (1994)

14. Granger, R., Hess, F., Oyono, R., Thériault, N., Vercauteren, F.: Ate Pairing on Hyperelliptic Curves. In: Naor, M. (ed.) EUROCRYPT 2007. LNCS, vol. 4515, pp. 430–447. Springer, Heidelberg (2007)

15. Hankerson, D., Menezes, A., Vanstone, S.: Guide to Elliptic Curve Cryptography. Springer, New York (2004)

16. Ó hÉigeartaigh, C., Scott, M.: Pairing Calculation on Supersingular Genus 2 Curves. In: Biham, E., Youssef, A.M. (eds.) SAC 2006. LNCS, vol. 4356, pp. 302–316. Springer, Heidelberg (2007)

17. Hess, F., Smart, N.P., Vercauteren, F.: The Eta Pairing Revisited. IEEE Transactions on Information Theory 52(10), 4595–4602 (2006)

18. Hitt, L.: Families of Genus 2 Curves with Small Embedding Degree, Cryptology ePrint Archive, Report 2007/001 (2007), http://eprint.iacr.org/2007/001

19. Joux, A.: A One-Round Protocol for Tripartite Diffie-Hellman. In: Bosma, W. (ed.) ANTS 2000. LNCS 1838, vol. 1838, pp. 385–394. Springer, Heidelberg (2000)

20. Karatsuba, A., Ofman, Y.: Multiplication of Multidigit Numbers on Automata. Soviet Physics Doklady (English Translation) 7(7), 595–596 (1963)

21. Kawazoe, M., Takahashi, T.: Pairing-friendly Hyperelliptic Curves of Type $y^2 = x^5 + ax$, Cryptology ePrint Archive, Report 2008/026 (2008), http://eprint.iacr.org/2008/026

22. Lange, T.: Formulae for Arithmetic on Genus 2 Hyperelliptic Curves. Applicable Algebra in Engineering, Communication and Computing 15(5), 295–328 (2005)
23. Lee, E., Lee, H.-S., Lee, Y.: Eta Pairing Computation on General Divisors over Hyperelliptic Curves $y^2 = x^7 - x \pm 1$. In: Takagi, T., Okamoto, T., Okamoto, E., Okamoto, T. (eds.) Pairing 2007. LNCS, vol. 4575, pp. 349–366. Springer, Heidelberg (2007)
24. Menezes, A., Okamoto, T., Vanstone, S.A.: Reducing Elliptic Curve Logarithms to a Finite Field. IEEE Transactions on Information Theory 39(5), 1639–1646 (1993)
25. Menezes, A., van Oorschot, P.C., Vanstone, S.A.: Handbook of Applied Cryptography. Chapman & Hall/CRC, Boca Raton (1997)
26. Miller, V.S.: Short Programs for Functions on Curves (Unpublished manuscript) (1986), http://crypto.stanford.edu/miller/miller.pdf
27. Miyamoto, Y., Doi, H., Matsuo, K., Chao, J., Tsujii, S.: A Fast Addition Algorithm of Genus Two Hyperelliptic Curve. In: The 2002 Symposium on Cryptography and Information Security - SCIS 2002, pp. 497–502 (2002) (in Japanese)
28. Mumford, D.: Tata Lectures on Theta II. In: Prog. Math., vol. 43. Birkhäuser (1984)
29. Sakai, R., Ohgishi, K., Kasahara, M.: Cryptosystems Based on Pairings. In: Proceedings of the 2000 Symposium on Cryptography and Information Security - SCIS 2002, Okinawa, Japan, pp. 26–28 (2000)
30. Scott, M.: MIRACL (Multiprecision Integer and Rational Arithmetic C/C++ Library), http://www.shamus.ie/
31. Scott, M.: Scaling Security in Pairing-based Protocols, Cryptology ePrint Archive, Report 2005/139 (2005), http://eprint.iacr.org/2005/139
32. Solinas, J.: Generalized Mersenne Primes, Centre for Applied Cryptographic Research (CACR) Technical Reports, CORR 99-39,
http://www.cacr.math.uwaterloo.ca/techreprots/1999/corr99-39.pdf

Appendix: Explicit Formulae for Genus 2 Curves over $\mathbb{F}_p$

In this appendix, we give efficient explicit formulae for group operations on genus 2 curves over $\mathbb{F}_p$ in projective coordinates in the context of pairing computations. Table 9 and Table 10 address the cases of projective coordinates. Given two divisor classes $\overline{E}_1$ and $\overline{E}_2$, Table 9 computes the divisor class $\overline{E}_3 = [u_3(x), v_3(x)]$ and the rational function $l(x)$ such that $E_1 + E_2 = E_3 + \text{div}\left(\frac{y-l(x)}{u_3(x)}\right)$ in the projective coordinate system, where $l(x) = \frac{s'_1}{r}x^3 + \frac{l_2}{rZ_2}x^2 + \frac{l_1}{rZ_2}x + \frac{l_0}{rZ_2}$. For doubling a reduced divisor class E_1, Table 10 calculates the divisor class $\overline{E}_3 = [u_3(x), v_3(x)]$ and the rational function $l(x)$ such that $2E_1 = E_3 + \text{div}\left(\frac{y-l(x)}{u_3(x)}\right)$ in projective coordinates, where $l(x) = \frac{s_1}{R'}x^3 + \frac{l_2}{R'Z_1}x^2 + \frac{l_1}{R'Z_1}x + \frac{l_0}{R'Z_1}$.

Table 9. Mixed-Addition Formula on a Genus 2 curve over $\mathbb{F}_p$ (Projective Coordinates)

Input	Genus 2 HEC $C : y^2 = x^5 + f_3 x^3 + f_2 x^2 + f_1 x + f_0$	
	$\overline{E}_1 = [U_{11}, U_{10}, V_{11}, V_{10}, 1]$ and $\overline{E}_2 = [U_{21}, U_{20}, V_{21}, V_{20}, Z_2]$	
Output	$\overline{E}_3 = [U_{31}, U_{30}, V_{31}, V_{30}, Z_3] = \overline{E}_1 \oplus \overline{E}_2$	
	$l(x)$ such that $E_1 + E_2 = E_3 + \mathrm{div}\left(\frac{y - l(x)}{u_3(x)}\right)$	
Step	Expression	Cost
1	**Compute resultant $r = \mathrm{Res}(u_1, u_2)$:**	$5M, 1S$
	$\tilde{U}_{11} = U_{11}Z_2, \tilde{U}_{10} = U_{10}Z_2, z_1 = \tilde{U}_{11} - U_{21}, z_2 = U_{20} - \tilde{U}_{10}$	
	$z_3 = U_{11}z_1, z_4 = z_2 + z_3, r = z_2 z_4 + z_1^2 U_{10}$	
2	**Compute almost inverse of $u_2 \bmod u_1$:**	–
	$inv_1 = z_1, inv_0 = z_4$	
3	**Compute s':**	$7M$
	$w_0 = V_{10}Z_2 - V_{20}, w_1 = V_{11}Z_2 - V_{21}, w_2 = inv_0 w_0$	
	$w_3 = inv_1 w_1, s_1' = z_1 w_0 + z_2 w_2, s_0' = w_2 - U_{10}w_3$	
4	**Precomputations:**	$4M, 1S$
	$R = r^2, \tilde{s}_0' = s_0' Z_2, \tilde{s}_1' = s_1' Z_2, S = s_1' \tilde{s}_1', \tilde{r} = r \tilde{s}_1'$	
5	**Compute l:**	$5M$
	$l_2 = s_1' U_{21} + \tilde{s}_0', l_0 = s_0' U_{20} + r V_{20}$	
	$l_1 = (s_1' + s_0')(U_{21} + U_{20}) - s_1' U_{21} - s_0' U_{20} + r V_{21}$	
6	**Compute U_3:**	$8M, 1S$
	$w_1 = \tilde{U}_{11} + U_{21}, U_{31} = s_1'(2\tilde{s}_0' - s_1' z_1) - RZ_2, l_1' = l_1 s_1'$	
	$U_{30} = \tilde{s}_0'(s_0' - 2s_1' U_{11}) + s_1'^2(z_3 - \tilde{U}_{10} - U_{20}) + 2l_1' + Rw_1$	
7	**Compute V_3:**	$6M$
	$w_1 = l_2 s_1' - U_{31}, V_{30} = U_{30}w_1 - S(l_0 s_1'), V_{31} = U_{31}w_1 + S(U_{30} - l_1')$	
8	**Adjust:**	$3M$
	$\overline{Z}_3 = \tilde{r}S, U_{31} = \tilde{r}U_{31}, U_{30} = \tilde{r}U_{30}$	
Sum		$38M, 3S$

Table 10. Doubling Formula on a Genus 2 Curve over $\mathbb{F}_p$ (Projective Coordinates)

Input	Genus 2 HEC $C : y^2 = x^5 + f_3 x^3 + f_2 x^2 + f_1 x + f_0$	
	$\overline{E}_1 = [U_{11}, U_{10}, V_{11}, V_{10}, Z_1]$	
Output	$\overline{E}_3 = [U_{31}, U_{30}, V_{31}, V_{30}, Z_3] = [2]\overline{E}_1$	
	$l(x)$ such that $2E_1 = E_3 + \mathrm{div}\left(\frac{y - l(x)}{u_3(x)}\right)$	
Step	Expression	Cost
1	**Compute resultant and precomputations:**	$4M, 3S$
	$\overline{Z}_2 = Z_1^2, \tilde{V}_{11} = 2V_{11}, \tilde{V}_{10} = 2V_{10}, w_0 = V_{11}^2, w_1 = U_{11}^2, w_2 = \tilde{V}_{10}Z_1$	
	$w_3 = 4w_0, w_4 = w_2 - U_{11}\tilde{V}_{11}, r = U_{10}w_3 + \tilde{V}_{10}w_4$	
2	**Compute almost inverse:**	–
	$inv_1' = -\tilde{V}_{11}, inv_0' = w_4$	
3	**Compute k':**	$5M$
	$w_3 = f_3 \overline{Z}_2 + w_1, w_4 = 2U_{10}, \tilde{w}_4 = w_4 Z_1, k_1' = 2w_1 + w_3 - \tilde{w}_4$	
	$k_0' = U_{11}(2\tilde{w}_4 - w_3) + Z_1(f_2 \overline{Z}_2 - w_0)$	
4	**Compute s':**	$7M$
	$w_0 = k_0' inv_0', w_1 = k_1' inv_1', s_2 = w_2 k_1' - \tilde{V}_{11}k_0'$	
	$s_1' = s_2 Z_1, s_0' = w_0 - Z_1 U_{10} w_1$	
5	**Precomputations:**	$6M, 2S$
	$R = r\overline{Z}_2, \tilde{R} = Rs_1', R' = Rs_2, S_0 = s_0'^2, S_1 = s_1'^2, S = s_0' s_1', s_0 = s_0' s_2, s_1 = s_1' s_2$	
6	**Compute l:**	$6M$
	$l_2 = s_1 U_{11} + s_0 Z_1, l_0 = s_0 U_{10} + R'V_{10}$	
	$l_1 = (s_1 + s_0)(U_{11} + U_{10}) - s_1 U_{11} - s_0 U_{10} + R'V_{11}$	
7	**Compute U_3:**	$4M, 1S$
	$U_{30} = S_0 + R(s_2 \tilde{V}_{11} + 2r Z_1 U_{11}), U_{31} = 2S - R^2$	
8	**Compute V_3:**	$4M$
	$w_1 = l_2 - U_{31}, w_2 = U_{30}w_1, w_3 = U_{31}w_1$	
	$V_{31} = w_3 + S_1(U_{30} - l_1), V_{30} = w_2 - S_1 l_0$	
9	**Adjust:**	$3M$
	$Z_3 = S_1 \tilde{R}, U_{31} = U_{31}\tilde{R}, U_{30} = U_{30}\tilde{R}$	
Sum		$39M, 6S$

On Software Parallel Implementation of Cryptographic Pairings⋆,⋆⋆

Philipp Grabher, Johann Großschädl, and Dan Page

University of Bristol, Merchant Venturers Building,
Woodland Road, Bristol, BS8 1UB, UK
{grabher,johann,page}@cs.bris.ac.uk

Abstract. A significant amount of research has focused on methods to improve the efficiency of cryptographic pairings; in part this work is motivated by the wide range of applications for such primitives. Although numerous hardware accelerators for pairing evaluation have used parallelism within extension field arithmetic to improve efficiency, thus far less emphasis has been placed on software exploitation of similar. In this paper we focus on parallelism within one pairing evaluation (intra-pairing), and parallelism between different pairing evaluations (inter-pairing). We identify several methods for exploiting such parallelism (extending previous results in the context of ECC) and show that it is possible to accelerate pairing evaluation by a significant factor in comparison to a naive approach.

1   Introduction

Generally speaking, one uses the term cryptographic pairing to describe a non-degenerate bilinear map of the form

$$e : \mathbb{G}_1 \times \mathbb{G}_2 \longrightarrow \mathbb{G}_T.$$

In this paper we focus on the Ate pairing which takes the concrete form

$$e : E(\mathbb{F}_p) \times \overline{E}(\mathbb{F}_{p^{k/6}}) \longrightarrow \mathbb{F}_{p^k}^{\times}$$

where $\overline{E}$ is the quadratic twist of an elliptic curve E defined over $\mathbb{F}_{p^{k/6}}$. The type and volume of applications enabled by pairings of this form has dictated that methods for their evaluation remain an ongoing research challenge. This is magnified by the fact that said applications have permeated both high-performance

⋆ The work described in this paper has been supported in part by the European Commission through the IST Programme under Contract IST-2002-507932 ECRYPT. The information in this document reflects only the author's views, is provided as is and no guarantee or warranty is given that the information is fit for any particular purpose. The user thereof uses the information at its sole risk and liability.
⋆⋆ The work described in this paper has been supported in part by EPSRC grant EP/E001556/1.

R. Avanzi, L. Keliher, and F. Sica (Eds.): SAC 2008, LNCS 5381, pp. 35–50, 2009.
© Springer-Verlag Berlin Heidelberg 2009

and embedded contexts: computational efficiency and storage footprint are both important. Improvements to high-level algorithms that relate to the pairing itself are clearly the most significant in terms of efficiency; for an overview of the evolution of this topic, see the excellent description by Scott [39]. In short, improvement of seminal but unpublished work by Miller [34] resulted in the first practical algorithms for evaluation of the Tate pairing [5,19]. These results were further optimised by Duursma and Lee [15] who developed an inexpensive, closed form for specific parameterisations later improved by Kwon [30]. Their techniques were generalised and extended to produce the Eta [4] and Ate [23] pairings, currently considered the fastest means of evaluation.

However, as well as the pairing itself, one depends on lower-level algorithms for arithmetic in the fields $\mathbb{F}_p$, $\mathbb{F}_{p^{k/6}}$ and $\mathbb{F}_{p^k}$. Previous results have reported on analysis and efficient realisation of said algorithms; see for example [28,20,13]. One can readily identify two types of parallelism within these algorithms and within pairing based cryptosystems more generally: that within a single pairing evaluation (intra-pairing) or between several pairing evaluations (inter-pairing). Put more simply, in the first case the aim is to compute $R = e(P,Q)$ for some P and Q from the appropriate groups; our focus is on parallelism within algorithms for the pairing and constituent arithmetic. Efficient implementation of pairings in hardware have used this feature to great effect; see [27] for an example design where extension field arithmetic is realised using several parallel computational units to reduce latency. In the second case, the aim is to compute several pairings $R_i = e(P_i, Q_i)$; our focus in on the fact that each R_i can be computed independently. Although Granger and Smart [21] describe a method to improve performance where the pairings form terms in a larger product, i.e. $R = \prod e(P_i, Q_i)$, actually capitalising on the parallelism between disjoint pairings is less well examined. This is despite the fact that numerous instances exist, verification of BLS signatures [9] to name one, where this could be useful.

Identifying parallelism in algorithms for the pairing and constituent arithmetic is only the first step: in order to exploit said parallelism, one must have effective methods to map an algorithm onto the capabilities of a given host platform. Often this mapping is difficult enough that any perceived advantage offered by parallelism is eliminated by implementation overhead, in other cases the correct choice of technique is limited by issues such as parameterisation and use of the pairing in real applications. Forthcoming work by Hankerson et al. [24] gives an excellent comparison between different algorithms and parameterisations, but does not investigate parallelism beyond that in $\mathbb{F}_p$. Our goal in this paper is to fill the resulting gap, focusing on parallelism realised using software techniques as a means of optimising concrete implementations of the Ate pairing.

We organise the paper as follows. In Section 2 we given an overview of the Ate pairing and standard methods for parameterisation and evaluation. Then, in Section 3, we make a detailed study of parallelism within algorithms for the pairing and constituent arithmetic. Section 4 describes details of our implementation including an efficient algorithm for parallel multiplication in $\mathbb{F}_p$. Using the identified techniques we present and analyse experimental results derived from

---

Algorithm 1. An algorithm to compute the Ate pairing

Input   : $Q \in E(\mathbb{F}_p)$, $P \in E(\mathbb{F}_{p^2})$, $s = t - 1 \in \mathbb{Z}$.
Output: $e(Q, P)$.

1 $f \leftarrow 1$, $T \leftarrow P$
2 **for** $i = |s| - 2$ **downto** 0 **do**
3     $f \leftarrow f^2 \cdot l_{T,T}(Q)$, $T \leftarrow 2 \cdot T$
4   **if** $s_i = 1$ **then**
5        $f \leftarrow f \cdot l_{T,P}(Q)$, $T \leftarrow T + P$
6 **return** $f^{(p^k - 1)/n}$

---

their implementation on Intel Core2 and Pentium 4 processors; this is captured in Section 5. Finally, we summarise our findings and conclude in Section 6.

2   The Ate Pairing

Successful parameterisation of the Ate pairing requires an elliptic curve $E(\mathbb{F}_p)$ whose order n is divisible by some large prime r. Let k, the embedding degree of the curve, be the smallest positive integer such that $r \mid p^k - 1$. A Barreto-Naehrig curve or BN-curve [6] of the form

$$E(\mathbb{F}_p) : y^2 = x^3 + b$$

where $b \neq 0$, satisfies these requirements. In particular, such a curve has prime order, i.e. $r = n$, and embedding degree $k = 12$. Additionally, the trace, curve order and characteristic of $\mathbb{F}_p$ can be parameterised by x as follows

$$
\begin{aligned}
t(x) &= 6x^2 + 1 \\
n(x) &= 36x^4 - 36x^3 + 18x^2 - 6x + 1 \\
p(x) &= 36x^4 - 36x^3 + 24x^2 - 6x + 1.
\end{aligned}
$$

We closely follow the excellent description of Devegili et al. [14] who show that by selecting $x = -6917529027641089837$ for example, one specifies a 256-bit value p and associated curve where n is of low Hamming weight. Selecting such an x makes the notation $t(x)$, for example, extraneous; using this specific value of x we simply write t instead. Since the associated p satisfies various congruences, it enables an efficient construction of extension field arithmetic using the tower $\mathbb{F}_{p^2} = \mathbb{F}_p[X]/(X^2 - \beta)$, $\mathbb{F}_{p^6} = \mathbb{F}_{p^2}[Y]/(Y^3 - \xi)$, $\mathbb{F}_{p^{12}} = \mathbb{F}_{p^6}[Z]/(Z^2 - \xi\prime)$ where $\beta = -2 \in \mathbb{F}_p$, $\xi = -1 - \sqrt{\beta} \in \mathbb{F}_{p^2}$ and $\xi\prime = \sqrt[3]{\xi} \in \mathbb{F}_{p^6}$.

Evaluation of the pairing is achieved using Algorithm 1 where $l_{A,B}(C)$ denotes the line function between points A and B evaluated at C. The selection of a sparse x allows for efficient realisation of the final exponentiation by $(p^k - 1)/n$ as described fully by Devegili et al. [14].

3   Exploitation of Parallelism

SIMD and SWAR. Many commodity processors now support SWAR (SIMD Within a Register), a form of vector processing; exemplar designs include several generations of SSE by Intel, VIS by Sun, 3DNow! by AMD, and AltiVec by Apple, IBM and Motorola. To utilise this feature, one packs say u sub-words, each v bits in size, into a large SWAR vector. Using such vectors one can permit SIMD style vector operations. Let $\overline{x}_i$ denote the i-th sub-word packed into vector $\overline{x} = (\overline{x}_0, \overline{x}_1, \ldots, \overline{x}_{u-1})_v$. Using such a representation, one can compute all u component-wise additions $\overline{r}_i = \overline{x}_i + \overline{y}_i$ with one operation. The choice of u and v, which dictate the number and size of sub-words that can be packed into a fixed vector length, depends on the application. Often an instruction set will support a (somewhat) orthogonal set of operations and choices of u and v. This approach has brought significant performance improvements in easily vectorised kernels such as those found in media processing; by making parallelism explicit the processor can maintain a high issue rate and ensure a good trade-off between provision of computational resources and their utilisation.

Use of SWAR style instruction sets have been successfully used to accelerate kernels in symmetric cryptography; for example [10,11,32,33,37]. Although exploiting parallelism within point multiplication in vanilla Elliptic Curve Cryptography (ECC) is possible [2,26], vectorisation of the public-key cryptography is often more problematic. Consider two n-bit multi-precision integers x and y represented by $l = \lceil n/w \rceil$ machine words where x_i denotes the i-th such w-bit word. Values represented as such are commonly manipulated within cryptosystems such as RSA and ECC. For the sake of clarity, imagine we set $n = 128$ and $u = l = 4$ such that $v = w = 32$. This implies that we can store x and y in one SWAR register each, i.e.

$$(x_0, x_1, x_2, x_3)_{32} \quad \text{and} \quad (y_0, y_1, y_2, y_3)_{32}$$

The problem is that to perform multi-precision addition, for example, one must deal with carry from one sub-word into another sub-word within the same vector. That is, say we want to compute $r = x + y$. The addition of x and y is not component-wise: for example, we need to take the carry produced by the primitive addition $x_0 + y_0$ and factor it into $x_1 + y_1$ thereby destroying the component-wise nature of computation and hence the SIMD style parallelism.

The flexibility of ECC parameterisations helps somewhat in resolving this problem. One might view specific field representations such as Residue Number Systems (RNS) and Optimal Extension Fields (OEF) [3] as more suitable for vectorisation; parameterisation and parallel implementation over $\mathbb{F}_{2^n}$ has also been effective [7] since carries are essentially eliminated by the nature of arithmetic. Motivated by application in RSA as well as ECC, there is a similar effort to accelerate arithmetic in $\mathbb{F}_p$ (or more exactly modulo some integer p). Work by Acar [1] and reports by Intel [25] and Apple [12] all investigate the use of SIMD parallelism for implementing multi-precision integer arithmetic. Acar states that his implementation of RSA on a processor with an MMX instruction set runs significantly slower due to a lack of unsigned 16-bit and 32-bit multiplication. Intel are more positive in their results that focus on the SSE2 instruction set. Their method applies a form of recoding into a representation with a smaller digit size; this allows fast combination of partial products without requiring carries at all. Hankerson et al. [22, Chapter 5.1.3] also discuss the same technique within the context of ECC.

This previous work offers a natural way to exploit intra-pairing parallelism: one simply accelerates arithmetic in $\mathbb{F}_p$ which, in turn, accelerates all higher layers of arithmetic and therefore the pairing evaluation itself.

Bit-slicing and Digit-slicing. Considering a scalar processor with a w-bit word size, let x_i denote the i-th bit of a machine word x where i is termed the index of the bit. Such a processor operates natively on word sized operands. For example, with a single operation one might perform addition of w-bit operands x and y to produce $r = x + y$, or component-wise XOR to produce $r_i = x_i \oplus y_i$ for all $0 \leq i < w$. This ability is restricted however when an algorithm is required to perform some operation involving different bits from the same word. For example one might be required to combine x_i and x_j, where $i \neq j$, using an XOR operation in order to compute the parity of x. In this situation one is required to shift (and potentially mask) the bits so they are aligned at the same index ready for combination via a native, component-wise XOR. The technique of bit-slicing, proposed by Biham for efficient implementation of DES [8], offers a way to reduce the associated overhead. Instead of representing the w-bit value x as one machine word, we represent x using w machine words where word i contains x_i aligned at the same fixed index j. As such, there is no need to align bits ready for use in a component-wise XOR operation. Additionally, since native word oriented logical operations in the processor operate on all w bits in parallel, one can pack w different values (say $x[k]$ for $0 \leq k < w$) into the w words and proceed using an analogy of SIMD style parallelism. Conversion to and from a bit-sliced representation can represent an overhead but this can be amortised if the cost of computation using the bit-sliced values is significant enough: Biham used this technique to extract a five-fold performance improvement from DES using a 64-bit Alpha processor.

Although it overloads the term somewhat, one might describe previous SWAR based implementations of public-key cryptography as digit-serial in the sense that they try to extract parallelism from a series of digits representing one value. An alternative approach, which one might describe as digit-sliced SWAR, represents the digit based analogy of the bit based slicing approach outlined above. This seems to have been first investigated by Montgomery in the context of ECM based factoring [36] and then rediscovered and applied in the context of RSA by Page and Smart [38]. Following the example above, the basic idea is that instead of representing an l-word multi-precision integer x by packing the digits x_i into one SWAR vector, we slice the digits into l separate SWAR vectors where vector i contains x_i aligned at the same fixed index j. For the case where $n = 128$, $u = l = 4$ and $v = w = 32$ we therefore represent x and y using four SWAR registers

$$
\begin{array}{ccc}
(x_0, \cdot, \cdot, \cdot)_{32} & & (y_0, \cdot, \cdot, \cdot)_{32} \\
(x_1, \cdot, \cdot, \cdot)_{32} & \text{and} & (y_1, \cdot, \cdot, \cdot)_{32} \\
(x_2, \cdot, \cdot, \cdot)_{32} & & (y_2, \cdot, \cdot, \cdot)_{32} \\
(x_3, \cdot, \cdot, \cdot)_{32} & & (y_3, \cdot, \cdot, \cdot)_{32}
\end{array}
$$

Algorithm 2. An algorithm to compute the Ate pairing

Input : $Q \in E(\mathbb{F}_p)$, $P \in E(\mathbb{F}_{p^2})$, $s = t - 1 \in \mathbb{Z}$.
Output: $e(Q, P)$.

1 $\tau_f[0] \leftarrow 1$
2 $\tau_T[0] \leftarrow P$

3 **for** $i = 1$ **upto** $|s| - 1$ **do**
4      $\tau_f[i] \leftarrow \tau_f[i-1]^2 \cdot l_{\tau_T[i-1], \tau_T[i-1]}(Q)$
5      $\tau_T[i] \leftarrow 2 \cdot \tau_T[i-1]$

6 $f_0 \leftarrow 1, T_0 \leftarrow \mathcal{O}$
7 $f_1 \leftarrow 1, T_1 \leftarrow \mathcal{O}$

8 **par**
9   **for** $i = 0$ **upto** $|s| - 1$ **do**
10   **if** $s_i = 1$ **and** $i = 0 \pmod 2$ **then**
11             $f_0 \leftarrow f_0 \cdot \tau_f[i] \cdot l_{T_0, \tau_T[i]}(Q)$, $T_0 \leftarrow T_0 + \tau_T[i]$
12   **for** $i = 0$ **upto** $|s| - 1$ **do**
13   **if** $s_i = 1$ **and** $i = 1 \pmod 2$ **then**
14                 $f_1 \leftarrow f_1 \cdot \tau_f[i] \cdot l_{T_1, \tau_T[i]}(Q)$, $T_1 \leftarrow T_1 + \tau_T[i]$

15 $f \leftarrow f_0 \cdot f_1 \cdot l_{T_0, T_1}(Q)$

16 **return** $f^{(p^k - 1)/n}$

---

where $\cdot$ denotes some arbitrary padding. The premise is that this makes carry easier to deal with: we are now faced with carries between sub-words of different vectors which are aligned at the same index rather than carries between sub-words in the same vector. As such and in a naive sense, one expects the amount of sub-word reorganisation, which represents a significant computational overhead, to be lower. Again there is an overhead in conversion to and from the digit-sliced representation. However, in common with the bit-slicing approach, we can operate on u packed values at the same time by replacing the padding (i.e. $\cdot$) with useful data. This essentially allows us to compute u separate multi-precision additions (say $x[k] + y[k]$ for $0 \leq k < u$), for example, at the same time. We call each such parallel digit-sliced operation a channel and term an implementation c-way digit-sliced if there are c channels utilised.

In terms of the pairing and constituent arithmetic, the technique of digit-slicing is potentially interesting. At any level, all the algorithms for arithmetic are (or are close to) control-flow invariant; for example for any given pairing evaluation using some fixed parameterisation, one performs the same operation at a given step so only the data values differ. As such, one can deploy digit-slicing to exploit intra-pairing parallelism (for example performing c multiplications in $\mathbb{F}_p$ at once to accelerate arithmetic in $\mathbb{F}_{p^2}$), or inter-pairing parallelism (for example evaluating c pairings at once).

Multi-core Processors. A modern trend in the design of microprocessors is that of multi-core, i.e. having many physical processor cores on a single die. This philosophy is in part guided by the need to make effective use of advances in fabrication which allow dies to house a huge number of transistors, and the so-called memory wall which posits that memory access dominates the performance of conventional single-core processors. In software, one can take advantage of multi-core processors using, for example, the OpenMP standard; with suitable compiler and operating system support this enables multiple code sequences to be executed in parallel, one on each core.

The use of multi-core processors is an emerging research topic in the context of cryptographic implementation, for example Fan et al. investigate modular multiplication [16] and ECC [17] on this type of platform. Intra-pairing parallelism is clearly possible at the field arithmetic level as evidenced by related hardware based approaches [27]. In software however, the overhead of thread management is a limiting factor: if the threads are too fine-grained then the cost of their management will dominate useful computation and eliminate the advantage of parallelism. An alternative, therefore, is to consider more coarse-grained parallelism. In this setting, inter-pairing parallelism is easy to exploit: we simply have each core compute a separate pairing. Exploiting coarse-grained intra-pairing parallelism requires more thought. For example, one might redesign Algorithm 1 to allow parallelism between point arithmetic or line function evaluations.

Consider Algorithm 2 which is derived from a specialisation of so-called fixed-base windowing [22, Algorithm 3.41] with window size 1. Use of the **par** keyword shows that after a precomputation phase comprised of point (resp. line) doublings, two threads can compute point (resp. line) additions in parallel (one thread deals with odd-indexed bits in s, the other even-indexed bits). The clear advantage of this approach is parallelism; the clear disadvantage is the significant memory overhead for tables τ_f and τ_T, and the fact that the point (resp. line) additions are now projective rather than mixed.

4 Implementation Details

In the following we elaborate on the concrete implementation of the field arithmetic using scalar (i.e. non-SIMD) as well as SIMD (i.e. MMX, SSE) instruction sets. Both implementations have in common that the modular multiplication (resp. squaring) operation is realised via Montgomery reduction [35]. The inversion is performed using the Extended Euclidean Algorithm (EEA).

4.1 Field Arithmetic with the IA-32/IA-64 Instruction Set

The IA-32 architecture provides an add-with-carry instruction (adc) and a 32-bit unsigned multiply instruction yielding a 64-bit result (mul). Thanks to the availability of these two instructions, the arithmetic operations in $\mathbb{F}_p$ can be implemented in a fairly straightforward way: a field element is simply represented in form of an array of single-precision (i.e. 32-bit) words and the software routines for addition and multiplication loop through these arrays and produce the result using the afore-mentioned instructions. Our implementation of the field

---

Algorithm 3. Montgomery multiplication (CIOS method)

Input : An s-word modulus $M = (m_{s-1}, \ldots, m_1, m_0)$, two operands
$A = (a_{s-1}, \ldots, a_1, a_0)$ and $B = (b_{s-1}, \ldots, b_1, b_0)$ with $A, B < M$, and
the constant $m_0' = -m_0^{-1} \bmod 2^w$.

Output: The Montgomery product $Z = A \cdot B \cdot 2^{-n} \bmod M$.

1   $Z \leftarrow 0$
2   **for** $i = 0$ **upto** $s - 1$ **do**
3      $u \leftarrow 0$
4   **for** $j = 0$ **upto** $s - 1$ **do**
5        $(u, v) \leftarrow a_j \times b_i + z_j + u$
6        $z_j \leftarrow v$
7      $(u, v) \leftarrow z_s + u, \; z_s \leftarrow v, \; z_{s+1} \leftarrow u$
8      $q \leftarrow z_0 \times m_0' \bmod 2^w$
9      $(u, v) \leftarrow z_0 + m_0 \times q$
10   **for** $j = 1$ **upto** $s - 1$ **do**
11       $(u, v) \leftarrow m_j \times q + z_j + u$
12       $z_{j-1} \leftarrow v$
13     $(u, v) \leftarrow z_s + u, \; z_{s-1} \leftarrow v, \; z_s \leftarrow z_{s+1} + u$
14 **if** $Z \geq M$ **then**
15     $Z \leftarrow Z - M$
16 **return** $Z = (z_{s-1}, \ldots, z_1, z_0)$

---

arithmetic is written in ANSI C and contains some hand-optimised assembly
language sections for the performance-critical inner-loop operations. As the size
of the fields used in pairing-based cryptography is relatively small, it is possible
to unroll the inner loops and gain some extra performance at the expense of a
slight increase in code footprint.

Algorithm 3 shows the Coarsely Integrated Operand Scanning (CIOS) method
for calculating the Montgomery product $Z = A \cdot B \cdot 2^{-n} \bmod M$ [29]. The n-bit
operands A, B, M are represented by arrays of s single-precision w-bit words.
The algorithm has a nested loop structure with two inner loops; the first con-
tributes to the calculation of the product $A \cdot B$ and the second implements the
modular reduction operation. Both inner loops perform the same operation: two
single-precision words are multiplied together, and then two other words are
added to the product. Therefore, each iteration of the inner loop executes a `mul`,
two `add`, and two `adc` instructions, respectively.

4.2   Field Arithmetic with the MMX/SSE Instruction Set

In order to accelerate the execution of multimedia kernels, Intel introduced the
MMX instruction set in 1997 as a SIMD extension to the IA-32 architecture.
MMX provides eight 64-bit registers and adds 57 new instructions. Most of these
instructions operate on packed data types, which means that a 64-bit MMX
operand can also be treated as either two 32-bit, four 16-bit, or eight 8-bit quantities. The Streaming SIMD Extensions (SSE) further enhance the capabilities

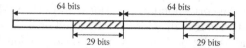

Fig. 1. The packed 29-bit digits within a single 128-bit SSE register, as detailed by [25]

Table 1. Timings for Montgomery multiplication and squaring (in cycles as reported by `rdtsc`) on a Pentium 4 processor for 256-bit, 384-bit and 512-bit operands

Implementation	256-bit	384-bit	512-bit
SIMD Montgomery Mul.	1182	2104	2978
GMP (`mpn_mul_n` + `redc`)	1171	2429	3700
SIMD Montgomery Sqr.	1063	1875	2523
GMP (`mpn_sqr_n` + `redc`)	1051	2257	3151

of the IA-32 architecture through the integration of eight 128-bit registers and appropriate instructions. For example, the SSE2 instruction `pmuludq` allows one to execute two 32×32-bit multiplications independently and in parallel, each yielding a 64-bit result. However, the main drawback of the MMX and SSE instruction sets in the context of multi-precision integer arithmetic is the lack of an add-with-carry instruction.

The fact that neither MMX nor SSE provide an add-with-carry instruction not only makes multiple-precision addition relatively costly, but also defines how multiple-precision multiplication must be implemented in order to exploit SIMD-level parallelism. In [25], Intel recommends that multi-precision integers should be represented as arrays of 29-bit words (instead of the more intuitive representation with 32-bit words) and to pack two such 29-bit words into a 128-bit quantity which can be loaded into SSE registers using the `movdqa` instruction. Two 29×29-bit multiplications can be executed in parallel and several 58-bit products can be accumulated without overflow. More precisely, the 29-bit representation eliminates the need to propagate carry bits from less to more significant words during a multiple-precision multiplication; a single carry propagation must be performed at the very end to obtain the correct result. We implemented the CIOS method for Montgomery multiplication following these guidelines which also allowed us to fuse the two inner loops. This loop fusion does not only reduces the loop overhead, but also eliminates a number of load/store instructions as, for example, the quantity z_j in Algorithm 3 needs to be loaded only once. However, a disadvantage of the multiplication technique described in [25] is that two arrays are necessary for storing the intermediate results during a Montgomery multiplication. This "redundant" representation makes the outer loop of Algorithm 3 relatively costly, in particular the calculation of the quotient q.

Table 1 compares the execution times (in clock cycles) of our Montgomery arithmetic implemented according to Algorithm 3 using the 29-bit representation detailed in [25], and the corresponding functions[1] from the GMP library version 4.2.2. Our implementation is slightly slower for 256-bit operands, but outperforms GMP for 384-bit and 512-bit operands. As mentioned previously, our implementation of Algorithm 3 is characterised by a relatively costly outer loop, while the inner loop is extremely efficient. However, for short operands, the operations in the outer loop dominate the execution time, which renders the 29-bit representation less attractive.

5   Implementation Results

In order to evaluate the options for exploiting parallelism introduced in previous sections, we used two experimental platforms; the rationale for their selection was that they represent previous (NetBurst) and current (Core2) generation micro-architectures in commodity microprocessors:

Platform A housed a 2.80GHz Intel Pentium 4 processor running a 32-bit installation of Linux including a 2.6.9 series kernel and 32-bit Intel C compiler version 10.1. The SIMD instruction set on this platform was limited to SSE2 series (and earlier) instructions only.

Platform B housed a 2.40GHz Intel Core2 Duo processor running a 64-bit installation of Linux including a 2.6.18 series kernel and 64-bit Intel C compiler version 10.1. The SIMD instruction set on this platform was limited to SSE3 series (and earlier) instructions only.

Since our goal is to highlight issues with existing processors, we do not investigate the impact of altering the number of execution pipelines within a particular micro-architecture (which could be interesting). Note that the second experimental platform includes a multi-core processor: it has two processor cores. Using the platforms we constructed eight separate implementations which represent a cross-section of the presented approaches to intra-pairing and inter-pairing parallelism (recalling that we have a fixed parameterisation where p is a 256-bit prime):

Implementation A uses the scalar (i.e. non-SIMD) instruction set and a 32-bit digit size; evaluates one pairing at a time using Algorithm 1.

Implementation B uses the scalar (i.e. non-SIMD) instruction set and a 64-bit digit size; evaluates one pairing at a time using Algorithm 1.

Implementation C uses the SIMD (i.e. SSE) instruction set and a 29-bit digit size to perform digit-serial $\mathbb{F}_p$ arithmetic; evaluates one pairing at a time using Algorithm 1.

---

[1] Note that GMP features a function for Montgomery reduction (redc), but not for Montgomery multiplication. Therefore, a Montgomery multiplication must be composed of mpn_mul_n and redc. We evaluated the execution times of mpn_mul_n, mpn_sqr_n, and redc with help of the speed program.

Table 2. Timings for major operations (in cycles as reported by `rdtsc`) on experimental platform A (Pentium 4). $\mathbb{F}_p$ is a 256-bit prime field.

	$\mathbb{F}_p$			$\mathbb{F}_{p^{12}}$			$e(P,Q)$
	Inv	Add	Mul	Inv	Add	Mul	
A	278754	188	5826	892508	1870	347249	177634471
B	–	–	–	–	–	–	–
C	271063	226	1182	624667	2144	174774	58266382
D	278012	186	5813	633801	1803	229323	127986142
E	–	–	–	–	–	–	–
F	–	–	–	–	–	–	–
G	299268	566	3444	818690	6134	312738	147441219
H	–	–	–	–	–	–	–

Table 3. Timings for major operations (in cycles as reported by `rdtsc`) on experimental platform B (Core2 Duo). $\mathbb{F}_p$ is a 256-bit prime field.

	$\mathbb{F}_p$			$\mathbb{F}_{p^{12}}$			$e(P,Q)$
	Inv	Add	Mul	Inv	Add	Mul	
A	156179	132	1117	287160	1061	76002	44814516
B	155567	107	395	208603	779	31484	23319673
C	155514	114	477	290536	842	64490	28452901
D	154295	132	1106	278503	1062	73336	35215963
E	154217	107	399	207236	787	24494	14429439
F	155567	108	394	208612	781	31491	25321173
G	157287	261	1444	390626	2705	137356	64879334
H	155567	108	390	208607	773	31485	25925534

Implementation D uses the SIMD (i.e. SSE) instruction set and a 32-bit digit size to perform 2-way digit-sliced $\mathbb{F}_{p^2}$ arithmetic (i.e. two $\mathbb{F}_p$ operations in parallel); evaluates one pairing at a time using Algorithm 1.

Implementation E takes Implementation B as a starting point, uses OpenMP to perform parallel $\mathbb{F}_{p^6}$ arithmetic within $\mathbb{F}_{p^{12}}$ and parallel $\mathbb{F}_{p^2}$ arithmetic within Algorithm 1 in order to evaluate one pairing at a time.

Implementation F takes Implementation B as a starting point, but uses OpenMP to implement Algorithm 2 and thereby evaluate one pairing at a time.

Implementation G uses the SIMD (i.e. SSE) instruction set and a 32-bit digit size to perform 2-way digit-sliced pairing evaluation (i.e. two $e(P,Q)$ operations in parallel) and therefore evaluates two pairings at a time.

Implementation H takes Implementation B as a starting point, but uses OpenMP to execute two instances of Algorithm 1 in parallel and therefore evaluate two pairings at a time.

5.1   Analysis of Results

Timings obtained by executing these implementation on the two experimental platforms are detailed in Tables 2 and 3. In each case the number of cycles (as reported by `rdtsc`) required for the entire operation is quoted. That is, if an operation generates n results in parallel then the tables quote the total time: the per-result time requires division by n.

Although our results are not exhaustive, for the given parameterisation they prompt some interesting conclusions. On the Pentium 4 based platform, if one is required to evaluate a single pairing then the best option is to parallelise arithmetic in $\mathbb{F}_p$ (Implementation C); if the requirement is for two pairing evaluations, the best option is actually two invocations of Implementation C. Hankerson et al. [24] obtain significantly better results for their Implementation A on the same platform (but different processor model), but arrive at the same overall conclusion.

On the Core2 based platform, if one is required to evaluate a single pairing then it makes more sense to use 64-bit scalar arithmetic in $\mathbb{F}_p$ and multi-core parallel arithmetic within $\mathbb{F}_{p^{12}}$ and the pairing itself (Implementation E) than consider SIMD parallelism; if the requirement is for two pairing evaluations, the slightly trivial conclusion is that one can perform one pairing on each core (Implementation H), doubling the performance versus two sequential invocations of any other method that does not already use multi-core parallelism internally.

5.2   Analysis of Platforms

Design of SIMD Instruction Sets. Interestingly, in early 2008 Intel announced an update to the SSE lineage of SIMD instruction sets, and a totally new instruction set specialised toward implementation of AES. Specifically, the Advanced Vector Extensions (AVX) includes the `pclmulqdq` instruction for carry-less multiplication that can be used to accelerate arithmetic in binary finite fields. In addition, the Advanced Encryption Standard Instructions Set (AES-NI) includes instructions that perform whole AES rounds with the view to improving performance and eliminating cache based side-channel attack.

In contrast with this new emphasis on supporting cryptography, our results show that on a current Core2 platform, a 64-bit implementation (Implementation B) is faster than that based on SIMD parallel techniques. In the short term, microprocessors with a 64-bit data-path width seem sure to be ubiquitous before longer operands (e.g. 512-bit). One might conclude that using current technology, non-parallel implementation is best; given the specific nature of the updates described above, it seems this will remain the fact in next-generation processors. This seems an unattractive conclusion since it implies that current support for SIMD parallelism is less effective that it could be for this particular domain. We posit that this problem demands research into more public-key cryptography centric SIMD instruction sets: in the longer term, the chance of the processor data-path width doubling (e.g. from 64-bit to 128-bit) is less likely than the operand length doubling and so effective use of parallelism is crucial to scalability.

In a sense, it is not a surprise that Implementation B outperforms C on the Core2 platform. For example, the SSE3 instruction set allows 2-way parallel 32×32-bit multiplication; the cost of such multiplication plus the overhead of data reorganisation will intuitively be greater than native 64×64-bit multiplication. Furthermore, the SSE3 instruction set lacks a method for performing an add-with-carry operation that exists in the scalar instruction set. As such, enhancements over SSE3 such as the pshufb instruction help to reduce said overhead but the instruction set still lacks features which could improve performance of our results. For example, the PLX [31] processor eases the issue of shuffles between sub-words by including odd and even multiplication, i.e.

$$\overline{r}_{2i+1...2i+0} = \overline{x}_{2i+0} \cdot \overline{y}_{2i+0}$$
$$\overline{r}_{2i+1...2i+0} = \overline{x}_{2i+1} \cdot \overline{y}_{2i+1}$$

for $i \in \{0, 1\}$. Another improvement would be provision of hardware support for add-with-carry via vector-carry registers; Fournier [18] investigates this approach within the context of a dedicated vector processor. The upcoming SSE5 instruction set offers an alternative approach by departing from purely 3-address instructions by adding support for a range of 4-address alternatives. In this context, it seems possible to extend the instruction set further and allow explicit specification of a vector-carry register rather than via an implicit, special purpose register as proposed by Fournier.

Effective Utilisation of Multi-core. Another interesting feature is that using the multi-core capabilities of the Core2 platform to evaluate one pairing, we are presented with two problems. Firstly, the overhead from use of OpenMP limits where we can exploit the inherent parallelism within field arithmetic; if the processor had a more light-weight means of managing fine-grained threads, Implementation E would potentially be even more lucrative. The results from using coarse-grained threads in Algorithm 2 are underwhelming. The low Hamming weight of s coupled with the significant overhead introduced by using projective rather than mixed point (resp. line) addition means it is slower than the non-parallel alternative. The first problem motivates research into fine-grained multi-core and multi-threaded processors; an exemplar design is the XCore. The second problem motivates research into forms of easily parallelised pairing algorithms.

6   Conclusions

The efficient evaluation of cryptographic pairings underpins a wide range of modern cryptographic applications. There are a wide range of parameterisation and implementation options to consider, in this paper we focused on the exploitation of parallelism in software. The capabilities of modern processors in this respect are diverse; the correct option and realisation in terms of implementation is therefore far from trivial. In particular we found that, unlike implementation in hardware, on a Pentium 4 based platform one should parallelise arithmetic in $\mathbb{F}_p$ rather than a higher level; on a Core2 based platform one should utilise native support for 64-bit arithmetic and then harness the multi-core features to parallelise arithmetic in $\mathbb{F}_{p^{12}}$ and the pairing itself. Although our results improve significantly on a naive approach, we identified areas for further improvement through study of new algorithm types and changes to processor architecture. The results for arithmetic in $\mathbb{F}_p$ have a direct implication for vanilla ECC in which it seems a similar argument wrt. implementation approach should apply.

Acknowledgements

The authors would like to thank Peter Schwabe for helping to correct some problems with initial performance results, and various anonymous reviewers for improving the clarity of discussion.

References

1. Acar, T.: High-Speed Algorithms & Architectures For Number-Theoretic Cryptosystems. PhD Thesis, Oregon State University (1997)
2. Aoki, K., Hoshino, F., Kobayashi, T., Oguro, H.: Elliptic curve arithmetic using SIMD. In: Davida, G.I., Frankel, Y. (eds.) ISC 2001. LNCS, vol. 2200, pp. 235–247. Springer, Heidelberg (2001)
3. Bailey, D.V., Paar, C.: Efficient Arithmetic in Finite Field Extensions with Application in Elliptic Curve Cryptography. Journal of Cryptology 14(3), 153–176 (2001)
4. Barreto, P.S.L.M., Galbraith, S., Ó hÉigeartaigh, C., Scott., M.: Efficient Pairing Computation on Supersingular Abelian Varieties. Designs, Codes and Cryptography 42(3), 239–271 (2007)
5. Barreto, P.S.L.M., Kim, H., Lynn, B., Scott, M.: Efficient Algorithms for Pairing-Based Cryptosystems. In: Yung, M. (ed.) CRYPTO 2002. LNCS, vol. 2442, pp. 354–368. Springer, Heidelberg (2002)
6. Barreto, P.S.L.M., Naehrig, M.: Pairing-friendly Elliptic Curves of Prime Order. In: Preneel, B., Tavares, S. (eds.) SAC 2005. LNCS, vol. 3897, pp. 319–331. Springer, Heidelberg (2006)
7. Bhaskar, R., Dubey, P.K., Kumar, V., Rudra, A., Sharma, A.: Efficient Galois Arithmetic on SIMD Architectures. In: ACM Symposium on Parallel Algorithms and Architectures, pp. 256–257. ACM Press, New York (2003)
8. Biham, E.: A fast new DES implementation in software. In: Biham, E. (ed.) FSE 1997. LNCS, vol. 1267, pp. 260–272. Springer, Heidelberg (1997)
9. Boneh, D., Lynn, B., Shacham, H.: Short signatures from the Weil pairing. Journal of Cryptology 17(4), 297–319 (2004)
10. Bosselaers, A., Govaerts, R., Vandewalle, J.: SHA: A design for parallel architectures? In: Fumy, W. (ed.) EUROCRYPT 1997. LNCS, vol. 1233, pp. 348–362. Springer, Heidelberg (1997)
11. Clapp, C.S.K.: Optimizing a Fast Stream Cipher for VLIW, SIMD, and Superscalar Processors. In: Biham, E. (ed.) FSE 1997. LNCS, vol. 1267, pp. 273–287. Springer, Heidelberg (1997)
12. Crandall, R., Klivington, J.: Vector Implementation of Multiprecision Arithmetic. Technical Report (1999)

13. Devegili, A.J., ÓhÉigeartaigh, C., Scott, M., Dahab, R.: Multiplication and Squaring on Pairing-Friendly Fields. Cryptology ePrint Archive, Report 2006/471 (2006)
14. Devegili, A.J., Scott, M., Dahab, R.: Implementing Cryptographic Pairings over Barreto-Naehrig Curves. In: Takagi, T., Okamoto, T., Okamoto, E., Okamoto, T. (eds.) Pairing 2007. LNCS, vol. 4575, pp. 197–207. Springer, Heidelberg (2007)
15. Duursma, I., Lee, H.: Tate Pairing Implementation for Hyperelliptic Curves $y^2 = x^p$ - x+d. In: Laih, C.-S. (ed.) ASIACRYPT 2003. LNCS, vol. 2894, pp. 111–123. Springer, Heidelberg (2003)
16. Fan, J., Sakiyama, K., Verbauwhede, I.: Montgomery Modular Multiplication Algorithm on Multi-Core Systems. In: Workshop on Signal Processing Systems: Design and Implementation (SIPS), pp. 261–266 (2007)
17. Fan, J., Sakiyama, K., Verbauwhede, I.: Elliptic Curve Cryptography on Embedded Multicore Systems. In: WESS 2007, pp. 17–22 (2007)
18. Fournier, J.J.A.: Vector Microprocessors for Cryptography. PhD Thesis, University of Cambridge (2007)
19. Galbraith, S.D., Harrison, K., Soldera, D.: Implementing the tate pairing. In: Fieker, C., Kohel, D.R. (eds.) ANTS 2002. LNCS, vol. 2369, pp. 324–337. Springer, Heidelberg (2002)
20. Granger, R., Page, D., Smart, N.P.: High security pairing-based cryptography revisited. In: Hess, F., Pauli, S., Pohst, M. (eds.) ANTS 2006. LNCS, vol. 4076, pp. 480–494. Springer, Heidelberg (2006)
21. Granger, R., Smart, N.P.: On Computing Products of Pairings. In: Cryptology ePrint Archive, Report 2006/172 (2006)
22. Hankerson, D., Menezes, A., Vanstone, S.: Guide to Elliptic Curve Cryptography. Springer, Heidelberg (2004)
23. Hess, F., Smart, N.P., Vercauteren, F.: The Eta Pairing Revisited. Transactions on Information Theory 52, 4595–4602 (2006)
24. Hankerson, D., Menezes, A.J., Scott, M.: Software Implementation of Pairings. To appear in Identity-Based Cryptography,
    http://www.math.uwaterloo.ca/~ajmeneze/research.html
25. Intel Cooperation. Using Streaming SIMD Extensions (SSE2) to Perform Big Multiplications. Technical Report (2000)
26. Izu, T., Takagi, T.: Fast elliptic curve multiplications with SIMD operations. In: Deng, R.H., Qing, S., Bao, F., Zhou, J. (eds.) ICICS 2002. LNCS, vol. 2513, pp. 217–230. Springer, Heidelberg (2002)
27. Kerins, T., Marnane, W.P., Popovici, E.M., Barreto, P.S.L.M.: Efficient Hardware for the Tate Pairing Calculation in Characteristic Three. In: Rao, J.R., Sunar, B. (eds.) CHES 2005. LNCS, vol. 3659, pp. 412–426. Springer, Heidelberg (2005)
28. Koblitz, N., Menezes, A.: Pairing-based Cryptography at High Security Levels. In: Smart, N.P. (ed.) Cryptography and Coding 2005. LNCS, vol. 3796, pp. 13–36. Springer, Heidelberg (2005)
29. Koc, C.K., Acar, T., Kaliski, B.S.: Analyzing and Comparing Montgomery Multiplication Algorithms. IEEE Micro 16(3), 26–33 (1996)
30. Kwon, S.: Efficient tate pairing computation for elliptic curves over binary fields. In: Boyd, C., González Nieto, J.M. (eds.) ACISP 2005. LNCS, vol. 3574, pp. 134–145. Springer, Heidelberg (2005)
31. Lee, R.B., Fiskiran, A.M.: PLX: A Fully Subword-Parallel Instruction Set Architecture for Fast Scalable Multimedia Processing. In: International Conference on Multimedia and Expo, pp. 117–120 (2002)
32. Lipmaa, H.: IDEA: A cipher for multimedia architectures? In: Tavares, S., Meijer, H. (eds.) SAC 1998. LNCS, vol. 1556, pp. 248–263. Springer, Heidelberg (1999)

33. Matsui, M., Nakajima, J.: On the power of bitslice implementation on intel core2 processor. In: Paillier, P., Verbauwhede, I. (eds.) CHES 2007. LNCS, vol. 4727, pp. 121–134. Springer, Heidelberg (2007)
34. Miller, V.: Short programs for functions on curves, http://crypto.stanford.edu/miller/miller.pdf
35. Montgomery, P.L.: Modular Multiplication without Trial Division. Mathematics of Computation 44(170), 519–521 (1985)
36. Montgomery, P.L.: Vectorization of the Elliptic Curve Method, ftp://ftp.cwi.nl/pub/pmontgom/ecmvec.psl.gz
37. Nakajima, J., Matsui, M.: Performance analysis and parallel implementation of dedicated hash functions. In: Knudsen, L.R. (ed.) EUROCRYPT 2002. LNCS, vol. 2332, pp. 165–180. Springer, Heidelberg (2002)
38. Page, D., Smart, N.P.: Parallel Cryptographic Arithmetic Using a Redundant Montgomery Representation. Transactions on Computers 53(11), 1474–1482 (2004)
39. Scott, M.: Implementing Cryptographic Pairings, ftp://ftp.computing.dcu.ie/pub/resources/crypto/pairings.pdf

The Cryptanalysis of Reduced-Round SMS4

Jonathan Etrog* and Matt J.B. Robshaw

Orange Labs
38–40 rue du Général Leclerc
92794 Issy les Moulineaux Cedex 9, France
{forename.surname}@orange-ftgroup.com

Abstract. In this paper we consider the cryptanalysis of the block cipher SMS4. The cipher has received much recent attention due its simplicity and prominence (it is used in wireless networks in China) and a range of differential attacks break up to 21 of the 32 rounds used in SMS4. Here we consider the application of linear cryptanalysis to the cipher and we demonstrate a simple attack on 22 rounds of SMS4. We also consider some advanced linear cryptanalytic techniques which, under the best conditions for the cryptanalyst, might (just) extend to 23 rounds.

1 Introduction

In this paper we consider the security of the block cipher SMS4 which is reputedly mandated for wireless networks in China [10]. A Chinese description of the cipher was made public in 2006 by the Chinese government, and the first analysis in the open community was published in 2007 [10]. The cipher takes a 128-bit block and key, and it consists of 32 simple rounds. Its intriguing design encourages analysis; something which is due in no small part to the fact that minor variants of the cipher are exceptionally weak.

The first open analysis of a reduced-round version of SMS4 examined the algebraic nature of the algorithm—thereby uncovering the construction of the S-box—and yielded a saturation attack over 13 rounds using 2^{16} chosen plaintext pairs and 2^{114} operations [10]. This was followed by a differential attack on 14 rounds and then by an impossible differential attack on 16 rounds with the claimed requirements of 2^{105} chosen plaintext pairs and 2^{107} operations [11]. These are rather complex attacks, and a more natural differential attack has been revealed that suggests that 21 rounds could be compromised using 2^{118} chosen plaintext pairs and $2^{126.6}$ operations [22]. This is the previous best known attack in the literature.

In this paper we present the first reported application of linear cryptanalysis to SMS4. Apart from DES [15], there are few ciphers for which linear cryptanalysis yields a more efficient attack than differential cryptanalysis. However, for SMS4 we propose an attack on 22 rounds of the cipher with less than 2^{119} known

---

* Partially supported by the national research project RFIDAP ANR-08-SESU-009-03.

R. Avanzi, L. Keliher, and F. Sica (Eds.): SAC 2008, LNCS 5381, pp. 51–65, 2009.

plaintexts and a work effort roughly equivalent to 2^{117} 22-round SMS encryptions. The attack can be clearly described and the necessary components have been experimentally verified. We also consider attacks on 23 rounds of SMS4 and highlight some future research directions.

2   Description of SMS4

We briefly describe the block cipher SMS4, but first we establish our notation.

Notation. For the most part we will be working with 32-bit words, though the context will be clear when we restrict ourselves to bytes. The left rotation (*resp.* right rotation) of a word x by b bit positions will be denoted $x \lll b$ (*resp.* $x \ggg b$). The remaining notation is standard in the cryptographic literature.

Encryption and Decryption. SMS4 is a 32-round block cipher with a 128-bit key and block. It is an unbalanced Feistel cipher, that repeatedly uses an 8-bit S-box S. This is described in the appendices and it is, by way of construction [10], closely related to the AES S-box [16]. We define the L function and the γ function as follows

$$L(x) = x \oplus (x \lll 2) \oplus (x \lll 10) \oplus (x \lll 18) \oplus (x \lll 24)$$
$$\gamma(x) = (S[x_{31\ldots24}] \,\|\, S[x_{23\ldots15}] \,\|\, S[x_{15\ldots8}] \,\|\, S[x_{7\ldots0}]).$$

The action of the round function f on input X_{i-1} to the ith round of SMS4 is given by $f(X_{i-1}) = L(\gamma(X_{i-1} \oplus k_i))$. Two rounds of SMS4 are shown in Figure 1.

The SMS4 S-box

	-0	-1	-2	-3	-4	-5	-6	-7	-8	-9	-a	-b	-c	-d	-e	-f
0-	d6	90	e9	fe	cc	e1	3d	b7	16	b6	14	c2	28	fb	2c	05
1-	2b	67	9a	76	2a	be	04	c3	aa	44	13	26	49	86	06	99
2-	9c	42	50	f4	91	ef	98	7a	33	54	0b	43	ed	cf	ac	62
3-	e4	b3	1c	a9	c9	08	e8	95	80	df	94	fa	75	8f	3f	a6
4-	47	07	a7	fc	f3	73	17	ba	83	59	3c	19	e6	85	4f	a8
5-	68	6b	81	b2	71	64	da	8b	f8	eb	0f	4b	70	56	9d	35
6-	1e	24	0e	5e	63	58	d1	a2	25	22	7c	3b	01	21	78	87
7-	d4	00	46	57	9f	d3	27	52	4c	36	02	e7	a0	c4	c8	9e
8-	ea	bf	8a	d2	40	c7	38	b5	a3	f7	f2	ce	f9	61	15	a1
9-	e0	ae	5d	a4	9b	34	1a	55	ad	93	32	30	f5	8c	b1	e3
a-	1d	f6	e2	2e	82	66	ca	60	c0	29	23	ab	0d	53	4e	6f
b-	d5	db	37	45	de	fd	8e	2f	03	ff	6a	72	6d	6c	5b	51
c-	8d	1b	af	92	bb	dd	bc	7f	11	d9	5c	41	1f	10	5a	d8
d-	0a	c1	31	88	a5	cd	7b	bd	2d	74	d0	12	b8	e5	b4	b0
e-	89	69	97	4a	0c	96	77	7e	65	b9	f1	09	c5	6e	c6	84
f-	18	f0	7d	ec	3a	dc	4d	20	79	ee	5f	3e	d7	cb	39	48

The Key Schedule. The key schedule is similar to the encryption function. Each subkey k_i is derived as one word from the output of a single round of SMS-like encryption where the "key" for each round i is a constant $g(i)$ (to be defined below). The plaintext for the start of the key generation is the 128-bit user-supplied key $K_{[127...0]}$. The round function for the SMS-like encryption is given by

$$L'(x) = x \oplus (x \lll 13) \oplus (x \lll 23)$$
$$\gamma(x) = (S[x_{31...24}] \,||\, S[x_{23...15}] \,||\, S[x_{15...8}] \,||\, S[x_{7...0}])$$

so only the L-function is changed in comparison with encryption. At the start, the user-supplied key is xor-ed with a constant

$$T = \texttt{0xa3b1bac6 0x56aa3350 0x677d9197 0xb27022dc},$$

and the initialization of the generation of the subkeys[1] is as follows:

$$k_{-3} = K_{[127...96]} \oplus T_{[127...96]}, \quad k_{-2} = K_{[95...64]} \oplus T_{[95...64]},$$
$$k_{-1} = K_{[63...32]} \oplus T_{[63...32]}, \quad k_0 = K_{[31...0]} \oplus T_{[31...0]}.$$

The key k_i for the ith round, for $1 \leq i \leq 32$ is computed as

$$k_i = k_{i-4} \oplus L'(\gamma(k_{i-3} \oplus k_{i-2} \oplus k_{i-1} \oplus g(i)))$$

where each constant $g(i)$ is defined by

$$g(i) = ((28 \times (i-1)) \,||\, (28 \times (i-1)+7) \,||\, (28 \times (i-1)+14) \,||\, (28 \times (i-1)+21)).$$

2.1   RED-SMS4: A Small Version of SMS4

We confirm some of the work in this paper with experiments, and for these we will need to define a reduced-version of SMS4. This will be a block cipher with a 64-bit key and block size which uses a 4-bit S-box S_r. For experiments we chose the S-box used in PRESENT [2]. We can define a reduced L_r function and a reduced γ_r function as follows:

$$L_r(x) = x \oplus (x \lll 8) \oplus (x \lll 10)$$
$$\gamma_r(x) = (S_r[x_{15...12}] \,||\, S_r[x_{11...8}] \,||\, S_r[x_{7...4}] \,||\, S_r[x_{3...0}]).$$

L_r was built using the rotations that appear in L modulo 16. In this way, the round function f_r used in the ith round of reduced-SMS4 is given by $f_r(X_{i-1}) = L_r(\gamma_r(X_{i-1} \oplus k_i)$. A reduced version of the key schedule requires us to change the linear function L' to L'_r just as we changed L to L_r in the encryption routine, and to revise the per-round constants to $g_r(i) = ((28 \times (i-1)) \,||\, (28 \times (i-1)+7))$.

---

[1] This is slightly different to other descriptions so as to accommodate the natural numbering of rounds starting with 1.

3   Linear Cryptanalysis

While linear cryptanalytic methods appeared in [21], the linear cryptanalytic attack and its application to DES was developed by Matsui [12,13]. The basic idea is to find a *linear approximation* to the action of the block cipher. By this we mean a linear equation that includes a bits of the plaintext $P_{r_1}, \ldots, P_{r_a}$, together with b bits of the ciphertext $C_{s_1}, \ldots, C_{s_b}$ and a single bit of key-related information κ. Borrowing the vector inner-product, we will use the notation $\alpha \cdot P$ to denote the sum of plaintext bits $P_{r_1} \oplus \ldots \oplus P_{r_a}$ where $\alpha = \sum_{j=1}^{a} 2^{r_j}$ and α is called a *linear mask*. We will then write a single linear approximation as

$$\alpha \cdot P \oplus \beta \cdot C = \kappa. \tag{1}$$

If κ (the exclusive-or of subkey bits) is fixed, then Equation 1 will be correct with probability $p = \frac{1}{2} + \epsilon$ and we say that the linear approximation has a bias of ϵ. Given a bias of sufficiently large absolute value $|\epsilon|$ and sufficiently many known plaintext/ciphertext pairs, the value of κ can be deduced thereby revealing one bit of key information. Throughout the paper the term "bias" will refer to its absolute value.

It is well-known that we can recover more bits of the key by using Matsui's *Algorithm 2* [12]. Here we use a linear approximation over several inner rounds, say rounds b to c of the r-round cipher, and this approximates one *inner* bit of key information (which is a function of the subkeys $k_b, \ldots, k_c$). Since the inputs to this linear approximation are a function of the plaintext, the ciphertext, and the outer subkeys $k_1, \ldots, k_{b-1}$ and $k_{c+1}, \ldots, k_r$, if we were to test for a bias as part of an exhaustive search over these *outer* key bits, then we would expect a bias to appear for the correct guess. In this way we can recover more key information and derive a more practical attack.

Clearly the basic building block to all these attacks will be the linear approximation, and to build a linear approximation we approximate individual components of the cipher and join these together. We will therefore use the following notation for the linear approximation of a component f, say, where we write $\alpha \xrightarrow{f} \beta$ if $\alpha \cdot X = \beta \cdot f(X)$ with some associated bias ϵ. Approximations to larger components of a block cipher, such as a round, can be written in the same way.

3.1   Linear Cryptanalysis and SMS4

To find a linear approximation of SMS4, we first compute the biases of all linear approximations $\alpha \xrightarrow{S} \beta$ to the S-box. Then we consider the evolution of a linear mask through the function L. For this, we define the function

$$L_2(x) = x \oplus (x \ggg 2) \oplus (x \ggg 10) \oplus (x \ggg 18) \oplus (x \ggg 24)$$

and we observe the following. Since for bit-wise rotations $\alpha \cdot (x \lll i) = (\alpha \ggg i) \cdot x$, we have for all 32-bit inputs x, and all linear masks α, that $\alpha \cdot L(x) = L_2(\alpha) \cdot x$.

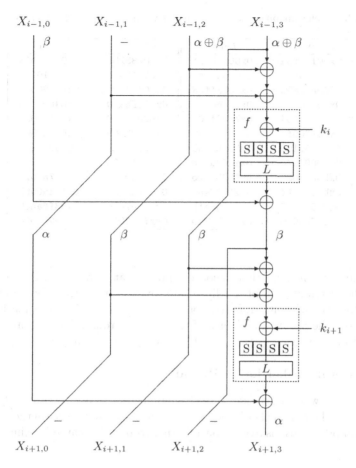

Fig. 1. Two rounds of SMS4 along with a two-round linear approximation using masks α and β. The input to round i is $X_{i-1,0} \parallel X_{i-1,1} \parallel X_{i-1,2} \parallel X_{i-1,3}$.

As can be seen from Figure 1, we can identify the potential for two-round linear characteristics of the following form:

$$(\beta, 0, \alpha \oplus \beta, \alpha \oplus \beta) \to (\alpha, \beta, \beta, \beta) \to (0, 0, 0, \alpha).$$

Such a linear approximation would require the approximation $\alpha \xrightarrow{f} \beta$ in the first round and $\beta \xrightarrow{f} \alpha$ in the second. Interestingly, by setting $\beta = \alpha$ this reduces to

$$(\alpha, 0, 0, 0) \to (\alpha, \alpha, \alpha, \alpha) \to (0, 0, 0, \alpha)$$

and by exploiting the structure of SMS4 in the preceeding three rounds, we derive a five-round iterative linear approximation, of which only the last two rounds are active

$$(0, 0, 0, \alpha) \to (0, 0, \alpha, 0) \to (0, \alpha, 0, 0) \to (\alpha, 0, 0, 0) \to (\alpha, \alpha, \alpha, \alpha) \to (0, 0, 0, \alpha).$$

Table 1. The relevant bitmasks for the iterative linear approximations in this paper

α	$L_2(\alpha)$	α	$L_2(\alpha)$
0x0011ffba	0x0084be2f	0x007852b3	0x00582b15
0x007905e1	0x005afbc6	0x00a1b433	0x00f1027a
0x00edca7c	0x0083ffaa	0x00fa7099	0x00d20b1d
0x05e10079	0xfbc6005a	0x11ffba00	0x84be2f00
0x3300a1b4	0x7a00f102	0x52b30078	0x2b150058
0x709900fa	0x0b1d00d2	0x7852b300	0x582b1500
0x7905e100	0x5afbc600	0x7c00edca	0xaa0083ff
0x9900fa70	0x1d00d20b	0xa1b43300	0xf1027a00
0xb3007852	0x1500582b	0xb43300a1	0x027a00f1
0xba0011ff	0x2f0084be	0xca7c00ed	0xffaa0083
0xe1007905	0xc6005afb	0xedca7c00	0x83ffaa00
0xfa709900	0xd20b1d00	0xffba0011	0xbe2f0084

To identify a bit-mask α that yields an approximation $\alpha \xrightarrow{f} \alpha$ with a good bias, we use the distribution table for linear approximations of the S-box. In this way we can list 24 different $(\alpha, L_2(\alpha))$ pairs, where $L_2(\alpha)$ gives the mask for the output from the S-boxes, and each of these 24 five-round linear approximations holds with a bias of $\frac{7}{32768} \approx 2^{-10.2}$. These are given in Table 1.

3.2   A Distinguisher for 18-Round SMS4

It is straightforward to see that a classical application of linear cryptanalysis gives us an 18-round distinguisher for SMS4. We can concatenate three of the five-round iterative approximations to give the following 18-round linear approximation with bias ϵ_1:

$$(0,0,0,\alpha) \xrightarrow{5 \ rounds} (0,0,0,\alpha) \xrightarrow{5 \ rounds} (0,0,0,\alpha)$$
$$\xrightarrow{5 \ rounds} (0,0,0,\alpha) \rightarrow (0,0,\alpha,0) \rightarrow (0,\alpha,0,0) \rightarrow (\alpha,0,0,0)$$

To combine linear approximations, and to estimate the resultant bias, it is typical to appeal to the so-called *piling-up lemma* [12]. The suitability of applying the piling-up lemma depends on the algorithms in question; for some, such as DES [15], it gives accurate results while for others, such as RC5 [18], an inter-round dependence means that the piling-up lemma can be misleading [19]. This problem can be particularly acute when we have two consecutive active rounds. However, experimental results below suggest that the piling-up lemma should remain a reasonable tool to use with SMS4. We therefore estimate the resultant bias of the 18-round linear approximation to be $\epsilon_1 = (2^{-10.2})^6 \times 2^5 = 2^{-56.2}$. This means that if we were to use $\epsilon_1^{-2} = 2^{112.4}$ known plaintexts then we would expect our distinguisher to identify non-ideal behaviour in the reduced-round SMS4 and/or to recover a single bit of key information with a success rate of 97.7% [12]. With regards to the work effort, we need to evaluate a single bit and increment a single counter $2^{112.4}$ times. This will be a fraction of the work required to exhaustively search a 128-bit key.

In what follows we will use the 18-round linear approximations of the form described above, of which there are 24 (see Table 1). It will therefore be convenient to refer to a generic approximation from this class as $\mathcal{A}_\alpha^{18}$.

Experimental Confirmation. To confirm the applicability of the piling-up lemma with the basic linear approximations that we will use, we consider the equivalent linear approximations in RED-SMS4. The bias of the best approximation over a single active round—for which the input and output mask is the same—is 2^{-5}. So over five rounds, of which two are active, the linear approximation $(0,0,0,\alpha) \to (0,0,0,\alpha)$ with $\alpha = \text{0x040c}$ would have a theoretical bias of 2^{-9} . We extend this to give a six-round approximation

$$(\alpha, \beta, \beta, \beta)\xrightarrow{\text{1 round}}(0,0,0,\alpha)\xrightarrow{\text{5 rounds}}(0,0,0,\alpha)$$

with $\beta = \text{0x0406}$ and a bias of $2^{-2.7}$ for $\beta\xrightarrow{f_r}\alpha$. The resultant six-round approximation has a theoretical bias of $2^{-10.7}$ and in experiments with 100 keys using the reduced key schedule and 2^{23} known plaintexts, the measured bias ranged between $2^{-10.0}$ and $2^{-12.1}$ with an average of $2^{-10.7}$.

On Extending to 19 Rounds. Taking $\mathcal{A}_\alpha^{18}$ we can prepend a single-round linear approximation of the form $(\alpha, \beta, \beta, \beta) \to (0,0,0,\alpha)$. Here we can choose β so as to maximise the bias of this extra round. For each of the valid $L_2(\alpha)$ that we identified in Section 3.1, we find that there are 125 possible values to β that give a maximum bias of 2^{-10} over a single round of S-box transformations. This means that there are $125 \times 24 = 3000$ 19-round linear approximations with a bias of $\epsilon_2 = (2^{-10.2})^6 \times 2^{-10} \times 2^6 = 2^{-65.2}$. While the bias means that such approximations aren't immediately useful to us, the large number of such approximations makes them a tempting object for more advanced analysis, see Section 5.1.

4   Advanced Techniques

We now use the 18-round approximations $\mathcal{A}_\alpha^{18}$ to recover the full 128-bit key. Standard techniques immediately compromise 20-round SMS4, while a novel extension of the work of Collard et al [4] extends this to 22 rounds. In the literature notation, this constitutes a 4R-attack for which there are few precedents.

4.1   An Attack on 20-Round SMS4

The classical approach to using an 18-round distinguisher is to recover key information from the two outer rounds of the cipher. We will use the linear approximations $\mathcal{A}_\alpha^{18}$ that have only three active S-boxes in an active round and we will need the following definition: Given mask α, denote the *restriction* of

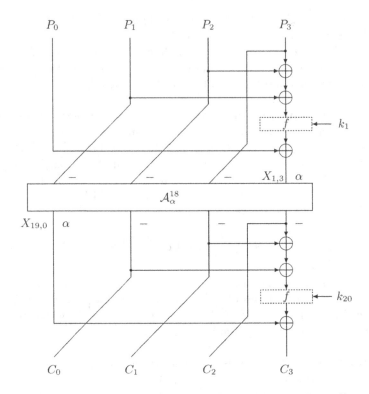

Fig. 2. Intermediate values for the 2R attack on 20-round SMS4

a 32-bit word y by α to be $R_\alpha(y)$ where $R_\alpha(y)$ consists of the deletion of bits corresponding to the inactive byte. For example, given $\alpha = $ 0x0011ffba then $R_\alpha(y) = $ y&0x00ffffff. Note that this can be viewed as a 24-bit quantity even when the inactive byte is not in the most significant position.

It is easy to verify the following (see Figure 2). For plaintext $P_0||P_1||P_2||P_3$, the bit value $\alpha \cdot X_{1,3}$ depends solely on $\alpha \cdot P_0$, $R_\alpha(P_1 \oplus P_2 \oplus P_3)$, and $R_\alpha(k_1)$. We can make a similar observation on the ciphertext, namely that the bit value $\alpha \cdot X_{19,0}$ depends solely on $\alpha \cdot C_3$, $R_\alpha(C_0 \oplus C_1 \oplus C_2)$, and $R_\alpha(k_{20})$. In our 2R-attack we will recover the values of $R_\alpha(k_1)$ and $R_\alpha(k_{20})$ giving 48 bits of key information. The rest of the key can be deduced using exhaustive search.

The data-related information that we need to evaluate the approximation is $\alpha \cdot P_0$, $R_\alpha(P_1 \oplus P_2 \oplus P_3)$, $\alpha \cdot C_3$, and $R_\alpha(C_0 \oplus C_1 \oplus C_2)$ and we can consider a plaintext-ciphertext as being in one of 2^{50} possible classes according to the values of these quantities. Note that under the same key guess $R_\alpha(k_1) || R_\alpha(k_{20})$, two plaintext/ciphertext pairs from the same class yield the same values to $\alpha \cdot X_{1,3}$ and $\alpha \cdot X_{19,0}$. In [4] an efficient 1R-attack is described. We extend this approach to give a 2R-attack recovering information from both outer rounds and adopting an optimisation that means we need only store 2^{48} rather than 2^{50} counters.

1. Take $N = 32\epsilon^{-2} = 2^{117.4}$ plaintext/ciphertext pairs.
2. Initialise a set of counters $A[0] \ldots A[2^{48} - 1]$ to zero.
3. For each plaintext/ciphertext pair, compute $b = \alpha \cdot P_0 \oplus \alpha \cdot C_3$ and increment $A[R_\alpha(P_1 \oplus P_2 \oplus P_3) || R_\alpha(C_0 \oplus C_1 \oplus C_2)]$ if $b = 0$ or decrement it if $b = 1$, *i.e.*

$$A[R_\alpha(P_1 \oplus P_2 \oplus P_3) || R_\alpha(C_0 \oplus C_1 \oplus C_2)] \mathrel{+}= (-1)^{(\alpha \cdot P_0 \oplus \alpha \cdot C_3)}.$$

4. For each key guess $k' = R_\alpha(k_1) || R_\alpha(k_{20})$ keep a counter, and compute the bias generated during the attack as follows:

   (a) Taking each $x = R_\alpha(P_1 \oplus P_2 \oplus P_3) || R_\alpha(C_0 \oplus C_1 \oplus C_2)$ in turn, where $0 \le x \le 2^{48} - 1$, compute the value

   $$c = (-1)^{(\alpha \cdot f(R_\alpha(k_1 \oplus P_1 \oplus P_2 \oplus P_3)) \oplus (\alpha \cdot f(R_\alpha(k_{20} \oplus C_0 \oplus C_1 \oplus C_2))}.$$

   (b) Add $c \times A[R_\alpha(P_1 \oplus P_2 \oplus P_3) || R_\alpha(C_0 \oplus C_1 \oplus C_2)]$ to the score for key guess k'.

5. After recovering the 48-bit k', perform exhaustive search on the remaining 80 bits of key.

We expect to recover the right value to the 48 bits of the key by identifying the guess which gives the highest score of absolute value; using [20] the correct key should be recovered with a probability of 99.9%

While the work effort for each plaintext/ciphertext pair in step 3 is much less than a round of SMS4, we might estimate the work effort for the first three steps to be equivalent to $2^{117.4} \times \frac{1}{20} \approx 2^{113.1}$ 20-round SMS4 computations. The work effort for finding the right 48 bits of key material in step 4 is $2^{48} \times 2^{48} = 2^{96}$ basic operations and the work to recover the rest of the key. is $2^{128-48} = 2^{80}$ reduced-round SMS encryptions. One point of detail: it is possible (see below) that several keys are identified along with the correct one. However this is not uncommon, and merely extends the search for the remainder of the key.

An Optimisation. Even though the work effort for Step 4 is lower than that for data processing, we can adapt techniques introduced in [4]. Consider initialising a $(2^{48} \times 2^{48})$ matrix M, where rows are indexed by $R_\alpha(k_1) || R_\alpha(k_{20})$ and the columns indexed by $R_\alpha(P_1 \oplus P_2 \oplus P_3) || R_\alpha(C_0 \oplus C_1 \oplus C_2)$. Then the bias for the ith guess of $R_\alpha(k_1) || R_\alpha(K_{20})$ is given by $\sum_{j=0}^{2^{48}-1} M_{(i,j)} x_j$ and we can view the counters $A[\cdot]$ as a column vector $\mathbf{x} = A^T$. Following [4], since entry $M_{(i,j)}$ is a function of $i \oplus j$ the entire matrix $M_{(i,j)}$ can be reconstructed from a single row or column, and it is possible to compute the product $M\mathbf{x} = \mathbf{e}$ with only three products between a Discrete Fourier Transform matrix and a vector [4].This means that the complexity of generating a set of final scores for each key, represented by a vector $\mathbf{e}$ is reduced from $O((2^{48})^2)$ to $3 \times O(2^{48} \log_2(2^{48}))$ [4]. The work effort for data analysis can therefore be estimated as $2^{48} \times 3 \times 48 \approx 2^{55.2}$ basic operations.

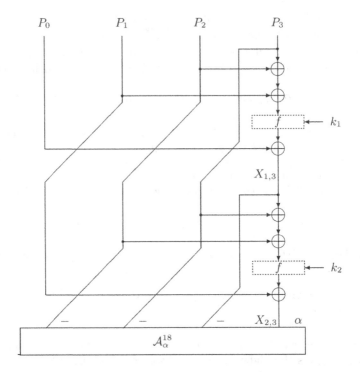

Fig. 3. The upper half of the 4R-attack on 22-round SMS4

Experimental Confirmation. To illustrate this more advanced linear attack, we use a 10-round version of RED-SMS4 with the same linear approximation as was used in the experiments of Section 3.2, namely $(\alpha, \beta, \beta, \beta) \rightarrow (\alpha, 0, 0, 0)$ over nine rounds. We will recover information about k_{10}, though for RED-SMS4 the mask α we use has two inactive bytes. Recalling the bias of the approximation is $2^{-10.7}$, we take $8 \times (2^{11})^2 = 2^{25}$ plaintexts. The data is separated according to the restriction $R_\alpha(C_0 \oplus C_1 \oplus C_2)$ and we perform key recovery as outlined in Section 4.1, though adapted to the 1R-scenario. For S_r there is a slight complication with the bit mask α since there are equivalent keys for one of the active nibbles. Experiments and analysis show that the best score applies to four equivalent key values, and so we recover at most six bits of k_{10}. With the plaintext amount we use, we theoretically have a probability of 99.9% to recover the right six bits of key [20]. In 100 experiments the correct set of keys was recovered 99 times.

4.2   An Attack on 22-Round SMS4

We use $\mathcal{A}_\alpha^{18}$ to make a 4R-attack on 22-round SMS4 (see Figure 3 for the plaintext side of the attack) and we aim to recover k_1, k_{22}, $R_\alpha(k_2)$ and $R_\alpha(k_{21})$. For the data analysis, we will appeal to the optimisation of Collard *et al.* [4] described in Section 4.1.

1. We take $N = 64\epsilon^{-2} = 2^{118.4}$ plaintext/ciphertext pairs.
2. View counters $A[0] \ldots A[2^{112} - 1]$ as a column vector $\mathbf{x}$ and set to zero.
3. For each plaintext/ciphertext pair, compute $b = \alpha \cdot P_1 \oplus \alpha \cdot C_2$ and increment $A[P_1 \oplus P_2 \oplus P_3 \, || \, R_\alpha(P_0 \oplus P_2 \oplus P_3) \, || \, R_\alpha(C_0 \oplus C_1 \oplus C_3) \, || \, C_0 \oplus C_1 \oplus C_2]$ if $b = 0$ or decrement it if $b = 1$.
4. Define (conceptually) the $(2^{112} \times 2^{112})$ matrix $M_{(i,j)}$ where rows are indexed by the key guess $k' = k_1 \, || \, R_\alpha(k_2) \, || \, R_\alpha(k_{21}) \, || \, k_{22}$ and columns indexed by $x = P_1 \oplus P_2 \oplus P_3 \, || \, R_\alpha(P_0 \oplus P_2 \oplus P_3) \, || \, R_\alpha(C_0 \oplus C_1 \oplus C_3) \, || \, C_0 \oplus C_1 \oplus C_2$. Recall we need only store the first column of this matrix $M_{(i,j)}$ since all values for subsequent computations can be reconstructed from a single row/column.

   (a) Compute the values in the first column as $(-1)^b$ where

$$b = \alpha \cdot f(R_\alpha(k_{21} \oplus C_0 \oplus C_1 \oplus C_3) \oplus f(k_{22} \oplus C_0 \oplus C_1 \oplus C_2))$$
$$\oplus \, \alpha \cdot f(R_\alpha(k_2 \oplus P_0 \oplus P_2 \oplus P_3) \oplus f(k_1 \oplus P_1 \oplus P_2 \oplus P_3))$$

   (b) Efficiently compute $M\mathbf{x} = \mathbf{e}$ using [4]. This gives the right result since

$$\alpha \cdot X_{2,3} = \alpha \cdot f(R_\alpha(k_2) \oplus R_\alpha(P_2 \oplus P_3 \oplus X_{1,3})) \oplus \alpha \cdot P_1$$
$$= \alpha \cdot f(R_\alpha(k_2) \oplus R_\alpha(P_0 \oplus P_2 \oplus P_3)$$
$$\oplus \, f(k_1 \oplus P_1 \oplus P_2 \oplus P_3)) \oplus \alpha \cdot P_1$$

   and we have a similar expression for the ciphertext side.
5. Recover the 112-bit k' from $\mathbf{e}$ and search the remaining bits of the key.

The only hypothesis needed to apply [4] to the 22-round attack is that the $(2^{112} \times 2^{112})$ matrix $M_{(i,j)}$ (see optimisation to Section 4.1) should only depend on $i \oplus j$, which is the case for the expression in Step 2. We expect to recover the right value to the 112 bits of the key from the guess with the highest score of absolute value. With $2^{118.4}$ plaintexts, the method of [20] suggests that we are very likely to recover the correct value, see Table 2. The work effort for Steps

Table 2. The estimated work efforts for a range of linear cryptanalytic attacks on r-round SMS4 for $19 \leq r \leq 22$. Work is estimated in terms of the number of r-round encryptions (for appropriate r) with that exceeding 2^{128} placed in parentheses.

r	texts	mem.	work steps 1-3	work step 4 (w/o [4])	work step 4 (w. [4])	work step 5	success (%)
19	$2^{116.4}$	2^{24}	$2^{112.2}$	$2^{43.8}$	2^{26}	2^{104}	99.5
20	$2^{117.4}$	2^{48}	$2^{113.1}$	$2^{91.7}$	$2^{50.9}$	2^{80}	99.9
21	$2^{117.4}$	2^{80}	$2^{113.0}$	$(2^{155.6})$	$2^{83.5}$	2^{48}	84.8
21	$2^{118.4}$	2^{80}	$2^{114.0}$	$(2^{155.6})$	$2^{83.5}$	2^{48}	99.9
22	$2^{117.4}$	2^{112}	$2^{112.9}$	$(2^{219.5})$	$2^{115.9}$	2^{16}	17.7
22	$2^{118.4}$	2^{112}	$2^{113.9}$	$(2^{219.5})$	$2^{115.9}$	2^{16}	99.9

1-3 can be estimated as $2^{118.4} \times \frac{1}{22} \approx 2^{113.9}$ 22-round SMS4 computations while the effort in Step 4 is approximately $2^{112} \times 3 \times 112 \times \frac{1}{22} \approx 2^{115.9}$ 22-round SMS4 computations, and this dominates the attack.

5   Ongoing and Future Research

It is natural to consider some more advanced techniques in trying to attack more rounds of SMS4. In this section we consider the use of multiple linear approximations as well as the use of chosen-plaintexts.

5.1   Multiple Linear Approximations

Multiple linear approximations were first proposed in [6,7] and they have been the subject of much recent analysis [3,5]. Here we take m different linear approximations, where we use κ_j to denote a single bit of key information,

$$\alpha_j \cdot P \oplus \beta_j \cdot C = \kappa_j.$$

The purpose is to use several approximations to reduce the number of plaintexts when keeping the same probability of sucess. Let ϵ^j denote the theoretical bias of the jth approximation and let $e^j_{k_{\text{outer}}}$ denote the experimental bias of the jth approximation observed when using the guess k_{outer} for the outer key bits[2]. If, with sufficiently many plaintexts, we compute

$$\min_{k_{\text{outer}}} \min_{(\kappa_1,\dots,\kappa_m)\in\{0,1\}^m} \sum_{j=1}^{m} \left(\epsilon^j - (-1)^{\kappa_j} e^j_{k_{\text{outer}}}\right)^2,$$

then the minimum value will be given by the correct values of k_{outer} and the correct values of the m bits of internal key represented by $(\kappa_1,\dots,\kappa_m)$.

A straightforward application of this method needs $2^{|k|+m}$ computations. However this can be reduced if we introduce σ^j_k where $\sigma^j_k = 1$ if $sgn(\epsilon^j) = sgn(e^j_k)$ and zero otherwise. Then we observe that, for each j,

$$\min_{\kappa_j\in\{0,1\}} \left(\epsilon^j - (-1)^{\kappa_j} e^j_k\right)^2 = \left(\epsilon^j - (-1)^{\sigma^j_k} e^j_k\right)^2$$

and so we have the equality

$$\min_{k_{\text{outer}}} \min_{(\kappa_1,\dots,\kappa_m)\in(0,1)^m} \sum_{j=1}^{m} \left(\epsilon^j - (-1)^{\kappa_j} e^j_{k_{\text{outer}}}\right)^2$$

$$= \min_{k_{\text{outer}}} \sum_{j=1}^{m} \left(\epsilon^j - (-1)^{\sigma^j_{k_{\text{outer}}}} e^j_{k_{\text{outer}}}\right)^2.$$

This requires $m2^{|k_{\text{outer}}|}$ computations, though we only recover the correct value to k_{outer}. However this is usually the most important block of key information to recover.

---

[2] An equivalent approach considers the *imbalance* which is double the bias [3].

Application to SMS4. To gauge the possible limits of linear cryptanalysis, we will optimistically assume that the gain that can be made when using multiple linear approximations is linear in the number of approximations. We will then use the techniques above to combine a set of different linear 19-round approximations and illustrate the basis for a possible attack on 23-round SMS4.

To do this we need a set of linear approximations and we will choose 125 19-round approximations $\mathcal{A}^{19}_{\alpha\beta}$ where these are the extensions of a given, fixed, 18-round distinguisher $\mathcal{A}^{18}_{\alpha}$ by the 125 choices for β. (These approximations were identified in Section 3.2). We denote by ϵ the theoretical bias of $2^{-65.2}$ which is the same for each of the $\mathcal{A}^{19}_{\alpha\beta}$.

1. Take $N = 2^{125.4}$ plaintext/ciphertext pairs.
2. For each β view counters $A[\beta][0], \ldots, A[\beta][2^{112} - 1]$ as a column vector $\mathbf{x}^\beta$ and set this to zero.
3. For each β and each plaintext/ciphertext pair, compute $b = \alpha \cdot P_2 \oplus \beta \cdot P_0 \oplus \beta \cdot P_1 \oplus \beta \cdot P_3 \oplus \alpha \cdot C_2$ and increment $A[\beta][P_1 \oplus P_2 \oplus P_3 \| R_\alpha(P_0 \oplus P_2 \oplus P_3) \| R_\alpha(C_0 \oplus C_1 \oplus C_3) \| C_0 \oplus C_1 \oplus C_2]$ if $b = 0$ or decrement it if $b = 1$.
4. Define for each β (conceptually) the $(2^{112} \times 2^{112})$ matrix $M^\beta_{(i,j)}$ where rows are indexed by the outer key guess $k_{outer} = k_1 \| R_\alpha(k_2) \| R_\alpha(k_{21}) \| k_{22}$ and columns indexed by $x = P_1 \oplus P_2 \oplus P_3 \| R_\alpha(P_0 \oplus P_2 \oplus P_3) \| R_\alpha(C_0 \oplus C_1 \oplus C_3) \| C_0 \oplus C_1 \oplus C_2$. Recall we need only store the first column of this matrix $M^\beta_{(i,j)}$ since all values for subsequent computations can be reconstructed from a single row/column.
5. Compute the values in the first column as $(-1)^b$ where

$$b = \alpha \cdot f(R_\alpha(k_{21} \oplus C_0 \oplus C_1 \oplus C_3) \oplus f(k_{22} \oplus C_0 \oplus C_1 \oplus C_2))$$
$$\oplus \beta \cdot f(R_\alpha(k_2 \oplus P_0 \oplus P_2 \oplus P_3) \oplus f(k_1 \oplus P_1 \oplus P_2 \oplus P_3))$$

6. Efficiently compute $M^\beta \mathbf{x}^\beta = \mathbf{e}^\beta$ using [4].
7. For each guess to k_{outer}, compute $\sum_\beta \left(\epsilon - (-1)^{\sigma^\beta_{k_{outer}}} e^\beta_{k_{outer}}\right)^2$.
8. Assume that the minimum value is given by the correct guess for the 112-bit k_{outer} and then search the remaining bits of the key.

The work effort for this attack is dominated by Step 3. To derive the maximum number of plaintexts we can use, we observe that the work effort of Step 3 can be expressed as $\frac{125 \times N}{23}$ 23-round SMS4 computations. To give an academic attack, we need this to be less than 2^{128} 23-round SMS4 computations and so we have $N \leq \frac{23 \times 2^{128}}{125} \approx 2^{125.4}$ for a valid attack.

However $2^{125.4}$ corresponds to around $4 \times \frac{(2^{65.2})^2}{125}$, but since we are recovering 112 bits of key information the success rate [20] will be almost negligible. Thus while it is conceivable that 23 rounds could be attacked (academically) we feel that this is somewhat optimistic.

Unfortunately the reduced version of SMS4 used earlier doesn't exhibit different linear approximations with the same bias. Intead we were able to experiment on a different reduced cipher design, but it was too far-removed from SMS4 for us to be able to draw any substantive conclusions. Our experiments demonstrated improvements to the number of plaintexts required in a successful attack, but the probability of success was somewhat less than anticipated by theory. We therefore leave it as an object of future research to provide a sound estimate for the effectiveness of multiple linear approximations on SMS4.

5.2   On Using Chosen Plaintext

Several extensions to linear cryptanalysis consider the use of chosen plaintext. One of these is described by Knudsen and Mathiassen [8]. In early work for this paper we considered using variants of this technique and at first sight it seemed to be well-suited to SMS4. However technical complications meant that it was hard to use these techniques directly with the 18-round distinguisher and we were unable to get any satisfactory advantages. We also considered using differential-linear cryptanalysis [9] but our preliminary conclusion was somewhat negative. We therefore leave it as an open problem to decide whether chosen plaintext can give any real advantage over the typical known plaintext approach.

6   Conclusions

In this paper we have considered the cryptanalysis of the block cipher SMS4. The cipher is both actively deployed and of an elegant and simple design, making it of considerable interest to the cryptanalyst. While much of the preceeding work is concentrated on the differential cryptanalysis of SMS4, by turning to linear cryptanalysis we have demonstrated some simple and effective attacks. These yield results which are superior to all previous claims and which, therefore, give the best current attacks on SMS4.

References

1. Biham, E., Shamir, A.: Differential Cryptanalysis of the Data Encryption Standard. Springer, Heidelberg (1993)
2. Bogdanov, A., Knudsen, L.R., Leander, G., Paar, C., Poschmann, A., Robshaw, M.J.B., Seurin, Y., Vikkelsoe, C.: PRESENT: An ultra-lightweight block cipher. In: Paillier, P., Verbauwhede, I. (eds.) CHES 2007. LNCS, vol. 4727, pp. 450–466. Springer, Heidelberg (2007)
3. Biryukov, A., De Cannière, C., Quisquater, M.: On multiple linear approximations. In: Franklin, M. (ed.) CRYPTO 2004. LNCS, vol. 3152, pp. 1–22. Springer, Heidelberg (2004)
4. Collard, B., Standaert, F.-X., Quisquater, J.-J.: Improving the time complexity of matsui's linear cryptanalysis. In: Nam, K.-H., Rhee, G. (eds.) ICISC 2007. LNCS, vol. 4817, pp. 77–88. Springer, Heidelberg (2007)
5. Collard, B., Standaert, F.-X., Quisquater, J.-J.: Experiments on the multiple linear cryptanalysis of reduced round serpent. In: Nyberg, K. (ed.) FSE 2008. LNCS, vol. 5086, pp. 382–397. Springer, Heidelberg (2008)

6. Kaliski, B.S., Robshaw, M.J.B.: Linear cryptanalysis using multiple approximations. In: Desmedt, Y.G. (ed.) CRYPTO 1994. LNCS, vol. 839, pp. 26–39. Springer,
   Heidelberg (1994)
7. Kaliski, B.S., Robshaw, M.J.B.: Linear Cryptanalysis and FEAL. In: Preneel, B.
   (ed.) FSE 1994. LNCS, vol. 1008, pp. 249–264. Springer, Heidelberg (1995)
8. Knudsen, L., Mathiassen, J.: A chosen-plaintext linear attack on DES. In: Schneier,
   B. (ed.) FSE 2000. LNCS, vol. 1978, pp. 262–272. Springer, Heidelberg (2001)
9. Langford, S.K., Hellman, M.E.: Differential-linear cryptanalysis. In: Desmedt, Y.G.
   (ed.) CRYPTO 1994. LNCS, vol. 839, pp. 17–25. Springer, Heidelberg (1994)
10. Liu, F., Ji, W., Hu, L., Ding, J., Lv, S., Pyshkin, A., Weinmann, R.-P.: Analysis
    of the SMS4 block cipher. In: Pieprzyk, J., Ghodosi, H., Dawson, E. (eds.) ACISP
    2007. LNCS, vol. 4586, pp. 158–170. Springer, Heidelberg (2007)
11. Lu, J.: Attacking reduced-round versions of the SMS4 block cipher in the chi-
    nese WAPI standard. In: Qing, S., Imai, H., Wang, G. (eds.) ICICS 2007. LNCS,
    vol. 4861, pp. 306–318. Springer, Heidelberg (2007)
12. Matsui, M.: Linear cryptanalysis method for DES cipher. In: Helleseth, T. (ed.)
    EUROCRYPT 1993. LNCS, vol. 765, pp. 386–397. Springer, Heidelberg (1994)
13. Matsui, M.: The first experimental cryptanalysis of the data encryption stan-
    dard. In: Desmedt, Y.G. (ed.) CRYPTO 1994. LNCS, vol. 839, pp. 1–11. Springer,
    Heidelberg (1994)
14. Murphy, S.: The Independence of Linear Approximations in Symmetric Cryptanal-
    ysis. IEEE Transactions on Information Theory 52, 5510–5518 (2006)
15. National Institute of Standards and Technology. FIPS 46-3: Data Encryption Stan-
    dard (November 1998), http://csrc.nist.gov
16. National Institute of Standards and Technology. FIPS 197: Advanced Encryption
    Standard (November 2001), http://csrc.nist.gov
17. Nyberg, K.: Linear approximation of block ciphers. In: De Santis, A. (ed.)
    EUROCRYPT 1994. LNCS, vol. 950, pp. 439–444. Springer, Heidelberg (1995)
18. Rivest, R.L.: The RC5 Encryption Algorithm. In: Preneel, B. (ed.) FSE 1994.
    LNCS, vol. 1008, pp. 86–96. Springer, Heidelberg (1995)
19. Selçuk, A.A.: New results in linear cryptanalysis of RC5. In: Vaudenay, S. (ed.)
    FSE 1998. LNCS, vol. 1372, pp. 1–16. Springer, Heidelberg (1998)
20. Selçuk, A.: On Probability of Success in Linear and Differential Cryptanalysis.
    Journal of Cryptology 21(1), 131–147 (2008)
21. Tardy-Corfdir, A., Gilbert, H.: A known plaintext attack of FEAL-4 and FEAL-6.
    In: Feigenbaum, J. (ed.) CRYPTO 1991. LNCS, vol. 576, pp. 172–182. Springer,
    Heidelberg (1992)
22. Zhang, L., Zhang, W., Wu, W.: Cryptanalysis of reduced-round SMS4 block ci-
    pher. In: Mu, Y., Susilo, W., Seberry, J. (eds.) ACISP 2008. LNCS, vol. 5107,
    pp. 216–229. Springer, Heidelberg (2008)

Building Secure Block Ciphers
on Generic Attacks Assumptions

Jacques Patarin[1] and Yannick Seurin[1,2]

[1] University of Versailles, France
[2] Orange Labs, Issy-les-Moulineaux, France
jacques.patarin@prism.uvsq.fr,
yannick.seurin@orange-ftgroup.com

Abstract. Up to now, the design of block ciphers has been mainly driven by heuristic arguments, and little theory is known to constitute a good guideline for the development of their architecture. Trying to remedy this situation, we introduce a new type of design for symmetric cryptographic primitives with high self-similarity. Our design strategy enables to give a reductionist security proof for the primitive based on plausible assumptions regarding the complexity of the best distinguishing attacks on random Feistel schemes or other ideal constructions. Under these assumptions, the cryptographic primitives we obtain are perfectly secure against any adversary with computational resources less than a given bound. By opposition, other provably secure symmetric primitives, as for example C [3] and KFC [4], designed using information-theoretic results, are only proved to resist a limited (though significant) range of attacks. Our construction strategy leads to a large expanded key size, though still usable in practice (around 1 MB).

Keywords: block ciphers, Feistel schemes, generic attacks, provable security.

1 Introduction

Provable Security. Building provably secure but still efficient block ciphers is certainly the most desired but also the most challenging goal of symmetric cryptography. In the area of asymmetric cryptography, "provable security" means that one is mathematically able to reduce the security of a primitive to a well studied and presumably difficult problem such as integer factorisation or discrete logarithm (see [17] for an overview but also a critical look at "provable security" in public key cryptography). The situation in symmetric cryptography is quite different: the security of the most widely deployed primitives often relies on heuristic arguments of one of the three following types:

- lack of known attacks whose complexity is less than "brute-force" attacks or less than the desired security level (typically 2^{80} operations nowadays).
- provable security against some classes of attacks, typically differential and linear cryptanalysis when dealing with block ciphers. For example, AES does possess such security arguments.

R. Avanzi, L. Keliher, and F. Sica (Eds.): SAC 2008, LNCS 5381, pp. 66–81, 2009.

– provable security when some components of the primitive are replaced by "ideal" ones. This kind of arguments apply for example for all Feistel ciphers such as DES, for which the celebrated result of Luby and Rackoff [19] shows that when the internal functions are pseudorandom, the cipher is secure in the sense that it is a pseudorandom permutation. This, however, does not yield any security proof for the real primitive, but only ensures that the general structure of the algorithm does not present intrinsic weaknesses.

Provable security in symmetric cryptography in the reductionist sense discussed for asymmetric cryptography is rather rare. Most notable examples include some number-theoretic hash functions like VSH [10] and the stream cipher QUAD [6] whose security relies on the difficulty of solving systems of multivariate quadratic equations. However, there is to the best of our knowledge no block cipher with security reduction to some hard problem proposed so far. More concernedly, no difficult problems have been identified as suitable for such a design goal. We will see that the problem of distinguishing a Feistel scheme from a random permutation could be a potential candidate.

The Proposal. We propose to build a block cipher whose security can be reduced to some simple and well studied problem. The hard problem we propose is not number-theoretic like for most schemes of asymmetric cryptography. We will use the problem of distinguishing a random Feistel scheme from a random permutation. The rational for such a choice is that Feistel schemes have been extensively studied in the cryptographic literature since the introduction of DES. Though most of this literature is primarily concerned with the information-theoretic properties of these schemes, some authors have studied the so-called "generic attacks" on them. The term generic attacks, introduced by Kilian and Rogaway in [16], means any attack performed on Feistel schemes instantiated using uniformly random and independent functions in each round (which we will name a "random Feistel scheme" in the following), and hence not making use of the underlying structure of the function generator of a real cipher such as DES. Though we will primarily use Feistel schemes, any well studied structure with similar properties could be used.

We propose to go beyond the intrinsic limitations of information-theoretic designs. For Feistel schemes, information theory is "stuck" at five rounds in the sense that increasing the number of rounds beyond five does not increase the number of queries needed by a computationally unbounded adversary to distinguish the Feistel scheme from a random permutation. Indeed, whatever the number $r \geq 5$ of rounds used in a random Feistel schemes from $2n$ bits to $2n$ bits, there is always an oracle adversary making $\Theta(2^n)$ queries and distinguishing a random Feistel scheme from a random permutation with high probability. However the *computational complexity* of this distinguisher can be extremely high. Taking the problem in the opposite way, we will make the hypothesis (and give arguments supporting it) that the best generic attacks described against Feistel schemes cannot be improved, and design a permutation generator such that any distinguishing attack against it would imply an improvement of the generic attacks against random Feistel schemes.

To achieve this goal, we will start from a Feistel scheme with r_1 rounds using random and independent functions at each round, and evaluate its security according to the best generic attacks. Then, rather than using independent and random functions directly as the key, we will instantiate each of these functions with independent Feistel schemes with r_2 rounds, and again estimate the security of the overall construction with respect to the best generic attacks. We will keep on using this recursive structure until the total size of the key (constituted of the random functions used at the innermost level of the construction) becomes practical. We name this design strategy the "Russian Dolls" construction. As we will see, the complexity of the best distinguisher described so far increases exponentially with the number of rounds of the Feistel scheme, so that using a reasonable number of rounds will be sufficient for a good level of security. Note that in the information-theoretic setting, the innermost Feistel schemes would be potentially weak as they have very small block size. However, any attack on the resulting block cipher would imply a better generic attack on random Feistel schemes at some level of the construction.

Related Work. There have been a number of "provably secure" block ciphers proposals. We review the most prominent of them. BEAR and LION were proposed by Anderson and Biham [2]. They are constructed from an ideal stream cipher and an ideal hash function, and the authors proved that attacking the block cipher would imply an attack on one of the underlying components. Later Pat Morin [22] identified some weaknesses in BEAR and LION and proposed AARDVARK, which is based on the same design strategy.

Zheng, Matsumoto and Imai [36] presented block ciphers built on so-called Generalized Type-2 transformations (which are kinds of generalized Feistel constructions). They analysed their constructions in the information-theoretic setting and gave evidence supporting the security of their primitives, but no formal security proof.

Baignères and Finiasz built on Vaudenay's *decorrelation* theory [35] to propose two block ciphers, C [3] and KFC [4], provably secure against a wide range of attacks. This is the logical continuation of the work initiated with the NUT family [35] (COCONUT, PEANUT) and the AES proposal DFC [13]. Again, their security proof relies on information-theoretic arguments. In particular, KFC is based on a 3-round Feistel scheme using round functions with a very low decorrelation bias and is proved resistant against "d-limited" adversaries making less than $d = 8$ or 70 queries, depending on the parameters. The security proof also handles so-called "iterated attacks" of order $d/2$, where the adversary repeats independent non-adaptive $d/2$-limited attacks. However, we note that as the Feistel scheme of KFC has only 3 rounds, it is vulnerable to a distinguishing attack making only 3 chosen plaintext-ciphertext queries (see Section 4.2).

Granboulan and Pornin [14] proposed an efficient way of generating perfectly random permutations (*i.e.* statistically very close to the uniform distribution, even for an attacker having the entire codebook) using a pseudorandom number generator, however their construction is only practical for small plaintext domains (typically less than 30-bit blocks).

The prior proposal which is the closest to our work was made by Blaze [7] but never published. He proposed the block cipher TURTLE and the stream cipher HAZE. TURTLE is simply the Russian Dolls construction where 4-rounds Feistel schemes are used at each stage, and HAZE is based on TURTLE in counter mode. Yet the security arguments proposed by Blaze are quite different from ours. He claims that retrieving the secret functions of an r-round Feistel scheme, $r \geq 3$, is NP-complete by reducing this problem to Numerical Matching with Target Sums (NMTS) [11]. However, keeping the number or rounds constant as the block-size decreases implies a dramatic loss of security.

Organization. Our paper is organized as follows. First we give our notations and some standard security definitions. Then, we describe the Russian Dolls design strategy in all generality and state theorems about its security. In Section 4 we analyse the Russian Dolls construction using balanced Feistel schemes. We highlight some promising possibilities for future work and draw our conclusions in Section 5.

2   Preliminaries

Notations. Throughout the whole paper, we will use the following notations. We will denote by $s \xleftarrow{\$} S$ the operation of selecting an element in the set S endowed with the uniform probability distribution. Func $(\mathcal{D}, \mathcal{R})$ will denote the set of all functions from $\mathcal{D}$ to $\mathcal{R}$, Perm $(\mathcal{D})$ the set of all permutations on $\mathcal{D}$, and Perm$^+$ $(\mathcal{D})$ the set of all permutations on $\mathcal{D}$ with an even signature. I_n will denote the set of binary strings of length n, and we will use Func (n, m), Perm (n) and Perm$^+$ (n) as shorthands for Func (I_n, I_m), Perm (I_n) and Perm$^+$ (I_n) respectively.

A family of functions from $\mathcal{D}$ to $\mathcal{R}$ indexed by key space $\mathcal{K}$ is a function $E : \mathcal{K} \times \mathcal{D} \to \mathcal{R}$. We will use the notation $E_K(X)$ as shorthand for $E(K, X)$. E is a family of permutations if $\mathcal{D} = \mathcal{R}$ and E_K is a permutation for each $K \in \mathcal{K}$. We will denote by E_K^{-1} the inverse of E_K. We will sometimes use the terms function or permutation generator instead of family of functions or permutations.

Given a function f of Func (n, n), the 1-round Feistel scheme Ψ_f is the element of Perm $(2n)$ defined by $\Psi_f(x) = x_\mathrm{R} \| x_\mathrm{L} \oplus f(x_\mathrm{R})$, where x_L and x_R denote respectively the left and right halves of the $2n$-bit string x. We will note $\Psi_{f_1,\ldots,f_r}$ the r-rounds Feistel scheme $\Psi_{f_r} \circ \ldots \circ \Psi_{f_1}$. Given two non null integers n and r, $\Psi^{(r)}(2n)$ will denote the permutation generator on I_{2n} with key space Func $(n, n)^r$, taking as arguments r functions $(f_1, \ldots, f_r)$ in Func (n, n) and $x \in I_{2n}$ and returning $\Psi_{f_1,\ldots,f_r}(x)$. When we omit the block-size, i.e. $\Psi^{(r)}$, it will implicitly be $2n$.

The adversaries we will consider are probabilistic. Implicitly, when we note $\Pr[s \xleftarrow{\$} S : \mathcal{A} = 1]$ the probability will always be on S and the internal randomness of $\mathcal{A}$.

Pseudorandom Functions and Permutations. The notion of pseudorandom function (PRF) was introduced by [12], and the notion of pseudorandom and strong (or super-) pseudorandom permutation (PRP and SPRP) by [18]. Informally, a PRF is a family of functions E indexed by a key space $\mathcal{K}$ such that any efficient adversary with access to an oracle can distinguish a function associated to a random key $K \xleftarrow{\$} \mathcal{K}$ from a uniformly random function only with negligible probability. The definition of a PRP is quite similar, except that the adversary tries to distinguish the permutation family from a uniformly random permutation. For a SPRP, the adversary is given access to two oracles, either E_K and E_K^{-1} for a random K, or G and G^{-1} for a uniformly random permutation G. Rather than using the usual asymptotic notions of PRF and PRP, we will use the concrete security approach introduced in [5] where the distinguishing advantage of an adversary is measured as a function of its resources (namely, runtime and number of oracle queries). We give now the following formal definitions.

Definition 1 (PRF). *Let $E : \mathcal{K} \times \mathcal{D} \to \mathcal{R}$ be a family of functions from $\mathcal{D}$ to $\mathcal{R}$ indexed by keys $\mathcal{K}$. An adversary $\mathcal{A}$ (ϵ, T)-distinguishes E as a PRF if it runs in time at most T and*

$$\mathrm{Adv}_E^{\mathrm{prf}}(\mathcal{A}) = \left| \Pr\left[K \xleftarrow{\$} \mathcal{K} : \mathcal{A}^{E_K} = 1 \right] \right.$$
$$\left. - \Pr\left[G \xleftarrow{\$} \mathrm{Func}\,(\mathcal{D}, \mathcal{R}) : \mathcal{A}^G = 1 \right] \right| \geq \epsilon \ .$$

We will say that E is an (ϵ, T)-secure PRF if no adversary is able to (ϵ, T)-distinguish it.

Definition 2 (PRP). *Let $E : \mathcal{K} \times \mathcal{D} \to \mathcal{D}$ be a family of permutations on $\mathcal{D}$ indexed by keys $\mathcal{K}$. An adversary $\mathcal{A}$ (ϵ, T)-distinguishes E as a PRP if it runs in time at most T and*

$$\mathrm{Adv}_E^{\mathrm{prp}}(\mathcal{A}) = \left| \Pr\left[K \xleftarrow{\$} \mathcal{K} : \mathcal{A}^{E_K} = 1 \right] \right.$$
$$\left. - \Pr\left[G \xleftarrow{\$} \mathrm{Perm}\,(\mathcal{D}) : \mathcal{A}^G = 1 \right] \right| \geq \epsilon \ .$$

We will say that E is an (ϵ, T)-secure PRP if no adversary is able to (ϵ, T)-distinguish it.

Definition 3 (SPRP). *Let $E : \mathcal{K} \times \mathcal{D} \to \mathcal{D}$ be a family of permutations on $\mathcal{D}$ indexed by keys $\mathcal{K}$. An adversary $\mathcal{A}$ (ϵ, T)-distinguishes E as a SPRP if it runs in time at most T and*

$$\mathrm{Adv}_E^{\mathrm{sprp}}(\mathcal{A}) = \left| \Pr\left[K \xleftarrow{\$} \mathcal{K} : \mathcal{A}^{E_K, E_K^{-1}} = 1 \right] \right.$$
$$\left. - \Pr\left[G \xleftarrow{\$} \mathrm{Perm}\,(\mathcal{D}) : \mathcal{A}^{G, G^{-1}} = 1 \right] \right| \geq \epsilon \ .$$

We will say that E is an (ϵ, T)-secure SPRP if no adversary is able to (ϵ, T)-distinguish it.

Alternatively, when a primitive is $(O(\frac{T}{f(n)}), T)$-secure for some parameter n, where O stands for some small constant independent of n, we will say that it is $\Omega(f(n))$-secure, meaning that a distinguisher must have runtime greater than $f(n)$ to have a non-negligible advantage. Note that all our definitions are stated in terms of runtime T of the adversary. The total number q of queries of the adversary to the oracle will only be constrained by the obvious inequality $q \leq T$.

As we will see later, it is always possible to distinguish a random Feistel scheme $\Psi^{(r)}(2n)$ from a uniformly random permutation with complexity $O(2^{2n})$. This comes from the fact that a Feistel scheme has always an even signature, whereas a random permutation has an even signature with probability $1/2$. We will therefore sometimes consider the difficulty of distinguishing a random Feistel scheme from a random permutation with an even signature. For this reason we also define the notion of (S)PRP$^+$ (strong pseudorandom *even* permutation) by simply substituting Perm$^+$ ($\mathcal{D}$) to Perm ($\mathcal{D}$) in the definitions of PRP and SPRP.

We will use sometimes the term CPA (Chosen Plaintext Attack) to qualify an adversary trying to break the pseudorandomness of a permutation generator, and CPCA (Chosen Plaintext-Ciphertext Attack) to qualify an adversary trying to break the strong pseudorandomness of a permutation generator. It will always imply *adaptive* attacks.

3   The Russian Dolls Construction

In this section we explain our design strategy in all generality. Assume one knows how to construct a secure (S)PRP E on $\mathcal{D}$ using a relatively large set of keys $\mathcal{K}$ structured as a direct product of smaller permutations spaces $\mathcal{K} = $ Perm $(\mathcal{D}_1) \times \ldots \times$ Perm $(\mathcal{D}_\lambda)$. Assume now that there exists secure PRPs $E^{(i)}$, $1 \leq i \leq \lambda$, on $\mathcal{D}_i$ with key spaces $\mathcal{K}_i$. Then it is possible to define a new (S)PRP E' on $\mathcal{D}$ with key space $\mathcal{K}' = \mathcal{K}_1 \times \ldots \times \mathcal{K}_\lambda$, by

$$E'_{(K_1, \ldots, K_\lambda)}(\cdot) = E_{(E^{(1)}_{K_1}, \ldots, E^{(\lambda)}_{K_\lambda})}(\cdot) \ . \tag{1}$$

For simplicity, we will make the assumption that when the $E^{(i)}$'s are given as oracles, ciphering or deciphering with E' requires only *direct* queries to the $E^{(i)}$'s. As will be clear from the proof of the theorem below, this enables to use only secure PRPs for the $E^{(i)}$'s. As soon as it requires access to the direct and the inverse oracle for some i, $E^{(i)}$ has to be a secure SPRP. The security of the new (S)PRP E' is characterized by the following theorem:

Theorem 1 (Security of the Russian Dolls construction). *Let E be an (ϵ, T)-secure PRP (resp. SPRP) on $\mathcal{D}$ indexed by key space $\mathcal{K} = $ Perm $(\mathcal{D}_1) \times \ldots \times$ Perm $(\mathcal{D}_\lambda)$. Let also $E^{(i)}$, $1 \leq i \leq \lambda$, be (ϵ_i, T)-secure PRPs on $\mathcal{D}_i$ with key spaces $\mathcal{K}_i$. Then the permutation generator E' defined by Equ. 1 is an $(\epsilon + \sum_{i=1}^{\lambda} \epsilon_i, T)$-secure PRP (resp. SPRP) on $\mathcal{D}$ with key space $\mathcal{K}' = \mathcal{K}_1 \times \ldots \times \mathcal{K}_\lambda$.*

Proof. The proof proceeds by a standard hybrid method. Let $\mathcal{A}$ be an oracle algorithm running in time T. We are interested in bounding its advantage in distinguishing the PRP E':

$$\left| \Pr\left[(K_1,\ldots,K_\lambda) \xleftarrow{\$} \mathcal{K}_1 \times \ldots \times \mathcal{K}_\lambda : \mathcal{A}^{E_{(E_{K_1}^{(1)},\ldots,E_{K_\lambda}^{(\lambda)})}} = 1 \right] \right.$$
$$\left. - \Pr\left[G \xleftarrow{\$} \mathrm{Perm}\,(\mathcal{D}) : \mathcal{A}^G = 1 \right] \right| .$$

This advantage is upper bounded through the triangular inequality by the sum of

$$\left| \Pr\left[(G_1,\ldots,G_\lambda) \xleftarrow{\$} \mathcal{K} : \mathcal{A}^{E_{(G_1,\ldots,G_\lambda)}} = 1 \right] \right.$$
$$\left. - \Pr\left[G \xleftarrow{\$} \mathrm{Perm}\,(\mathcal{D}) : \mathcal{A}^G = 1 \right] \right|$$

and the sum for $i = 1$ to λ of the following quantities (where by convention for $i = 1$ (resp. $i = \lambda$), the expressions were $i-1$ (resp. $i+1$) appears are discarded):

$$\left| \Pr\left[(K_1,\ldots,K_i) \xleftarrow{\$} \mathcal{K}_1 \times \ldots \times \mathcal{K}_i, \right.\right.$$
$$(G_{i+1},\ldots,G_\lambda) \xleftarrow{\$} \mathrm{Perm}\,(\mathcal{D}_{i+1}) \times \ldots \times \mathrm{Perm}\,(\mathcal{D}_\lambda) :$$
$$\left. \mathcal{A}^{E_{(E_{K_1}^{(1)},\ldots,E_{K_i}^{(i)},G_{i+1},\ldots,G_\lambda)}} = 1 \right]$$
$$- \Pr\left[(K_1,\ldots,K_{i-1}) \xleftarrow{\$} \mathcal{K}_1 \times \ldots \times \mathcal{K}_{i-1}, \right.$$
$$(G_i,\ldots,G_\lambda) \xleftarrow{\$} \mathrm{Perm}\,(\mathcal{D}_i) \times \ldots \times \mathrm{Perm}\,(\mathcal{D}_\lambda) :$$
$$\left.\left. \mathcal{A}^{E_{(E_{K_1}^{(1)},\ldots,E_{K_{i-1}}^{(i-1)},G_i,\ldots,G_\lambda)}} = 1 \right] \right|$$

The first term is upper bounded by definition by ϵ as E is an (ϵ,T)-secure PRP. The i-th of the λ other terms is upper bounded by ϵ_i. Indeed, one can build a probabilistic distinguisher $\mathcal{A}_i$ for $E^{(i)}$ as follows. Let F be the oracle to which $\mathcal{A}_i$ has access. $\mathcal{A}_i$ draws random keys $(K_1,\ldots,K_{i-1})$ and random permutations $(G_{i+1},\ldots,G_\lambda)$ and runs $\mathcal{A}$, answering each of its queries with $E_{(E_{K_1}^{(1)},\ldots,E_{K_{i-1}}^{(i-1)},F,G_{i+1},\ldots,G_\lambda)}$. Then $\mathcal{A}_i$ runs in time T and its advantage is exactly the quantity above. Hence by hypothesis on $E^{(i)}$ it cannot be greater than ϵ_i. The theorem follows. The SPRP case is handled in a similar way. $\square$

More restricted versions of this theorem in the information-theoretic setting can be found in [20, Theorem 1] and [35, Lemma 20]. When the key spaces $\mathcal{K}_i$ are themselves permutations spaces, the construction can be iterated to decrease the key size of the outermost PRP. This construction may use functions instead of permutations or even a mix of functions and permutations. However, we will be primarily interested in permutations. We will now see how to use the Russian Dolls construction with concrete PRP schemes.

4 Constructions with Balanced Feistel Schemes

Two main lines of research have been explored concerning Feistel schemes: one aims at giving security bounds against information-theoretic adversaries, the other tries to describe generic attacks on random Feistel schemes. We sum up some known results about these two domains.

4.1   Information-Theoretic Bounds

First, we review the security results on random Feistel schemes holding in the information-theoretic setting, *i.e.* against computationally unbounded adversaries. All these results are purely combinatorial and can be restated in terms of *statistical closeness* between the output of a Feistel permutation and the output of a uniformly random permutation. Though we restate them in terms of computational runtime T, it is essential to note that they are in fact all true in terms of number of oracle queries q. The computational statement simply stems from $q \leq T$.

Luby and Rackoff started the subject by proving [19] that $\Psi^{(3)}(2n)$ is a $\Omega(2^{\frac{n}{2}})$-secure PRP, and claiming (without proof) that $\Psi^{(4)}(2n)$ is a $\Omega(2^{\frac{n}{2}})$-secure SPRP. The later was proved by Patarin in [23]. The first improvements beyond the so-called "birthday bound" (namely, $\Omega(2^{\frac{n}{2}})$-security) came from Patarin who proved respectively in [25] and [26] that $\Psi^{(5)}$ is a $\Omega(2^{\frac{2n}{3}})$-secure PRP and $\Psi^{(6)}$ is a $\Omega(2^{\frac{3n}{4}})$-secure PRP. Maurer and Pietrzak showed [21] that for r sufficiently large, $\Psi^{(r)}$ is a $\Omega(2^{(1-O(\frac{1}{r}))n})$-secure SPRP. Finally, Patarin proved in [28,29] that the information-theoretic optimal security is obtained for 5 rounds in a CPA attack (*i.e.* $\Psi^{(5)}$ is a $\Omega(2^n)$-secure PRP) and 6 rounds for a CPCA attack (*i.e.* $\Psi^{(6)}$ is a $\Omega(2^n)$-secure SPRP). It is still an open problem to improve the bound for $\Psi^{(5)}$ in a CPCA attack (for now it is only known that $\Psi^{(5)}$ is a $\Omega(2^{\frac{n}{2}})$-secure SPRP).

However, building on these results doesn't enable to construct secure schemes using the Russian Dolls construction as the security decreases with the block size. We will see in the following how we can circumvent this problem by making hypotheses on the best generic attacks on random Feistel schemes.

4.2   Generic Attacks on Feistel Schemes

Generic Attacks on $\Psi^{(3)}$ and $\Psi^{(4)}$. Generic attacks on $\Psi^{(3)}$ and $\Psi^{(4)}$ matching the information-theoretic security bounds were described in [25] and later independently in [1]. In the 3-round case, for a CPA attack, the adversary gets m values $y_i = E(x_i)$ and counts the number of (i,j), $i < j$, such that $x_{iR} \oplus y_{iL} = x_{jR} \oplus y_{jL}$. It can be proved that this number will be about twice greater in the case of $\Psi^{(3)}$ than for a random permutation, and this leads to an attack with $O(2^{\frac{n}{2}})$ queries and runtime. However, there is a very efficient CPCA attack with only 3 queries: $\mathcal{A}$ asks for $y_1 = E(x_1)$ and $y_2 = E(x_2)$ where $x_{1R} = x_{2R}$. Then, it asks for $x_3 = E^{-1}(y_{2L}||y_{2R} \oplus x_{1L} \oplus x_{2L})$ and checks whether $x_{3R} = x_{1R} \oplus y_{1L} \oplus y_{2L}$. This will always be the case for $\Psi^{(3)}$ but will happen only with probability $1/2^n$ for a random permutation. We note that this attack applies to KFC as it is based on a 3-round Feistel scheme. However KFC was explicitly designed to resist only chosen-plaintext attacks.

In the 4-round case, there is the following CPA attack: the adversary gets m values $y_i = E(x_i)$ such that x_{iR} is constant and counts the number of (i,j), $i < j$, such that $x_{iL} \oplus y_{iL} = x_{jL} \oplus y_{jL}$. Again, it can be proved that this number will be about twice greater in the case of $\Psi^{(4)}$ than for a random permutation, and this leads to an attack with $O(2^{\frac{n}{2}})$ queries and runtime.

Brute Force Attacks. We state the following result concerning brute force attacks on Feistel schemes, valid for any number of rounds.

Claim. Let r, n be non null integers, r fixed. Then there exists an oracle adversary, running in time $\Theta(2^{rn2^n})$ and distinguishing $\Psi^{(r)}(2n)$ from a random permutation with overwhelming probability.

A rigorous proof of this claim can be found in [24]. Note that a simple entropy argument [21, footnote 2] shows that the number of oracle queries required is only $r \cdot 2^n$, which is in $O(2^n)$ for any fixed r. The adversary proceeds by making an exhaustive search on the key space $\mathrm{Func}\,(n, n)^r$ to see if there is one for which all queries match. It is however highly non trivial to reduce the complexity of the distinguisher described in the above claim in the case $r \geq 5$, as we will see now.

Attacks "By The Signature". As noticed by Patarin in [27], there are better attacks than the exhaustive search described above taking advantage of the fact that Feistel schemes lie in a proper subgroup of $\mathrm{Perm}\,(2n)$, namely $\mathrm{Perm}^+\,(2n)$. Indeed, it can easily be checked (see [27]) that a Feistel scheme has always an even signature. Clearly, the signature of a permutation $E \in \mathrm{Perm}\,(2n)$ can be computed in time $O(2^{2n})$ when all the cipherbook is available. As a random permutation has an even signature with probability $\frac{1}{2}$, we have the following claim:

Claim. Let r, n be non null integers. Then there exists an oracle adversary, running in time $\Theta(2^{2n})$ and distinguishing $\Psi^{(r)}(2n)$ from a uniformly random permutation with probability $\frac{1}{2}$.

However, as we will see in the following, it is much harder to distinguish $\Psi^{(r)}$ when this "global" property is suppressed, *i.e.* when the adversary tries to distinguish $\Psi^{(r)}$ from a random permutation with an even signature.

Best Known Attacks against $\Psi^{(r)}$ as an SPRP$^+$ When $r \geq 5$. The best generic attacks for distinguishing $\Psi^{(r)}$ from a random even permutation fall in the class of iterated attacks of order 2. The notion of iterated distinguisher of order d has been defined by Vaudenay [34,35]. Roughly, such a distinguisher obtains a number d of plaintext-ciphertext pairs (x_j, y_j), takes a binary decision γ_i depending on $\boldsymbol{x} = (x_1, \ldots, x_d)$ and $\boldsymbol{y} = (y_1, \ldots, y_d)$, and after N repetitions of this, outputs 0 or 1 depending on $(\gamma_1, \ldots, \gamma_N)$. At each iteration i, the d-tuple of plaintext-ciphertext pairs that is tested is determined, possibly adaptively, and possibly in a probabilistic way[1] by the adversary, by making only queries to E for a CPA attack, or to E and E^{-1} for a CPCA attack. It is important however that the decision function Γ such that $\gamma_i = \Gamma(\boldsymbol{x}, \boldsymbol{y})$ is fixed during all the attack. In particular, it must not depend on the previously tested d-tuples and previous decisions. Indeed, if it were the case, the i-th decision γ_i of the adversary would in

---

[1] Indeed, as we consider computationally bounded adversaries, there may be an advantage for the adversary to be probabilistic.

Parameters: number of iterations N, decision function $\Gamma : \mathcal{D}^d \times \mathcal{D}^d \to \{0,1\}$,
acceptance set $S \subset \{0,1\}^N$
Oracle: a permutation $E \in \mathrm{Perm}\,(\mathcal{D})$ (and possibly its inverse E^{-1})

1: for $i = 1$ to N do
2:     for $j = 1$ to d do
3:         select $x_j \in \mathcal{D}$ and get $y_j = E(x_j)$ or select $y_j \in \mathcal{D}$ and get $x_j = E^{-1}(y_j)$
4:     end for
5:     set $\gamma_i = \Gamma(\boldsymbol{x}, \boldsymbol{y})$, where $\boldsymbol{x} = (x_1, \ldots, x_d)$ and $\boldsymbol{y} = (y_1, \ldots, y_d)$
6: end for
7: if $(\gamma_1, \ldots, \gamma_N) \in S$ then output 1 else output 0

Fig. 1. Iterated attack of order d

fact depend on all previous d-tuples already tested and the distinguisher would in fact be a classical d'-limited adversary with $d' > d$. Note that this is only a logical description. In particular the total runtime of the adversary can be less than N. For example, the generic attack described previously on $\Psi^{(4)}$ is an iterated attack of order 2 where the attacker makes $N = m(m-1)$ tests in time m by storing the m values of $x_{iL} \oplus y_{iL}$ and counting the number of collisions. The total runtime of the adversary is thus $T = \sqrt{N}$. It is evident that making the same test more than one time does not increase the advantage of the adversary, hence we will assume that the distinguisher never makes twice the same test. Thus, the total number of possible tests is $2^{2n}(2^{2n} - 1) \cdots (2^{2n} - d + 1)$. Note that the outcomes of the tests are of course not independent.

Up to now, the best distinguishing attacks on Feistel schemes with $r \geq 5$ rounds, described in [28], are iterated attacks of order 2. They follow the general description of Fig. 1. We describe the case r even; the case r odd is handled in a similar way. The attacks need only to access the direct oracle E. To understand how these attacks work, we introduce the d-ary transition probabilities associated to a permutation generator E on $\mathcal{D}$ with key space $\mathcal{K}$ defined for any pairs of d-tuples $\boldsymbol{x} = (x_1, \ldots, x_d)$, $\boldsymbol{y} = (y_1, \ldots, y_d)$ of distinct elements of $\mathcal{D}$ by

$$\Pr[\boldsymbol{x} \xrightarrow{E_K} \boldsymbol{y}] = \Pr\left[K \xleftarrow{\$} \mathcal{K} : E_K(x_i) = y_i \text{ for all } i \in [1..d]\right] . \qquad (2)$$

These quantities were introduced and extensively studied by Patarin in [24,23] and are fundamental in upper bounding the advantage of information-theoretic adversaries making less than d queries and trying to distinguish E_K from a uniformly random permutation on $\mathcal{D}$. In particular, closed formula were given in the binary case $d = 2$, for any number of rounds r. Let $\Pr^* = \frac{1}{2^{2n}(2^{2n}-1)}$ denote the binary transition probability for a random even permutation for any $\boldsymbol{x}$ and $\boldsymbol{y}$. We will simply note $\Pr$ for $\Pr[\boldsymbol{x} \xrightarrow{\Psi^{(r)}} \boldsymbol{y}]$. For r even, when $x_{1R} = x_{2R}$, then depending on (y_1, y_2) the transition probabilities have the following values:

1. when $y_{1L} = y_{2L}$, $\Pr = \Pr^* \left(1 - \frac{1}{2^{(r-2)n}}\right)$

2. when $y_{1L} \neq y_{2L}$ and $x_{1L} \oplus y_{1L} \neq x_{2L} \oplus y_{2L}$, $\Pr \simeq \Pr^* \left(1 - \frac{1}{2^{\left(\frac{r}{2}-1\right)n}}\right)$

3. when $y_{1L} \neq y_{2L}$ and $x_{1L} \oplus y_{1L} = x_{2L} \oplus y_{2L}$, $\Pr \simeq \Pr^* \left(1 + \frac{1}{2^{(\frac{r}{2}-2)n}}\right)$

With these notations the attack proceeds as follows. The adversary tests N pairs (x_1, y_1), (x_2, y_2) such that $x_{1R} = x_{2R}$. The decision function is defined by

$$\Gamma(\boldsymbol{x}, \boldsymbol{y}) = \begin{cases} 0 \text{ if } \Pr \leq \Pr^* & \text{(cases 1 and 2)} \\ 1 \text{ if } \Pr > \Pr^* & \text{(case 3)} \end{cases}$$

Let X be the random variable defined by $X = \sum_{i=1}^{N} \gamma_i$. Let $E(X)$ and $\sigma(X)$ (resp. $E^*(X)$ and $\sigma^*(X)$) be the expected value and the standard deviation of X for a random Feistel scheme (resp. a random even permutation). One can easily check that $E^*(X) \simeq \frac{N}{2^n}$ and $E(X) \simeq \frac{N}{2^n}\left(1 + \frac{1}{2^{(\frac{r}{2}-2)n}}\right)$, and it can be proved that $\sigma^*(X) \simeq \frac{\sqrt{N}}{2^{\frac{n}{2}}}$ and $\sigma(X) \simeq \frac{\sqrt{N}}{2^{\frac{n}{2}}}$. If we let the acceptance set be $S = \{(\gamma_1, \ldots, \gamma_N) \mid \sum_{i=1}^{N} \gamma_i \geq \tau\}$ for $\tau = (E(X) - E^*(X))/2$, the adversary will have a noticeable advantage as soon as τ is larger than $\sigma(X)$ and $\sigma^*(X)$. This implies the condition $N \geq 2^{(r-3)n}$.

Because of the constraint $x_{1R} = x_{2R}$, the number of possible tests is only 2^{3n}. So in order to have a meaningful attack for $r \geq 7$ we have to broaden slightly the security model by letting the adversary interact with $\mu > 1$ permutations randomly outputted by the generator. The adversary will have to repeat the test on $\mu = 2^{(r-6)n}$ permutations. For each permutation, the 2^{3n} tests can in fact be implemented in time 2^{2n} by building, for each possible value of x_R, the list of the 2^n values for $x_{iL} \oplus y_{iL}$ and counting the number of collisions. Hence the total runtime of $\mathcal{A}$ is $\mathrm{T} = \mu 2^{2n} = 2^{(r-4)n}$. Note that originally Patarin [28] described a known plaintext attack with roughly the same complexity.

We will take these best known generic attacks as a starting point to build secure PRPs by making the following conjecture:

Conjecture 1. Let $n > 1$ be an integer, r be an integer ≥ 5. Then $\Psi^{(r)}(2n)$ is a $(O(\frac{T}{2^{(r-4)n}}), T)$-secure SPRP$^+$.

Evidence in favour of this conjecture is that the best distinguishing attacks for 3 and 4 rounds, matching the information-theoretic bounds, are iterated attacks of order 2. Hence this conjecture may be viewed as a natural generalization to $r \geq 5$ of a provable result for $r < 5$. We also conjecture that for a fixed d, iterated attacks of order d are not more efficient than the best iterated attack of order 2 for sufficiently large n. Hence improving the attacks described above would require to handle large d-tuples of plaintext-ciphertext pairs, which appears to be intractable as the computation of the transition probabilities for random Feistel schemes becomes very involved as soon as $d \geq 3$.

4.3   The Russian Dolls Construction with Balanced Feistel Schemes

We now concretely describe how to construct a secure SPRP using the Russian Dolls construction and Conjecture 1. The parameters of the construction will be as follows:

- the block size of the SPRP will be $2n$,
- s will denote the number of iterations of the Russian Dolls construction,
- $r_1, r_2, \ldots, r_s =$ will denote the number of rounds of the Feistel schemes used at the i-th iteration of the process.

We start with the outermost Feistel scheme, which will have r_1 rounds. If it were to be instantiated with r_1 random functions, the obtained permutation generator would be a $(O(\frac{T}{2^{(r_1-4)n}}), T)$-secure SPRP$^+$. However, the size of the key would be $r_1 n 2^n$ bits, which is impractical for usual values of n. Using the Russian Dolls construction, one can decrease the size of the key while maintaining a good level of security by instantiating each function inside the Feistel scheme $\Psi^{(r_1)}$ with independent Feistel schemes with r_2 rounds. Again, each function used in the r_1 Feistel schemes $\Psi^{(r_2)}$ can be instantiated using independent Feistel schemes with r_3 rounds, and so on... Note that we implicitly make here the assumption that the security of a Feistel scheme with internal random permutations is close to the security obtained when using internal random functions. A security proof by Piret [33] as well as preliminary results on generic attacks on Feistel schemes with internal permutations [32] point towards the validity of this assumption.

Consider the permutation generator obtained after s iterations of the nesting process. The innermost Feistel schemes use random functions from $\frac{n}{2^{s-1}}$ bits to $\frac{n}{2^{s-1}}$ bits which will constitute the key for the global permutation generator. It can easily be seen that the total number of functions needed to define the global permutation is $r_1 \cdot r_2 \ldots r_s$. Hence the size of the key defining a permutation is

$$\log_2(|\mathcal{K}|) = r_1 \cdot r_2 \cdots r_s \cdot \frac{n}{2^{s-1}} \cdot 2^{\frac{n}{2^{s-1}}} .$$

Suppose now that the numbers of rounds r_i were chosen as the minimal integers to satisfy, for some α, the following inequality:

$$(r_i - 4)\frac{n}{2^{i-1}} \geq \alpha \quad i.e. \quad r_i = \left\lceil \frac{2^{i-1}\alpha}{n} + 4 \right\rceil . \tag{3}$$

According to Conjecture 1, any Feistel scheme used in the construction is a $(\frac{T}{2^\alpha}, T)$-secure SPRP. Then, according to Theorem 1, any adversary running in time T and trying to distinguish a permutation resulting from the overall construction from a uniformly random even permutation has an advantage upper bounded by

$$\left(\frac{T}{2^\alpha} + r_1 \left(\frac{T}{2^\alpha} + r_2 \left(\cdots \left(\frac{T}{2^\alpha} + r_s \cdot \frac{T}{2^\alpha} \right) \cdots \right) \right) \right) = \left(1 + \sum_{i=1}^{s} \prod_{j=1}^{i} r_j \right) \frac{T}{2^\alpha} .$$

Suppose that n is a power of 2. From an asymptotic point of view, if we set $\alpha = \text{poly}(n)$, Equation 3 shows that for a logarithmic number on iterations $s = \log_2(n) - c$, for some constant c, (which means that the key is constituted of functions from 2^{c+1} bits to 2^{c+1} bits), the numbers of rounds r_i will all be polynomials in n. Hence the size of the key will be in $\text{poly}(n)^{\log n} = e^{O((\log n)^2)}$, which is quasi-polynomial, whereas the security is in $(e^{O((\log n)^2)} \frac{T}{2^{\text{poly}(n)}}, T)$. So the Russian Dolls construction will be quite efficient *and* secure.

In practice, the optimal number of iterations is determined the following way. Assume that s iterations have been made, and we want to know whether the following iteration will increase or decrease the size of the key (we suppose that the loss of security coming from the next iteration is negligible). Up to now, the number of bits needed to store one of the functions constituting the key is $\frac{n}{2^{s-1}} \cdot 2^{\frac{n}{2^{s-1}}}$. Iterating the construction one more time would require to instantiate each of these functions with Feistel schemes with r_{s+1} rounds, where r_{s+1} verifies Equ. 3. Hence the storage requirements for each function would become $r_{s+1} \cdot \frac{n}{2^s} \cdot 2^{\frac{n}{2^s}}$. Consequently, it is unfavourable to iterate again as soon as

$$r_{s+1} \cdot \frac{n}{2^s} \cdot 2^{\frac{n}{2^s}} \geq \frac{n}{2^{s-1}} \cdot 2^{\frac{n}{2^{s-1}}}, \quad i.e. \quad r_{s+1} \geq 2^{\frac{n}{2^s}+1} .$$

4.4   Concrete Instantiations

We give now some concrete values for the parameters (n, s, r_i). We describe a block cipher with 128-bit blocks, hence $n = 64$. We aim roughly at 80-bit security, meaning that the cipher has to be a $(T/2^{80}, T)$-secure SPRP. After some optimizations, one can verify that $s = 5$ iterations, with the following number of rounds: $r_1 = 6$, $r_2 = 7$, $r_3 = 10$, $r_4 = 16$ and $r_5 = 28$, is optimal and gives the desired level of security. The size of the expanded key, constituted of functions from 4 bits to 4 bits, is

$$\log_2(|\mathcal{K}|) = 6 \times 7 \times 10 \times 16 \times 28 \times 4 \times 2^4 \simeq 1.5 \text{ MB} ,$$

which is quite practical. Note however that stopping at $s = 4$ iterations (with the same number of rounds r_1 to r_4) yields an expanded key size of $\simeq 1.7$ MB, which is close to the previous size. Yet the resulting block cipher would be much faster as the number of table accesses to encrypt or decrypt one plaintext would only be $6 \times 7 \times 10 \times 16 = 6,720$ instead of $6 \times 7 \times 10 \times 16 \times 28 = 188,160$, which shows that trade-offs are possible.

Key Schedule. It is arguable that such a block cipher as we just described would be implemented using pseudorandom bits for the expanded key. We did not consider this problem in details and expect that a provably secure pseudorandom number generator, such as BBS [8] or QUAD [6] would be used to expand a smaller key. It may even be possible to design a key expansion procedure relying itself on the Russian Dolls construction with PRFs rather than PRPs. Besides, we'd like to underline that the nonexistence of short keys may be turned into an advantage in some cases, particularly in a white-box context of operation [9]. We leave this as topics for further research.

5   Conclusion and Further Work

We described a general recursive strategy enabling to build secure PRFs or PRPs and applied this design approach with random balanced Feistel schemes in order to obtain symmetric primitives provably secure under plausible conjectures about generic attacks on random Feistel schemes. The schemes we obtain look very promising: the size of the expanded key required for our proposed constructions is of the order of 1 MB, and hence compares very favorably with other proposals of provably secure block ciphers such as KFC which may require in extreme cases up to 4 GB of expanded key. Moreover our schemes should be very fast in software as they require only XOR operations and table look-ups.

Other structures are potentially very interesting to use inside the Russian Dolls construction. In the case of PRP constructions, unbalanced Feistel schemes could be suitable. They have been studied in [15,30,31] and could lead to expanded key size savings and efficiency improvements. Such schemes are currently under investigation.

Finally, proving results in the vein of Conjecture 1 may be very difficult because of its connexions with the "P vs. NP" problem. However it may be possible to obtain more restricted security results by considering weaker models of adversary (such as iterated attacks of order d). Such results would greatly reinforce the confidence in the primitives based on the Russian Dolls construction. Exploring new kinds of attacks on random Feistel schemes (e.g., by studying the cycle structure of the permutation) might also be a fruitful avenue of research.

References

1. Aiello, W., Venkatesan, R.: Foiling birthday attacks in length-doubling transformations. In: Maurer, U.M. (ed.) EUROCRYPT 1996. LNCS, vol. 1070, pp. 307–320. Springer, Heidelberg (1996)
2. Anderson, R.J., Biham, E.: Two Practical and Provably Secure Block Ciphers: BEAR and LION. In: Gollmann, D. (ed.) FSE 1996. LNCS, vol. 1039, pp. 113–120. Springer, Heidelberg (1996)
3. Baignères, T., Finiasz, M.: Dial C for cipher. In: Biham, E., Youssef, A.M. (eds.) SAC 2006. LNCS, vol. 4356, pp. 76–95. Springer, Heidelberg (2007)
4. Baignères, T., Finiasz, M.: KFC - the krazy feistel cipher. In: Lai, X., Chen, K. (eds.) ASIACRYPT 2006. LNCS, vol. 4284, pp. 380–395. Springer, Heidelberg (2006)
5. Bellare, M., Kilian, J., Rogaway, P.: The Security of the Cipher Block Chaining Message Authentication Code. J. Comput. Syst. Sci. 61(3), 362–399 (2000)
6. Berbain, C., Gilbert, H., Patarin, J.: QUAD: A practical stream cipher with provable security. In: Vaudenay, S. (ed.) EUROCRYPT 2006. LNCS, vol. 4004, pp. 109–128. Springer, Heidelberg (2006)
7. Blaze, M.: Efficient Symmetric-Key Ciphers Based on an NP-Complete Subproblem (1996), http://www.crypto.com/papers/turtle.pdf
8. Blum, L., Blum, M., Shub, M.: A Simple Unpredictable Pseudo-Random Number Generator. SIAM J. Comput. 15(2), 364–383 (1986)
9. Chow, S., Eisen, P.A., Johnson, H., van Oorschot, P.C.: White-box cryptography and an AES implementation. In: Nyberg, K., Heys, H.M. (eds.) SAC 2002. LNCS, vol. 2595, pp. 250–270. Springer, Heidelberg (2003)
10. Contini, S., Lenstra, A.K., Steinfeld, R.: VSH, an efficient and provable collision-resistant hash function. In: Vaudenay, S. (ed.) EUROCRYPT 2006. LNCS, vol. 4004, pp. 165–182. Springer, Heidelberg (2006)

11. Garey, M.R., Johnson, D.S.: Computers and Intractability: A Guide to the Theory of NP-Completeness. W. H. Freeman, New York (1979)
12. Goldreich, O., Goldwasser, S., Micali, S.: How to Construct Random Functions. J. ACM 33(4), 792–807 (1986)
13. Granboulan, L., Nguyên, P.Q., Noilhan, F., Vaudenay, S.: DFCv2. In: Stinson, D.R., Tavares, S. (eds.) SAC 2000. LNCS, vol. 2012, pp. 57–71. Springer, Heidelberg (2001)
14. Granboulan, L., Pornin, T.: Perfect block ciphers with small blocks. In: Biryukov, A. (ed.) FSE 2007. LNCS, vol. 4593, pp. 452–465. Springer, Heidelberg (2007)
15. Jutla, C.S.: Generalized birthday attacks on unbalanced feistel networks. In: Krawczyk, H. (ed.) CRYPTO 1998, vol. 1462, pp. 186–199. Springer, Heidelberg (1998)
16. Kilian, J., Rogaway, P.: How to protect DES against exhaustive key search. In: Koblitz, N. (ed.) CRYPTO 1996. LNCS, vol. 1109, pp. 252–267. Springer, Heidelberg (1996)
17. Koblitz, N., Menezes, A.: Another Look at Provable Cryptography. J. Cryptology 20(1), 3–37 (2007)
18. Luby, M., Rackoff, C.: Pseudo-random Permutation Generators and Cryptographic Composition. In: STOC, pp. 356–363. ACM, New York (1986)
19. Luby, M., Rackoff, C.: How to Construct Pseudorandom Permutations from Pseudorandom Functions. SIAM J. Comput. 17(2), 373–386 (1988)
20. Maurer, U.M.: A simplified and generalized treatment of luby-rackoff pseudorandom permutation generators. In: Rueppel, R.A. (ed.) EUROCRYPT 1992. LNCS, vol. 658, pp. 239–255. Springer, Heidelberg (1993)
21. Maurer, U.M., Pietrzak, K.: The Security of Many-Round Luby-Rackoff Pseudo-Random Permutations. In: Biham, E. (ed.) EUROCRYPT 2003. LNCS, vol. 2656, pp. 544–561. Springer, Heidelberg (2003)
22. Morin, P.: Provably Secure and Efficient Block Ciphers. In: Selected Areas in Cryptography - SAC 1996, pp. 30–37 (1996)
23. Patarin, J.: Pseudorandom Permutations Based on the DES Scheme. In: Charpin, P., Cohen, G. (eds.) EUROCODE 1990. LNCS, vol. 514, pp. 193–204. Springer, Heidelberg (1991)
24. Patarin, J.: Etude des générteurs de permutations basés sur le schéma du DES, Ph.D. thesis, INRIA, Domaine de Voluceau, Le Chesnay, France (1991)
25. Patarin, J.: New results on pseudorandom permutation generators based on the DES scheme. In: Feigenbaum, J. (ed.) CRYPTO 1991. LNCS, vol. 576, pp. 301–312. Springer, Heidelberg (1992)
26. Patarin, J.: About feistel schemes with six (or more) rounds. In: Vaudenay, S. (ed.) FSE 1998. LNCS, vol. 1372, pp. 103–121. Springer, Heidelberg (1998)
27. Patarin, J.: Generic attacks on feistel schemes. In: Boyd, C. (ed.) ASIACRYPT 2001. LNCS, vol. 2248, pp. 222–238. Springer, Heidelberg (2001)
28. Patarin, J.: Security of random feistel schemes with 5 or more rounds. In: Franklin, M. (ed.) CRYPTO 2004. LNCS, vol. 3152, pp. 106–122. Springer, Heidelberg (2004)
29. Patarin, J.: On linear systems of equations with distinct variables and small block size. In: Won, D.H., Kim, S. (eds.) ICISC 2005. LNCS, vol. 3935, pp. 299–321. Springer, Heidelberg (2006)
30. Patarin, J., Nachef, V., Berbain, C.: Generic attacks on unbalanced feistel schemes with contracting functions. In: Lai, X., Chen, K. (eds.) ASIACRYPT 2006. LNCS, vol. 4284, pp. 396–411. Springer, Heidelberg (2006)

31. Patarin, J., Nachef, V., Berbain, C.: Generic attacks on unbalanced feistel schemes with expanding functions. In: Kurosawa, K. (ed.) ASIACRYPT 2007. LNCS, vol. 4833, pp. 325–341. Springer, Heidelberg (2007)
32. Patarin, J., Treger, J.: Generic Attacks on Feistel Networks with Internal Permutations (2008) (in submission)
33. Piret, G.: Luby-Rackoff Revisited: On the Use of Permutations as Inner Functions of a Feistel Scheme. Des. Codes Cryptography 39(2), 233–245 (2006)
34. Vaudenay, S.: Resistance against general iterated attacks. In: Stern, J. (ed.) EUROCRYPT 1999. LNCS, vol. 1592, pp. 255–271. Springer, Heidelberg (1999)
35. Vaudenay, S.: Decorrelation: A Theory for Block Cipher Security. J. Cryptology 16(4), 249–286 (2003)
36. Zheng, Y., Matsumoto, T., Imai, H.: On the construction of block ciphers provably secure and not relying on any unproved hypotheses. In: Brassard, G. (ed.) CRYPTO 1989. LNCS, vol. 435, pp. 461–480. Springer, Heidelberg (1990)

Lifting and Elliptic Curve Discrete Logarithms

Joseph H. Silverman*

Mathematics Department, Brown University, Providence, RI 02912 USA
jhs@math.brown.edu

Abstract. The difficulty of the elliptic curve discrete logarithm problem (ECDLP) underlies the attractiveness of elliptic curves for use in cryptography. The index calculus is a lifting algorithm that solves the classical finite field discrete logarithm problem in subexponential time, but no such algorithm is known in general for elliptic curves. It turns out that there are four distinct lifting scenarios that one can use in attempting to solve the ECDLP; the lifting field may be a local field or a global field, and the lifted points may be torsion points or nontorsion points. These choices lead to four quite different ways to try to solve the ECDLP via lifting. None of these approaches has led to a solution to the ECDLP, but each method has its own reasons for failing to work. In this article I survey the four ways of lifting the ECDLP, explain their similarities and their differences, and describe the distinct roadblocks that arise in each case.

Introduction

The *elliptic curve discrete logarithm problem* (ECDLP) has attracted considerable attention since Neal Koblitz [14] and Victor Miller [20] independently proposed its use as the basis for crytography. To date, the best general algorithms for ECDLP are no better than the square root algorithms which are known to be best possible for the discrete logarithm problem in a generic group. This is in marked contrast to the discrete logarithm problem in the multiplicative group of a finite field, for which subexponential algorithms are known.

A number of writers have considered the possibility of solving the ECDLP by lifting to either a p-adic (complete local) field such as $\mathbb{Q}_p$ or to a global field such as $\mathbb{Q}$, see for example [4,5,12,13,26,27,34,35,36]. In this paper we consider the general question of lifting as it relates to the ECDLP. In particular, we observe that there are four, quite distinct, lifting scenarios, depending on whether the lifting field is local or global and whether the lifted point is torsion or nontorsion. This leads to four surprisingly different ways to try to solve ECDLP via lifting. (Actually, five different methods, because the global/nontorsion approach comes in two different flavors.) As we will see, none of these approaches has led to a solution to ECDLP, but as we will also see, the reasons for their failures are quite varied.

* Research supported by NSA H98230-04-1-0064 and NSF DMS-0650017. This article is an expanded version of talks presented at ECC 2007 and SAC 2008.

Disclaimer. This article is a survey that draws together a number of threads and attempts to present them in a unified and coherent manner. I have endeavored to give credit as appropriate and I apologize to anyone who may feel slighted. Uncredited results are mostly elementary, well-known to experts in the field, and have undoubtedly been discovered and rediscovered by numerous researchers over the past couple of decades, although many have not previously been published.

1   ECDLP and Lifting Problems

In this section we state the ECDLP and various sorts of lifting problems and briefly indicate how each lifting problem might be used to solve ECDLP and why each turns out not to give a practical algorithm. The remainder of this article gives further details and works out several numerical examples that illustrate what is realistically computable and what is not. For ease of exposition, these examples are done with small numbers (e.g., we often consider the ECDLP over $\mathbb{F}_{257}$), but except as noted, all computations that we perform over $\mathbb{F}_{257}$ can be done for cryptographically useful finite fields containing between 2^{130} and 2^{400} elements.

We do not review the basic theory of elliptic curves or elliptic curve cryptography. The reader will find this material amply covered in [2,6,11,21,25,37] and in numerous other books and articles.

We generally let k denote a finite field and K a local or global field to which we lift. Continuing this convention, we use lower case letters to denote quantities defined over the finite field k and the corresponding upper case letters to denote the lifted quantities defined over the local or global field K.

We now state the two problems whose interconnections lie at the heart of our investigation.

Definition 1. *Let e be an elliptic curve defined over a finite field k and let s and t be points in $e(k)$. Assuming that t is in the group generated by s, the Elliptic Curve Discrete Logarithm Problem (ECDLP) is the problem of finding an integer m such that $t = ms$.*

Definition 2. *Let e/k be an elliptic curve and let $s_1, \ldots, s_r \in e(k)$. The Lifting Problem for $(k, e, s_1, \ldots, s_r)$ is the problem of finding the following quantities:*

- *a field K with subring R.*
- *a maximal ideal $\mathfrak{p}$ of R satisfying $R/\mathfrak{p} \cong k$.*
- *an elliptic curve E/K satisfying $E \bmod \mathfrak{p} \cong e$.*
- *points $S_1, \ldots, S_r \in E(K)$ satisfying $S_i \bmod \mathfrak{p} = s_i$ for $1 \leq i \leq r$.*

Remark 1. There are many variants of the lifting problem, including:

A. Given e/k, find a lift E/K of e/k and an algorithm that is able to (efficiently) lift some sizable collection of the points in $e(k)$ to points in $E(K)$.

Table 1. ECDLP Options for Lifting Points

	Lift to Torsion Point	Lift to Nontorsion Point
p-**adic**	• preserves unique relation • computationally feasible • cannot move to formal group	• does not preserve relation • can move to formal group • easy to compute
Global	• preserves unique relation • computationally infeasible • can move to complex numbers	• can find $E(K) \twoheadrightarrow E(k)$, hard to lift • can lift (up to 9) points, hard to make them dependent

B. Lifting $(k, e, s_1, \ldots, s_r)$ with the added restriction that $E(K)$ is a finitely generated group of rank strictly less than r.

C. Lifting $(k, e, s_1, \ldots, s_r)$ with the added restriction that the lifted points $S_1, \ldots, S_r$ are torsion points.

Roughly speaking, we can divide the lifting problem into four cases, depending on whether the field K is local or global and depending on whether the lifted point(s) are torsion or nontorsion. This separation into four cases may appear, at first glance, to be somewhat artificial, but as we shall see, each case offers a different path leading to a solution of ECDLP.

Thus suppose that we are given an ECDLP (k, e, s, t) whose solution is $t = ms$. It is very easy to find a lift (K, E, S, T) to a *local* field K with *torsion* points S and T that satisfy the same relation $T = mS$, but this does not seem to help in finding m. On the other hand, if we could instead lift to *torsion* points in a *global* field, then we could solve ECDLP using Diophantine approximation. Unfortunately (or fortunately, depending on your point of view), it does not seem to be feasible to lift to torsion points in a global field because the degree of the field will necessarily be very large.

It is also quite easy to lift to *nontorsion* points S and T in a *local* field. The problem with this scenario is that there are many different lifts and it appears to be hard to lift while preserving the relation $T = mS$. If we could find a relation-preserving lift to nontorsion points, then it would be easy to solve ECDLP by moving to the formal group. (More precisely, let n be the order of the point $s \in e(k)$ and rewrite the relation as $nT = mnS$, then nS and nT are in the formal group, so it is easy to find m using the formal logarithm.) Finally, there are two approaches to lifting to *nontorsion* points over a *global* field K. First, we can easily find E/K with a lift of s to $S \in E(K)$. If we could also find $T \in E(K)$ satisfying $T = mS$, then we could use height functions or descent theory to recover m and solve ECDLP. But it appears to be very difficult to find T. Second, we can easily find a lift of (k, e, s, t) to (K, E, S, T), but then it will almost always be true that S and T are independent points in $E(K)$, so they do not satisfy the relation $T = mS$ and cannot be used to solve ECDLP.

The preceding discussion is summarized in Table 1. The remainder of this article is devoted to expanding on these preliminary remarks.

2 Lifting to a p-adic Nontorsion Point

Let e/k be an elliptic curve defined over a finite field k and let $s \in e(k)$. Let K be a complete local field with ring of integers R, maximal ideal $\mathfrak{p}$, and residue field $R/\mathfrak{p} = k$. The reduction map $E(K) \to e(k)$ is surjective [30, VII.2.1] , and indeed there is an exact sequence

$$0 \longrightarrow E_1(K) \longrightarrow E(K) \longrightarrow e(k) \longrightarrow 0. \tag{1}$$

Hensel's lemma provides an efficient method to calculate a lift of s to $E(K)$. (There are even more efficient methods, but Hensel's lemma suffices for our purposes.) Here is the basic idea.

Hensel's Lemma: Let e/k and E/K be elliptic curves given by Weierstrass equations

$$e : f(x, y) = y^2 + a_1 xy + a_3 y - x^3 - a_2 x^2 - a_4 x - a_6 = 0,$$
$$E : F(X, Y) = Y^2 + A_1 XY + A_3 Y - X^3 - A_2 X^2 - A_4 X - A_6 = 0,$$

with $E \bmod \mathfrak{p} = e$, and let $s = (x_1, y_1) \in e(k)$ be a point to be lifted. Also let π be a generator of the ideal $\mathfrak{p}$. Define a sequence of points (X_i, Y_i) satisfying

$$F(X_i, Y_i) \equiv 0 \pmod{\mathfrak{p}^i}, \qquad i = 1, 2, 3, \ldots, \tag{2}$$

as follows:

- Choose any $(X_1, Y_1) \in R^2$ satisfying $X_1 \bmod \mathfrak{p} = x_1$ and $Y_1 \bmod \mathfrak{p} = y_1$. Note that (X_1, Y_1) satisfies (2).
- Suppose that (X_i, Y_i) has been chosen and satisfies (2). Choose $u, v \in R$ satisfying

$$\frac{F(X_i, Y_i)}{\pi^i} + \frac{\partial F}{\partial X}(X_i, Y_i)u + \frac{\partial F}{\partial Y}(X_i, Y_i)v \equiv 0 \pmod{\mathfrak{p}} \tag{3}$$

  and set

$$X_{i+1} = X_i + \pi^i u \qquad \text{and} \qquad Y_{i+1} = Y_i + \pi^i v.$$

  The nonsingularity of e ensures that one of the partial derivatives is nonzero modulo $\mathfrak{p}$, so there will be *many* choices for $u, v \in R$.
- Repeat the previous step to construct a sequence of points (X_i, Y_i) that reduce modulo $\mathfrak{p}$ to s, that satisfy $F(X_i, Y_i) \equiv 0 \pmod{\mathfrak{p}^i}$, and that converge to a point $S \in E(K)$ lifting s.

Remark 2. Notice that the Hensel construction does not yield a particular lift S of s. Instead, at each step, it is necessary to choose values $u, v \in R$ satisfying (3). In practice, only the values of u and v modulo $\mathfrak{p}$ matter, and the value of $v \bmod \mathfrak{p}$ is determined by the value of u. Thus for each $S_i = (X_i, Y_i)$, there is one lift S_{i+1} for each value of u in $R/\mathfrak{p}$. In other words, the set of lifts of $s \in e(k)$ to $S \in E(K)$ is parameterized by R, with the lifts modulo $\mathfrak{p}^i$ being parametrized by $R/\mathfrak{p}^i$. In particular, if k is a large field, then even the set of lifts modulo $\mathfrak{p}^2$ is large.

Remark 3. Suppose that $t = ms$ and that we are searching for the value of m. As explained above, we can lift e, s, and t to a curve E/K and points $S, T \in E(K)$. Suppose that we manage to do this while maintaining the initial relation, i.e., $T = mS$. The kernel of the reduction-modulo-$\mathfrak{p}$ map, which we denote $E_1(K)$, is called the *formal group*, and there is an exact sequence

$$0 \longrightarrow E_1(K) \longrightarrow E(K) \xrightarrow{\text{red mod } \mathfrak{p}} e(k) \longrightarrow 0.$$

We know that $ns = nt = 0$, so nS and nT are in $E_1(K)$. The significance of this lies in the fact that the formal group comes equipped with an easily computable formal logarithm homomorphism

$$\log_E^f : E_1(K) \longrightarrow K^+.$$

(See [30, chapter IV] for information about formal groups and formal logarithms.)

We apply the formal logarithm to the relation $T = mS$. Since $\log_E^f$ is a homomorphism, we find that

$$\log_E^f(T) = m \log_E^f(S).$$

This allows us to solve for the discrete logarithm

$$m = \frac{\log_E^f(T)}{\log_E^f(S)},$$

unless we are unlucky and $\log_E^f(S) = 0$. Further, it is not hard to prove that for a given s, most lifts S satisfy $\log_E^f(S) \neq 0$.

So why does local-nontorsion lifting fail to solve the ECDLP? The answer lies in our requirement that the lifted points S and T satisfy $T = mS$. Assume for the moment that we have already lifted s to a point modulo $\mathfrak{p}^2$. Then among the many possible lifts of t modulo $\mathfrak{p}^2$, only one of them satisfies the relation $T \equiv mS \pmod{\mathfrak{p}^2}$. So the difficulty of using local-nontorsion lifts to solve the ECDLP is that there are too many lifts, and there is no way known[1] to consistently lift two points so as to preserve the desired relation.

We illustrate with a small numerical example.

Example 1. We let $p = 257$ and consider the field $k = \mathbb{F}_{257}$ and the elliptic curve and point

$$e : y^2 = x^3 + 23x + 11, \qquad s = (7, 1) \in e(k).$$

It is easy to check that $\#e(k) = 249 = 3 \cdot 83$ and that s has order $n = 83$ in $e(k)$. We lift e to a p-adic curve E in the obvious way,

$$E : Y^2 = X^3 + 23X + 11.$$

We will lift s to a point $S \bmod p^2$.

---

[1] This is not strictly true; see Section 3 on local-torsion lifts. What we should say is that there is no way known to lift to nontorsion points satisfying $T = mS$.

We write S in the form

$$S = (7 + pu, 1 + pv) \bmod p^2.$$

In order for S to represent a point on E modulo p^2, we need

$$(1 + pv)^2 \equiv (7 + pu)^3 + 23(7 + pu) + 11 \pmod{p^2}.$$

Expanding this gives

$$1 + 2pv \equiv 515 + 170pu \pmod{p^2},$$

so we find that $v \equiv 85u + 1 \pmod{p}$. Thus S has the form

$$S = (7 + pu, 258 + 85pu) = (7 + 257u, 258 + 21845u) \pmod{257^2}. \quad (4)$$

This gives the complete set of lifts of s modulo p^2 with each value of $u \bmod p$ giving a distinct lift.

Now fix a particular mod 257^2 lift of s, say $S = (7, 258)$, and consider a second point $t = (150, 14) \in e(k)$. The ECDLP for (e, s, t) asks us for the integer $0 \le m < 83$ such that $t = ms$ in $e(k)$. (The solution turns out to be $m = 54$, but we will suppose we do not know the answer.) The lifts of t modulo 257^2 are given by the formula

$$T = (150 + pu, 61694 + 72pu) = (150 + 257u, 61694 + 18247u) \pmod{257^2}, \quad (5)$$

where we are free to choose any $u \bmod 257$. Unfortunately, for most choices of u we have $T \ne mS \pmod{257^2}$, so for most choices of u we lose the relation that we are seeking. For example, if we take the obvious lifts $S = (7, 258)$ and $T = (150, 61694)$, then

$$54S \equiv (24565, 25971) \not\equiv T \pmod{257^2}.$$

Indeed, for these lifts the smallest solution to $T \equiv mS \pmod{257^2}$ is $T = 11093S$ $\pmod{257^2}$.

It turns out that the "correct" choice for u in (5) is $u = 95$. Then we get the point $T = (15570, 33681) \bmod 257^2$, and this point T does satisfy $T = 54S$ $\pmod{257^2}$. Further, if we know the point $T = (15570, 33681)$, we can use the formal logarithm to compute as follows.

It turns out to be easier to do computations if we make the change of variables $Z = -X/Y$ and $W = -1/Y$. This brings the identity element to $(Z, W) = (0, 0)$, and the equation of our curve becomes $W = Z^3 + 23ZW^2 + 11W^3$. The formal logarithm for E starts $\log_E^f(Z) = Z + 23Z^7 + \cdots$, so since we are working modulo 257^2, it suffices to use $\log_E^f(Z) \approx Z$. We first compute (in (Z, W) coordinates)

$$83S = (24 \cdot 257 \bmod 257^2, 203 \cdot 257^3 \bmod 257^4),$$
$$83T = (11 \cdot 257 \bmod 257^2, 46 \cdot 257^3 \bmod 257^4).$$

Then

$$\frac{\log_E^f([83]T)}{\log_E^f([83]S)} = \frac{11 \cdot 257}{24 \cdot 257} \equiv 54 \pmod{257}$$

yields the discrete logarithm $m = 54$ that solves $t = ms$.

Of course, for $p = 257$ it was not hard to find the right value for u. But if $p = 257$ is replaced by a large prime, say $p \approx 2^{160}$, then there is no efficient algorithm known for selecting a "correct" value of u, i.e., a value of u for which $T = mS \pmod{p^2}$.

Remark 4. We have just said that there is no efficient way to lift modulo p^2 to points S and T while maintaining the relation $T = mS$, but this is not entirely true. There are actually two situations in which we can perform this lift . The first is when we lift to points S and T that have the same order modulo p^2 as s and t have modulo p. We discuss this situation in more detail in the next section, but we note here that this is the one case in which we cannot multiply S and T by n and still get useful information.

The other situation in which nontorsion local lifting does work and leads to an essentially linear-time solution to the ECDLP is the case that $n = \#E(\mathbb{F}_p) = p$. Elliptic curves with this property are called *anomalous*. In this case s and t have order p, but if we lift them to points S and T modulo p^2 such that S and T do not have order p, then it turns out that $[p]S$ and $[p]T$ automatically satisfy $T = [m]S \pmod{p^2}$. The reason is that if S' and T' are some other lifts, then $S - S'$ and $T - T'$ are in the formal group, so

$$[p](S - S') \equiv O \pmod{p^2} \qquad \text{and} \qquad [p](T - T') \equiv O \pmod{p^2}.$$

For details see [26,27,36]. Thus anomalous curves are not suitable for use in cryptography.

Remark 5. In Example 1 we considered the problem of lifting modulo p^2. For ease of exposition we restricted attention to p^2, but we note that it is easy to repeat the process and lift to higher powers of p. For example, if we take $u = 0$ in (4), then $S = (7, 258)$ and we can work modulo 257^3 to find the collection of lifts

$$S' = (7 + p^2 u, 8454530 + 85p^2 u) \pmod{257^3}.$$

Continuing in this way, we can lift s to any desired level 257^i as long as we can perform basic arithmetic with numbers of size 257^i. At each stage we have 257 choices for the next lift.

3   Lifting to a p-adic Torsion Point

Let e/k be an elliptic curve defined over a finite field k and let $s \in e(k)$. Let K be a complete local field with ring of integers R, maximal ideal $\mathfrak{p}$, and residue field $R/\mathfrak{p} = k$. The order n of the point s divides $\#e(k)$. In this section we consider the question of lifting s to a point $S \in E(K)$ that also has finite order n. We have seen in Section 2 that there are many ways of lifting s to $E(K)$, so there are questions of both existence and uniqueness. Both are answered by the following well-known result.

Theorem 1. *Let e/k, E/K, and $s \in e(k)$ be as above, and assume that the order n of s is not divisible by the characteristic p of k. Then there exists a unique n-torsion point $S \in E(K)$ satisfying S mod $\mathfrak{p} = s$.*

Proof. We begin with uniqueness. Suppose that $S, S' \in E(K)$ are both n-torsion points that lift s. Then $T = S - S'$ is an n-torsion point that reduces of zero, and now our assumption that $p \nmid n$ implies that $T = O$; see [30, VII.3.1b]. Hence $S = S'$.

Let E be given by a minimal Weierstrass equation

$$E : F(X, Y) = y^2 + A_1 xy + A_3 y - x^3 - A_2 x^2 - A_4 x - A_6 = 0.$$

Thus $A_1, \ldots, A_6 \in R$ and the discriminant $\Delta \in R^*$, since by assumption the reduction E mod $\mathfrak{p} = e$ is nonsingular. The nth division polynomial of E is a polynomial

$$\psi_n(X) = n^2 X^{(n^2-1)/2} + \cdots \in \mathbb{Z}[A_1, \ldots, A_6][X]$$

whose roots are the x-coordinates of the n-torsion points of E. (Strictly speaking, this is only true if n is odd, otherwise a polynomial of a slightly different form is required.) Further, the discriminant of $\psi_n(X)$ has the form $n^\alpha \Delta^\beta$, so in particular $\mathrm{Disc}(\psi_n) \in R^*$. (Another way to see this last fact is to use the earlier observation that the n-torsion of E injects under reduction, hence the roots of $\psi_n(X)$ remain distinct modulo $\mathfrak{p}'$ for every extension K' of K, hence its discriminant is relatively prime to $\mathfrak{p}$.)

We are given that $s = (x_0, y_0)$ is an n-torsion point in $e(k)$, so x_0 is a root of $\psi_n(X) \equiv 0 \pmod{\mathfrak{p}}$. Further, it is a simple root (i.e., $\psi_n'(x_0) \not\equiv 0 \pmod{\mathfrak{p}}$), so Hensel's lemma tells us that there is a (unique) $X_0 \in R$ satisfying

$$X_0 \equiv x_0 \pmod{\mathfrak{p}} \qquad \text{and} \qquad \psi_n(X_0) = 0.$$

Finally, we use the fact that $F(X_0, Y) \equiv 0 \pmod{\mathfrak{p}}$ has the root $Y \equiv y_0 \pmod{\mathfrak{p}}$ and use Hensel's lemma to find a $Y_0 \in R$ satisfying $Y_0 \equiv y_0 \pmod{\mathfrak{p}}$ and $F(X_0, Y_0) = 0$. Then $S = (X_0, Y_0)$ is an n-torsion point in $E(K)$ lifting $s \in E(k)$. □

Remark 6. Theorem 1 assures us that every $s \in e(k)$ of order n can be lifted to an n-torsion point $S \in E(K)$, where K is a complete local field with residue field k. However, the proof relies on properties of the division polynomial $\psi_n(X)$. If n is large, then it is not feasible to explicitly compute ψ_n, since ψ_n has degree $(n^2 - 1)/2$. Luckily, there is a more direct way to compute the lift. Roughly speaking, we look at the one-parameter family of lifts, and then the condition that S have finite order gives a linear equation for the parameter. The next example illustrates the process.

Example 2. We continue with Example 1. The formula for S,

$$S = (7 + pu, 258 + 85pu) = (7 + 257u, 258 + 21845u) \pmod{257^2},$$

gives a one-parameter collection of lifts of s. That is, we get one lift modulo 257^2 for each value of u. We now add the condition that $83S \equiv 0 \pmod{p^2}$ and use it to pin down a precise value for u. The easiest way to exploit this condition is to write it as $41S \equiv -42S \pmod{p^2}$. It may seem difficult to compute a large multiple of S when S involves the indeterminate quantity u. However, the variable u appears as pu, so its square modulo p^2 is 0. Hence we never need to deal with general polynomails in u. The only expressions that appear have the form $\alpha + \beta pu$ with $0 \le \alpha < p^2$ and $0 \le \beta < p$. Thus the usual elliptic curve addition formula and general methods for computing large multiples (e.g., by binary expansion of the multiplier) work quite well. The results in our case are

$$41S \equiv (59609 + 12336u, 39178 + 44718u) \pmod{257^2},$$
$$-42S \equiv (40334 + 24415u, 27099 + 63736u) \pmod{257^2}.$$

The congruence $41S \equiv -42S \pmod{257^2}$ leads to the two congruences

$$\frac{x(41S) - x(-42S)}{257} \equiv -47u + 75 \equiv 0 \pmod{257},$$
$$\frac{y(41S) - y(-42S)}{257} \equiv -74u + 47 \equiv 0 \pmod{257}.$$

These congruences have the solution $u \equiv 18 \pmod{257}$. Of course, it is no coincidence that there is a simultaneous solution. Substituting into the formula for S yields the unique point

$$S = (4633, 63223) = (7 + 18 \cdot 257, 1 + 246 \cdot 257) \pmod{257^2}$$

satisfying

$$S \in E \pmod{257^2}, \qquad S \equiv (7, 1) \pmod{257}, \qquad 83S \equiv 0 \pmod{257^2}.$$

We apply the same process to the point $t = (150, 14)$ from Example 1. We find that $T = (150 + pu, 14 + pv)$ is on the curve modulo p^2 if and only if $v = 71u + 240$, so the full set of lifts is $T = (257u + 150, 18247u + 61694)$. We now impose the condition $83T \equiv O \pmod{p^2}$ in the form $41T \equiv -42T$. This leads to $u = 47$ and $T = (12229, 60666)$. Then T has order 83 modulo p^2, and one can check that $T \equiv 54S \pmod{257^2}$. However, we cannot ascertain this last formula by computing $83S$ and $83T$ and working in the formal group, because both $83S$ and $83T$ are zero modulo 257^2. And there is no known way to efficiently use the mod 275^2 lifts to compute the discrete logarithm 54 without first moving into the formal group.

4   Lifting to a Global Torsion Point

As we saw in Section 3, if we lift points $s, t \in e(k)$ to torsion points $S, T \in E(K)$, then the relation $t = ms$ is preserved as $T = mS$. However, lifitng to a local field did not help to compute m. In this section we observe that if we can lift to torsion points defined over a global field, for example $\mathbb{Q}$ or a number field, then it is compartively easy to find m from S and T. For example, we can reduce modulo many small primes q and find the value of m modulo $\#E(\mathbb{F}_q)$.

However if we try to lift to torsion points defined over $\mathbb{Q}$ or a number field, then we run into a severe restriction.

Theorem 2. (Mazur [17], Merel [18]) *Let $E/\mathbb{Q}$ be an elliptic curve and let $P \in E(\mathbb{Q})_{tors}$. Then P has order at most 12.*

More generally, for any $d \geq 1$ there is a bound $C(d)$ so that if $K/\mathbb{Q}$ is a number field of degree d and E/K is an elliptic curve with torsion point $P \in E(K)_{tors}$, then P has order at most $C(d)$.

We can also turn the question around. Thus we lift e to an elliptc curve E, say defined over $\mathbb{Q}$, and we ask how large a number field K is needed in order to get an n-torsion point in $E(K)$. The asymptotic answer is provided by a theorem of Serre.

Theorem 3. (Serre [28,29]) *Let $E/\mathbb{Q}$ be an elliptic curve. There is a constant $c = c(E) > 0$ so that for all integers $n \geq 2$ and any number field K such that $E(K)$ has a torsion point of exact order n, we have*

$$[K : \mathbb{Q}] \geq c \# \mathrm{GL}_2(\mathbb{Z}/n\mathbb{Z}) \approx cn^4.$$

Since for cryptographic applications we need points whose order is between 2^{160} and 2^{320}, the theorems of Mazur, Merel, and Serre make it unlikely that lifting to torsion points over global fields will lead to a workable attack on the ECDLP, since it is not feasible to write down such points or to work in the fields over which they are defined.

5   Lifting to a Global Nontorsion Point

The index calculus is the most powerful method known for solving the classical discrete logarithm problem in the multiplicative group of a finite field. Miller's original article [20] briefly mentions some of the difficulties in extending the index calculus to elliptic curve groups. A more detailed analysis is given in [35]. We briefly summarize the results in Section 5.2. An alternative approach to solving the ECDLP tries to force curves to have low rank rather than using curves of high rank. This method, dubbed the xedni calculus, is described in Section 5.3. In the final section we briefly mention another unsuccessful global nontorsion lifting method that exploits the covering of elliptic curves by modular curves and the existence of special points called Heegner points on modular curves.

5.1   Canonical Heights and Global Lifting

The group of rational points $E(K)$ on an elliptic curve over a number field has a canonical height function

$$\hat{h} : E(K) \longrightarrow [0, \infty)$$

possessing a number of very nice properties. (See, e.g., [30, VIII §9] for basic material on $\hat{h}$.) It has been suggested that the existence of the canonical height in some way protects the ECDLP from index calculus methods. In this section we explain why we feel that the mere existence of the canonical height does not, in and of itself, imply that ECDLP should be hard. Our reason for this assertion is that canonical heights exist in a wide variety of situations, including some such as the classical DLP for which the index calculus does work. We then briefly indicate what we feel is the real reason that the index calculus does not work on elliptic curves. A more complete discussion is given in the next section.

In general terms, a *canonical height* on an abelian group G is a function

$$\hat{h} : G \longrightarrow [0, \infty)$$

with the following four properties:

Power Rule. There is a constant $d > 0$ such that

$$\hat{h}(n\alpha) = |n|^d \hat{h}(\alpha) \qquad \text{for all } n \in \mathbb{Z} \text{ and } \alpha \in G.$$

Addition Rule. There is a constant $c_1 = c_1(G) > 0$ such that

$$\hat{h}(\alpha + \beta) \leq c_1\big(\hat{h}(\alpha) + \hat{h}(\beta)\big) \qquad \text{for all } \alpha, \beta \in G.$$

Normalization. For any $\alpha \in G$, let $|\alpha|_H$ denote the number of bits it takes to describe the element α. (The "H" stands for Hamming weight.) There are constants $c_2 = c_2(G) > 0$ and $c_3 = c_3(G) > 0$ such that

$$c_2|\alpha|_H \leq \hat{h}(\alpha) \leq c_3|\alpha|_H \qquad \text{for all } \alpha \in G.$$

Finiteness. For any bound B, the set $\{\alpha \in G : \hat{h}(\alpha) < B\}$ is finite.

We have the following well-known result, which is a version of Fermat's method of descent.

Proposition 1. *Let G be an abelian group. Suppose that there exists a canonical height on G. Further suppose that the quotient group G/nG is finite for some integer $n \geq 2$. Then the group G is finitely generated.*

More precisely, suppose that G/nG is finite for some integer n satisfying $n^d > 2c_1$, where d and c_1 are the constants appearing in the power rule and the addition rule for $\hat{h}$, respectively. Choose coset representatives

$$\alpha_1, \alpha_2, \ldots, \alpha_t \quad \text{for } G/nG.$$

Then the finite set

$$\left\{ \alpha \in G : \hat{h}(\alpha) \leq \max_{1 \leq i \leq t} \hat{h}(\alpha_i) \right\} \tag{6}$$

generates G.

Table 2. Canonical height on multiplicative and elliptic curve groups

	Multiplicative Group	**Elliptic Curve**
Power Rule	$\hat{h}(\alpha^n) = |n|\hat{h}(\alpha)$	$\hat{h}(nP) = n^2\hat{h}(P)$
Addition Rule	$\hat{h}(\alpha\beta) \le \hat{h}(\alpha) + \hat{h}(\beta)$	$\hat{h}(P+Q) \le 2\hat{h}(P) + 2\hat{h}(Q)$
Normalization	Standard	Standard

Proof. We prove the second part and leave the first part as an exercise (or see any standard text). Let S denote the set (6), let G_S denote the subgroup of G generated by S, and let $M = \max_i \hat{h}(\alpha_i)$. We suppose that $G_S \ne G$ and we choose an element $\alpha \in G \setminus G_S$ of minimal canonical height. Notice in particular that $\hat{h}(\alpha) > M$. The image of $-\alpha$ in G/nG is represented by one of the α_i's, say

$$-\alpha \equiv \alpha_k \pmod{nG}.$$

This means that $-\alpha = \alpha_k - n\beta$ for some $\beta \in G$. We compute

$$n^d\hat{h}(\beta) = \hat{h}(n\beta) = \hat{h}(\alpha + \alpha_k) \le c_1\big(\hat{h}(\alpha) + \hat{h}(\alpha_k)\big) \le c_1\big(\hat{h}(\alpha) + M\big) \le 2c_1\hat{h}(\alpha).$$

Thus $\hat{h}(\beta) \le (2c_1/n^d)\hat{h}(\alpha) < \hat{h}(\alpha)$, so by assumption we have $\beta \in G_S$. Thus $\beta = \sum_i m_i\alpha_i$ for some $m_i \in \mathbb{Z}$, which in turn implies that

$$\alpha = n\beta - \alpha_k = n\sum_i m_i\alpha_i - \alpha_k \in G_S,$$

contradicting the assumption that $\alpha \notin G_S$. This proves that $G_S = G$, and hence that S generates G. $\qquad\square$

The part of the definition that makes the height "canonical" is the power rule. If the height function only satisfies $\hat{h}(g^n) \gg\ll |n|^d\hat{h}(g)$, one can still easily deduce the conclusion of Proposition 1 that G/nG is finitely generated, although the actual set of generators will be somewhat different.

It is not only elliptic curves that have a canonical height. The ordinary multiplicative group of $\mathbb{Q}$, or more generally of a number field, also has a canonical height. Indeed, the standard Weil height

$$h\left(\frac{a}{b}\right) = \log\max\{|a|, |b|\}, \qquad a, b \in \mathbb{Z}, \ \gcd(a,b) = 1,$$

is a canonical height on $\mathbb{Q}^*$. Table 2 gives a point-by-point comparision of the canonical heights on multiplicative groups and elliptic curves and shows how they are analogous.

Since we know that the index calculus works when we lift from the multiplicative group $\mathbb{F}_p^*$ to (finitely generated subgroups of) the multiplicative group of $\mathbb{Q}^*$, it does not seem that the mere existence of the canonical height prevents index calculus methods from working on elliptic curves. We must look elsewhere for the reason. We give a detailed discussion in the next section, but briefly we point out here the following dichotemy that gives a clear distinction between the two situations.

Consider first a finitely generated subgroup of $\mathbb{Q}^*$, say the subgroup

$$G = \langle p_1, p_2, p_3, \ldots, p_r \rangle$$

generated by the first r primes. Similarly, let $E/\mathbb{Q}$ be an elliptic curve whose Mordell–Weil group is given by

$$E(\mathbb{Q}) = \langle P_1, P_2, \ldots, P_r \rangle.$$

In both situations, the canonical height lets us work with linear combinations of the generators. However, when we look at the actual sizes of the generators, we find (assuming some standard conjectures) a striking difference:

$$\max_{1 \le i \le r} \hat{h}(p_i) \approx \log r \quad \text{and} \quad \max_{1 \le i \le r} \hat{h}(P_i) \gg r \log r.$$

Thus it is quite reasonable to work with subgroups of $\mathbb{Q}^*$ of rank (say) 10^6 or 10^7, since the Hamming weight of the generators is not very large. But it would be difficult to deal with elliptic curves of such high rank, even if one knew how to find them. Further, there are other conjectures predicting that for "most" elliptic curve, $\max_i \hat{h}(P_i)$ actually grows exponentially in r. Thus it is not the canonical height, per se, that "protects" elliptic curves from the index calculus. Rather, it is the fact that generating sets for elliptic curve groups have heights (i.e., Hamming weights) that are at least exponentially larger than those for multiplicative groups.

5.2   Elliptic Curves and the Index Calculus (Hard Lift Method)

Let (k, e, s, t) be an ECDLP whose solution $t = ms$ we seek. We also let n denote the order of s (and t) in the group $e(k)$. The idea of the index calculus is to find a number field K and a lift (K, E) of (k, e) for which it is possible to solve the lifting problem

$$E(K) \longrightarrow e(k)$$

for some reasonable fraction of the points in $e(k)$. One way to do this might be to choose E so that the group $E(K)$ has large rank. Of course, the Mordell–Weil theorem tells us that $E(K)$ has finite rank, and indeed for any given field K, it appears to be very difficult to find elliptic curves of very large rank. In any case, we start by showing that if this lifting problem can be solved, then ECDLP can similarly be solved.

Theorem 4. *Let K be a number field and let (K, E) be a lift of (k, e). Let $\mathcal{A}$ be an algorithm that, given a point $u \in e(k)$, has an ϵ-probability of finding a lift $U \in E(K)$ of u. Then there is an algorithm that solves the ECDLP for $e(k)$ in time $O(\epsilon^{-1})$. (The implied constants depend on K and E.)*

Proof. Let r be an upper bound for the rank of the Mordell–Weil group $E(K)$. An upper bound can be given explicitly in terms of the coefficients of E and the discriminant of the field K. (The methods in [30, chapter 10] can easily be used to derive such a bound, or see [23] for a general formulation. The upper bound for r is logarithmic in the coefficients and discriminant of K, so tends to be fairly small.)

Let (k, e, s, t) be an ECDLP to be solved. Choose at random $2(r+1)\epsilon^{-1}$ pairs of integers (a, b) and for each pair compute the point

$$as - bt \in e(k).$$

Applying the algorithm $\mathcal{A}$ to each of these points, we expect to lift at least $r+1$ of them. Let

$$u_i = a_i s - b_i t \in e(k), \qquad 0 \le i \le r,$$

be the points that we are able to lift and let $U_i \in E(K)$ be the lift of u_i.

The fact that $E(K)$ has rank at most r implies that the points $U_0, \ldots, U_r$ are dependent. Further, it is generally possible to find an equation of dependency

$$m_0 U_0 + m_1 U_1 + \cdots + m_r U_r = 0 \qquad \text{in } E(K). \tag{7}$$

More precisely, we can find the dependence relation using either the theory of canonical heights or the theory of descent. See [34] and the references listed there for details, but as a practical matter we observe that the method will generally work in time that is polynomial in the number of bits in the description of E, K, and $U_0, \ldots, U_r$.

Having produced a relation (7) over the global field K, we use the reduction map $E(K) \to e(k)$ to deduce the relation

$$m_0 u_0 + m_1 u_1 + \cdots + m_r u_r = 0 \qquad \text{in } e(K).$$

Substituting $U_i = a_i s - b_i t$ and rearranging terms gives

$$\left(\sum_{i=0}^{r} m_i a_i\right) s = \left(\sum_{i=0}^{r} m_i b_i\right) t \qquad \text{in } e(k).$$

In other words, we have a relation $As = Bt$ with $A, B \in \mathbb{Z}$. Further, there is a reasonable probability that B will be relatively prime to n. (In practice, n will be a large prime, in which case B is almost certainly prime to n.) Multiplying the relation $As = Bt$ by $B^{-1} \bmod n$ yields the solution $ms = t$ to the ECDLP. $\square$

We next formulate a general notion of an index calculus for a group and relate it to the ideas described in the proof of Theorem 4.

Definition 3. *Let G be a (finitely generated abelian) group. The* relation problem *on G is the problem of finding, for a given set $\{U_0, U_1, \ldots, U_r\}$ of dependent elements of G, a nontrivial relation*

$$m_0 U_0 + m_1 U_1 + \cdots + m_r U_r = 0.$$

Definition 4. *An* index calculus *for a finite group g is a finitely generated group G for which the relation problem can be efficiently solved, a (surjective) homomorphism*

$$\pi : G \longrightarrow g,$$

and an algorithm that has an ϵ-probability of lifting an element of g to an element of G.

Example 3. Let G is the subgroup of $\mathbb{Q}^*$ generated by the first n primes, say for $n = 10^5$ or $n = 10^6$. It is relatively easy to check if an element of $\mathbb{Q}^*$ is in G, and it is also not hard to solve the relation problem for elements of G. Finally, let $\mathbb{F}_p$ be a finite field with p elements and consider the reduction map

$$G \longrightarrow \mathbb{F}_p^*.$$

For primes p of an appropriate size, there is a nontrivial probability that elements of $\mathbb{F}_p^*$, lifted into the interval $[0, p-1]$, will lie in G. Thus there is an index calculus for $\mathbb{F}_p^*$.

Example 4. We let $p = 257$ and consider the curve and points

$$e : y^2 = x^3 + 23x + 11, \quad s = (7, 1) \in e(\mathbb{F}_{257}), \quad t = (140, 71) \in e(\mathbb{F}_{257}).$$

It is not hard to find a lift $E/\mathbb{Q}$ such that $S = (7, 1) \in E(\mathbb{Q})$, for example

$$E : Y^2 = X^3 + 23X - 503, \quad S = (7, 1) \in \hat{E}(\mathbb{Q}).$$

In this example it is likely that rank $E(\mathbb{Q}) = 1$ and that the reduction map $E(\mathbb{Q}) \to e(\mathbb{F}_{257})$ is surjective, so there are points $T \in E(\mathbb{Q})$ whose reduction is $t = (140, 71)$. If we can find such a T, then it is relatively easy to express T as a multiple of S, and hence to solve the ECDLP for s and t. However, although such T exist, there are no known algorithms that efficiently find a T. For this example it turns out that the least complicated value of $T \in E(\mathbb{Q})$ satisfying $T \equiv t \pmod{257}$ is the point

$$T = \left(\frac{6239431086988004986355 9}{8736078981416085105625}, \frac{413066569237376536975672924043787 7}{8165350423947492616771476241718 75} \right).$$

Further, we are lucky that T is so uncomplicated, since it happens that $T = 5S$. If instead T were equal to, say, $51S$, then its coördinates would require numbers with thousands of digits.

Now let G be the group of points $E(\mathbb{Q})$ of an elliptic curve. As we observed during the proof of Theorem 4, there are efficient algorithms based on canonical heights and on descent theory for solving the relation problem in $E(\mathbb{Q})$. Further, for a given elliptic curve $e/\mathbb{F}_p$, it is not hard to find a lift $E/\mathbb{Q}$ such that the reduction map is surjective, and one can even force the rank of $E(\mathbb{Q})$ to be larger than one. (However, it is not known if the rank can be arbitrarily large; the current record for the rank of $E(\mathbb{Q})$ is less than 30.) Thus if one could find an efficient algorithm $\mathcal{A}$ that had an ϵ-probability of lifting the map $E(\mathbb{Q}) \to e(\mathbb{F}_p)$, then one would have an index calculus and be able to solve ECDLP.

We now briefly sketch the reasons why such an algorithm is unlikely to exist. Our material is taken from [20] and [35] and we refer the reader to those sources for further details. For simplicity, we restrict attention to an elliptic curve e defined over a finite field $\mathbb{F}_p$ and a lift of e to an elliptic curve $E/\mathbb{Q}$. In order to have an index calculus, we need to find an efficient algorithm $\mathcal{A}$ that lifts a significant number of the points of $e(\mathbb{F}_p)$ to points in $E(\mathbb{Q})$.

The complexity of a point $P \in E(\mathbb{Q})$ is measured by its canonical height $\hat{h}(P)$, so we suppose that $\mathcal{A}$ lifts points in $e(\mathbb{F}_p)$ into the set

$$E_B(\mathbb{Q}) = \{P \in E(\mathbb{Q}) : \hat{h}(P) \leq B\}.$$

(See [30, VIII §9] for basic material on canonical heights and [31,33] for computational methods.) A conjecture of Lang (proven in many cases, see [10]) says that $\hat{h}(P)$ cannot be too small. A theoretical and experimental analysis given in [35] shows that at best we can expect

$$\#E_B(\mathbb{Q}) \gg\ll \left(\frac{c \log B}{r \cdot \log |\Delta(E)|} \right)^{r/2},$$

where r is the rank of $E(\mathbb{Q})$, $\Delta(E)$ is the discriminat of E, and c is an explicit constant.

In order to lift points from $e(\mathbb{F}_p)$ into $E_B(\mathbb{Q})$, we need $\#E_B(\mathbb{Q})$ to be a nontrivial fractional multiple of p. On the other hand, the fact that E is a lift of e means that $\log |\Delta(E)| \gg \log p$, and a theorem of Mestre [19] (conditional on various standard conjectures) implies that $\log |\Delta(E)| \gg r \log r$. The calculations in [35] then show that if $p \approx 2^{160}$ and if we want $\#E_B(\mathbb{Q}) \geq p/2^{10}$, then we probably need $r \approx 180$ and $B \approx 2^{7830} \approx p^{49}$. The first problem would be to merely find a curve of rank 180. (Mestre's work says roughly that the coefficients of a curve of rank r will be larger than $r^{c'r}$.) However, even if this problem could be solved, we still have no way of lifting points from $e(\mathbb{F}_p)$ to points in $E_B(\mathbb{Q})$.

Remark 7. Although the index calculus does not work on elliptic curves, we mention that it does work on hyperelliptic Jacobian varieties when the genus is sufficiently large compared to the order of the field; see [1] for details.

5.3   Elliptic Curves and the Xedni Calculus (Easy Lift Method)

As described in the previous section, an index calculus for a group g involves a lifting homomorphism $G \to g$ such that G is finitely generated and such that there is an efficient algorithm for lifting many elements of g to elements of G. Thus in the index calculus scenario, we start with the homomorphism $G \to g$ and then select points to lift. In this section we consider the reverse scenario, which we dub the *xedni calculus* (xedni is index reversed). The idea is to first select the points to be lifted, and then to find an appropriate group into which to lift them. We begin with an abstract formulation.

Definition 5. *A* xedni calculus *for a finite abelian group g is an algorithm that has an ε-probability of taking a set of elements $u_0, \ldots, u_r \in g$ and efficiently finding a finitely generated group G of rank at most r for which the relation problem can be efficiently solved, a (surjective) homomorphism*

$$\pi : G \longrightarrow g,$$

and points $U_0, \ldots, U_r \in G$ satisfying $\pi(U_i) = u_i$.

Proposition 2. *Let g be a finite abelian group for which there is a xedni calculus. Then there is an algorithm to solve the discrete logarithm problem on g.*

Proof. Let $s, t \in g$ be a discrete logarithm problem to be solved. Choose at random integers $a_0, \ldots, a_r, b_0, \ldots, b_r$ and apply the given xedni calculus algorithm to the points

$$u_i = a_i s - b_i t \in g, \qquad 0 \leq i \leq r.$$

The algorithm will probably be successful in fewer than $2/\epsilon$ attempts. Let $U_0, \ldots, U_r \in G$ be the lifts of $u_0, \ldots, u_r$ found by the algorithm. The group G has rank at most r, so $U_0, \ldots, U_r$ are dependent; and by assumption there is an efficient method for finding a relation $m_0 U_0 + \cdots + m_r U_r = 0$.

The remainder of the proof is the same as the proof of Theorem 4, so we just briefly sketch. Substituting and rearranging yields $(\sum_i m_i a_i)s = (\sum_i m_i b_i)t$. Then multiplying by the inverse of $\sum_i m_i b_i$ modulo the order of s and t gives the desired relation $ms = t$. □

There is a natural way to try to use the xedni calculus to solve the ECDLP. Thus let $e/\mathbb{F}_p$ be an elliptic curve and let $u_0, \ldots, u_r \in e(\mathbb{F}_p)$. Writing $u_i = (x_i, y_i) \in \mathbb{F}_p^2$, we lift the u_i to points $U_i = (X_i, Y_i) \in \mathbb{Z}^2$ without regard to the curve.

Suppose that e is given by that Weierstrass equation

$$e : f(x, y) = y^2 + a_1 xy + a_3 y - x^3 - a_2 x^2 - a_4 x - a_6 = 0.$$

Then lifts of e to $\mathbb{Q}$ are given by Weierstrass equations

$$E : F(X, Y) = Y^2 + A_1 XY + A_3 Y - X^3 - A_2 X^2 - A_4 X - A_6 = 0$$

whose coefficients $A_1, \ldots, A_6 \in \mathbb{Q}$ are required to satisfy

$$A_1 \equiv a_1, \ A_2 \equiv a_2, \ A_3 \equiv a_3, \ A_4 \equiv a_4, \ A_6 \equiv a_6 \pmod{p}.$$

The formulas $F(X_i, Y_i) = 0$ for $0 \leq i \leq r$ give $r+1$ linear equations for $A_1, \ldots, A_6$, so as long as $r \leq 4$, there is a solution $A_1, \ldots, A_6 \in \mathbb{Q}$. Further, the fact that $f(x_i, y_i) = 0$ in $\mathbb{F}_p$ means that we can find a solution with $A_i \equiv a_i \pmod{p}$. Then the curve $E/\mathbb{Q}$ defined by $F(X, Y) = 0$ is a lift of e, and we have arranged matters so that the points $u_i \in e(\mathbb{F}_p)$ have lifts to points $U_i \in E(\mathbb{Q})$.

More generally, we can lift e using a general cubic polynomial of two variables, $F(X, Y) = \sum_{j+k \leq 3} A_{jk} X^j Y^k$. There are 10 coefficients A_{jk}, so using only linear algebra, we can lift $e/\mathbb{F}_p$ and up to 9 points $u_i \in e(\mathbb{F}_p)$ to an elliptic curve $E/\mathbb{Q}$ and points $U_i \in E(\mathbb{Q})$. If it turns out that (with non-negligible probability) the rank of $E(\mathbb{Q})$ is smaller than the number of lifted points, then the xedni calculus succeeds.

Example 5. We let $p = 257$ and consider the curve and points

$$e : y^2 = x^3 + 23x + 11, \quad s = (7,1) \in e(\mathbb{F}_{257}), \quad t = (110, 15) \in e(\mathbb{F}_{257}).$$

We write the lifts $E/\mathbb{Q}$ of $e/\mathbb{F}_{257}$ as

$$E : Y^2 = X^3 + (23 + 257\alpha)X + (11 + 257\beta).$$

Substituting $S = (7,1)$ and $T = (110, 15)$ yields two equations for α and β whose solution gives

$$E : Y^2 = X^3 - \frac{1330433}{103}X + \frac{9277805}{103},$$
$$S = (7,1) \in E(\mathbb{Q}) \quad \text{and} \quad T = (110, 15) \in E(\mathbb{Q}).$$

However, the points S and T are linearly independent in $E(\mathbb{Q})$, so they cannot be used to solve the ECDLP for s and t in $e(\mathbb{F}_{257})$.

We may view this naive xedni approach to the ECDLP as a specialization process. Thus if we write $U_i = (X_i, Y_i)$ and treat the coordinates X_i and Y_i as indeterminates, then we can create an elliptic curve $\mathcal{E}$ whose coefficients are in the field of rational functions $\mathcal{K} = \mathbb{Q}(X_0, \ldots, X_r, Y_0, \ldots, Y_r)$ and such that $U_i \in \mathcal{E}(\mathcal{K})$. Then the above process involves substituting in particular values for the X_i and Y_i. It is not hard to see that before we substitute values, the points $U_0, \ldots, U_r$ are independent in the group $\mathcal{E}(\mathcal{K})$. Then results of Néron and Masser, as described in the following result, say that most substitutions give specialized points that are independent.

Theorem 5. (Néron [22], Masser [16]) *Let $\mathcal{E}_Z$ be a parameterized family of elliptic curves, where $Z = (Z_1, \ldots, Z_n)$, and let $U_{0,Z}, \ldots, U_{r,Z}$ be parameterized families of points that are linearly independent. Then*

$$\left\{ z \in \mathbb{Q}^n : Q_{1,z}, \ldots, Q_{r,z} \text{ are dependent in } \mathcal{E}_z(\mathbb{Q}) \right\}$$

is a small set (a set of density 0).

If we view the coordinates of the points as being the parameters, then the precise statement of Masser's theorem says that the probability that lifted (i.e., specialized) points are linearly dependent is at most $O(1/p)$. Hence the probability that this naive version of the xedni calculus succeeds is negligible.

The reason that the naive xedni calculus does not work is because the lifted points tend to be independent. This suggests imposing further conditions on the lifts in order to make them more likely to be dependent. Mestre [19] has a method, based on the Birch–Swinnerton-Dyer conjecture, for influencing elliptic curves to have *higher* ranker than expected. His idea is to impose congruence conditions on the coefficients of $E/\mathbb{Q}$ for small primes $\ell \leq L$ in order to force $\#E(\mathbb{F}_\ell)$ to be large. Since we know that $\#E(\mathbb{F}_\ell) = \ell + 1 - a_\ell$ with $|a_\ell| \leq 2\sqrt{\ell}$, Mestre's idea is to require that a_ℓ be close to $-2\sqrt{\ell}$. Mestre used this idea to produce an elliptic curve with rank $E(\mathbb{Q}) = 15$, and his idea is still used in algorithms to find curves of high rank.

This led the author to suggest using Mestre's method in reverse to try to influence the lifted curve E to have smaller than expected rank [34]. Thus we impose both the mod p condition that $E/\mathbb{Q}$ is a lift of $e/\mathbb{F}_p$, and also mod ℓ conditions for small primes in order to force $\#E(\mathbb{F}_\ell)$ to be small, i.e., for a_ℓ to be close to $2\sqrt{\ell}$. The hope was that this would cause $E(\mathbb{Q})$ to have smaller rank than expected, which would allow the xedni calculus to succeed.

However, as described in [13], it turns out that there are two difficulties that cause this approach to fail. First, asymptotically one can show using canonical heights, a height specialization theorem [32, III §11] and Lang's height lower bound conjecture [15, page 78] that the lifted points are independent. Second, even for numbers of cryptographic size, experiments show that the rank lowering effect of the small primes is offset by the increased size of the coefficients of E, which negates the (heuristic) application of the Birch–Swinnerton-Dyer conjecture.

5.4   Elliptic Curves and Heegner Point Lifts

We conclude by briefly describing another global lifting method based on entirely different ideas. Suppose that $e/\mathbb{F}_p$ can be lifted to a curve $E/\mathbb{Q}$ with small coefficients. Then we can exploit the fact (Wiles et.al. [3,38,39]) that E is covered by a modular curve, $X_0(N) \to E$, where N is the conductor of E. The curve $X_0(N)$ has special points called Heegner points that are constructed using the theory of complex multiplication, and Deuring's work on CM [7] explains how to lift points in $X_0(\mathbb{F}_p)$ to Heegner points in $X_0(K)$ for certain number fields K. If these Heegner points have a non-negligible probabiilty of being dependent, then one might use their modular interpretation and height formulas of Gross, Zagier, and Kohnen [8,9] to find explicit dependencies without having to explicitly determine the coordinates of the points. This would give a xedni calculus solution to the ECDLP. However, it turns out that the Heegner point lifts are (almost) always independent, although proving their independence is far from trivial. See [24] for details.

6   Summary and Final Remarks

In this paper we have outlined four lifting methods for the ECDLP:

Local-Nontorsion. Lift to nontorsion points in $E(\mathbb{Q}_p)$.
  Fails because we lose the relationship $T = mS$.

Local-Torsion. Lift to torsion points in $E(\mathbb{Q}_p)_{\text{tors}}$
  The relation $T = mS$ is true, but the method fails because we cannot move into the formal group, and there is no known way to determine m without moving into the formal group.

Global-Torsion. Lift to points in $E(\mathbb{Q})_{\text{tors}}$ or $E(K)_{\text{tors}}$
Fails because $E(\mathbb{Q})_{\text{tors}}$ is too small and $[K : \mathbb{Q}]$ is too large.

Global-Nontorsion. Lift to nontorsion points in $E(\mathbb{Q})$.
Hard Lift Method (index calculus):
Fails because there is no known method to lift additional points.
Easy Lift Method (xedni calculus):
Fails because the lifted points are independent.

Acknowledgements. I would like to thank Jeff Achter for his comment on lifting the ECDLP to global torsion points, a remark that led me to consider anew the overall question of lifting and the ECDLP.

References

1. Adleman, L.M., DeMarrais, J., Huang, M.-D.A.: A subexponential algorithm for discrete logarithms over hyperelliptic curves of large genus over $GF_{(q)}$. Theoret. Comput. Sci. 226(1-2), 7–18 (1999)
2. Blake, I.F., Seroussi, G., Smart, N.P.: Elliptic Curves in Cryptography. Cambridge University Press, Cambridge (1999)
3. Breuil, C., Conrad, B., Diamond, F., Taylor, R.: On the modularity of elliptic curves over $\mathbb{Q}$: wild 3-adic exercises. J. Amer. Math. Soc. 14, 843–939 (2001)
4. Cheng, Q., Huang, M.-D.: Partial lifting and the elliptic curve discrete logarithm problem. Algorithmica 46(1), 59–68 (2006)
5. Kim, H.J., Cheon, J.H., Hahn, S.G.: On remarks on lifting problems for elliptic curves. Adv. Stud. Contemp. Math (Pusan) 2, 21–36 (2000)
6. Cohen, H., Frey, G., Avanzi, R., Doche, C., Lange, T., Nguyen, K., Vercauteren, F.: Handbook of Elliptic and Hyperelliptic Curve Cryptography. Discrete Mathematics and Its Applications (Boca Raton). Chapman & Hall/CRC, Boca Raton (2006)
7. Deuring, M.: Die Typen der Multiplikatorenringe elliptischer Funktionenkörper. Abh. Math. Sem. Hansischen Univ. 14, 197–272 (1941)
8. Gross, B., Kohnen, W., Zagier, D.: Heegner points and derivatives of L-series. II. Math. Ann. 278, 497–562 (1987)
9. Gross, B.H., Zagier, D.B.: Heegner points and derivatives of L-series. Invent. Math. 84, 225–320 (1986)
10. Hindry, M., Silverman, J.H.: The canonical height and integral points on elliptic curves. Invent. Math. 93, 419–450 (1988)
11. Hoffstein, J., Pipher, J., Silverman, J.H.: An Introduction to Mathematical Cryptography, UTM. Springer, New York (2008)
12. Huang, M.-D., Kueh, K.L., Tan, K.-S.: Lifting elliptic curves and solving the elliptic curve discrete logarithm problem. In: Bosma, W. (ed.) ANTS 2000. LNCS, vol. 1838, pp. 377–384. Springer, Heidelberg (2000)
13. Jacobson, M.J., Koblitz, N., Silverman, J.H., Stein, A., Teske, E.: Analysis of the xedni calculus attack. Designs, Codes and Cryptography 20, 41–64 (2000)
14. Koblitz, N.: Elliptic curve cryptosystems. Mathematics of Computation 48, 203–209 (1987)
15. Lang, S.: Elliptic Curves: Diophantine Analysis. In: Grund. Math. Wiss., vol. 231. Springer, Berlin (1978)

16. Masser, D.: Specializations of finitely generated subgroups of abelian varieties. Trans. Amer. Math. Soc. 311, 413–424 (1989)
17. Mazur, B.: Modular curves and the Eisenstein ideal. Inst. Hautes Études Sci. Publ. Math 47, 33–186 (1977)
18. Merel, L.: Bornes pour la torsion des courbes elliptiques sur les corps de nombres. Invent. Math. 124, 437–449 (1996)
19. Mestre, J.-F.: Formules explicites et minoration de conducteurs de variétés algébriques. Compositio Math. 58, 209–232 (1986)
20. Miller, V.S.: Use of elliptic curves in cryptography. In: Williams, H.C. (ed.) CRYPTO 1985. LNCS, vol. 218, pp. 417–426. Springer, Heidelberg (1986)
21. Menezes, A.J., van Oorschot, P.C., Vanstone, S.A.: Handbook of Applied Cryptography. CRC Press, Boca Raton (1996)
22. Néron, A.: Problèmes arithmétiques et géométriques rattachés à la notion de rang d'une courbe algébrique dans un corps. Bull. Soc. Math. France 80, 101–166 (1952)
23. Ooe, T., Top, J.: On the Mordell–Weil rank of an abelian variety over a number field. J. Pure Appl. Algebra 58(3), 261–265 (1989)
24. Rosen, M., Silverman, J.H.: On the independence of Heegner points associated to distinct quadratic imaginary fields. Journal of Number Theory 127, 10–36 (2007)
25. Rosing, M.: Implementing Elliptic Curve Cryptography. Manning Publications (1998)
26. Satoh, T., Araki, K.: Fermat quotients and the polynomial time discrete log algorithm for anomalous elliptic curves. Commentarii Math. Univ. St. Pauli 47, 81–92 (1998); Errata. 48, 211–213 (1999)
27. Semaev, I.A.: Evaluation of discrete logarithms in a group of p-torsion points of an elliptic curves in characteristic p. Math. Comp. 67, 353–356 (1998)
28. Serre, J.-P.: Abelian l-adic representations and elliptic curves. In: Research Notes in Mathematics, vol. 7. A K Peters Ltd, Wellesley (1998)
29. Serre, J.-P.: Propriétés galoisiennes des points d'ordre fini des courbes elliptiques. Invent. Math. 15, 259–331 (1972)
30. Silverman, J.H.: The Arithemtic of Elliptic Curves. In: Graduate Texts in Mathematics, vol. 106. Springer, Heidelberg (1986)
31. Serre, J.-P.: Computing heights on elliptic curves. Math. Comp. 51, 339–358 (1988)
32. Serre, J.-P.: Advanced Topics in the Arithemtic of Elliptic Curves. Graduate Texts in Mathematics, vol. 151. Springer, Heidelberg (1994)
33. Serre, J.-P.: Computing canonical heights with little (or no) factorization. Math. Comp. 66, 787–805 (1997)
34. Serre, J.-P.: The xedni calculus and the elliptic curve discrete logarithm problem. Designs, Codes and Cryptography 20, 5–40 (2000)
35. Silverman, J.H., Suzuki, J.: Elliptic curve discrete logarithms and the index calculus. In: Ohta, K., Pei, D. (eds.) ASIACRYPT 1998. LNCS, vol. 1514, pp. 110–125. Springer, Heidelberg (1998)
36. Smart, N.P.: The discrete logarithm problem on elliptic curves of trace one. J. Cryptology 12, 193–196 (1999)
37. Stinson, D.: Cryptography: Theory and Practice. CRC Press, Boca Raton (1997)
38. Taylor, R., Wiles, A.: Ring-theoretic properties of certain Hecke algebras. Ann. of Math. 141, 553–572 (1995)
39. Wiles, A.: Modular elliptic curves and Fermat's last theorem. Ann. of Math. 141, 443–551 (1995)

Preimage Attacks on One-Block MD4, 63-Step MD5 and More

Kazumaro Aoki and Yu Sasaki

NTT, 3-9-11 Midoricho, Musashino-shi, Tokyo, 180-8585 Japan

Abstract. This paper shows preimage attacks on one-block MD4 and MD5 reduced to 63 (out of 64) steps. Our attacks are based on the meet-in-the-middle attack, and many additional improvements make the preimage computable faster than that of the brute-force attack, 2^{128} hash computation. A preimage of one-block MD4 can be computed in the complexity of the 2^{107} MD4 compression function computation, and a preimage of MD5 reduced to 63 steps can be computed in the complexity of the 2^{121} MD5 compression function computation. Moreover, we optimize the computational order of the brute-force attack against MD5, and a preimage of full-round MD5 can be computed in the complexity of the 2^{127} MD5 compression function computation.

Keywords: MD5, MD4, meet-in-the-middle, local collision, one-way, preimage.

1 Introduction

A cryptographic hash function is an important primitive of cryptographic techniques. There are many applications to make a scheme secure using a hash function: message compression in digital signatures and message authentication, for example. However, surprisingly, unlike block ciphers, there are not many concrete instantiations of hash functions. MD5 [12] and SHA-1 [14] are the de facto standards of a hash function and their security is not analyzed well.

A hash function should have several security properties such as collision resistance and one-wayness. After the breakthrough of Wang's work [15], a lot of study has been applied to collision resistance of hash functions. However, the one-wayness of hash functions is not analyzed much.

At FSE 2008, Leurent showed that a preimage attack of MD4, which is a predecessor of MD5 and consists of 48 steps, can be computed in the complexity of the $2^{100.5}$ MD4 compression function [10]. (Hereafter, we omit the unit of complexity, which is the computational complexity of the compression function of the corresponding hash function.) The attack is based on the pioneering work by Dobbertin [4] and its extension [8]. The techniques used in that paper made extensive use of the property of MD4 such as simple step function, not

R. Avanzi, L. Keliher, and F. Sica (Eds.): SAC 2008, LNCS 5381, pp. 103–119, 2009.
© Springer-Verlag Berlin Heidelberg 2009

well-mixed message expansion, and so on. Therefore, applying those techniques to MD5 directly seems difficult. Recently, [13] have tried to compute a preimage of MD5, which consists of 64 steps, utilizing the techniques in [10]. However, [13] can compute a preimage of reduced variants of MD5 up to only 44 steps faster than the brute-force attack. While De et al. proposed preimage attacks on reduced variants of MD4 and MD5 based on SAT-solver [3].

This paper applies the meet-in-the-middle attack to MD5. With newly developed techniques, a preimage of MD5 reduced to 63 steps can be computed in 2^{121}, and a pseudo-preimage of full-round MD5 can be computed in $2^{125.7}$, which is faster than the brute-force attack. On the concrete preimage of full-round MD5, we develop a clever brute-force algorithm, and it finds a preimage of full-round MD5 in 2^{127}. Moreover, utilizing our technique with absorption properties of Boolean functions used in MD4, we can compute a one-block preimage of MD4 in 2^{107}, while [10] computes a preimage of more than 1 block.

A summary of our results and previously published results is shown in Table 1[1]. Note that we do not think that our attack can be used to practically compute a preimage by using currently available resources, since all of our attacks need very high complexity. Since the storage requirements for our attacks are 2^{32} blocks at most, we do not mention the precise memory requirement in this paper.

Table 1. Comparison of preimage attacks against MD4 and MD5

Target	Attack	Attacked steps	Complexity	
			Pseudo-preimage	Preimage
MD4	[4]	32	2^{32} †	
(Total 48 steps)	[8]	32	2^{32} †	
	[3]	39	Not given (8 hours) †	
	[10]	48 (Full)	2^{96}	$2^{100.5}$
	Our result (Sect. 5.2)	48 (Full)	2^{107} †	
MD5	[3]	26	Not given	
(Total 64 steps)	[13]	44	2^{96} †	
	[1]	47	2^{96}	2^{102}
	Our result (Sect. 3.2)	55	2^{96}	2^{113}
	Our result (Sect. 3.3)	59	2^{96}	*
	Our result (Sect. 3.4)	63	2^{112}	2^{121}
	Our result (Sect. 4)	64 (Full)	$2^{125.7}$ ‡	2^{127} †‡

† One-block attack.
‡ The attack is just the brute-force attack, but the computation is optimized.
* This attack only computes a pseudo-preimage. If a very long preimage is accepted, the attack can be converted to a preimage attack whose preimage length is $\approx 2^{64}$ blocks and computed in 2^{113}.

---

[1] Aumasson et al. independently shows a preimage attack in [1]. We refer their result in the table for convenience.

2 Description of MD5 and MD4

2.1 MD5 Specification and Its Properties

This section briefly describes the specification of MD5. Refer to details in [12].

MD5 is one of the Merkle-Damgård hash functions, that is, the hash value is computed as follows:

$$\begin{cases} H_0 \leftarrow IV, \\ H_{i+1} \leftarrow \mathrm{md5}(H_i, M_i) & \text{for } i = 0, 1, \ldots, n-1, \end{cases} \tag{1}$$

where IV is the initial value defined in the specification, md5: $\{0,1\}^{128} \times \{0,1\}^{512} \rightarrow \{0,1\}^{128}$ is the compression function of MD5, and the output of the hash function is H_n. Before applying (1), the messages string M is processed as follows:

- The messages are padded in 512-bit multiples.
- The padded string includes the length of the message, which is represented by 64-bits, and the length string is placed at the end of the padding.

After the process, the message string is divided into 512-bit blocks, M_i ($i = 0, 1, \ldots, n-1$).

The compression function $H_{i+1} \leftarrow \mathrm{md5}(H_i, M_i)$ is computed as follows.

1. M_i is divided into 32-bit message words m_j ($j = 0, 1, \ldots, 15$).
2. Do the following recurrence:

$$\begin{cases} p_0 \leftarrow H_i, \\ p_{j+1} \leftarrow R_j^{\mathrm{MD5}}(p_j, m_{\pi^{\mathrm{MD5}}(j)}) & \text{for } j = 0, 1, \ldots, 63. \end{cases}$$

3. Output H_{i+1} ($= p_{64} + H_i$), where "$+$" denotes 32-bit word-wise addition. In this paper, we similarly use "$-$" to denote 32-bit word-wise subtraction.

R_j^{MD5} is the step function for Step j. Let Q_j be a 32-bit value that satisfies $p_j = (Q_{j-3}\|Q_j\|Q_{j-1}\|Q_{j-2})$. R_j^{MD5} is defined as follows:

$$R_j^{\mathrm{MD5}}(p_j, m_{\pi^{\mathrm{MD5}}(j)}) = (Q_{j-2}\|Q_{j+1}\|Q_j\|Q_{j-1}), \quad \text{where } Q_{j+1}$$
$$= Q_j + (Q_{j-3} + \Phi_j(Q_j, Q_{j-1}, Q_{j-2}) + m_{\pi^{\mathrm{MD5}}(j)} + k_j) \lll s_j, \tag{2}$$

where Φ_j, k_j, and s_j are bitwise Boolean function, constant value, and left rotation defined in the specification. $\pi^{\mathrm{MD5}}(j)$ is a function for MD5 message expansion shown in Table 2. Note that $(R_j^{\mathrm{MD5}})^{-1}(\cdot, m_{\pi^{\mathrm{MD5}}(j)})$ can be computed in almost the same complexity as that of R_j^{MD5}.

2.2 MD4 Specification and Its Properties

The structure of MD4 is similar to that of MD5. The compression function of MD4 consists of 48 steps. The step function R_j^{MD4} for Step j is defined as follows:

$$Q_{j+1} = (Q_{j-3} + \Phi_j(Q_j, Q_{j-1}, Q_{j-2}) + m_{\pi^{\mathrm{MD4}}(j)} + k_j) \lll s_j, \tag{3}$$

Table 2. MD5 message expansion

$\pi^{\mathrm{MD5}}(0), \pi^{\mathrm{MD5}}(1), \ldots, \pi^{\mathrm{MD5}}(15)$	0	1	2	3	4	5	6	7	8	9	10	11	12	13	14	15
$\pi^{\mathrm{MD5}}(16), \pi^{\mathrm{MD5}}(17), \ldots, \pi^{\mathrm{MD5}}(31)$	1	6	11	0	5	10	15	4	9	14	3	8	13	2	7	12
$\pi^{\mathrm{MD5}}(32), \pi^{\mathrm{MD5}}(33), \ldots, \pi^{\mathrm{MD5}}(47)$	5	8	11	14	1	4	7	10	13	0	3	6	9	12	15	2
$\pi^{\mathrm{MD5}}(48), \pi^{\mathrm{MD5}}(49), \ldots, \pi^{\mathrm{MD5}}(63)$	0	7	14	5	12	3	10	1	8	15	6	13	4	11	2	9

Table 3. MD4 Boolean functions and message expansion

$\Phi_j(X,Y,Z),\ 0 \le j \le 15$	$(X \wedge Y) \vee (\neg X \wedge Z)$
$\Phi_j(X,Y,Z), 16 \le j \le 31$	$(X \wedge Y) \vee (Y \wedge Z) \vee (X \wedge Z)$
$\Phi_j(X,Y,Z), 32 \le j \le 47$	$X \oplus Y \oplus Z$

$\pi^{\mathrm{MD4}}(0), \pi^{\mathrm{MD4}}(1), \ldots, \pi^{\mathrm{MD4}}(15)$	0	1	2	3	4	5	6	7	8	9	10	11	12	13	14	15
$\pi^{\mathrm{MD4}}(16), \pi^{\mathrm{MD4}}(17), \ldots, \pi^{\mathrm{MD4}}(31)$	0	4	8	12	1	5	9	13	2	6	10	14	3	7	11	15
$\pi^{\mathrm{MD4}}(32), \pi^{\mathrm{MD4}}(33), \ldots, \pi^{\mathrm{MD4}}(47)$	0	8	4	12	2	10	6	14	1	9	5	13	3	11	7	15

where Φ_j, k_j, s_j, and $\pi^{\mathrm{MD4}}(j)$ are defined *differently* than in MD5. Φ_j and $\pi^{\mathrm{MD4}}(j)$ are shown in Table 3. Note that $(R_j^{\mathrm{MD4}})^{-1}(\cdot, m_{\pi^{\mathrm{MD4}}(j)})$ can be computed in almost the same complexity as that of R_j^{MD4}.

Hereafter, we omit superscripts of $R_j^{\mathrm{MD5}}, R_j^{\mathrm{MD4}}, \pi^{\mathrm{MD5}}$, and π^{MD4} if the hash function discussed is obvious from the context.

3   Preimage Attacks against Reduced MD5

3.1   Converting Pseudo-preimages to a Preimage

First, we describe the generic algorithm that converts pseudo-preimages to a preimage [11, Fact 9.99]. Assume that there is an algorithm that finds $(H_1, (M_1, M_2, \ldots, M_{n-1}))$ such that $H_{i+1} = \mathrm{md5}(H_i, M_i)$ $(i = 1, 2, \ldots, n-1)$ in the complexity of 2^x and H_1 looks random. Prepare a table that includes $2^{64-x/2}$ entries of $(H_1, (M_1, M_2, \ldots, M_{n-1}))$. Compute $2^{64+x/2}$ md5(H_0, M_0) for random M_0, then one of them agrees with one of the entries in the table with high probability. The required complexity of the attack is about $2^{65+x/2}$. Therefore, showing how to compute (H_1, M_1) from a given hash value within 2^x where $x < 126$ is enough for theoretical preimage attack.

3.2   A Preimage Attack against MD5 Reduced to 55 Steps

The proposed attack finds a preimage of MD5 reduced to 55 steps from a given hash value H_n. Our attack target is MD5 reduced to 55 steps, and the steps lie from Step 5 to Step 59[2]. We propose a new technique called the splice-and-cut technique.

---

[2] We confirmed that Step 5 to Step 59, which are 55 steps in total, are the longest section that can be attacked with only the splice-and-cut technique.

Technique 1: Splice-and-Cut

We consider the first and last steps of the attack target as consecutive steps. Then, we divide the attack target into two chunks of steps so that each chunk includes at least one message word that is independent from the other chunk. We call such message words "neutral words." Then, we find pseudo-preimages by the meet-in-the-middle approach.

Observe the message expansion described in Table 2 and notice that Steps 23-37 do not contain m_0, m_6, m_{10}, m_{15}, and Steps 5-22 and 38-59 do not contain m_4 as shown in Fig. 1.

Our attack finds a 2-block preimage, so first, the appropriate padding strings for 2-block messages are set in m_{13}, m_{14}, and m_{15}. For a given H_2, an attack procedure is given below.

Attack Procedure

1. Choose m_i ($i \notin \{4, 6, 13, 14, 15\}$) and p_{38} randomly.
2. For all m_6, do the following:

$$\begin{cases} p_{j+1} \leftarrow R_j(p_j, m_{\pi(j)}) & \text{for } j = 38, 39, \ldots, 59, \\ p_5 \leftarrow H_2 - p_{60}, \\ p_{j+1} \leftarrow R_j(p_j, m_{\pi(j)}) & \text{for } j = 5, 6, \ldots, 22. \end{cases}$$

3. Make a table of (m_6, p_{23})s which are computed in the last step.
4. For all m_4, do the following:

$$p_j \leftarrow R_j^{-1}(p_{j+1}, m_{\pi(j)}) \qquad \text{for } j = 37, 36, \ldots, 23,$$

and examine that the computed p_{23} is in the table made by the previous step. If p_{23} is in the table, the corresponding message and H_1 is just a pseudo-preimage of H_2.

Note that p_5 in the attack is just H_1.

The computational complexity of the above attack procedure is about 2^{32} ($= 2^{32}\frac{40}{55} + 2^{32}\frac{15}{55}$), and the success probability is about 2^{-64} ($= 2^{32} \cdot 2^{32}/2^{128}$). Thus, by iterating the above procedure 2^{64} times, we expect to find one pseudo-preimage (H_1, M_1), and its complexity is about 2^{96} ($= 2^{64} \cdot 2^{32}$). By applying the technique in Section 3.1, we expect that a preimage of MD5 reduced to 55 steps can be computed in 2^{113} ($= 2^{65+96/2}$).

3.3   A Preimage Attack against MD5 Reduced to 59 Steps

We propose an attack that finds a preimage of MD5 reduced to 59 steps starting from Step 3 and ending with Step 61. We notice that this attack cannot deal with the message padding, therefore, the attack can only find a pseudo-preimage.

In the attack against MD5 reduced to 55 steps, two chunks reach the same p_i, and we examine 128-bit matching. Here, we do not have to check all 128 bits, but we check part of them e.g. only 32 bits.

Assume that one chunk produces 2^{32} p_is and the other chunk produces 2^{32} p_{i-3}s. Since $p_i = (Q_{i-3} \| Q_i \| Q_{i-1} \| Q_{i-2})$ and $p_{i-3} = (Q_{i-6} \| Q_{i-3} \| Q_{i-4} \| Q_{i-5})$,

Step	0	1	2	3	4	5	6	7	8	9	10	11	12	13	14	15
index	⓪	1	2	3	④	5	⑥	7	8	9	⑩	11	12	13	14	⑮
	excluded					first chunk										

Step	16	17	18	19	20	21	22	23	24	25	26	27	28	29	30	31
index	1	⑥	11	⓪	5	⑩	⑮	④	9	14	3	8	13	2	7	12
	first chunk							second chunk								

Step	32	33	34	35	36	37	38	39	40	41	42	43	44	45	46	47
index	5	8	11	14	1	④	7	⑩	13	⓪	3	⑥	9	12	⑮	2
	second chunk						first chunk									

Step	48	49	50	51	52	53	54	55	56	57	58	59	60	61	62	63
index	⓪	7	14	5	12	3	⑩	1	8	⑮	⑥	13	④	11	2	9
	first chunk												excluded			

Fig. 1. Message word distribution in MD5 observed in 55-step attack

Step	0	1	2	3	4	5	6	7	8	9	10	11	12	13	14	15
index	0	1	②	3	4	5	6	7	8	9	10	11	12	13	14	⑮
	excluded			first chunk												

Step	16	17	18	19	20	21	22	23	24	25	26	27	28	29	30	31
index	1	6	11	0	5	10	⑮	4	9	14	3	8	13	②	7	12
	first chunk							second chunk								

Step	32	33	34	35	36	37	38	39	40	41	42	43	44	45	46	47
index	5	8	11	14	1	4	7	10	13	0	3	6	9	12	⑮	②
	second chunk													skip		

Step	48	49	50	51	52	53	54	55	56	57	58	59	60	61	62	63
index	0	7	14	5	12	3	10	1	8	⑮	6	13	4	11	②	9
	first chunk														excluded	

Fig. 2. Message word distribution in MD5 observed in 59-step attack

we can examine 32-bit matching without computing three steps. This enables us to find longer sections that are vulnerable against our attack.

Technique 2: Partial Matching
By executing only one-word matching instead of all-word matching, up to three consecutive steps can be skipped from the attack target.

Observe the message expansion described in Table 2 and notice that Steps 23-44 do not contain m_{15}, and Steps 3-22 and 48-61 do not contain m_2 as shown in Fig. 2. For a given H_2, the rough sketch of the attack procedure is as follows[3].

Attack Procedure
1. Choose m_i ($i \notin \{2, 15\}$) and p_{23} randomly.
2. For all m_{15}, do the following:

---

[3] In this attack, skipping two steps is enough. However, we explain the attack procedure for skipping three steps to show the generality of our attack.

$$\begin{cases} p_j \ \leftarrow R_j^{-1}(p_{j+1}, m_{\pi(j)}) & \text{for } j = 22, 21, \ldots, 3, \\ p_{62} \leftarrow H_2 - p_3, \\ p_j \ \leftarrow R_j^{-1}(p_{j+1}, m_{\pi(j)}) & \text{for } j = 61, 60, \ldots, 48, \end{cases}$$

and store (m_{15}, p_{48})s in a table.

3. For all m_2, do the following:

$$p_{j+1} \leftarrow R_j(p_j, m_{\pi(j)}) \qquad \text{for } j = 23, 24, \ldots, 44.$$

Since $p_{48} = (Q_{45}\|Q_{48}\|Q_{47}\|Q_{46})$ and $p_{45} = (Q_{42}\|Q_{45}\|Q_{44}\|Q_{43})$ we can examine Q_{45} is in the table. If Q_{45} is in the table, we compute Q_{46} to Q_{48} by the corresponding m_i, and check whether all of Q_{46} to Q_{48} are matched.

The computational complexity and the success probability are almost the same with the attack against MD5 reduced to 55 steps. Therefore, a pseudo-preimage of MD5 reduced to 59 steps can be found at the complexity of 2^{96}. Note that we can compute a very long preimage by using the technique in Section 3.1 and expandable message introduced in [6], where the length is determined by m_{15}.

Note that we can attack MD5 reduced up to 50 steps even if we restrict that the reduced MD5 should start with the first step (Step 0). The first chunk starts with Step 17 and is 19 steps long, and the second chunk starts with Step 36 and is 28 steps long. The neutral words are m_1 and m_{14}.

3.4   A Preimage Attack against MD5 Reduced to 63 Steps

We propose an attack that finds a preimage of the last 63 steps of MD5. In addition to the splice-and-cut and partial-matching techniques, we use partial-fixing technique.

In previous attack variants, neutral words are totally free when we execute the meet-in-the-middle attack, and thus, both chunks can produce 2^{32} outputs. In this attack, we fix the lower 16 bits of a neutral word. By this effort, computation of one chunk can be partially continued even if the message word for the other chunk appears.

Let us see the inversion of the step function R_j^{-1}. $R_j^{-1}(\cdot, m_{\pi(j)})$ is written by using Q_j as follows:

$$Q_{j-3} = ((Q_{j+1} - Q_j) \ggg s_j) - \Phi_j(Q_j, Q_{j-1}, Q_{j-2}) - m_{\pi(j)} - k_j. \qquad (4)$$

When the lower n bits of Q_{j-1}, Q_{j-2}, and $m_{\pi(j)}$ are fixed and other variables are fully fixed, we can compute the lower n bits of $R_j^{-1}(\cdot, m_{\pi(j)})$ independently from the higher $32 - n$ bits of Q_{j-1}, Q_{j-2}, and $m_{\pi(j)}$. As a consequence, we can partially compute 3 more steps if neutral words are partially fixed. This is graphically explained in Appendix A.

Technique 3: Partial Fixing
By partially fixing neutral words in chunks, up to three consecutive steps can be additionally skipped from the attack target.

Observe the message expansion described in Table 2 and notice that Steps 19-42 do not contain m_6, and Steps 1-18 and 49-63 do not contain m_0 as shown in Fig. 3.

For a given H_2, the rough sketch of the attack procedure is as follows.

Attack Procedure
1. Set m_{13}, m_{14}, and m_{15} to appropriate padding for 2-block messages.
2. Choose m_i ($i \notin \{0,6\}$), p_{19}, and the lower 16 bits of m_0, randomly.
3. For all higher 16 bits of m_0, do the following:

$$p_{j+1} \leftarrow R_j(p_j, m_{\pi(j)}) \qquad \text{for } j = 19, 20, \dots, 42,$$

and store (m_0, p_{43})s in a table, where $p_{43} = (Q_{40} \| Q_{43} \| Q_{42} \| Q_{41})$.
4. (a) For all m_6, do the following:

$$\begin{cases} p_j \ \leftarrow R_j^{-1}(p_{j+1}, m_{\pi(j)}) & \text{for } j = 18, 17, \dots, 1, \\ p_{64} \leftarrow H_2 - p_1, \\ p_j \ \leftarrow R_j^{-1}(p_{j+1}, m_{\pi(j)}) & \text{for } j = 63, 62, \dots, 49. \end{cases}$$

   (b) From obtained $p_{49} = (Q_{46} \| Q_{49} \| Q_{48} \| Q_{47})$, by the partial-fixing technique, we can compute the lower 16 bits of Q_{45}, Q_{44}, and Q_{43}.

   (c) From the partial-matching technique described in Section 3.3, we can examine 16-bit matching by Q_{43}.

Step 3 of the above procedure needs the complexity of 2^{16}, and steps 4(a) and 4(b) need the complexity of 2^{32}. Therefore, the total complexity is 2^{32}. At step 4(c), we examine 16-bit matching for 2^{48} pairs, and we obtain $2^{48} \times 2^{-16} = 2^{32}$ pairs whose 16 bits are matched. Finally, by repeating the above procedure 2^{80} times, we obtain a pair, where all 128 bits are matched. Therefore, the final complexity of the pseudo-preimage attack is $2^{32} \times 2^{80} = 2^{112}$, and this is converted to a preimage attack whose complexity is 2^{121}.

Step	0	1	2	3	4	5	6	7	8	9	10	11	12	13	14	15
index	⓪	1	2	3	4	5	⑥	7	8	9	10	11	12	13	14	15
excluded							first chunk									

Step	16	17	18	19	20	21	22	23	24	25	26	27	28	29	30	31
index	1	⑥	11	⓪	5	10	15	4	9	14	3	8	13	2	7	12
	first chunk			second chunk												

Step	32	33	34	35	36	37	38	39	40	41	42	43	44	45	46	47
index	5	8	11	14	1	4	7	10	13	⓪	3	⑥	9	12	15	2
	second chunk											skip				

Step	48	49	50	51	52	53	54	55	56	57	58	59	60	61	62	63
index	⓪	7	14	5	12	3	10	1	8	15	⑥	13	4	11	2	9
	skip	first chunk														

Fig. 3. Message word distribution in MD5 observed in 63-step attack

4 Notes on Preimage Attack against Full-Round MD5

This section studies the preimage resistance against full-round MD5. We, unfortunately, cannot find any "cryptanalytic attacks" against full-round MD5. While, we find clever technique to perform a brute-force attack. Using this technique, we can find a pseudo-preimage of MD5 at the complexity of 2^{126}, and a preimage of MD5 at the complexity of about 2^{127}.

4.1 Finding Pseudo-preimage of MD5

As we learned by the partial-matching and partial-fixing techniques, a few steps can be skipped from the attack target. Based on this finding, we searched for the minimum number of steps that must be skipped to attack the full-round MD5. The best selection of two chunks where the number of skipped steps is 19 is shown in Fig. 4. (Only this pattern allows skipped steps to be less than 20.)

Since the number of skipped steps is large, we cannot find an efficient way to check whether results from both chunks are matched or not. In this attack, we exhaustively search for the pair that can be matched. Assume we obtain values of p_{14}, p_{33}, and all message words. Whether the computation for that message from p_{14} reaches p_{33} can be checked at the complexity of computing only 13 steps with negligible cost since the complexity of computing 6 steps can be saved by the partial-matching and partial-fixing techniques.

When we only consider pseudo-preimage of the compression function md5, the attack procedure becomes very simple. However, we later want to discuss the conversion from pseudo-preimage(s) to a preimage in Section 4.3. So, we stress that m_{14} is selected as a neutral word, Therefore, some effort is necessary to adjust the padding part. Since m_5 is selected as a neutral word, the last message block must be longer than or equal to 192 bits. As explained later, this attack needs at least a 2-block message. Therefore, we fix 9 bits of m_{14} to guarantee that the value of m_{14} is $192 + 512n, n \geq 1$ for any choice of other bits. Details of messages we select are as follows:

Step	0	1	2	3	4	5	6	7	8	9	10	11	12	13	14	15
index	0	1	2	3	4	⑤	6	7	8	9	10	11	12	13	⑭	15
	first chunk														skip	

Step	16	17	18	19	20	21	22	23	24	25	26	27	28	29	30	31
index	1	6	11	0	⑤	10	15	4	9	⑭	3	8	13	2	7	12
	skip															

Step	32	33	34	35	36	37	38	39	40	41	42	43	44	45	46	47
index	⑤	8	11	⑭	1	4	7	10	13	0	3	6	9	12	15	2
	skip		second chunk													

Step	48	49	50	51	52	53	54	55	56	57	58	59	60	61	62	63
index	0	7	⑭	⑤	12	3	10	1	8	15	6	13	4	11	2	9
	2nd chunk			first chunk												

Fig. 4. Message word distribution in MD5 observed in full-round MD5

- $m_0, \ldots, m_4 \leftarrow$ Randomly chosen fixed value,
- Lower 16 bits of $m_5 \leftarrow$ Randomly chosen fixed value,
- $m_6 \leftarrow$ 0x00000080,
- $m_7, \ldots, m_{13} \leftarrow$ 0x00000000,
- m_{14} is chosen to be $192 + 512n$, $n \geq 1$,
- $m_{15} \leftarrow$ 0x00000000.

For a given hash value H_n, the attack procedure is as follows.

Attack Procedure

1. Set messages as explained above and choose p_{51} randomly.
2. For all the 23 free-bits of m_{14}, do the following:

$$\begin{cases} p_j \leftarrow R_j^{-1}(p_{j+1}, m_{\pi(j)}) \text{ for } j = 50, 49, \ldots, 33, \\ \text{Partially compute } Q_{29}, Q_{28}, \text{ and } Q_{27} \text{ by the partial-fixing technique,} \end{cases}$$

and store $(m_{14}, p_{33}, \text{partial } Q_{29}, \text{partial } Q_{28}, \text{partial } Q_{27})$s in a table.
3. For all of higher 16-bits of m_5, do the following:

$$\begin{cases} p_{j+1} \leftarrow R_j(p_j, m_{\pi(j)}) & \text{for } j = 51, 52, \ldots, 63, \\ p_0 \quad \leftarrow H_n - p_{64}, \\ p_{j+1} \leftarrow R_j(p_j, m_{\pi(j)}) & \text{for } j = 0, 1, \ldots, 13, \end{cases}$$

and keep the values of (m_5, p_{14}).

(a) For all $(m_{14}, p_{33}, \text{partial } Q_{29}, \text{partial } Q_{28}, \text{partial } Q_{27})$ stored in a table, do the following:

$$p_{j+1} \leftarrow R_j(p_j, m_{\pi(j)}) \qquad \text{for } j = 14, 15, \ldots, 26,$$

and examine the lower 16-bit match of Q_{27}.
(b) If lower 16 bits of Q_{27} are matched, compute all bits of Q_{29}, Q_{28}, and Q_{27} by using p_{33} and m_5. Then, examine the higher 16-bit match of Q_{27}.
(c) If higher 16 bits of Q_{27} are matched, compute $p_{28} = R_{27}(p_{27}, m_{\pi(27)})$, and examine the match of Q_{28}.
(d) If Q_{28} is matched, compute $p_{29} = R_{28}(p_{28}, m_{\pi(28)})$, and examine the match of Q_{29}.
(e) If Q_{29} is matched, compute $p_{30} = R_{29}(p_{29}, m_{\pi(29)})$, and examine the match of Q_{30}. If matched, corresponding (p_0, M) is a pseudo-preimage.

Step 2 of the above procedure needs the complexity of $2^{23}\frac{21}{64}$. For each m_5, the first 4 lines of step 3 need the complexity of $\frac{27}{64}$. Step 3(a) needs the complexity of $2^{23}\frac{13}{64}$. As a result of lower 16-bit match of Q_{27}, $2^{23} \times 2^{-16} = 2^7$ pairs are expected to be remained. Step 3(b) needs the complexity of $2^7\frac{3}{64}$. As a result of higher 16-bit match of Q_{27}, $2^7 \times 2^{-16} = 2^{-9}$ pair is expected to be remained. Step 3(c) needs the complexity of $2^{-9}\frac{1}{64}$. After the match of Q_{28}, $2^{-9} \times 2^{-32} = 2^{-41}$ pair is expected to be remained. The complexities of 3(d) and 3(e) are negligible. Hence, the complexity of step 3 is $2^{23}\frac{21}{64} + 2^{16}\frac{27}{64} + 2^{16}(2^{23}\frac{13}{64} + 2^7\frac{3}{64} + 2^{-9}\frac{1}{64}) < 2^{37}$.

Finally, by repeating this procedure 2^{89} times, we obtain a pair, where all 128 bits are matched. Therefore, the final complexity of the pseudo-preimage attack is $2^{89} \times (2^{16} \times 2^{23}\frac{13}{64}) < 2^{126}$.

4.2   Increase the Speed of the Naive Search

When we compute hash values of 2^{128} different messages, we do not have to compute 2^{128} times of md5. For example, two different messages whose m_0 to m_{14} are the same and whose m_{15} are different will have the same computation result after the first 15 steps. This saves us the cost of computing the first 15 steps of the second message. By extending this idea, the complexity of computing hash values of 2^{128} different messages becomes $2^{127} = 2^{128}\frac{32}{64}$.

We use a technique named *Q4 Tunnel* by Klima [7], which enables us to compute md5 from an intermediate step. In this technique, the value of m_3 in the first round is changed, however, any change in m_3 can be offset by modifying m_4 and m_7 so that all other chaining variables in the first round are kept unchanged. Since m_3, m_4, and m_7 appear in Steps 23, 26, and 30, respectively in the second round, any choice of m_3 does not impact on chaining variables up to Step 22. Therefore, this technique saves us the cost for computing the first 23 steps. Moreover, we can save the complexity of a few more steps:

1. Since the initial value and hash value are fixed and message words used in Steps 61, 62, and 63 are not m_3, m_4, and m_7, we can compute p_{63}, p_{62}, and p_{61} independently of m_3, m_4, and m_7. This saves us the complexity of three steps.
2. The partial-matching and partial-fixing techniques described in Section 3.3 save us the complexity of six steps.

Finally, the complexity to compute a hash value becomes $64 - 23 - 3 - 6 = 32$ steps, we can compute hash values of 2^{128} messages at the complexity of $2^{128}\frac{32}{64}$.

4.3   Discussion on Converting a Pseudo-preimage to a Preimage

Let $E(x) = 1 - \exp(-x)$, then the success probability of a brute-force attack for computing preimage is $b = E(2^{128}/2^{128}) \approx 0.63$. The attack described in Section 4.2 finds a pseudo-preimage at the complexity of $2^{125.70}$ $(= 2^{128}\frac{13}{64})$ with probability b. If we directly use the conversion described in Section 3.1, a preimage will be found at the complexity of $2^{127.85}$ with probability b^2. This complexity is higher than $2^{127.00}$ $(= 2^{128}\frac{32}{64})$ described in Section 4.2. Applying the technique in Section 4.2 to the computation of md5(H_0, M_0) in the conversion described in Section 3.1, resulting preimage attack still requires $2^{127.39}$ with probability b^2.

Using the idea of expandable message [6] as in [10], one preimage is enough to compute a preimage[4]. This attack requires $2^{126.86}$ $(= 2^{128}\frac{13}{64} + 2^{128}\frac{32}{64}/2)$. However, the success probability is b^2, which is lower than that of the brute-force attack, b. Spending $c_1 2^{128}\frac{13}{64}$ work for computing a pseudo-preimage and $c_2 2^{128}\frac{32}{64}/2$ for brute-force attack, the success probability of the attack is $E(c_1)$ $E(c_2)$ and the complexity is $2^{128}\frac{13c_1+16c_2}{64}$. To achieve the same success probability $b = E(c_1)E(c_2)$, the attack requires $2^{127.52}$, where $c_1 \approx 1.672$ and $c_2 \approx 1.507$.

---

[4] Appendix B shows extensions of the attack.

Even if computing a pseudo-preimage fails, we can continue to seek a preimage using brute-force attack. The complexity of the attack is also $2^{128} \frac{13c_1 + 16c_2}{64}$, and the success probability increases to $E(c_1)E(c_2) + (1 - E(c_1))E(c_2/2)$. To achieve the same success probability b, the attack requires $2^{126.94}$, where $c_1 \approx 0.354$ and $c_2 \approx 1.636$.

5   Preimage Attacks against MD4

The first preimage attack against full-round MD4 was proposed by Leurent [10]. It finds a preimage of MD4 with the complexity of 2^{102} by using messages of 34 blocks[5]. Therefore, no one has succeeded in attacking MD4 by using one-block messages. A one-block attack is particularly interesting since an attacker cannot use the characteristics of the Merkle-Damgård structure. A one-block attack analyzes the security of the compression function md4.

In this section, we first show a preimage attack using messages of 2 blocks to show that the splice-and-cut approach can be also applied to MD4. This attack finds a preimage of MD4 at the complexity of 2^{121}. Second, we show a one-block attack that finds a preimage at the complexity of 2^{107}.

5.1   Two-Block Preimage Attack against MD4

MD4 can be analyzed in a manner similar to MD5. By using the splice-and-cut, partial-matching, and partial-fixing techniques, we can find a pseudo-preimage at the complexity of 2^{112}, and this attack is converted to a preimage attack at the complexity of 2^{121}. The selection of two chunks is shown in Fig. 5.

5.2   One-Block Preimage Attack against MD4

By checking the details of the step function of MD4, we can find a preimage that consists of a one-block message. The key idea is fixing the value of p_0 to the original MD4 IV when we compute a chunk. To achieve this, we use a local-collision approach. By this approach, the value of p_0 can be kept unchanged even if the value of a neutral word in a chunk is changed. (The similar idea is used by Sasaki et al. [13] to analyze MD5.) We thus search for a pair of chunks in which one chunk includes one neutral word and the other chunk includes two neutral words where changes of one neutral word can be offset by changing the other neutral word. The selected chunks are shown in Fig. 6.

When we compute the first chunk by changing the value of m_7, the corresponding chaining variable Q_4 is updated according to the selection of m_7. In this attack, by selecting m_3 adaptively and fixing p_7, m_0 to m_2, and m_4 to m_6 in advance, Q_0 to Q_{-3} can be fixed to the original IV of MD4 for any m_7. This attack heavily uses the absorption properties of Boolean functions of MD4, so readers who are not familiar with them are recommended to read [10, Section 2.1]. The method to

---

[5] This attack can be easily converted to an attack that finds a preimage with the complexity of 2^{113} by using messages of 2 blocks.

Step	0 1 2 3 4 5 6 7	8 9 10 11 12 13 14 15
index	0 1 2 3 4 5 6 ⑦	⑧ 9 10 11 12 13 14 15
	first chunk	second chunk
Step	16 17 18 19 20 21 22 23 24 25 26 27 28	29 30 31
index	0 4 ⑧ 12 1 5 9 13 2 6 10 14 3	⑦ 11 15
	second chunk	skip
Step	32 33	34 35 36 37 38 39 40 41 42 43 44 45 46 47
index	0 ⑧	4 12 2 10 6 14 1 9 5 13 3 11 ⑦ 15
	skip	first chunk

Fig. 5. Message word distribution in MD4 observed in 2-block attack

Step	0 1 2 3 4 5 6 7	8 9 10 11 12 13 14 15
index	0 1 2 ③ 4 5 6 ⑦	⑧ 9 10 11 12 13 14 15
	first chunk	second chunk
Step	16 17 18 19 20 21 22 23 24 25 26 27	28 29 30 31
index	0 4 ⑧ 12 1 5 9 13 2 6 10 14	③ ⑦ 11 15
	second chunk	skip
Step	32 33	34 35 36 37 38 39 40 41 42 43 44 45 46 47
index	0 ⑧	4 12 2 10 6 14 1 9 5 13 ③ 11 ⑦ 15
	skip	first chunk

Fig. 6. Message word distribution in MD4 observed in 1-block attack

select p_7 and m_0 to m_7 is shown in Table 4. **0**, **1**, **C_i**, and $*$ denote 0x00000000, 0xffffffff, a randomly fixed value, and a flexible value which depends on the value of m_7, respectively.

The attack procedure is as follows:

Precomputation

1. Set the values of chaining variables Q_j as shown in Table 4. Note the value of $*$ is left undetermined.
2. Compute $m_j, j \in \{0, 1, 2, 4, 5, 6\}$ by the following equation:

$$m_{\pi(j)} = (Q_{j+1} \ggg s_j) - Q_{j-3} - \Phi_j(Q_j, Q_{j-1}, Q_{j-2}) - k_j. \tag{5}$$

Computation of the first chunk including m_7 and m_3

3. For all 32-bits of m_7, compute the value of $*$.
4. For each m_7, compute m_3 by the following equation:

$$m_{\pi(3)} = (Q_4 \ggg s_3) - Q_0 - \Phi_3(Q_3, Q_2, Q_1) - k_3. \tag{6}$$

Table 4. Fixed values for MD4 one-block preimage attack

step j	$m_{\pi(j)}$	Q_{j-2}	Q_{j+1}	Q_j	Q_{j-1}
0	m_0	Q_{-3}	Q_0	Q_{-1}	Q_{-2}
1	m_1	Q_{-2}	$\mathbf{C_4}$	Q_0	Q_{-1}
2	m_2	Q_{-1}	$\mathbf{C_3}$	$\mathbf{C_4}$	Q_0
3	$\textit{m}_3$	Q_0	$\mathbf{C_3}$	$\mathbf{C_3}$	$\mathbf{C_4}$
4	m_4	$\mathbf{C_4}$	$*$	$\mathbf{C_3}$	$\mathbf{C_3}$
5	m_5	$\mathbf{C_3}$	0	$*$	$\mathbf{C_3}$
6	m_6	$\mathbf{C_3}$	1	0	$*$
7	$\textit{m}_7$	$*$	$\mathbf{C_2}$	1	0
8		0	$\mathbf{C_1}$	$\mathbf{C_2}$	1

At Step 2 of the above procedure, the value of $*$ is involved in the computation for m_4, m_5, and m_6. However, due to the absorption properties, m_4, m_5, and m_6 can be computed independently on $*$.

In this attack, we fix lower 11-bits of m_8 to an arbitrary value. Then, we compute the second chunk for all the remaining 21-bits of m_8 and store the results. After that, we compute the first chunk for all m_7, then check whether they are matched with stored items by comparing the lower 11-bits of Q_{27} and Q_{28}[6]. Finally, we can find a one-block preimage at the complexity of 2^{107}.

6   Conclusion

This paper has shown the preimage attacks of one-block MD4 and MD5 reduced to 63 (out of 64) steps. A preimage of MD5 reduced to 63 steps can be computed in 2^{121} MD5 computations, which is faster than the brute-force attack, and a pseudo-preimage of full-round MD5 can be computed in $2^{125.7}$ MD5 computations. On a preimage of full-round MD5, we optimize the computational order of the brute-force attack against MD5, and a preimage of full-round MD5 can be computed in the complexity of the 2^{127} MD5 compression function computation. Moreover, a one-block preimage of MD4 can be computed in 2^{107} MD4 computations, while the previous work [10] computes a preimage of more than 1 block. The key idea of our attacks, which are based on the meet-in-the-middle technique, is quite simple, but very effective for preimage attacks. We left the application of our attack to other hash functions as a problem.

Acknowledgments

The authors wish to thank Christophe De Cannière and Christian Rechberger for providing [2] and anonymous referees for many useful comments on this paper.

---

[6] In MD4, up to four steps can be additionally skipped by the partial-fixing technique.

References

1. Aumasson, J.-P., Meier, W., Mendel, F.: Preimage attacks on 3-pass HAVAL and step-reduced MD5. In: Avanzi, R., Keliher, L., Sica, F. (eds.) Selected Areas in Cryptography — Workshop Records of 15th Annual International Workshop, SAC 2008, Sackville, New Brunswick, Canada, pp. 99–114 (2008); also appeared in IACR Cryptology ePrint Archive: Report 2008/183 http://eprint.iacr.org/2008/183

2. De Cannière, C., Rechberger, C.: Preimages for reduced SHA-0 and SHA-1. In: Wagner, D. (ed.) CRYPTO 2008. LNCS, vol. 5157, pp. 179–202. Springer, Heidelberg (2008); slides on preliminary results were appeared at ESC 2008 seminar, http://wiki.uni.lu/esc/

3. De, D., Kumarasubramanian, A., Venkatesan, R.: Inversion attacks on secure hash functions using SAT solvers. In: Marques-Silva, J., Sakallah, K.A. (eds.) SAT 2007. LNCS, vol. 4501, pp. 377–382. Springer, Heidelberg (2007)

4. Dobbertin, H.: The first two rounds of MD4 are not one-way. In: Vaudenay, S. (ed.) FSE 1998. LNCS, vol. 1372, pp. 284–292. Springer, Heidelberg (1998)

5. Flajolet, P., Odlyzko, A.M.: Random mapping statistics. In: Quisquater, J.-J., Vandewalle, J. (eds.) EUROCRYPT 1989. LNCS, vol. 434, pp. 329–354. Springer, Heidelberg (1990)

6. Kelsey, J., Schneier, B.: Second preimages on n-bit hash functions for much less than 2^n work. In: Cramer, R. (ed.) EUROCRYPT 2005. LNCS, vol. 3494, pp. 474–490. Springer, Heidelberg (2005)

7. Klima, V.: Tunnels in hash functions: MD5 collisions within a minute (IACR Cryptology ePrint Archive: Report 2006/105) (2006),
http://eprint.iacr.org/2006/105

8. Kuwakado, H., Tanaka, H.: New algorithm for finding preimages in a reduced version of the MD4 compression function. IEICE Transactions Fundamentals of Electronics, Communications and Computer Sciences (Japan) E83-A(1), 97–100 (2000)

9. Lai, X., Massey, J.L.: Hash functions based on block ciphers. In: Rueppel, R.A. (ed.) EUROCRYPT 1992. LNCS, vol. 658, pp. 55–70. Springer, Heidelberg (1993)

10. Leurent, G.: MD4 is not one-way. In: Nyberg, K. (ed.) FSE 2008. LNCS, vol. 5086, pp. 412–428. Springer, Heidelberg (2008)

11. Menezes, A.J., van Oorschot, P.C., Vanstone, S.A.: Handbook of applied cryptography. CRC Press, Boca Raton (1997)

12. Rivest, R.L.: Request for Comments 1321: The MD5 Message Digest Algorithm. The Internet Engineering Task Force (1992),
http://www.ietf.org/rfc/rfc1321.txt

13. Sasaki, Y., Aoki, K.: Preimage attacks on step-reduced MD5. In: Mu, Y., Susilo, W., Seberry, J. (eds.) ACISP 2008. LNCS, vol. 5107, pp. 282–296. Springer, Heidelberg (2008)

14. U.S. Department of Commerce, National Institute of Standards and Technology. Announcing the SECURE HASH STANDARD (Federal Information Processing Standards Publication 180-2) (2002),
http://csrc.nist.gov/publications/fips/fips180-2/
fips180-2withchangenotice.pdf

15. Wang, X., Yu, H.: How to break MD5 and other hash functions. In: Cramer, R. (ed.) EUROCRYPT 2005. LNCS, vol. 3494, pp. 19–35. Springer, Heidelberg (2005)

A   Graphical Explanation of Partial-Matching and Partial-Fixing Techniques

The way partial-matching and partial-fixing techniques work when we skip 6 steps from the attack target are shown in Fig. 7. The numbers in Fig. 7 denote the number of bits that can be computed independently of the neutral word for the other chunk.

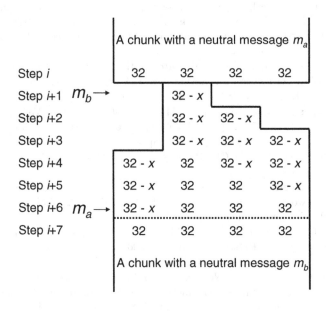

Fig. 7. Graphical explanation of the partial-matching and partial-fixing techniques

First, we store results of the computation of the chunk including m_a for all possible values of m_a. Note that to use the partial-fixing technique, we fix the lower x-bits of m_a to any value. Since m_b is used in Step $i + 1$, this computation can be carried out until Step i. Then, we compute the chunk including m_b for all possible values of m_b. Until Step $i + 7$, all 128-bit values can be computed independently of m_a.

Since the lower x-bits of m_a are fixed by the partial-fixing technique, we can partially execute inverse computation in Step $i + 6$ independently of the higher $(32 - x)$-bits of m_a. A similar situation occurs in Steps $i + 5$ and $i + 4$. Finally, by applying the partial-matching technique, we can compare $32 - x$ bits of results of the two chunks.

B   Notes on MD-Strengthening

Section 3.1 describes how to convert pseudo-images to a preimage on Merkle-Damgård structure. When the length of preimages is not fixed and MD-strengthening [9] is used in the target hash function, the method cannot be applied. The problem can be solved using "expandable message" introduced in [6]. An (a, b)-expandable message inputs a fixed chaining value, and can take any message length between a and b blocks, and outputs the same chaining value. The expandable message to adjust the length of preimage is already applied to MD4 [10]. Actually, when the compression function is constructed by Davies-Meyer, an expandable message is easily found, and [6] showed how to construct $(n, n+2^n-1)$-expandable message for a generic compression function. However, when the computational cost of computing preimages is nearly the complexity of brute-force attack, [6] may not be efficient to compute a preimage. The following algorithm efficiently produce $(k + 1, k + n)$-expandable message for given k. Note that the algorithm is not efficient when n is large compared with [6].

Assume we need to generate a multi-collision that consists of messages whose length are $(k + 1)$-block, $(k + 2)$-block, ..., $(k + n)$-block. Such a multi-collision is generated as follows:

1. Randomly generate a k-block message M_k and compute $H_k = h(H_0, M_k)$.
2. Randomly generate a 1-block message M_{k+1} and compute $H_{k+1} = h(H_k, M_{k+1})$.
3. For $i = 1, 2, \ldots, n - 1$, search for a 1-block message M_{k+i+1} such that $h(H_k, M_{k+i+1}) = h(H_{k+i}, M_{k+i+1})$. Let the generated value be H_{k+i+1}.
4. Finally, $(M_k \| M_{k+n})$, $(M_k \| M_{k+n-1} \| M_{k+n})$, ..., $(M_k \| M_{k+1} \| \cdots \| M_{k+n-1} \| M_{k+n})$ are multi-collision messages of $(k + 1)$, $(k + 2)$, ..., $(k + n)$ blocks.

In the above procedure, generating M_{k+i+1} costs the complexity of the birthday paradox, which is sufficiently low in the preimage attacks.

Very recently, [2] introduced the use of P^3graph. When a random directed graph has n nodes which are a part of chaining values, $2n$ edges are sufficient to connect from IV to a given hash value, and the path from IV to the given hash value can take any length if the length is large enough. On the other hand, we know n edges are sufficient to connect to the given hash value with high probability, and we conjectured that there exists paths to the given hash value from about $\sqrt{n}$ nodes. The conjecture is true for the case of random map [5], and we examine the conjecture by computer simulations. Though we only examined that the number of nodes is less than 4096, about $1.1n$ edges make $\sqrt{n}$ nodes connect to the given hash value. Followed by the idea in Section 3.1 with expandable message and above conjecture, a preimage can be computed and its complexity is about half compared with that in [2]. More precisely, the number of nodes in P^3graph is small, a preimage can be computed more efficiently.

Preimage Attacks on 3-Pass HAVAL and Step-Reduced MD5[*]

Jean-Philippe Aumasson[1,**], Willi Meier[1,***], and Florian Mendel[2]

[1] FHNW, Windisch, Switzerland
[2] IAIK, Graz University of Technology, Graz, Austria

Abstract. This paper presents preimage attacks on the hash functions 3-pass HAVAL and step-reduced MD5. Introduced in 1992 and 1991 respectively, these functions underwent severe collision attacks, but no preimage attack. We describe two preimage attacks on the compression function of 3-pass HAVAL. The attacks have a complexity of about 2^{224} compression function evaluations instead of 2^{256}. We present several preimage attacks on the MD5 compression function that invert up to 47 steps (out of 64) within 2^{96} trials instead of 2^{128}. Although our attacks are not practical, they show that the security margin of 3-pass HAVAL and step-reduced MD5 with respect to preimage attacks is not as high as expected.

Keywords: cryptanalysis, hash function, preimage attack.

1 Introduction

A cryptographic hash function h maps a message M of arbitrary length to a fixed-length hash value H and has to fulfill the following security requirements:

- *Collision resistance:* it is infeasible to find two messages M and $M^\star$, with $M^\star \neq M$, such that $h(M) = h(M^\star)$.
- *Second preimage resistance:* for a given message M, it is infeasible to find a second message $M^\star \neq M$ such that $h(M) = h(M^\star)$.
- *Preimage resistance:* for a given hash value H, it is infeasible to find a message M such that $h(M) = H$.

The resistance of a hash function to collision and (second) preimage attacks depends in the first place on the length n of the hash value. Regardless of how a hash function is designed, an adversary will always be able to find preimages or second preimages after trying out about 2^n different messages. Finding collisions requires a much smaller number of trials: about $2^{n/2}$ due to the birthday paradox. A function is said to achieve *ideal security* if these bounds are guaranteed.

---

[*] The work in this paper was supported in part by the Austrian Science Fund (FWF), project no. P19863.
[**] Supported by the Swiss National Science Foundation, project no. 113329.
[***] Supported by GEBERT RÜF STIFTUNG, project no. GRS-069/07.

R. Avanzi, L. Keliher, and F. Sica (Eds.): SAC 2008, LNCS 5381, pp. 120–135, 2009.
© Springer-Verlag Berlin Heidelberg 2009

Recent cryptanalytic results on hash functions mainly focus on collision attacks but only few results with respect to preimages have been published to date. In this article, we analyze the preimage resistance of the hash functions MD5 and HAVAL. Both are iterated hash functions based on the Merkle-Damgård design principle. MD4 and MD5 both underwent critical collision attacks [4, 7, 8, 17, 18, 19], and hence should not be used anymore. But in practice MD5 is still widespread and remains secure for applications that do not require collision resistance. While three preimage attacks on MD4 are known [3, 5, 6], the picture is different for MD5: using a SAT-solver De et al. [3] inverted 26 (out of 64) steps of MD5, and no analytical attack is known to date. Idem for HAVAL: while several collision attacks [7, 13, 20, 21] and even a second preimage attack [9] were published, no preimage attack is known.

Independent Work. Sasaki and Aoki discovered preimage attacks on round-reordered and step-reduced MD5 [14]: their best attack with original round-ordering inverts 44 steps of the compression function within 2^{96} trials, starting at the step 3 and ending at step 46. They subsequently improved this result in a paper presented at this workshop [15].

Our Contribution. First, we invert the compression function of MD5 reduced to 45 steps by using a meet-in-the-middle approach. The attack makes about 2^{100} compression function evaluations and needs negligible memory. Second, we exploit special properties of the permutations used in the compression function to extend this attack to 47 steps (out of 64). The attack has a complexity of 2^{96} compressions and memory requirements of 2^{36} bytes. Third, we extend the attacks on the compression function to the hash function by using a meet-in-the-middle and tree-based approach. With this method we can construct preimages for MD5 reduced to 45 and 47 steps with a complexity of about 2^{106} and 2^{102} compression function evaluations and memory requirements of 2^{39} bytes.

Similar strategies can be applied to the compression function of HAVAL. We can invert the compression function of 3-pass HAVAL with a complexity of about 2^{224} compression function evaluations and memory requirements of 2^{69} bytes. We can turn the attack on the compression function into a preimage attack on the hash function with a complexity of about 2^{230} compression function evaluations and memory requirements of 2^{70} bytes.

Outline. The article is structured as follows. §2 presents two methods to invert to compression function of MD5 reduced to 45 and 47 steps. We use the same methods to invert the compression function of 3-pass HAVAL in §3. In §4, we show how the attacks on the compression function of MD5 and HAVAL can be extended to preimage attacks on the hash function, and §5 concludes.

2   Preimage Attacks on Step-Reduced MD5

This section presents two techniques to invert the MD5 compression function. The first attack on 45 steps is based on a standard meet-in-the-middle (MITM) and requires about 2^{100} trials. The second attack inverts up to 47 steps, and exploits special properties of the message ordering. Combined with a MITM, we construct a preimage attack with complexity about 2^{96} trials. But prior to that, we provide a brief description of MD5 and illustrate the basic idea of our attacks over 32 steps.

2.1   Short Description of MD5

The MD5 compression function takes as input a 512-bit message block and a 128-bit chain value and outputs another 128-bit chain value.

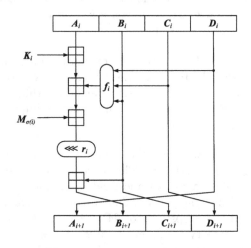

Fig. 1. The step function of MD5

The input chain value $H_0 \ldots H_3$ is first copied into registers $A_0 \ldots D_0$:

$$(A_0, B_0, C_0, D_0) \leftarrow (H_0, H_1, H_2, H_3). \tag{1}$$

This inner state is then transformed by a series of 64 steps and the output is

$$(H_0^\star, H_1^\star, H_2^\star, H_3^\star) = (A_{64} + A_0, B_{64} + B_0, C_{64} + C_0, D_{64} + D_0). \tag{2}$$

where $A_{64} \ldots D_{64}$ are defined by the recursion below:

$$\begin{aligned} A_i &= D_{i-1} \\ B_i &= B_{i-1} + (A_{i-1} + f_i(B_{i-1}, C_{i-1}, D_{i-1}) + M_{\sigma(i)} + K_i) \lll r_i \\ C_i &= B_{i-1} \\ D_i &= C_{i-1} \end{aligned} \tag{3}$$

The K_i's and r_i's are predefined constants and $\sigma(i)$'s are in Table 1. The function f_i is defined as

$$\begin{aligned} f_i(B, C, D) &= (B \wedge C) \vee (\neg B \wedge D) & \text{if } 0 < i \le 16 \\ f_i(B, C, D) &= (D \wedge B) \vee (\neg D \wedge C) & \text{if } 16 < i \le 32 \\ f_i(B, C, D) &= B \oplus C \oplus D & \text{if } 32 < i \le 48 \\ f_i(B, C, D) &= C \oplus (B \vee \neg D) & \text{if } 48 < i \le 64 \end{aligned} \tag{4}$$

Table 1. Values of $\sigma(i)$ in MD5 for $i = 1, \ldots, 64$ (we boldface the M_2 key inputs used in the attacks on 32 and 47 steps, and the M_6 and M_9 key inputs used in the attack on 45 steps)

Step index i	1	2	3	4	5	6	7	8	9	10	11	12	13	14	15	16
Message word $\sigma(i)$	0	1	**2**	3	4	5	6	7	8	9	10	11	12	13	14	15
Step index i	17	18	19	20	21	22	23	24	25	26	27	28	29	30	31	32
Message word $\sigma(i)$	1	6	11	0	5	10	15	4	9	14	3	8	13	**2**	7	12
Step index i	33	34	35	36	37	38	39	40	41	42	43	44	45	46	47	48
Message word $\sigma(i)$	5	8	11	14	1	4	7	10	13	0	3	**6**	**9**	12	15	**2**
Step index i	49	50	51	52	53	54	55	56	57	58	59	60	61	62	63	64
Message word $\sigma(i)$	0	7	14	5	12	3	10	1	8	15	6	13	4	11	2	9

Fig. 1 gives a schematic view of the step function, and [12] gives a complete specification.

Fact 1. *At step i only B_i is a really new value, the others are just shifted as in a feedback shift register. Hence for $i = 0, \ldots, 60$ we have $B_i = C_{i+1} = D_{i+2} = A_{i+3}$.*

Fact 2. *The step function is* invertible, *i.e. from $A_i \ldots D_i$ and $M_{\sigma(i)}$ we can always compute $A_{i-1} \ldots D_{i-1}$. Removing the feedforward by $H_0 \ldots H_3$ in Eq. (2) would thus make the compression function trivially invertible.*

2.2   Preimage Attack on 32 Steps

This attack computes preimages for the 32-step compression function within about 2^{96} trials (instead of 2^{128}). It introduces two tricks used in the 45- and 47-step attacks: absorption of changes in C_0 and exploitation of the ordering of the message words.

Key Facts. Observe in Table 1 that M_2 is only input at the very beginning and the very end of 32-step MD5, namely at steps 3 and 30. Hence, if we could pick a message and freely modify M_2 such that B_3 stays unchanged, we would be able to "choose" $B_{30} = C_{31} = D_{32}$ (cf. Fact 1). A key observation is that the function f_i can either preserve or absorb an input difference: indeed for $0 < i \leq 16$ and any C and D we have

$$f_i(\texttt{0x00000000}, C, D) = (0 \wedge C) \vee (\texttt{0xffffffff} \wedge D) = D \qquad (5)$$

$$f_i(\texttt{0xffffffff}, 0, D) = (\texttt{0xffffffff} \wedge 0) \vee (0 \wedge D) = 0 \qquad (6)$$

These properties will be used to "absorb" a change in $C_0 = D_1 = A_2$ at steps 1 and 2. More precisely, we need that $B_0 = 0$ to absorb the changes of C_0 at step 1. And to absorb the change in $D_1 = C_0$ we need that $B_1 = \texttt{0xffffffff}$. We can now sketch the attack:

1. pick a chain value $H_0 \ldots H_3 = A_0 \ldots D_0$ (with certain constraints)
2. pick a message $M_0 \ldots M_{15}$ (with certain constraints)
3. modify M_2 to choose $B_{30} = C_{31} = D_{32}$
4. modify $H_2 = C_0$ such that the change in M_2 doesn't alter subsequent $A_i \ldots D_i$

Our strategy is inspired from Leurent's MD4 inversion [6]; the main difference is that [6] exploits absorption in the second round, whereas we use it in the early steps.

Description of the Attack. Suppose we seek a preimage of $\tilde{H} = \tilde{H}_0 \ldots \tilde{H}_3$. The algorithm below first sets $B_0 = 0$ and $B_1 = \texttt{0xffffffff}$, to guarantee that a change in C_0 will only affect A_2. Then, from an arbitrarily chosen message, Algorithm 1 modifies M_2 in order to "meet in the middle". Finally, C_0 corrects the change in M_2, and this new value of C_0 does not affect the initial steps of the function.

Algorithm 1. Preimage attack on 32-step MD5

---

1. set $B_0 = 0$ and A_0, C_0, D_0 to arbitrary values
2. **repeat**
3.   pick M_0 such that $B_1 = \texttt{0xffffffff}$
4.   pick arbitrary values for $M_1 \ldots M_{15}$
5.   compute $A_{30} \ldots D_{30}$
6.   modify M_2 to get $B_{30} = D_{32} = \tilde{H} - D_0$
7.   correct C_0 to keep B_3 unchanged
8.   compute the final hash value $H^* = H_0^* \ldots H_3^*$
9.   **if** $H^* = \tilde{H}$ **then**
10.   **return** $A_0 \ldots D_0$ and $M_0 \ldots M_{15}$

Algorithm 1 makes about 2^{96} trials by choosing 32 bits in the 128-bit image and bruteforcing the 96 remaining bits. (We denote $H^* = H_0^* \ldots H_3^*$ a final hash value, so our goal is to have in the end $H^* = \tilde{H}$.)

Correctness of the Attack. We now explain in details why the attack works. First, the operation at line 3 of our algorithm is feasible because it corresponds to setting

$$M_0 = \texttt{0xffffffff} - A_0 - D_0 - K_0. \tag{7}$$

Then right after line 4 we have for any choice of C_0:

1. $f_1(B_0, C_0, D_0) = f_1(0, C_0, D_0) = D_0$
2. $f_2(B_1, C_1, D_1) = f_2(\texttt{0xffffffff}, C_1 = B_0, D_1) = 0$

In other words, the first two steps are *independent of* C_0. This will allow us to modify $C_0 = D_1 = A_2$—to correct a change in M_2—without altering $A_i \ldots D_i$ between steps 4 and 30.

Now, at line 6 we set

$$M_2 = (\tilde{H}_3 - D_0 - B_{29}) \ggg 9 - G(B_{29}, C_{29}, D_{29}) - A_{29} - K_{30} \qquad (8)$$

With this new value of M_2 we get in the end $H_3^\star = \tilde{H}_3$.

Finally we "correct" this change by setting

$$C_0 = (B_3 - B_2) \ggg r_3 - f_3(B_2, C_2, D_2) - M_2 - K_2. \qquad (9)$$

With this new value of $C_0 = A_2$ we keep the same B_3 as with the original choice of M_2.

We can thus choose the output value $H_3^\star$ by modifying M_2 and "correcting" C_0. However, $H_0^\star$, $H_1^\star$ and $H_2^\star$ are random for the attacker. Hence, 96 bits have to be bruteforced to invert the 32-step function. This gives a total cost of 2^{96} trials.

We experimentally verified the correctness of our algorithm by searching for inputs that give $H_2^\star = H_3^\star = 0$ (see Appendix A).

2.3   Preimage Attack on 45 Steps

We present here an attack that computes 45-step preimages within 2^{100} trials and negligible memory. This combines a MITM with a conditional linear approximation of the step function. In short, the attack is based on the fact that M_2 appears at the very beginning and that M_6 and M_9 appear at the very end of 45-step MD5. Another key observation is that M_2 is used only once in the first 25 steps, and M_6 and M_9 are used only once after step 25. Algorithm 2 describes the attack for finding a preimage of $\tilde{H}_0 \ldots \tilde{H}_3$.

Correctness of the Attack. First, we use again (at line 1) the trick to absorb the modification of C_0, necessary to keep the forward stage unchanged with the new value of M_2. Then, observe that

- between steps 25 and 45, M_6 and M_9 are input at steps 44 and 45 (cf. Table 1)
- at line 7 we use values of M_6 and M_9 distinct from the ones used in the forward stage (line 5)

Hence, by setting M_6 and M_9 to the values chosen the matching L entry, we would expect different values of $B_{44} = C_{45}$ and B_{45} than the (zero) ones used for the backward computation. Recall (cf. line 1) that we need $A_{45} = 0$, $B_{45} = \tilde{H}_1$, $D_{45} = 0$, hence the values of C_{45} will not matter; we would however expect a random B_{45} from the new values of M_6 and M_9.

The trick used here is that the condition imposed on M_6 and M_9 at line 5 implies that the new B_{45} equals the original $H_1^\star = \tilde{H}_1$ with probability 2^{-4} instead of 2^{-32} for random values (see below). The attack thus succeeds to find a 96-bit preimage when the MITM succeeds *and* $B_{45} = \tilde{H}_1$, that is with probability $2^{-64} \times 2^{-4} = 2^{-68}$. Storage for 2^{68} bytes is required for the MITM. For full (128-bit preimage) we bruteforce the 32 remaining bits thus the costs grows to 2^{100} trials.

Algorithm 2. Preimage attack on 45-step MD5

1. set $A_0 = \tilde{H}_0$, $B_0 = 0$, $D_0 = \tilde{H}_3$
   (We thus need $A_{45} = 0$, $B_{45} = H_1^*$, $D_{45} = 0$. Note that we'll have $f_{45}(B_{44}, C_{44}, D_{44}) = f_{45}(C_{45}, D_{45}, A_{45}) = C_{45}$.)
2. **repeat**
3.   pick M_0 such that $B_1 = $ `0xffffffff`
4.   set arbitrary values to the remaining M_i's except M_6 and M_9
5.   **for** all 2^{64} choices of C_0 and (M_6, M_9) such that

$$M_9 = -((M_6 \lll 19) + (M_6 \lll 23))$$

    (Here 23 coincides with r_{44} and $19 = r_{44} - r_{45}$)
6.   compute $A_{25} \ldots D_{25}$, store it in a list L
7.   **for** $M_6 = M_9 = 0$ and all 2^{64} choices of C_{45} and M_2
8.   compute $A_{25} \ldots D_{25}$
9.   **if** this $A_{25} \ldots D_{25}$ matches an entry in L **then**
10.   correct C_0 to keep B_3 unchanged
11.   **return** $A_0 \ldots D_0$ and $M_0 \ldots M_{15}$
    (Here the message contains the M_2, M_6, M_9 corresponding to the matching entries)

Reducing the Memory Requirements. By using a cycle-finding algorithm (as for instance [16,11]) the memory requirements of the meet-in-the-middle step of the attack can be significantly reduced. Hence, we can find a preimage for 45-step MD5 with a complexity of about 2^{100} and negligible memory requirements.

On the Choice of M_6 and M_9. We explain here why the condition

$$M_9 = -(M_6 \lll 19 + M_6 \lll 23) \tag{10}$$

gives $B_{45} = \tilde{H}_1$ with high probability.

Consider the last two steps (44 and 45): because $A_{45} = D_{45} = 0$ we have $C_{44} = D_{44} = 0$ and $B_{43} = C_{43} = 0$. Hence we have

$$f_i(B, C, D) = B \oplus C \oplus D = B + C + D \tag{11}$$

in these two steps.

Note that A_{43} and D_{43} depend on the C_{45} used for the backward computation. Now we can compute B_{44} and B_{45} (note $r_{44} = 23, r_{45} = 9$)

$$B_{44} = (A_{43} + D_{43} + K_{43} + M_6) \lll 23 \tag{12}$$

$$B_{45} = (A_{44} + B_{44} + K_{44} + M_9) \lll 4 + B_{44} \tag{13}$$

For simplicity we rewrite

$$B_{44} = (X + M_6) \lll 23 \tag{14}$$

$$B_{45} = ((Y + B_{44} + M_9) \lll 4) + B_{44} \tag{15}$$

Now we can express B_{45}:

$$B_{45} = ((Y + ((X + M_6) \lll 23) + M_9) \lll 4) + ((X + M_6) \lll 23) \quad (16)$$

Since (cf. line 7 of the algorithm) we chose $(M_6, M_9) = (0, 0)$ this simplifies to

$$B_{45} = ((Y + (X \lll 23)) \lll 4) + (X \lll 23) \quad (17)$$

Consider now the case $M_9 = -(M_6 \lll 19 + M_6 \lll 23)$; Eq. (16) becomes:

$$B_{45} = ((Y + ((X + M_6) \lll 23) - ((M_6 \lll 19) + (M_6 \lll 23))) \lll 4) \quad (18)$$
$$+((X + M_6) \lll 23)$$

We will simplify this equation by using the generic approximation:

$$(A + B) \lll k = A \lll k + B \lll k \quad (19)$$

Daum showed [2, §4.1.3] that Eq. (19) holds with probability about 2^{-2} for random A and B. We first use this approximation to replace $(X + M_6) \lll 23$ by

$$(X \lll 23) + (M_6 \lll 23). \quad (20)$$

Thus Eq. (18) yields

$$B_{45} = ((Y + (X \lll 23) - (M_6 \lll 19)) \lll 4) + (X \lll 23) \quad (21)$$
$$+(M_6 \lll 23)$$

Finally we approximate $(Y + (X \lll 23) - (M_6 \lll 19)) \lll 4$ by

$$((Y + (X \lll 23)) \lll 4) - ((M_6 \lll 19) \lll 4) \quad (22)$$

and Eq. (21) becomes

$$B_{45} = ((Y + X \lll 23) \lll 4) + (X \lll 23) \quad (23)$$

Note that this is the same equation as for $(M_6, M_9) = (0, 0)$ in Eq. (17). Hence, we get the correct value in B_{45} with a probability of 2^{-4}, since we used two approximations[1].

Delayed-Start Attack. This attack strategy can be applied to invert the 47 steps from step 16 to 62, using M_6 in place of M_2, and the pair (M_4, M_{11}) instead of (M_6, M_9).

2.4   Preimage Attack on 47 Steps

In the following we will show how to construct a preimage for the compression function of 47-step MD5 with a complexity of about 2^{96}. This attack combines the 32-step attack with a meet-in-the-middle (MITM) strategy. The latter is made possible by the invertibility of the step function.

---

[1] The exact probability is $2^{-3.9097}$ according to Daum's formulas.

The attack on 47-step MD5 can be summarized as follows:

1. set initial state variable to absorb a change in C_0, as in the 32-step attack
2. compute $A_{29} \ldots D_{29}$ for all 2^{32} choices of C_0 and save the result in a list L
3. compute $A_{30} \ldots D_{30}$ for all 2^{32} choices of C_{47} and "meet in the middle" by finding a matching entry in L

Algorithm 3 describes the attack more formally.

Algorithm 3. Preimage attack on 47-step MD5

---

1. set $B_0 = 0$ and A_0, C_0, D_0 to arbitrary values
2. **repeat**
3.   pick M_0 such that $B_1 = \texttt{0xffffffff}$
4.   pick arbitrary values for $M_1 \ldots M_{15}$
5.   **for all** 2^{32} choices of C_0
6.   compute $A_{29} \ldots D_{29}$, store it in a list L
7.   set $A_{47} = \tilde{H}_0 - A_0$, $B_{47} = \tilde{H}_1 - B_0$, $D_{47} = \tilde{H}_3 - D_0$
8.   **for all** 2^{32} choices of C_{47}
9.   compute (backwards) $A_{30} \ldots D_{30}$
10.   **if** L contains an entry $A_{30} = D_{29}, C_{30} = B_{29}, D_{30} = C_{29}$ **then**
11.   modify M_2 to have

$$B_{30} = ((A_{29} + f(B_{29}, C_{29}, D_{29}) + M_2 + K_{29}) \lll 9) + B_{29}$$

12.   correct C_0 to keep B_3 unchanged
13.   compute the final hash value $H_0^\star \ldots H_3^\star$
14.   **return** $A_0 \ldots D_0$ and $M_1 \ldots M_{15}$

Again this attack essentially exploits the "absorption" of 32 bits during the early steps to save a 2^{32} complexity factor. Note that when the MITM succeeds, i.e. when the line 10 predicate holds, we only have a 96-bit preimage because $H_2^\star = C_{47} + C_0$ is random. This is because both C_0 and C_{47} are random for the attacker.

Each **repeat** loop hence succeeds in finding a 96-bit preimage with probability 2^{-32}, and costs 2^{32} trials. This is respectively because

1. we have $2^{32} \times 2^{32} = 2^{64}$ candidate pairs that each match with probability 2^{-96}
2. the cost of the two **for** loops amounts to 2^{32} computations of the compression function

The total cost for finding a 128-bit preimage is thus $2^{32} \times 2^{32} \times 2^{32} = 2^{96}$, with a required storage of 2^{36} bytes (64 Gb) for the MITM. This allows us to find preimages on the 47-step MD5 compression function 2^{32} times faster than bruteforce. However it doesn't directly give a preimage attack for the hash function because the initial value is here partially random, whereas in the hash function it is fixed.

3   Preimage Attacks on 3-Pass HAVAL

HAVAL was proposed with either 3, 4, or 5 passes, i.e. 96, 128, or 160 steps. It has message blocks and hash values twice as large as MD5, i.e. 1024 bits (32 words) and 256 bits (8 words) respectively. In the following, we present two methods to invert the compression function of 3-pass HAVAL. Both attacks have a complexity of about 2^{224} compression function evaluations. Like in the attacks on step-reduced MD5, we combine a generic MITM with weaknesses in the design of the compression function. In detail, we exploit the properties of the Boolean functions to absorb differences in its input and special properties of the message ordering in 3-pass HAVAL. But before describing the attacks, we give a short description of 3-pass HAVAL.

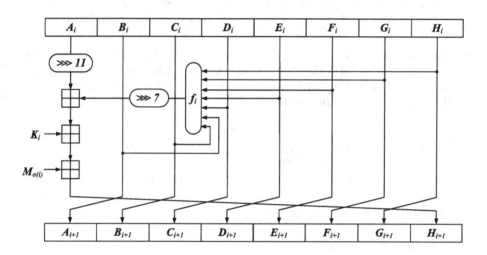

Fig. 2. The step function of HAVAL

3.1   Short Description of 3-Pass HAVAL

The structure of HAVAL is similar to that of MD5: registers $A_0, B_0, \ldots, G_0, H_0$ are initialized to the input chain values and finally the function returns

$$(H_0^\star, \ldots, H_7^\star) = (A_{96} + A_0, B_{96} + B_0, \ldots, G_{96} + G_0, H_{96} + H_0) \qquad (24)$$

after 96 steps that set

$A_i = B_{i-1},$
$B_i = C_{i-1}$
$\ldots \qquad \ldots$
$G_i = H_{i-1}$
$H_i = A_{i-1} \ggg 11 + f_i(B_{i-1}, C_{i-1}, D_{i-1}, E_{i-1}, F_{i-1}, G_{i-1}, H_{i-1}) \ggg 7 + K_i + M_{\sigma(i)}$
$$(25)$$

Table 2. Values of $\sigma(i)$ in 3-pass HAVAL for $i = 1, \ldots, 96$ (we boldface the key inputs of M_5 and M_6)

Step index i	1	**2**	3	4	5	6	7	8	9	10	11	12	13	14	15	16
Message word $\sigma(i)$	0	1	2	3	4	**5**	**6**	7	8	9	10	11	12	13	14	15
Step index i	17	18	19	20	21	22	23	24	25	26	27	28	29	30	31	32
Message word $\sigma(i)$	16	17	18	19	20	21	22	23	24	25	26	27	28	29	30	31
Step index i	33	34	35	36	37	38	39	40	41	42	43	44	45	46	47	48
Message word $\sigma(i)$	**5**	14	26	18	11	28	7	16	0	23	20	22	1	10	4	8
Step index i	49	50	51	52	53	54	55	56	57	58	59	60	61	62	63	64
Message word $\sigma(i)$	30	3	21	9	17	24	29	**6**	19	12	15	13	**2**	25	31	27
Step index i	65	66	67	68	69	70	77	72	73	74	75	76	77	78	79	80
Message word $\sigma(i)$	19	9	4	20	28	17	8	22	29	14	25	12	24	30	16	26
Step index i	81	82	83	84	85	86	87	88	89	90	91	92	93	94	95	96
Message word $\sigma(i)$	31	15	7	3	1	0	18	27	13	**6**	21	10	23	11	**5**	**2**

We thus have $H_i = G_{i+1} = F_{i+2} = E_{i+3} = D_{i+4} = C_{i+5} = B_{i+6} = A_{i+7}$ for $i = 0 \ldots 89$. Like in MD5 the step function is invertible, and uses step-specific constants, Boolean functions f_i, and message words $M_{\sigma(i)}$. The step functions are defined as (with e.g. $BC = (B \wedge C)$):

$$
\begin{aligned}
f_i(B, C, \ldots, H) &= FE \oplus BH \oplus CG \oplus DF \oplus D && \text{if } 0 < i \leq 32 \\
f_i(B, C, \ldots, H) &= ECH \oplus CGH \oplus CE \oplus EG \oplus CD \oplus FH \\
&\quad \oplus GF \oplus BC \oplus B && \text{if } 32 < i \leq 64 \\
f_i(B, C, \ldots, H) &= CDE \oplus CF \oplus DG \oplus EB \oplus EH \oplus H && \text{if } 64 < i \leq 96
\end{aligned}
\tag{26}
$$

The $\sigma(i)$'s are in Table 2. See [22] or [20] for a complete specification.

3.2 Preimage Attack A

Suppose we seek a preimage of $\tilde{H}_0 \ldots \tilde{H}_7$ with an arbitrary value for $\tilde{H}_6$; that is, we only want a 224-bit preimage. In the attack below we exploit the properties of the Boolean function f_i to absorb a difference in the input, and combine it with a MITM to improve on bruteforce search. Algorithm 4 describes the attack in detail. In the end the computed image $H^\star$ is the same as the image sought $\tilde{H}$ except (with probability $1 - 2^{-32}$) for $H_6^\star = G_{96} + G_0$. Here M_5 and M_6 are used as "neutral words", respectively in the second and the first part of the attack; the change in G_0 will correct the change in M_6, while being absorbed during the first six steps. Furthermore, if the MITM condition at line 8 is satisfied then we directly get a 224-bit preimage, because at line 6 we choose $A_{96} \ldots F_{96} H_{96}$.

Indeed we have 2^{64} candidates for $A_{48}, \ldots, H_{48}$ resulting from the forward computation and 2^{64} candidates resulting from the backward computation, so we'll find a match and thus a partial preimage with probability 2^{-128}. Hence, by repeating the attack 2^{128} times we'll find a 224-bit preimage with about $2^{128} \times 2^{64} = 2^{192}$ compression function evaluations. We need storage for 2^{69} bytes to perform the MITM. Note that a full (256-bit) preimage is obtained by bruteforcing the 32 remaining bits, increasing the cost to 2^{224} trials.

Algorithm 4. Preimage attack A on 3-pass HAVAL

---

1. set $C_0 = 0$, $D_0 = \tilde{H}_3 - \texttt{0xffffffff}$, $E_0 = F_0$, $H_0 = 0$, and arbitrary $A_0B_0G_0$
   (We need to assume $D_{96} = \texttt{0xffffffff}$ for our attack to work)
2. **repeat**
3.   choose an arbitrary message for which $H_1 = \texttt{0xffffffff}$ and $H_3 = H_5 = 0$
       (This guarantees that differences in G_0 will be absorbed in the first 6 rounds)
4.   **for all** 2^{64} choices of G_0 and M_5
       (A difference in M_5 only changes G_{96} after step 48)
5.   compute $A_{48} \ldots H_{48}$ and store it in a list L.
6.   set $A_{96} = \tilde{H}_0 - A_0, \ldots, H_{96} = \tilde{H}_7 - H_0$
7.   **for all** 2^{64} choices of G_{96} and M_6
8.   compute $A_{48} \ldots H_{48}$ by going backwards
9.   **if** this $A_{48} \ldots H_{48}$ matches an entry in L **then**
10.   correct G_0 such that $A_7 \ldots H_7$ remains unchanged
11.   **return** $A_0 \ldots H_0$ and $M_0 \ldots M_{31}$

3.3   Preimage Attack B

This attack exploits the fact that M_2 appears at the very beginning in the first pass and at the very end in the last pass. By combining this with absorption of the Boolean function in the early steps (similarly to our attack on 47-step MD5), we can construct a 192-bit preimage within about 2^{160} trials. By repeating the attack about 2^{64} times we can construct a preimage for the compression function with complexity of about 2^{224} instead of the expected 2^{256} compression function evaluations. Algorithm 5 computes a preimage of $\tilde{H}_0 \ldots \tilde{H}_7$ where all $\tilde{H}_i$'s are fixed but $\tilde{H}_2$ and $\tilde{H}_6$ (i.e. a 192-bit preimage):

Algorithm 5. Preimage attack B on 3-pass HAVAL

---

1. set $A_0 = \tilde{H}_0$, $B_0 = \tilde{H}_1$, $D_0 = \tilde{H}_3$, $E_0 = \tilde{H}_4$, $F_0 = \tilde{H}_5$, $G_0 = 0$.
   (To get a 192-bit preimage we thus need $A_{96} = B_{96} = 0$, $D_{96} = E_{96} = F_{96} = 0$, $G_{96} = \tilde{H}_7$)
2. **repeat**
3.   pick an arbitrary message for which the state variable $H_1 = 0$.
       (This guarantees that a change in C_0 will only affect A_2)
4.   **for all** 2^{64} choices of C_0 and H_0
5.   compute $A_{60} \ldots H_{60}$ and store it in a list L.
6.   **for all** 2^{64} choices of C_{96} and H_{96}
7.   compute $A_{61} \ldots H_{61}$
8.   **if** L contains a tuple such that $A_{61} = B_{60}, \ldots, G_{61} = H_{60}$ **then**
9.   modify M_2 to have

$$H_{61} = (A_{60} \ggg 11) + (f_{61}(\ldots) \ggg 7) + M_2 + K_{61}$$

10.   correct C_0 and H_{96} accordingly
11.   **return** $A_0 \ldots H_0$ and $M_0 \ldots M_{15}$

The MITM will succeed (line 8 of Algorithm 5) with probability $2^{-96} = 2^{64} \times 2^{64}/2^{224}$, hence $2^{96} \times 2^{64} = 2^{160}$ trials are required to get a 192-bit preimage (and storage 2^{69} bytes). A full (256-bit) preimage is obtained by bruteforcing the 64 remaining bits, which increases the cost to 2^{224} trials.

4   Extension to the Hash Functions

In this section, we will show how to extend the preimage attacks on the compression of step-reduced MD5 and 3-pass HAVAL to the hash function. The extension of the attacks to the hash function is constrained by the padding rule and the the predefined IV. The padding rule of MD5 and HAVAL forces the last bits of the message to encode its length. Thus a preimage attack should find messages that match this constraint. In our attacks we have no restrictions on the last message words and hence the padding rule is no problem; in each of the attacks proposed, we shall simply choose the end of the message to be of the form $100\cdots 0\langle\ell\rangle$, where $\langle\ell\rangle$ represents the bitlength of the original message (without the padding bits).

However, the IV of our preimages for the compression function is different from the fixed one; e.g. in the attack on MD5 reduced to 47-steps we require $B_0 = 0$, and get a random value for C_0. There are several methods to turn our attacks into preimage attacks starting from the predefined IV, as described in the next two sections; the general idea will be to find many preimages (with partially random initial value) and to find many images of the fixed IV, and then combine them to "bridge the gap" between the IV and the image.

4.1   Basic Meet-in-the-Middle

Suppose we want a preimage of H. This attack sets a parameter $0 < x < n$, and first computes 2^x preimages $(\tilde{H}_i, \tilde{M}_i)$, $i = 0, \ldots, 2^x - 1$, that is, such that $f(\tilde{H}_i, \tilde{M}) = H$; the $\tilde{M}_i$'s are chosen to have convenient padding bits. Then the attack computes 2^{n-x} random images $H_j = f(IV, M_j)$, $j = 0, \ldots, 2^{n-x} - 1$, for random M_j's and the IV specified for the function. Finally we find a pair (i, j) such that $\tilde{H}_i = H_j$, and return the message $M = M_j \| \tilde{M}_i$ as a preimage of H. Because there's in total 2^n pairs (i, j), the attack will work with high probability.

For reduced-step MD5 with the optimal x we compute forward 2^{112} random chain values and compute backward 2^{16} preimages within $2^{96} \times 2^{16} = 2^{112}$ trials. The total cost of the 47-step preimage attack is thus about 2^{113} trials and memory for a preimage attack. For 3-pass HAVAL we compute forward 2^{240} chain values and backward 2^{16} preimages within $2^{224} \times 2^{16} = 2^{240}$ trials. The total cost is 2^{241} trials plus memory for a preimage attack.

4.2   Tree Approach

This attack is an improved version of the meet-in-the-middle above. It is based on the finding of multi-target preimages, and the construction of a tree whose root is the target image. This is exactly the technique described in [6], (a similar approach was published before by Mendel and Rijmen in [10]). To summarize, we proceed in two stages

1. Backward stage: use a tree-based technique to compute a set S of multi-block preimages
2. Forward stage: compute images of random message blocks with the predefined IV until one lies in S

For MD5 the forward stage costs 2^{96} trials and the backward stages costs $32 \times 2^{97} = 2^{102}$ trials to compute 32-block preimages, plus storage for 2^{33} message blocks (i.e. 2^{39} bytes). Applied to 3-pass HAVAL we get a preimage attack that makes 2^{230} trials and needs 2^{71} bytes of storage.

5 Conclusion

We presented the first preimage attacks for the hash functions 3-pass HAVAL and step-reduced MD5: we described several preimage attacks on the MD5 compression function that invert up to 47 (out of 64) steps within 2^{96} compression function evaluations, instead of the expected 2^{128}, and two preimage attacks on the 3-pass HAVAL compression function that cost 2^{224} compression function evaluations instead of 2^{256}. We extended our best attacks to the hash functions (with padding and fixed IV) for a cost of 2^{230} and 2^{102} trials, respectively. Although these attacks are not practical (notably due to large memory requirements), they show that the security margin of 3-pass HAVAL and step-reduced MD5 with respect to preimage attacks is not as high as expected.

Acknowledgments

We would like to thank Kazumaro Aoki and Yu Sasaki for communicating us theirs results on MD5 and making helpful comments.

References

1. Cramer, R. (ed.): Advances in Cryptology - EUROCRYPT 2005, 24th Annual International Conference on the Theory and Applications of Cryptographic Techniques, Proceedings, Aarhus, Denmark, May 22-26, 2005. LNCS, vol. 3494, pp. 22–26. Springer, Heidelberg (2005)
2. Daum, M.: Cryptanalysis of Hash Functions of the MD4-Family. PhD thesis, Ruhr Universität Bochum (2005)
3. De, D., Kumarasubramanian, A., Venkatesan, R.: Inversion attacks on secure hash functions using SAT solvers. In: Marques-Silva, J., Sakallah, K.A. (eds.) SAT 2007. LNCS, vol. 4501, pp. 377–382. Springer, Heidelberg (2007)
4. den Boer, B., Bosselaers, A.: Collisions for the compression function of MD-5. In: Helleseth, T. (ed.) EUROCRYPT 1993. LNCS, vol. 765, pp. 293–304. Springer, Heidelberg (1994)

5. Dobbertin, H.: The first two rounds of MD4 are not one-way. In: Vaudenay, S. (ed.) FSE 1998. LNCS, vol. 1372, pp. 284–292. Springer, Heidelberg (1998)

6. Leurent, G.: MD4 is not one-way. In: Nyberg, K. (ed.) FSE 2008. LNCS, vol. 5086, pp. 412–428. Springer, Heidelberg (2008)

7. Kim, J.-S., Biryukov, A., Preneel, B., Lee, S.-J.: On the security of encryption modes of MD4, MD5 and HAVAL. In: Qing, S., Mao, W., López, J., Wang, G. (eds.) ICICS 2005. LNCS, vol. 3783, pp. 147–158. Springer, Heidelberg (2005)

8. Klima, V.: Tunnels in hash functions: MD5 collisions within a minute. Cryptology ePrint Archive, Report 2006/105 (2006), http://eprint.iacr.org/

9. Lee, E., Kim, J., Chang, D., Sung, J., Hong, S.: Second preimage attack on 3-pass HAVAL and partial key-recovery attacks on NMAC/HMAC-3-pass HAVAL (to appear) (2008)

10. Mendel, F., Rijmen, V.: Weaknesses in the HAS-V compression function. In: Nam, K.-H., Rhee, G. (eds.) ICISC 2007. LNCS, vol. 4817, pp. 335–345. Springer, Heidelberg (2007)

11. Quisquater, J.-J., Delescaille, J.-P.: How easy is collision search? Application to DES. In: Quisquater, J.-J., Vandewalle, J. (eds.) EUROCRYPT 1989. LNCS, vol. 434, pp. 429–434. Springer, Heidelberg (1990)

12. Rivest, R.: RFC 1321 - The MD5 Message-Digest Algorithm (1992)

13. Van Rompay, B., Biryukov, A., Preneel, B., Vandewalle, J.: Cryptanalysis of 3-pass HAVAL. In: Laih, C.-S. (ed.) ASIACRYPT 2003. LNCS, vol. 2894, pp. 228–245. Springer, Heidelberg (2003)

14. Sasaki, Y., Aoki, K.: Preimage attacks on step-reduced MD5. In: Mu, Y., Susilo, W., Seberry, J. (eds.) ACISP 2008. LNCS, vol. 5107, pp. 282–296. Springer, Heidelberg (2008)

15. Sasaki, Y., Aoki, K.: Preimage attacks on one-block MD4, 63-step MD5 and more. In: Avanzi, R., Keliher, L., Sica, F. (eds.) SAC 2008. LNCS, vol. 5381, pp. 103–119. Springer, Heidelberg (2009)

16. Sedgewick, R., Szymanski, T.G., Yao, A.C.-C.: The complexity of finding cycles in periodic functions. SIAM Journal of Computing 11(2), 376–390 (1982)

17. Stevens, M., Lenstra, A.K., de Weger, B.: Chosen-prefix collisions for MD5 and colliding X.509 certificates for different identities. In: Naor, M. (ed.) EUROCRYPT 2007. LNCS, vol. 4515, pp. 1–22. Springer, Heidelberg (2007)

18. X. Wang, X. Lai, D. Feng, H. Chen, X. Yu.: Cryptanalysis of the hash functions MD4 and RIPEMD. In: Cramer [1], pp. 1–18 (2005)

19. Wang, X., Yu, H.: How to break MD5 and other hash functions. In: Cramer [1], pp. 19–35 (2005)

20. Yoshida, H., Biryukov, A., De Cannière, C., Lano, J., Preneel, B.: Non-randomness of the full 4 and 5-pass HAVAL. In: Blundo, C., Cimato, S. (eds.) SCN 2004. LNCS, vol. 3352, pp. 324–336. Springer, Heidelberg (2005)

21. Yu, H., Wang, X., Yun, A., Park, S.: Cryptanalysis of the full HAVAL with 4 and 5 passes. In: Robshaw, M.J.B. (ed.) FSE 2006. LNCS, vol. 4047, pp. 89–110. Springer, Heidelberg (2006)

22. Zheng, Y., Pieprzyk, J., Seberry, J.: HAVAL - a one-way hashing algorithm with variable length of output. In: Zheng, Y., Seberry, J. (eds.) AUSCRYPT 1992. LNCS, vol. 718, pp. 83–104. Springer, Heidelberg (1993)

A Partial Preimage for 32-Step MD5

With the IV

$$H_0 = 0x67452301 \qquad H_2 = 0x382ca539$$
$$H_1 = 0x00000000 \qquad H_3 = 0x10325476$$

and the message

$$M_0 = 0xb11de410 \quad M_4 = 0x792a351e \quad M_8 = 0x6d32a030 \quad M_{12} = 0x1dd5ec6d$$
$$M_1 = 0x5c0cd1ec \quad M_5 = 0x420582b7 \quad M_9 = 0x16b2e752 \quad M_{13} = 0x4794f768$$
$$M_2 = 0xd7d35ac7 \quad M_6 = 0x77v8de3d \quad M_{10} = 0x3b70c422 \quad M_{14} = 0x04fef18f$$
$$M_3 = 0x5704c13b \quad M_7 = 0x2476b43b \quad M_{11} = 0x685cb2aa \quad M_{15} = 0x00000000$$

we get the image

$$H_0^\star = 0xb4df93c9 \qquad H_2^\star = 0x00000000$$
$$H_1^\star = 0x3348e3f2 \qquad H_3^\star = 0x00000000$$

This was found in fewer than five minutes on our 2.4 GHz Core 2 Duo, whereas brute force would take about 2^{64} trials (thousands of years on the same computer).

Cryptanalysis of Tweaked Versions of SMASH and Reparation

Pierre-Alain Fouque, Jacques Stern,
and Sébastien Zimmer

CNRS-École normale supérieure-INRIA, Paris, France
{Pierre-Alain.Fouque,Jacques.Stern,Sebastien.Zimmer}@ens.fr

Abstract. In this paper, we study the security of permutation based hash functions, *i.e.* blockcipher based hash functions with fixed keys. SMASH is such a hash function proposed by Knudsen in 2005 and broken the same year by Pramstaller *et al.* Here we show that the two tweaked versions, proposed soon after by Knudsen to thwart the attack, can also be attacked in collision in time $\mathcal{O}(n2^{n/3})$. This time complexity can be reduced to $\mathcal{O}(2^{2\sqrt{n}})$ for the first tweak version, which means an attack against SMASH-256 in $c \cdot 2^{32}$ for a small constant c. Then, we show that an efficient generalization of SMASH, using two permutations instead of one, can be proved secure against collision in the ideal-cipher model in $\Omega(2^{n/4})$ queries to the permutations. In order to analyze the tightness of our proof, we devise a non-trivial attack in $\mathcal{O}(2^{3n/8})$ queries. Finally, we also prove that our construction is preimage resistant in $\Omega(2^{n/2})$ queries, which the best security level that can be reached for 2-permutation based hash functions, as proved in [12].

1 Introduction

Hash functions have recently been the subject of many attacks, revealing weaknesses in widely trusted hash functions such as MD5 or SHA-1. For this reason, some recent papers deal with new designs for hash functions such as SMASH [6] or Radiogatún [2,3]. Most of previous constructions of hash functions use blockciphers, since we know how to build such secure and efficient primitives and since good constructions of compression functions based on them are known. However, in the proofs of classical constructions of compression function, such as Davies-Meyer used in MD5 and SHA-1, the assumption made on the blockcipher is very strong, namely that *for each key*, or message for hash function, *the blockcipher acts as a random permutation*. Such an assumption, which has been introduced by Shannon and formalized in the ideal-cipher model in 1998 by Bellare *et al.* in [1], is impossible to check. For instance, it is possible that among the 2^{512} possible keys of the SHACAL blockcipher used for SHA-1, some weak keys exist, which could be used by an attacker. One solution to restrict the power of the adversary is to fix the key as it is the case in SMASH.

In order to study blockcipher for such constructions, Knudsen and Rijmen at Asiacrypt last year [7] proposed to use the notion of known-key distinguisher.

R. Avanzi, L. Keliher, and F. Sica (Eds.): SAC 2008, LNCS 5381, pp. 136–150, 2009.

The latter is an adversary which tries to distinguish a blockcipher from random permutation when the key is known. This model seems to be less permissive than the ideal-cipher model required to study the Davies-Meyer construction. Knudsen and Rijmen mention that this approach could be used to analyze hash function constructions, but the security model is not formally defined, seems to be hard to formalize, and to take into account for a security proof.

Even if there is no security model well adapted to study this alternate construction mode, it is particularly interesting since the assumption on the blockcipher seems to be more realistic. We refer to *permutation* based hash functions to precise that we do not use the flexibility of having many permutations using a blockcipher. In the constructions we are interested in, we only require one or two permutations to behave as random permutations, thus the probability to have a weak key is low and cannot be used by the adversary. Finally, another practical advantage of such constructions is that the key schedule of some blockciphers is more costly than the encryption processes and so avoiding the key schedule algorithm is interesting in term of speed and in term of space for hardware implementation.

1.1   Related Work

At FSE 2005, Knudsen [6] proposed a design for a compression function using only *one* permutation and a particular instance called SMASH. Soon after, Pramstaller *et al.* [10] broke it in collision very efficiently and Lamberger *et al.* [9] broke it in second preimage. That is why Knudsen proposed two tweaks to avoid the attacks, so that the expected complexity of any collision attack is still $\mathcal{O}(2^{n/2})$.

The security of 1-permutation based hash functions has been studied at Eurocrypt 2005 by Black *et al.* [4]. They show a very interesting impossibility result: in the ideal-cipher model, the number of queries to the permutation required to attack in collision the hash function is very low, linear in the bitsize of the input/output permutation. This result seems to rule out the construction of compression function using *one* permutation. However, it has one drawback which is very important in practice: even if the number of queries is low, the overall time complexity of the attack presented is very high, namely $O(n2^n)$. Therefore, in a computational model which would take into accounts the time or space complexity of the attack, such a construction could be possible. That is why the result of [4] does not completely rules out the construction of hash function based on one permutation such as SMASH. Finally, in the same vein, Steinberger and Rogaway at Eurocrypt'08 extend this result for many permutations against collision and preimage attacks. The main result interesting for us, is that with two permutations the preimage and collision resistance cannot be proved if more than $O(2^{n/2})$ queries are made to the permutations.

1.2   Our Results

In this paper, we first exhibit a new collision attack against the two tweaked versions of SMASH with complexity in time and memory of order $O(n2^{n/3})$, generating a 2-block collision. For the first tweak version, the attack can be improved and the complexity reduced to approximately 2^{32} for $n = 256$. To avoid our attack, we propose to replace one special operation, namely the multiplication by a constant in an extension field of GF(2), by a strong permutation. This modification has already been proposed by Thomsen [13], but has never been analyzed. We prove that a collision attack against this new scheme requires at least $2^{n/4}$ queries to the permutations. In order to better evaluate its collision resistance in term of number of queries, we devise an attack that requires $2^{3n/8}$ queries but needs $O(2^{3n/4})$ in time and works only if the Merkle-Damgård strengthening is not used. Finally, we prove that the number of queries required to attack the preimage is at least of $2^{n/2}$. Note that this latter bound is optimal according to Steinberger and Rogaway attack. Therefore our construction has also a theoretical interest since it is the first 2-permutation based hash function provably collision and preimage resistant. It gives a lower bound for the best collision resistance that we can obtain with such a construction and proves that the best preimage collision resistance of these schemes is in $\Theta(2^{n/2})$ queries.

Remark that, even if we are not able to prove better bounds, this does not say that our function is weak since we are not aware of an attack requiring less than the birthday attack for collision if the Merkle-Damgård strengthening used. For the preimage, since the attack of Steinberger and Rogaway requires $O(2^n)$ time complexity, we propose an attack requiring $O(2^{n/2})$ time complexity for the compression function and $O(2^{3n/4})$ for the full hash function.

1.3   Organization of the Paper

In section 2, we recall the security model and the designs of SMASH and of our generalization. We propose our collision attack on SMASH in section 3 and study the resistance of our new hash function against collision attacks in section 4. Finally, in section 5, we study the resistance against preimage of our construction.

2   Construction and Security Model

2.1   Security Model

THE IDEAL-CIPHER MODEL. To model blockciphers, we use the ideal-cipher model introduced by Shannon. In this model, the adversary is not computationally limited and the blockcipher is viewed as a family of functions $E : \{0,1\}^\kappa \times \{0,1\}^n \to \{0,1\}^n$ such that for each k, $E(k, \cdot)$ is a permutation on $\{0,1\}^n$. For every key k, $E(k, \cdot)$ is chosen uniformly at random in the set of all permutations on n bits. This implies that, for the adversary, for each key $k \in \{0,1\}^\kappa$, $E(k, \cdot)$ is a random and independent permutation.

The adversary A is given access to the oracles E and E^{-1}, which is denoted by $A^{E,E^{-1}}$: it can ask at most Q oracle queries to either E or E^{-1} and the answer of a query (K_i, X_i) for E is $Y_i = E(K_i, X_i)$ and the answer of a query (K_i, Y_i) for E^{-1} is $X_i = E^{-1}(K_i, Y_i)$.

REMARKS ON THE SECURITY MODEL. In our 2-permutation based construction, the keys k_1 and k_2 chosen for the construction are public and given to the adversary. As the permutations $E(k, \cdot)$ for $k \neq k_1, k_2$ are independent of $E(k_1, \cdot)$ and $E(k_2, \cdot)$, we assume w.l.o.g that the adversary does not ask oracle queries (k, x) to E or (k, y) to E^{-1} with $k \neq k_1, k_2$. For the sake of simplicity we denote $\pi_1 = E(k_1, \cdot)$ and $\pi_2 = E(k_2, \cdot)$ and give oracle access to π_1, π_1^{-1}, π_2 and π_2^{-1}. Note that in this case we do not lean upon the whole power of the ideal-cipher model, we only require that π_1 and π_2 were chosen independently and uniformly at random in the set of all permutations. We do not use the fact that for every $k \neq k_1, k_2$, $E(k, \cdot)$ is a permutation chosen uniformly at random in the set of all the permutations.

COLLISION RESISTANCE. If H is a hash function, the goal of the adversary is to break the collision resistance of H, that is to find two different messages (M, M') such that $H(M) = H(M')$. The ability of the adversary to break H collision resistance is denoted $\mathsf{adv}_H^{\mathsf{Coll}}(A)$ and is equal to:

$$\Pr\left[H(M) = H(M') \wedge M \neq M' | A^{E, E^{-1}} \Rightarrow (M, M')\right]$$

The probability is taken over the random coins of A and over all the possible blockcipher E where E is generated as specified above. The notation $A^{E, E^{-1}} \Rightarrow (M, M')$ means that A, after at most Q queries to E or E^{-1}, outputs (M, M'). We denote by $\mathsf{adv}_H^{\mathsf{Coll}}(Q)$ the maximum of $\mathsf{adv}_H^{\mathsf{Coll}}(A)$ over all the adversaries A which can make at most Q queries.

ASSUMPTIONS. We assume that the adversary does not ask a query for which it already knows the answer; namely, it does not ask the same query twice or if it asks (k, x) to E, which returns y, it does not ask (k, y) to E^{-1}, and vice versa. Furthermore, we assume that when an adversary outputs (M, M'), it has already computed $H(M)$ and $H(M')$, i.e. it has already made all the oracle queries required to compute $H(M)$ and $H(M')$.

MEASURES OF THE COMPLEXITY. There are two classical ways to measure the complexity of the attack. On one hand, one can say that this complexity is equal to the time complexity of the adversary. This is the complexity we are interested in, in practice, and this is the complexity that Knudsen consider in his paper about SMASH [6] and that we consider in our attack of SMASH. We refer to this complexity as the *practical complexity*. On the other hand, the attack complexity can be measured by the number of queries made to the oracles. This is the complexity oftenly used in proofs [11,5], or on the contrary to show that proofs cannot be established [4,12]. We use this complexity in our security proofs. It is refered in the following as the *query complexity*. Note that the practical complexity is always greater than the query complexity.

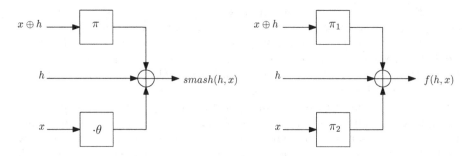

Fig. 1. SMASH compression function and our 2-permutation based compression function

2.2   SMASH Construction and Generalization

In this subsection, we introduce successively the original operating mode of SMASH, the modifications proposed by Knudsen and our new construction which is a generalization of the SMASH design.

Smash. Firstly we present the original version of SMASH. Let $\pi = E(0^n, \cdot)$ be a random permutations, $IV \in \{0,1\}^n$ be a fixed string, $\theta \neq 0, 1$ be a fixed element of $GF(2^n)$, the finite field of 2^n elements, and $smash \colon \{0,1\}^n \times \{0,1\}^n \to \{0,1\}^n$ be the function defined by:

$$smash^\pi(h, x) = \pi(h \oplus x) \oplus h \oplus \theta \cdot x,$$

where $\cdot$ denotes the multiplication in $GF(2^n)$.

Given (IV, π, θ), the hash of a message $x = (x_1, \ldots, x_\ell) \in \{0,1\}^{n \cdot \ell}$ is given by $SMASH(x) = h_{\ell+1}$ where $h_0 = \pi(IV) \oplus IV = smash^\pi(IV, 0^n)$, $h_k = smash^\pi(h_{k-1}, x_k)$, for all $1 \leq k \leq \ell$, and $h_{\ell+1} = \pi(h_\ell) \oplus h_\ell = smash^\pi(h_\ell, 0^n)$.

Tweaked Versions of Smash. After the attack of [10], it has been proposed two ways to modify the scheme [10,6], namely: "One is to use different permutations π for every iteration. Another is to use a secure compression function $(\ldots)$ after the processing of every t blocks of the message for, say $t = 8$ or $t = 16$".

We call the modification which consists in using a different permutation for every iteration, the *first modification* and the modification which consists in using a secure compression function (as $\pi(h) \oplus h$ for example) after the processing of every t blocks of the message, the *second modification*.

Our Generalized Construction. Let $IV \in \{0,1\}^n$ be a fixed string, $\pi_1 = E(0^n, \cdot)$ and $\pi_2 = E(1^n, \cdot)$ be two random permutations, and $f \colon \{0,1\}^n \times \{0,1\}^n \to \{0,1\}^n$ be the function defined by:

$$f(h, x) = \pi_1(x) \oplus \pi_2(x \oplus h) \oplus h$$

Given (IV, π_1, π_2), the hash of a message $x = (x_1, \ldots, x_\ell) \in \{0,1\}^{n \cdot \ell}$ is $H(x) = h_\ell$ where $h_0 = IV$, $h_k = f(h_{k-1}, x_k)$ for all $1 \leq k \leq \ell$.

PADDING. The constructions introduced before require that the message length is a multiple of a fixed integer which depends on the block size. To extend this construction to arbitrary length messages, one can add an injective padding to the message, such as the classical padding proposed for SMASH: add a '1' and as many '0' as required. In SMASH, it is also required to add the so-called Merkle-Damgård strenghtening, that is to concatenate the encoded length of the message at the end of the message. We also advice to add the Merkle-Damgård strenghtening for our construction, since, even if the security proof we are able to establish does not require it, the best known collision attacks against the construction without strengthening are strictly more efficient than the best known collision attacks against the construction with the strengthening.

3   A Collision Attack against All Versions of SMASH

In this section, we present a collision attack against SMASH in $\mathcal{O}(n2^{n/3})$. Since it generates a 2-block collision, it can be mounted against the two modifications of SMASH, as long as $t \geq 3$ (we remind that t denotes the number of iterations using the classical SMASH compression function before the use of an alternate secure compression function). Then we present an improvement of the attack which can be used to reduce the complexity of the attack. It can be applied against the first modification and then the attack generates two $2^{\sqrt{n}-1}$-block long collision messages and has a practical complexity of $\mathcal{O}(2^{2\sqrt{n}})$. This means that for $n = 256$, there is an attack in $c \cdot 2^{32}$ where c is a small constant. It also can be applied against the second modification if $t = 8$ or $t = 16$, but its impact is more limited.

3.1   Generic Attack

This subsection describes an attack against the collision resistance of the two modifications of SMASH with practical complexity of $\mathcal{O}(n2^{n/3})$. It generates a 2-block collision.

Note that the generic collision attack presented in [4] by Black et al. also applies to SMASH used with the first modification and finds a collision with a query complexity of at most $\mathcal{O}(2(n + 1))$ but a practical complexity greater than $\mathcal{O}(2^n)$. Therefore, this attack does not negate the security level expected by Knudsen [6], namely a practical security of $\mathcal{O}(2^{n/2})$. The attack presented in [10] by Pramstaller et al. is very efficient against the original version of SMASH, but as they precise in their paper, it does not apply to the two modifications.

In the following, we use the notations already introduced in subsection 2.2. Let π and π' be the two permutations used respectively in the first and in the second iteration. Let (α_1, β_1) and (α'_1, β'_1) be 2 pairs such that $\pi(\alpha_1) = \beta_1$ and $\pi'(\alpha'_1) = \beta'_1$. Let us define $\gamma_1 = \beta_1 \oplus \theta \cdot \alpha_1$, $\gamma'_1 = \beta'_1 \oplus \theta \cdot \alpha'_1$, $x_1 = \alpha_1 \oplus h_0$, $h_1 = smash^{\pi}(h_0, x_1) = \beta_1 \oplus \theta \cdot \alpha_1 \oplus (\theta + 1) \cdot h_0$, and $x'_1 = \alpha'_1 \oplus h_1$. Consequently, for $h_2 = smash^{\pi'}(h_1, x'_1)$, we get:

$$h_2 = \gamma'_1 \oplus (\theta + 1) \cdot \gamma_1 \oplus (\theta + 1)^2 \cdot h_0.$$

Let (α_2, β_2) and (α'_2, β'_2) be 2 other pairs such that $\pi(\alpha_2) = \beta_2$ and $\pi'(\alpha'_2) = \beta'_2$. Let us define similarly as above $\gamma_2 = \beta_2 \oplus \theta \cdot \alpha_2$, $\gamma'_2 = \beta'_2 \oplus \theta \cdot \alpha'_2$, $x_2 = \alpha_2 \oplus h_0$, $h'_1 = smash^\pi(h_0, x_2) = \beta_2 \oplus \theta \cdot \alpha_2 \oplus (\theta + 1) \cdot h_0$ and $x'_2 = \alpha'_2 \oplus h'_1$. For $h'_2 = smash^{\pi'}(h'_1, x'_2)$, we get:

$$h'_2 = \gamma'_2 \oplus (\theta + 1) \cdot \gamma_2 \oplus (\theta + 1)^2 \cdot h_0.$$

First, notice that if $h_2 = h'_2$, then $SMASH(x_1, x'_1) = SMASH(x_2, x'_2)$. We have $h_2 = h'_2$ if and only if $\gamma'_1 \oplus (\theta + 1) \cdot \gamma_1$ equals $\gamma'_2 \oplus (\theta + 1) \cdot \gamma_2$, which is equivalent to:

$$(\theta + 1) \cdot \gamma_1 \oplus (\theta + 1) \cdot \gamma_2 \oplus \gamma'_1 \oplus \gamma'_2 = 0. \tag{1}$$

The attack can be easily deduced from this relation.

Let us makes $2q$ queries to π to generate 2 sequences with q elements $(\alpha_{1,i}, \beta_{1,i})$ and $(\alpha_{2,i}, \beta_{2,i})$ and $2q$ queries to π' to generate 2 sequences with q elements $(\alpha'_{1,i}, \beta'_{1,i})$ and $(\alpha'_{2,i}, \beta'_{2,i})$. Let us compute the associated $\gamma_{j,i} = \beta_{j,i} \oplus \theta \cdot \alpha_{j,i}$ and $\gamma'_{j,i} = \beta'_{j,i} \oplus \theta \cdot \alpha'_{j,i}$, for $j = 1, 2$ and $1 \le i \le q$.

If $q = 2^{n/4}$, the birthday paradox says that with high probability there exists a quadruple $(\gamma_{1,a}, \gamma'_{1,b}, \gamma_{2,c}, \gamma'_{2,d})$ such that equation (1) is true. However finding such a quadruple requires a time complexity of $\mathcal{O}(n2^{n/2})$. For $q = 2^{n/3}$, the algorithm presented in [14] allows to find such a quadruple in time $\mathcal{O}(n2^{n/3})$ and space $\mathcal{O}(2^{n/3})$. Therefore, using this algorithm, we can mount an attack with query complexity of $\mathcal{O}(2^{n/3})$ and practical complexity of $\mathcal{O}(n2^{n/3})$ which is much smaller than the practical complexity of $\mathcal{O}(2^{n/2})$ that one could expect.

3.2   Improvements of the Attack

The improvement presented in this subsection comes from the generalization presented in [14] of the 4-list algorithm. The more lists there are, the smaller the practical complexity is. The main drawback of this improvement is that it generates longer colliding messages and therefore cannot be used completely against the second modification.

Let assume that instead of searching for 2-block colliding messages, we are searching for 3-block colliding messages. Using the same notations as above, let us introduce π'' the permutation used in the third iteration and (α''_1, β''_1) and (α''_2, β''_2) two pairs such that $\pi''(\alpha''_1) = \beta''_1$ and $\pi''(\alpha''_2) = \beta''_2$. If we define similarly as above $x''_1 = \alpha''_1 \oplus h_2$ and $x''_2 = \alpha''_2 \oplus h'_2$, and generalize previous notations, we have that

$$h_3 = \gamma''_1 \oplus (\theta + 1) \cdot \gamma'_1 \oplus (\theta + 1)^2 \cdot \gamma_1 \oplus (\theta + 1)^3 \cdot h_0$$
$$h'_3 = \gamma''_2 \oplus (\theta + 2) \cdot \gamma'_2 \oplus (\theta + 2)^2 \cdot \gamma_2 \oplus (\theta + 1)^3 \cdot h_0$$

Therefore, $h_3 = h'_3$ if and only if $(\theta+1)^2 \cdot (\gamma_1 \oplus \gamma_2) \oplus (\theta+1) \cdot (\gamma'_1 \oplus \gamma'_2) \oplus \gamma''_1 \oplus \gamma''_2 = 0$.

This leads to an attack which generates 6 lists and tries to find one element in every list such that the xor of theses elements is equal to 0. This can be generalized to k-block long messages. We can show that $h_k = h'_k$ if and only if:

$$\bigoplus_{i=0}^{k} (\theta + 1)^{k-i} \cdot (\gamma_1^{(i)} \oplus \gamma_2^{(i)}) = 0. \tag{2}$$

The algorithm in [14] finds such a $2k$-tuple in time $\mathcal{O}(k \cdot 2^{n/(1+\log_2(2k))})$ and requires $2k$ lists of size $\mathcal{O}(2^{n/(1+\log_2(2k))})$, therefore it requires to make $\mathcal{O}(k \cdot 2^{n/(1+\log_2(2k))})$ queries to generate all these lists. The complexity of the attack is optimal for $2k = 2^{\sqrt{n}}$ and in this case the practical complexity is equal to $\mathcal{O}(2^{2\sqrt{n}})$.

This improvement can be applied for all values of k when the first modification is used and therefore this version of SMASH can be attack in $\mathcal{O}(2^{2\sqrt{n}})$, generating messages of $2^{\sqrt{n}-1}$ blocks. For $n = 256$, this means a complexity of $c \cdot 2^{32}$ for a small constant c and messages of 2^{15} 256-bit blocks, that is of 1 Mo.

However, it can be applied only for $k \leq t - 1$ when the second modification is used. Therefore against this modification, the improved attack has a practical complexity of $\mathcal{O}(t \cdot 2^{n/(2+\log_2(t-1))})$, that is $\mathcal{O}(2^{n/4})$ and $\mathcal{O}(2^{n/5})$ for $t = 8$ and $t = 16$ respectively, as proposed by Knudsen [6] (we remind that t denotes the number of iterations using the classical SMASH compression function before the use of an alternate secure compression function). For $n = 256$, this gives a complexity of approximately 2^{64} and 2^{52}.

4   Collision Resistance of the Generalized Design

Now, we examine the collision resistance of the generalized version we propose. Firstly, we prove that a collision attack requires at least $\Omega(2^{n/4})$ queries to succeed with good probability. Secondly, we give a collision attack against our scheme with a query complexity of $\mathcal{O}(2^{3n/8})$, but a practical complexity of $\mathcal{O}(2^{3n/4})$. Most often this attack generates two messages of different length and therefore does not work anymore if the Merkle-Damgård strengthening is used. In this latter case, the best attack we have against our scheme is the birthday paradox attack with $\mathcal{O}(2^{n/2})$ queries and a practical complexity of $\mathcal{O}(n2^{n/2})$.

4.1   Security Proof

The attack presented in [4] shows in particular that one cannot expect to *prove* the collision resistance of SMASH if more than $\mathcal{O}(n)$ queries are made. On the contrary, we prove here that if we replace the multiplication by θ by a strong permutation (modelized by an ideal cipher), then one can prove that at least $\Omega(2^{n/4})$ queries are required to break collision resistance, and therefore that such an attack has a practical complexity greater than $2^{n/4}$. This proof is valid even if the Merkle-Damgård strengthening is not used.

Theorem 1. *Let A be a computationally unbounded adversary which makes at most Q queries. Its advantage in breaking H collision resistance is upper bounded by:*

$$\mathsf{adv}_H^{\mathsf{Coll}}(A) \leq \frac{2Q^4}{2^n}.$$

Proof. A collision adversary is allowed to make at most Q queries to either π_1, π_2, π_1^{-1}, or π_2^{-1}. We show that the probability that the adversary finds a collision

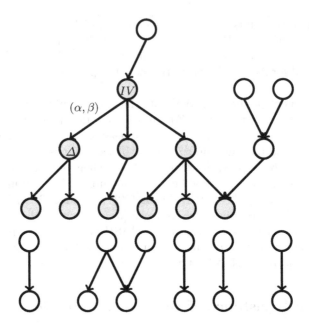

Fig. 2. An example of graph. In gray is the tree T.

for H is upper bounded by $Q^4/2^n$. The permutations π_1, π_2 and the initial value IV are chosen randomly.

THE GRAPH CONSTRUCTION. First, we introduce the following graph construction. Let $R_1 = \{(\alpha_i, \beta_i)_{1 \leq i \leq q_1}\}$ be q_1 pairs such that $\pi_1(\alpha_i) = \beta_i$ and $R_2 = \{(\alpha'_j, \beta'_j)_{1 \leq j \leq q_2}\}$ be q_2 pairs such that $\pi_2(\alpha'_j) = \beta'_j$. We define $\Delta_{i,j} = \alpha_i \oplus \alpha'_j$ and $\tilde{\Delta}_{i,j} = \beta_i \oplus \beta'_j \oplus \alpha_i \oplus \alpha'_j$ for $1 \leq i \leq q_1$ and $1 \leq j \leq q_2$. We construct a labelled directed graph $G = (V, E)$. The set of vertices V contains the bit strings $\Delta_{i,j}$, $\tilde{\Delta}_{i,j}$ and IV (that is at most $2q_1 \cdot q_2 + 1$ nodes). The set of edges E contains the directed edges $(\Delta_{i,j}, \tilde{\Delta}_{i,j})$ labelled with (α_i, β_i) denoted $\left((\Delta_{i,j}, \tilde{\Delta}_{i,j}), \alpha_i, \beta_i \right)$ (there are exactly $q_1 \cdot q_2$ labelled directed edges, possibly several edges between the same pair of nodes).

We define a path in the graph G as a sequence of edges $p = (e_1, \ldots, e_\ell)$ such that for each of its edge e_i, $1 \leq i \leq \ell - 1$ the output vertex is equal to the input vertex of e_{i+1}. Let us denote $\Delta \overset{p}{\leadsto} \Delta'$ which means that either $\Delta = \Delta'$ (and p is empty) or there exists a path $p = (e_1, \ldots, e_\ell)$ for which the input vertex of e_1 is Δ and the output vertex of e_ℓ is Δ'.

CORRESPONDENCE BETWEEN THE HASH FUNCTION AND THE GRAPH CONSTRUCTION. A message $x = (x_1, \ldots, x_\ell)$ is said to be *valid* if one can compute its digest value thanks to the already made requests, that is if and only if: $h_0 = IV$ and for every $k \geq 1$, $(x_k, \pi_1(x_k)) \in R_1$ and $(x_k \oplus h_{k-1}, \pi_2(x_k \oplus h_{k-1})) \in R_2$, with $h_k = \pi_1(x_k) \oplus \pi_2(x_k \oplus h_{k-1}) \oplus h_{k-1}$. Let us denote by M the set of all the valid messages. Let P be the set of all non-empty paths in G with IV as input node, that is $P = \{p \neq \emptyset \mid \exists \Delta \in V, IV \overset{p}{\leadsto} \Delta\}$. We now show that there is a bijection between P and M.

Let $p = (e_1, \ldots, e_\ell)$ be a non-empty path from IV to a node Δ. For this path p we construct a message $x = (x_1, \ldots, x_\ell)$, such that $H(x) = \Delta$, where x is defined as follows. For the kth edge e_k, by construction, there exists (a unique) (i_k, j_k) such that $e_k = \left((\Delta_{i_k, j_k}, \tilde{\Delta}_{i_k, j_k}), \alpha_{i_k}, \beta_{i_k}\right)$, and we define $x_k = \alpha_{i_k}$. Using the same notations as in the definition of H one can easily check that $h_0 = IV = \Delta_{i_1, j_1}$, and for all other $1 \leq k \leq \ell$, $h_k = \Delta_{i_{k+1}, j_{k+1}} = \tilde{\Delta}_{i_k, j_k}$:

$$h_k = f(h_{k-1}, x_k) = \pi_1(x_k) \oplus \pi_2(x_k \oplus h_{k-1}) \oplus h_{k-1}$$
$$= \pi_1(\alpha_{i_k}) \oplus \pi_2(\alpha_{i_k} \oplus \Delta_{i_k, j_k}) \oplus \Delta_{i_k, j_k} = \pi_1(\alpha_{i_k}) \oplus \pi_2(\alpha'_{j_k}) \oplus \alpha_{i_k} \oplus \alpha'_{j_k}$$
$$= \beta_{i_k} \oplus \beta'_{j_k} \oplus \alpha_{i_k} \oplus \alpha'_{j_k} = \tilde{\Delta}_{i_k, j_k} = \Delta_{i_{k+1}, j_{k+1}}.$$

Therefore x is valid and $H(x) = h_\ell = \tilde{\Delta}_{i_\ell, j_\ell} = \Delta$. We say that p induces the message x. One can check easily that if $p \neq p'$ induce respectively x and x', then $x \neq x'$.

Conversely, let $x = (x_1, \ldots, x_\ell)$ be a valid message and p be the path defined as $p = (e_1, \ldots, e_\ell)$ with $e_k = ((h_{k-1}, h_k), x_k, \pi_1(x_k))$ (we remind that $h_0 = IV$ and $h_k = f(h_{k-1}, x_k)$). The path p is clearly in P. We say that x induces p. One can check easily that if $x \neq x'$ induce respectively a path p and p' in G then $p \neq p'$.

Therefore, finding two colliding messages in M is equivalent to find two paths in P with the same output nodes. We say that these two paths collide and that there is a collision in G.

UPPER BOUND OF THE COLLISION PROBABILITY. Consider now the collision adversary. Let us assume that it has already made q_1 queries to π_1 or π_1^{-1} and q_2 queries to π_2 or π_2^{-1}. These queries induce two sets R_1 and R_2, and a graph G defined as above. We also introduce the following sets:

$$T = \{\Delta \in V \mid \exists p, IV \overset{p}{\leadsto} \Delta\}$$
$$A = \{\alpha \mid \exists 1 \leq j \leq q_2, \exists \Delta \in T, \alpha'_j \oplus \Delta = \alpha\}$$
$$B = \{\gamma \mid \exists 1 \leq j \leq q_2, \exists \Delta' \in V, \beta'_j \oplus \alpha'_j \oplus \Delta' = \gamma\}$$

Without loss of generality, we can assume that the adversary is ready to make a query to π_1 or π_1^{-1}. Let us denote by $(\tilde{\alpha}, \tilde{\beta} = \pi_1(\tilde{\alpha}))$ the pair induced by this query. With this query the graph G expands, new edges are generated. Let us denote by $\tilde{G}$ the graph after this expansion and similarly $\tilde{T}$ the expansion of T and $\tilde{P}$ the expansion of P.

We now show that if there is a collision in $\tilde{G}$, then $\tilde{\beta} \oplus \tilde{\alpha} \in B$ and $\tilde{\alpha} \in A$ with high probability. Assume that there is a collision in $\tilde{G}$ but not in G. Let Δ be a node in $\tilde{G}$, let p, p' be two paths in $\tilde{P}$ such that $p \neq p'$, $IV \overset{p}{\leadsto} \Delta$ and $IV \overset{p'}{\leadsto} \Delta$ in $\tilde{G}$. Let us denote $(IV, \Delta_1, \ldots, \Delta_\ell = \Delta)$ the sequence of vertices crossed by p

in $\tilde{G}$ and $(IV, \Delta'_1, \ldots, \Delta'_m = \Delta)$ the sequence of vertices crossed by p' in $\tilde{G}$. As there is not any collision in G, then either p or p' is not in P. Let us say it is p.

Note that with high probability Δ is already in G and was not generated by the expansion. If it were not the case, then there would be $i \neq j$ such that $\Delta = \tilde{\beta} \oplus \beta'_i \oplus \alpha'_i = \tilde{\beta} \oplus \beta'_j \oplus \alpha'_j$. This implies that $\beta'_i \oplus \alpha'_i = \beta'_j \oplus \alpha'_j$. The probability that there exists such a pair (i, j) is upper bounded by $q_2^2/2^n$. Let us assume that such a pair does not exist and therefore that Δ is already in G.

Let a be the smallest integer such that there exists r suffix of p with $\Delta_a \overset{r}{\rightsquigarrow} \Delta_\ell$ in G (hence $\Delta_a \in V$), that is r exists before the expansion. Due to the previous remark, a exists and $a \leq \ell$. As $\Delta_{a-1} \notin r$, it means that the edge (Δ_{a-1}, Δ_a) is generated by the expansion, that is there exists j such that $\Delta_{a-1} = \tilde{\alpha} \oplus \alpha'_j$ and $\Delta_a = \tilde{\beta} \oplus \tilde{\alpha} \oplus \beta'_j \oplus \alpha'_j$. Therefore we have $\tilde{\beta} \oplus \tilde{\alpha} \in B$.

Similarly, let b be the greatest integer such that, there exists r' prefix of p with $IV \overset{r'}{\rightsquigarrow} \Delta_b$ in G (hence $\Delta_b \in T$). As $\Delta_{b+1} \notin r'$, it means that the edge (Δ_b, Δ_{b+1}) is generated by the expansion, that is there exists j such that $\Delta_b = \tilde{\alpha} \oplus \alpha'_j$ and $\Delta_{b+1} = \tilde{\beta} \oplus \tilde{\alpha} \oplus \beta'_j \oplus \alpha'_j$. Therefore we have $\tilde{\alpha} \in A$.

If it is π_1 which was queried by $\tilde{\alpha}$, then the collision probability is upper bounded by:

$$\frac{\#B}{2^n - q_1} \leq \frac{\#V \cdot q_2}{2^n - q} \leq \frac{2(2q_1q_2 + 1)q_2}{2^n} \leq \frac{2q^3}{3 \cdot 2^n} \leq \frac{q^3}{2^n},$$

where $q = q_1 + q_2$. The last inequality is true because the function $x \mapsto 2(2(q - x)x + 1)x$ reaches its maximum for $x \approx 2q/3$ and is smaller than $2q^3/3$ at this point. The collision probability can be similarly upper bounded by $q^3/2^n$ if it is π_1^{-1} which was queried.

Therefore, at the qth iteration, the success probability is lower than $2q^3/3 \cdot 2^n + q^2/2^n$, and at the end the success probability is lower than

$$\sum_{q=1}^{Q} \left(2q^3/3 \cdot 2^n + q^2/2^n \right) \leq Q^4/2^n. \qquad \square$$

4.2   Attacks

Now we present an attack against the entire hash function, this gives upper bounds of its collision resistance. Before that, note that the birthday paradox allows to easily construct a collision attack which succeeds with probability nearly 1 with $\mathcal{O}(2^{n/2})$ queries to π_1 and π_2 and time complexity of $\mathcal{O}(n2^{n/2})$. This attack generates two 1-block messages which collide and therefore works even if the Merkle-Damgård strengthening is used (see the full version of the paper for a description of this attack).

We present now an attack which succeeds with probability nearly 1. It is a better attack than the birthday attack since its query complexity is only equal to $2 \cdot 2^{3n/8}$, but it has a practical complexity of $\mathcal{O}(n2^{3n/4})$. Moreover, one does not control the size of the two messages generated during the attack and most probably they won't have the same size. Therefore the Merkle-Damgård strengthening allows to thwart the attack. Our analysis of this latter is heuristic and not proved.

However, we have tested the attack for several values of n up to $n = 40$ and it turned out to work well in practice.

Proposition 1. *For $Q \geq 2^{3n/8+2}$ there is a computationally unbounded collision adversary with high success probability.*

SKETCH OF THE ATTACK. In the sequel, first we explain how we make the queries, then we informally evaluate the expected number of messages for which we are able to compute the hash. For a precise algorithm, see the full version of the paper. Note that the following attack is inspired from the way we have proved the collision resistance : we introduce the same tree T and try to make it grow as much as possible, so that it quickly contains $2^{n/2}$ vertices.

Let α_0 and β_0 be two random n-bit strings such that $\alpha_0 \oplus \beta_0 = IV$. Let q be an integer. We generate the sequences $(\alpha_i)_{0 \leq i \leq q}$, and $(\beta_i)_{0 \leq i \leq q}$ such that for all $1 \leq i \leq q$, $\alpha_i = \pi_1(\alpha_{i-1}) \oplus \alpha_{i-1}$, and $\beta_i = \pi_2(\beta_{i-1}) \oplus \beta_{i-1}$. For all $0 \leq i, j \leq q$, let us define $\Delta_{i,j} = \alpha_i \oplus \beta_j$. Note that we make $Q = 2q$ queries to π_1 and π_2, and we generate about $(q+1)^2$ different $\Delta_{i,j}$. We generate the sequences this way, because we have the following interesting property: for all (i,j), $f(\Delta_{i,j}, \alpha_i) = \Delta_{i+1,j+1}$. Therefore, for all $1 \leq \ell \leq q$, the message $\alpha_0 \| \alpha_1 \| \ldots \| \alpha_{\ell-1}$ hashes to $\Delta_{\ell,\ell}$ and if $\Delta_{k,k} = \Delta_{i,j}$, then for all $1 \leq \ell \leq q - \max(i,j)$ the message $M_{k,i,j,\ell} = \alpha_0 \| \alpha_1 \| \ldots \| \alpha_{k-1} \| \alpha_i \| \alpha_{i+1} \| \ldots \| \alpha_{i+\ell-1}$ hashes to $\Delta_{i+\ell,j+\ell}$. Such a triplet (k,i,j) is called a *colliding triplet* and the message $M_{k,i,j,\ell}$ is a preimage of $\Delta_{i+\ell,j+\ell}$ for H.

If there are many different colliding triplets (i,j,k), so we are able to find a preimage for many different values $\Delta_{i+\ell,j+\ell}$. Let us introduce the graph $T = (V, E)$ where:

$$V = \{\Delta_{a,b} | \exists \text{ a colliding triplet } (k,i,j) \text{ s.t. } a - i = b - j \geq 0\}$$
$$\cup \{\Delta_{a,a}, 0 \leq a \leq q\}$$
$$E = \{(\Delta_{a,b}, \Delta_{a+1,b+1}) \text{ s.t. } 0 \leq a, b \leq q - 1, \Delta_{a,b} \in V\}.$$

Note that we are able to find a preimage for all $\Delta_{a,b} \in V$ and that T is a tree if and only if there is no collision (otherwise we are able to find a cycle in T and there are two ways to reach some $\Delta_{a,b} \in V$). Therefore, our goal is to make V grow up to a size of about $2^{n/2}$ vertices so that a collision occurs. In the following we explain informally why this happens with high probability for $q = 2^{3n/8+1}$. This analysis considers that the α_i and β_j are uniformly distributed, which is of course not the case. However, we expect that the analysis gives a good intuition of what happens.

Let us evaluate roughly the expected value of T size, denoted t. Note that T contains at least the $q + 1$ vertices $\Delta_{a,a}$. If (i,j,k) is a collision triplet with $i \neq j$, then all the $\Delta_{i+\ell,j+\ell}$, with $0 \leq \ell \leq q - \max(i,j)$, are added to T. Thus, if $Set = \{(i,j,k) \text{ s.t. } i \neq j, i \neq k, j \neq k\}$, we have:

$$t \approx 1 + q + \sum_{(i,j,k) \in Set} \mathbb{1}_{\{\Delta_{k,k} = \Delta_{i,j}\}}(q - \max(i,j)),$$

therefore, $\mathbb{E}(t) \approx 1 + q + \sum_{(i,j,k) \in Set} \Pr[\Delta_{k,k} = \Delta_{i,j}](q - \max(i,j)).$

n	Q	number of experiments	size of T	percentage of success
36	2^{15}	10000	$2^{18} \leq \cdot \leq 2^{19}$	52%
40	$2^{16.5}$	1000	$2^{20} \leq \cdot \leq 2^{21}$	59%

Fig. 3. Experimental results

where $\mathbb{1}$ denotes the characteristic function. If the α_i and β_j were uniformly distributed (which is *not* the case) then we would have that $\Pr[\Delta_{k,k} = \Delta_{i,j}] = 1/2^n$ and therefore that

$$\mathbb{E}(t) \approx 1 + q + \frac{1}{2^n} \sum_{(i,j,k) \in Set} (q - \max(i,j))$$

$$= 1 + q + \frac{q(q+1)(q-1)(q-2)}{3 \cdot 2^n} \approx \frac{q^4}{3 \cdot 2^n}.$$

We can conclude that for $q = 2^{3n/8}$, we can expect that T contains more than $2^{n/2}$ vertices and, in this case, hope that the birthday paradox applies here so that two of these vertices collide. As already stated, if there is such a collision the attack is finished, we are able to find two messages which collide for H. □

COMPLEXITY OF THE ATTACK AND EXPERIMENTAL RESULTS. The precise algorithm is described in the full version of the paper. The attack requires $\mathcal{O}(2^{3n/8})$ queries to π_1 and π_2, $\mathcal{O}(2^{n/2})$ in space (to store T) and $\mathcal{O}(n2^{3n/4})$ in time (because we have to search for *all* the colliding triplets (i,j,k) with $1 \leq i,j,k \leq 2^{3n/8}$, that is all the triplets (i,j,k) such that $\Delta_{k,k} = \alpha_i \oplus \beta_j$).

We have run several tests for n equals 36 and 40. For that, we have used the blockcipher RC5 with two random keys and with a random IV. The results are summarized in figure 3. It appears that for $Q = 2\sqrt{2} \cdot 2^{3n/8}$, in all experiments the tree T contains between $2^{n/2}$ and $2^{n/2+1}$ vertices and a collision is found at least half the time. This validates our heuristic analysis of the attack.

NOTE. We have studied some other constructions of a compression function using only two permutations and some "xor". Some lead to hash functions which are trivially breakable, for all the others a variant of this attack could be applied (sometimes this variant is tricky and requires to make oracle queries to π_1^{-1} or π_2^{-1}).

5 Security against Preimage of the Generalized Construction

5.1 Security Proof

In this section we prove that the preimage resistance of our construction is provably in $\mathcal{O}(2^{n/2})$ queries. Note that in the design of the compression function, we have added a feed-forward (more precisely, we xor the chaining value to the output of the two permutations) exclusively in order to prevent trivial preimage attacks against the compression function (removing this feed-forward does not alter the collision resistance). Thus, the compression function is provably collision resistant up to $\mathcal{O}(2^{n/2})$ queries. That is what we show in the following. Since finding a preimage for the whole hash function implies finding a preimage for the compression function, this implies that the whole hash function is provably preimage resistant as long as less than $\mathcal{O}(2^{n/2})$ queries have been made.

Proposition 2. *Let A be a computationally unbounded adversary which makes at most Q queries, its advantage in breaking f preimage resistance is upper bounded by $Q^2/2^n$.*

Proof. Let α and α' be two queries made respectively to π_1 and π_2 and let β and β' be the respective answers (that is we have $\beta = \pi_1(\alpha)$ and $\beta' = \pi_2(\alpha')$). Let x be the value for which a preimage is searched. We have $\Pr[x = \alpha \oplus \beta \oplus \alpha' \oplus \beta'] = \sum_y \Pr[\beta = x \oplus \alpha \oplus \alpha' \oplus y]\Pr[\beta' = y] = 1/2^n$. The result is the same if π^{-1} is queried by β or if π^{-2} is queried by β'. Therefore, if A makes q_1 queries to π_1 or π_1^{-1} and q_2 queries to π_2 or π_1^{-2} (such that $q_1 + q_2 = Q$) and obtains the pairs (α_i, β_i) and (α'_j, β'_j) respectively, the union bound says that the probability that there exists a pair (i,j) such that $x = \alpha_i \oplus \beta_i \oplus \alpha'_j \oplus \beta'_j$ is upper bounded by $q_1 q_2/2^n$, and thus by $Q^2/2^n$. □

5.2 Optimality of the Proof and Attacks

In [12], Rogaway and Steinberger present a generic $\mathcal{O}(n2^{n/2})$ preimage attack against any 2-permutation based hash function. This means, as already stated, that our construction reaches the best security level against preimage that we can expect, namely $\mathcal{O}(2^{n/2})$ queries; in this sense, the construction is optimal.

Besides, the attack of [12] against the whole hash function requires $\mathcal{O}(n2^{n/2})$ queries, but the exact practical complexity is not established in general. However, in our case, its practical complexity seems greater than 2^n. This leads us to wonder what is the attack with the lowest practical complexity. Since finding a preimage for the compression function requires $\mathcal{O}(n2^{n/2})$ in time, the Lai and Massey attack [8] can be used. This attack is an unbalanced meet-in-the-middle attack: we compute $2^{n/4}$ preimages, hash $2^{3n/4}$ messages and meet in the middle using the birthday paradox. This requires to make $\mathcal{O}(2^{3n/4})$ queries and to make $\mathcal{O}(n2^{3n/4})$ computations. This is still greater than $\mathcal{O}(n2^{n/2})$ and it is an open problem to decrease the practical complexity of a preimage attack against the whole hash function.

Acknowledgment. The authors would like to thank Lars Knudsen for his useful comments. This work has been partially supported by the European Commission through the IST Program under Contract IST-2002-507932 ECRYPT, and the French RNRT/ANR SAPHIR Project.

References

1. Bellare, M., Krovetz, T., Rogaway, P.: Luby-rackoff backwards: Increasing security by making block ciphers non-invertible. In: Nyberg, K. (ed.) EUROCRYPT 1998. LNCS, vol. 1403, pp. 266–280. Springer, Heidelberg (1998)

2. Bertoni, G., Daemen, J., Peeters, M., Van Assche, G.: Radiogatùn, a belt-and-mill hash function. In: ECRYPT Hash Workshop (2007)
3. Bertoni, G., Daemen, J., Peeters, M., Van Assche, G.: On the indifferentiability of the sponge construction. In: Smart, N.P. (ed.) EUROCRYPT 2008. LNCS, vol. 4965, pp. 181–197. Springer, Heidelberg (2008)
4. Black, J.A., Cochran, M., Shrimpton, T.: On the impossibility of highly-efficient blockcipher-based hash functions. In: Cramer, R. (ed.) EUROCRYPT 2005. LNCS, vol. 3494, pp. 526–541. Springer, Heidelberg (2005)
5. Black, J.A., Rogaway, P., Shrimpton, T.: Black-box analysis of the block-cipher-based hash-function constructions from PGV. In: Yung, M. (ed.) CRYPTO 2002. LNCS, vol. 2442, pp. 320–335. Springer, Heidelberg (2002)
6. Knudsen, L.R.: SMASH - A cryptographic hash function. In: Gilbert, H., Handschuh, H. (eds.) FSE 2005. LNCS, vol. 3557, pp. 228–242. Springer, Heidelberg (2005)
7. Knudsen, L.R., Rijmen, V.: Known-key distinguishers for some block ciphers. In: Kurosawa, K. (ed.) ASIACRYPT 2007. LNCS, vol. 4833, pp. 315–324. Springer, Heidelberg (2007)
8. Lai, X., Massey, J.L.: Hash functions based on block ciphers. In: Rueppel, R.A. (ed.) EUROCRYPT 1992. LNCS, vol. 658, pp. 55–70. Springer, Heidelberg (1993)
9. Lamberger, M., Pramstaller, N., Rechberger, C., Rijmen, V.: Second preimages for SMASH. In: Abe, M. (ed.) CT-RSA 2007. LNCS, vol. 4377, pp. 101–111. Springer, Heidelberg (2006)
10. Pramstaller, N., Rechberger, C., Rijmen, V.: Breaking a new hash function design strategy called SMASH. In: Preneel, B., Tavares, S. (eds.) SAC 2005. LNCS, vol. 3897, pp. 233–244. Springer, Heidelberg (2006)
11. Preneel, B., Govaerts, R., Vandewalle, J.: Hash functions based on block ciphers: A synthetic approach. In: Stinson, D.R. (ed.) CRYPTO 1993. LNCS, vol. 773, pp. 368–378. Springer, Heidelberg (1994)
12. Rogaway, P., Steinberger, J.P.: Security/Efficiency tradeoffs for permutation-based hashing. In: Smart, N.P. (ed.) EUROCRYPT 2008. LNCS, vol. 4965, pp. 220–236. Springer, Heidelberg (2008)
13. Thomsen, S.S.: Cryptographic Hash Functions: PhD thesis, Technical University of Denmark (2005)
14. Wagner, D.: A generalized birthday problem. In: Yung, M. (ed.) CRYPTO 2002. LNCS, vol. 2442, p. 288. Springer, Heidelberg (2002)

Counting Functions for the k-Error Linear Complexity of 2^n-Periodic Binary Sequences[*]

Ramakanth Kavuluru and Andrew Klapper

Department of Computer Science, University of Kentucky,
Lexington, KY 40506, USA
{rvkavu2,klapper}@cs.uky.edu

Abstract. Linear complexity is an important measure of the cryptographic strength of key streams used in stream ciphers. The linear complexity of a sequence can decrease drastically when a few symbols are changed. Hence there has been considerable interest in the k-error linear complexity of sequences which measures this instability in linear complexity. For 2^n-periodic sequences it is known that minimum number of changes needed per period to lower the linear complexity is the same for sequences with fixed linear complexity. In this paper we derive an expression to enumerate all possible values for the k-error linear complexity of 2^n-periodic binary sequences with fixed linear complexity L, when k equals the minimum number of changes needed to lower the linear complexity below L. For some of these values we derive the expression for the corresponding number of 2^n-periodic binary sequences with fixed linear complexity and k-error linear complexity when k equals the minimum number of changes needed to lower the linear complexity. These results are of importance to compute some statistical properties concerning the stability of linear complexity of 2^n-periodic binary sequences.

Keywords: Periodic sequence, linear complexity, k-error linear complexity.

1 Introduction

The linear complexity of a sequence is the length of the shortest linear feedback shift register (LFSR) that can generate the sequence. The LFSR that generates a given sequence can be determined using the Berlekamp-Massey algorithm using only the first $2L$ elements of the sequence, where L is the linear complexity of the sequence. The typical assumption in the analysis of the security of stream ciphers is that the attacker has access to a part of the key stream and wants to use this to predict the remainder of the key stream. Thus the problem of designing a good stream cipher is reduced to the problem of designing a fast key stream

---

[*] This material is based upon work supported by the National Science Foundation under Grant No. CCF-0514660. Any opinions, findings, and conclusions or recommendations expressed in this material are those of the authors and do not necessarily reflect the views of the National Science Foundation.

R. Avanzi, L. Keliher, and F. Sica (Eds.): SAC 2008, LNCS 5381, pp. 151–164, 2009.

generator whose output is hard to predict from a prefix of the the output. Hence for cryptographic purposes sequences with high linear complexity are essential as an adversary would then need large initial segments of the sequences to recover the LFSRs that generate them using the Berlekamp-Massey algorithm.

A system is insecure if all but a few symbols of the key stream can be extracted. So for a cryptographically strong sequence, the linear complexity should not decrease drastically if a few symbols are changed. If it did, an attacker could modify the known prefix of the key stream and try to decrypt the result using the Berlekamp-Massey algorithm. If the resulting sequence differed from the actual key stream by only a few symbols, the attacker could extract most of the message. This observation gives rise to k-error linear complexity of sequences introduced in [7]. The k-error linear complexity of a periodic sequence is the smallest linear complexity achieved by making k or fewer changes per period. In addition to having large linear complexity, cryptographically strong sequences should, thus, also have large k-error linear complexity at least for small k.

Let $\mathbf{S} = (s_0, s_1, \cdots, s_{T-1})^\infty$ be a periodic binary sequence with period T. We associate the polynomial $\mathbf{S}(x) = s_0 + s_1 x + \cdots + s_{T-1} x^{T-1}$ and the corresponding T-tuple $\mathbf{S}^{(T)} = (s_0, s_1, \cdots, s_{T-1})$ to $\mathbf{S}$. The relationship between the linear complexity, denoted $L(\mathbf{S})$, of $\mathbf{S}$ and the associated polynomial $\mathbf{S}(x)$ is given by

$$L(\mathbf{S}) = T - \deg(\gcd(x^T - 1, \mathbf{S}(x))). \tag{1}$$

Let $w_H(\mathbf{S})$ denote the Hamming weight of the T-tuple $\mathbf{S}^{(T)}$. For $0 \le k \le T$, the k-error linear complexity of $\mathbf{S}$, denoted $L_k(\mathbf{S})$, is given by

$$L_k(\mathbf{S}) = \min_{\mathbf{E}} L(\mathbf{S} + \mathbf{E}), \tag{2}$$

where the minimum is over all T-periodic binary sequences $\mathbf{E}$ such that $w_H(\mathbf{E}) \le k$. Since we consider only 2^n-periodic sequences, we use $T = 2^n$ and the observation

$$x^T - 1 = x^{2^n} - 1 = (x-1)^{2^n} \tag{3}$$

for the rest of the paper.

Let $merr(\mathbf{S})$ denote the minimum value k such that the k-error linear complexity of a 2^n-periodic sequence $\mathbf{S}$ is strictly less than its linear complexity. That is

$$merr(\mathbf{S}) = \min\{k : L_k(\mathbf{S}) < L(\mathbf{S})\}. \tag{4}$$

Kurosawa et al. [3] derived the formula for the exact value of $merr(\mathbf{S})$.

Lemma 1. *For any nonzero 2^n-periodic sequence $\mathbf{S}$, we have*

$$merr(\mathbf{S}) = 2^{w_H(2^n - L(\mathbf{S}))},$$

where $w_H(j)$, $0 \le j \le 2^n - 1$, denotes the Hamming weight of the binary representation of j.

The counting function of a sequence complexity measure gives the number of sequences with a given complexity measure value. Rueppel [6] determined the counting function of linear complexity for 2^n-periodic binary sequences. Using equations (1) and (3) it is straightforward to obtain the number of 2^n-periodic binary sequences with fixed linear complexity. For the rest of the paper let $\mathcal{N}(L)$ and $\mathcal{A}(L)$ denote, respectively, the number of and the set of 2^n-periodic binary sequences with given linear complexity L, $0 \leq L \leq 2^n$. Rueppel [6] showed that

$$\mathcal{N}(0) = 1 \text{ and } \mathcal{N}(L) = 2^{L-1} \text{ for } 1 \leq L \leq 2^n. \tag{5}$$

Recently, using efficient algorithms to compute the linear complexity of p^n periodic sequences over $\mathbb{F}_p$, Meidl [4] obtained the counting function and the expected value for the 1-error linear complexity of 2^n-periodic binary sequences. Meidl and Venkateswarlu [5] extended these results to p^n-periodic sequences over $\mathbb{F}_p$. Fengxiang and Wenfeng [1] used Meidl's [4] approach of analyzing Games-Chan algorithm to obtain the counting functions and gave the exact expression for the expected value of the 2-error linear complexity of a random 2^n-periodic binary sequence with linear complexity $2^n - 1$.

In this paper we perform a more rigorous analysis of Games-Chan algorithm to enumerate all the possible values of k-error linear complexity of sequences in $\mathcal{A}(L)$ for $k = 2^{w_H(2^n - L)}$, that is when k is the minimum number of changes needed to lower the linear complexity below L. For certain sets of these values, we also derive the corresponding number of sequences in $\mathcal{A}(L)$ whose k-error linear complexity equals the values in those sets. For the rest of the paper by $k_{min}(L)$ denote the minimum number of changes needed to lower the linear complexity of sequences in $\mathcal{A}(L)$, that is $k_{min}(L) = 2^{w_H(2^n - L)}$.

2 Games-Chan Algorithm

In this section we describe the Games-Chan algorithm and list some results using its analysis.

By Lemma 1 for any 2^n-periodic sequence $\mathbf{S}$ with $merr(\mathbf{S}) = 2^m$, $m \in \{0, \cdots, n\}$, the linear complexity $L(\mathbf{S})$ can be uniquely expressed as

$$L(\mathbf{S}) = 2^n \text{ or } L(\mathbf{S}) = 2^n - \sum_{i=1}^{m} 2^{n-r_i},$$

where $0 < r_1 < \cdots < r_m \leq n$.

The Games-Chan algorithm [2] is a fast algorithm for computing the linear complexity of a 2^n-periodic binary sequence. For any $\mathbf{S} \in \mathcal{A}(L)$ with period $\mathbf{S}^{(2^n)} = (s_0, \cdots, s_{2^n-1})$, denote the left and right halves of $\mathbf{S}^{(2^n)}$ by

$$\mathbf{S}_L^{(2^{n-1})} = (s_0, \cdots, s_{2^{n-1}-1}) \text{ and } \mathbf{S}_R^{(2^{n-1})} = (s_{2^{n-1}}, \cdots, s_{2^n-1}).$$

Let $\mathbf{S}_L$ and $\mathbf{S}_R$ denote the 2^{n-1} periodic sequences

$$\mathbf{S}_L = (s_0, \cdots, s_{2^{n-1}-1})^\infty \text{ and } \mathbf{S}_R = (s_{2^{n-1}}, \cdots, s_{2^n-1})^\infty. \tag{6}$$

Games-Chan Algorithm. Let **S** be 2^n-periodic binary sequence.

(i) If $\mathbf{S}_L^{(2^{n-1})} = \mathbf{S}_R^{(2^{n-1})}$, then $L(\mathbf{S}) = L(\mathbf{S}_L)$.

(ii) If $\mathbf{S}_L^{(2^{n-1})} \neq \mathbf{S}_R^{(2^{n-1})}$, then $L(\mathbf{S}) = 2^{n-1} + L(\mathbf{S}_L + \mathbf{S}_R)$.

(iii) Apply the above procedure recursively to the 2^{n-1}-periodic binary sequence $\mathbf{S}_L$ in (i), or the 2^{n-1}-periodic binary sequence $\mathbf{S}_L + \mathbf{S}_R$ in (ii).

We make some observations and establish notation we use for the rest of the paper. We note that the procedure of the Games-Chan algorithm as stated here is executed a total of n times to compute the linear complexity of any $\mathbf{S} \in \mathcal{A}(L)$. In the ith step, $i = 0, \cdots, n-1$, the algorithm computes the linear complexity of a 2^{n-i}-periodic binary sequence. Let $\psi^i(\mathbf{S})$, $i = 0, \cdots, n-1$, denote the first period of the 2^{n-i}-periodic binary sequence considered in the ith step of the algorithm when run with input sequence $\mathbf{S}$. Let $\psi_L^i(\mathbf{S})$ and $\psi_R^i(\mathbf{S})$ denote, respectively, the left and right halves of $\psi^i(\mathbf{S})$. Let $m^i(\mathbf{S})$ denote the total value contributed to $L(\mathbf{S})$ in the algorithm during the execution from the 0-th step to the i-th step of the algorithm. For any two finite binary sequences, $\mathbf{S}$ and $\mathbf{S}'$, of same length let $d_H(\mathbf{S}, \mathbf{S}')$ denote the Hamming distance between $\mathbf{S}$ and $\mathbf{S}'$. We slightly abuse the notation because we also use $d_H(\mathbf{S}, \mathbf{S}')$ to denote the Hamming distance between the first periods of $\mathbf{S}, \mathbf{S}' \in A(L)$. It is straightforward to derive the following lemma from the Games-Chan algorithm.

Lemma 2. *Let* **S** *be a* 2^n*-periodic binary sequence. For any* t *integers* $r_1, \cdots, r_t$ *such that* $0 < r_1 < r_2 < \cdots < r_t \leq n$, *we have*

$$L(\mathbf{S}) = 2^n - (2^{n-r_1} + 2^{n-r_2} + \cdots + 2^{n-r_t}) \tag{7}$$

if and only if

$$\psi_L^{u-1}(\mathbf{S}) = \psi_R^{u-1}(\mathbf{S}) \quad \text{exactly when} \quad u \in \{r_1, \cdots, r_t\}.$$

For any $\mathbf{S} \in \mathcal{A}(L)$ where L is as in equation (7), the following properties of vectors $\psi^l(\mathbf{S})$, $0 \leq l \leq n$, are straightforward to obtain.

P1: If $l = r_i - 1$, for some $i \in \{1, \cdots, t\}$, then $w_H(\psi^l(\mathbf{S})) = 2 \cdot w_H(\psi^{l+1}(\mathbf{S}))$.

P2: For any $l \neq r_i - 1$, for all $i \in \{1, \cdots, t\}$, we have $w_H(\psi^l(\mathbf{S})) \geq w_H(\psi^{l+1}(\mathbf{S}))$.

By $\mathcal{P}_l$, $0 \leq l \leq n$, denote the number of distinct possibilities, over all sequences in $\mathcal{A}(L)$, for the 2^{n-l}-vector during the l-th step such that the 2^{n-l-1}-vector during the $(l+1)$-th step is fixed. It is straightforward to get the following properties.

P3: If $l = r_i - 1$, for some $i \in \{1, \cdots, t\}$, then $\mathcal{P}_l = 1$.

P4: For any $l \neq r_i - 1$, for all $i \in \{1, \cdots, t\}$, we have $\mathcal{P}_l = 2^{2^{n-l-1}}$.

We also use the following result in the next section. It can be proved using the procedure of Games-Chan algorithm and Lemma 2.

Lemma 3. *Let* $\mathbf{S} \in \mathcal{A}(L)$ *with* $L \neq 0$ *represented as*

$$L(\mathbf{S}) = 2^n - (2^{n-r_1} + 2^{n-r_2} + \cdots + 2^{n-r_t}), \tag{8}$$

where $0 < r_1 < r_2 < \cdots < r_t \leq n$. Let $\mathbf{S}' \neq \mathbf{S}$ be any other 2^n-periodic binary sequence such that $m^{l-1}(\mathbf{S}) = m^{l-1}(\mathbf{S}')$ for some $l \in \{1, \cdots, n\}$. If $d_H(\psi^l(\mathbf{S}), \psi^l(\mathbf{S}')) \neq 0$, then

$$d_H(\mathbf{S}, \mathbf{S}') \geq 2^b \cdot d_H(\psi^l(\mathbf{S}), \psi^l(\mathbf{S}')), \tag{9}$$

where b, $1 \leq b \leq t$, is the unique integer determined by the inequality $r_b \leq l < r_{b+1}$ assuming $r_0 = 0$ and $r_{t+1} = n + 1$.

3   Expression for $k_{min}(L)$-Error Linear Complexity

In this section we analyze the structure of the Games-Chan algorithm to derive an expression to enumerate all possible values of $k_{min}(L)$-error linear complexity of sequences in $\mathcal{A}(L)$ in terms the coefficients in the binary expansion of $2^n - L$. We handle the case when $1 < L < 2^n$ as the results are simple when $L = 0$ or 1 and as we already know the results when $L = 2^n$ [4]. We need the following generalization of [1, Lemma 2] whose proof is similar to that of Lemma 2 in [1].

Lemma 4. *For any sequence $\mathbf{S} = (s_0, \cdots, s_{2^n-1})^\infty \in \mathcal{A}(L)$, we have $L \leq 2^n - 2^{n-r}$, $r = 1, \cdots, n$, if and only if*

$$\sum_{i=0}^{2^r-1} s_{j+i \cdot 2^{n-r}} = 0 \text{ for } j = 0, \cdots, 2^{n-r} - 1.$$

We prove an auxiliary result that is used in the main result of this section.

Lemma 5. *Let $\mathbf{S} \in \mathcal{A}(L)$ with $1 < L < 2^n$. Consider the representation of L as*

$$L = 2^n - (2^{n-r_1} + 2^{n-r_2} + \cdots + 2^{n-r_t}), \tag{10}$$

where $r_0 = 0 < r_1 < r_2 < \cdots < r_t < n + 1 = r_{t+1}$ and $1 \leq t \leq n - 1$. Let $\mathbf{S}'$ be any 2^n-periodic binary sequence such that $d_H(\mathbf{S}, \mathbf{S}') = k_{min}(L) = 2^t$ and $L(\mathbf{S}') = L_{2^t}(\mathbf{S})$. Define the two integers

$$l_1 = \min\{i : 0 \leq i \leq n - 1 \quad and \quad m^i(\mathbf{S}') \neq m^i(\mathbf{S})\} \tag{11}$$

and

$$l_2 = \min\{i : 0 \leq i \leq n - 1 \quad and \quad d_H(\psi_L^i(\mathbf{S}), \psi_R^i(\mathbf{S})) = 2^{t-j}$$
$$with \quad r_j \leq i < r_{j+1}\}. \tag{12}$$

Then we have $l_1 = l_2$.

Proof. From Lemma 1 we know $k_{min}(L) = 2^t$ which implies $L(\mathbf{S}') < L(\mathbf{S})$. We note that there exists at least one integer i, $0 \leq i \leq n - 1$, such that $m^i(\mathbf{S}') \neq m^i(\mathbf{S})$ since otherwise $L(\mathbf{S}) = L(\mathbf{S}')$. Hence the set on the right hand side of equation (11) is not empty. From the procedure of the Games-Chan algorithm and using the fact $L(\mathbf{S}') < L(\mathbf{S})$ equation (11) implies

$$\psi_L^{l_1}(\mathbf{S}) \neq \psi_R^{l_1}(\mathbf{S}) \quad \text{and} \quad \psi_L^{l_1}(\mathbf{S}') = \psi_R^{l_1}(\mathbf{S}'). \tag{13}$$

From equation (13) we get

$$d_H(\psi^{l_1}(\mathbf{S}), \psi^{l_1}(\mathbf{S}')) \geq d_H(\psi_L^{l_1}(\mathbf{S}), \psi_R^{l_1}(\mathbf{S})). \tag{14}$$

Let b be the unique integer determined by the inequality $r_b \leq l_1 < r_{b+1}$. Since $\psi_L^{r_t-1}(\mathbf{S}) = \psi_R^{r_t-1}(\mathbf{S})$ and because the vectors considered during all the steps, except the last one, of the Games-Chan algorithm have nonzero Hamming weight, we have $w_H(\psi^{r_t-1}(\mathbf{S})) \geq 2$. So using properties P1 and P2 we get $w_H(\psi^{l_1+1}(\mathbf{S})) \geq 2^{t-b}$ and thus

$$d_H(\psi_L^{l_1}(\mathbf{S}), \psi_R^{l_1}(\mathbf{S})) \geq 2^{t-b}. \tag{15}$$

Now we show that $d_H(\psi_L^{l_1}(\mathbf{S}), \psi_R^{l_1}(\mathbf{S})) = 2^{t-b}$. If not, from equation (15) we have $d_H(\psi_L^{l_1}(\mathbf{S}), \psi_R^{l_1}(\mathbf{S})) > 2^{t-b}$. By equation (14) this implies

$$d_H(\psi^{l_1}(\mathbf{S}), \psi^{l_1}(\mathbf{S}')) > 2^{t-b}. \tag{16}$$

But from Lemma 3 we know $d_H(\mathbf{S}, \mathbf{S}') \geq 2^b \cdot d_H(\psi^{l_1}(\mathbf{S}), \psi^{l_1}(\mathbf{S}'))$, which implies $d_H(\psi^{l_1}(\mathbf{S}), \psi^{l_1}(\mathbf{S}')) \leq 2^{t-b}$ since $d_H(\mathbf{S}, \mathbf{S}') = 2^t$. This contradicts inequality in (16). Thus we have

$$d_H(\psi_L^{l_1}(\mathbf{S}), \psi_R^{l_1}(\mathbf{S})) = 2^{t-b}. \tag{17}$$

From equation (17) we know that the set on the right hand side of equation (12) is not empty and $l_2 \leq l_1$. By a denote the unique integer determined by the inequality $r_a \leq l_2 < r_{a+1}$. Because there are a steps before the l_2-th step where the left and right halves are equal it is evident from equation (21) that altering $\psi^{l_2}(\mathbf{S})$ such that $\psi_L^{l_2}(\mathbf{S}) = \psi_R^{l_2}(\mathbf{S})$ and propagating these changes to the 0-th step of the Games-Chan algorithm will require exactly $2^a \cdot 2^{t-a} = 2^t$ changes in $\mathbf{S}^{(2^n)}$. But if $l_2 < l_1$, forcing $\psi_L^{l_2}(\mathbf{S}) = \psi_R^{l_2}(\mathbf{S})$ will result in a 2^n-periodic binary sequence $\mathbf{S}''$ such that $d_H(\mathbf{S}, \mathbf{S}'') = 2^t$ and $L(\mathbf{S}'') < L(\mathbf{S}')$. This contradicts the fact that $L(\mathbf{S}') = L_{2^t}(\mathbf{S})$. Thus we have $l_2 = l_1$. □

Theorem 1. Let $\mathbf{S} \in \mathcal{A}(L)$ with $1 < L < 2^n$. Consider the representation of L as

$$L = 2^n - (2^{n-r_1} + 2^{n-r_2} + \cdots + 2^{n-r_t}), \tag{18}$$

where $r_0 = 0 < r_1 < r_2 < \cdots < r_t < n+1 = r_{t+1}$ and $1 \leq t \leq n-1$. Define the integer $w = \min\{i : r_i = n+i-t, 1 \leq i \leq t+1\}$. Then $L_{k_{min}(L)}(\mathbf{S})$ is 0 or is in one of the two forms

$$L_{j,l,C} := 2^n - \sum_{i=1}^{j-1} 2^{n-r_i} - 2^{n-l} + C, \quad 1 \leq j \leq w-1, \tag{19}$$

$$r_{j-1} \leq l \leq r_j - 2, \quad \text{and} \quad 1 \leq C \leq 2^{n-l-1} - 1,$$

or

$$L_{w,l,C} := 2^n - \sum_{i=1}^{w-1} 2^{n-r_i} - 2^{n-l} + C,$$

$$r_{w-1} \leq l \leq r_w - 3 \quad and \quad 1 \leq C \leq 2^{n-l-1} - 2^{t-w+1}.$$

(20)

Proof. From Lemma 1 and equation (18) $merr(\mathbf{S}) = k_{min}(L) = 2^t$. The sequences in $\mathcal{A}(L)$ whose 2^t-error linear complexity is 0 are those with exactly 2^t 1s per period. For any other sequence $\mathbf{S}$ in $\mathcal{A}(L)$ we show that the 2^t-error linear complexity is in one of the forms as stated in the theorem.

Define the integer l as in equation (12). That is

$$l = \min\{i : 0 \leq i \leq n-1 \quad and \quad d_H(\psi_L^i(\mathbf{S}), \psi_R^i(\mathbf{S})) = 2^{t-j}$$
$$with \quad r_j \leq i < r_{j+1}\}.$$

(21)

We already know that the set on the right hand side of equation (21) is not empty due to the intermediate findings of Lemma 5. By b denote the unique integer determined by the inequality $r_b \leq l < r_{b+1}$. From the proof of Lemma 5 we know that altering $\psi^l(\mathbf{S})$ such that $\psi_L^l(\mathbf{S}) = \psi_R^l(\mathbf{S})$ and propagating these changes to the 0-th step of the Games-Chan algorithm will require exactly 2^t changes in $\mathbf{S}^{(2^n)}$. We also see that it is necessary to alter $\psi^l(\mathbf{S})$ so that $\psi_L^l(\mathbf{S}) = \psi_R^l(\mathbf{S})$ to achieve the smallest linear complexity that can be obtained by making exactly 2^t errors in $\mathbf{S}^{(2^n)}$ since the remaining $n-l$ steps can only add a maximum of 2^{n-l-1} to the linear complexity of the modified sequence.

Note that $l \neq r_j - 1$, $j = 1, \cdots, t$, since $\psi_L^{r_j-1}(\mathbf{S}) = \psi_R^{r_j-1}(\mathbf{S})$, $j = 1, \cdots, t$. Next we show that

$$\forall \quad l+1 \leq i \leq n-1, \quad w_H(\psi^i(\mathbf{S})) = 2^{t-j} \quad with \quad r_j \leq i < r_{j+1}.$$ (22)

If equation (22) does not hold, then let m be any integer such that $l+1 \leq m \leq n-1$ and $w_H(\psi^m(\mathbf{S})) \neq 2^{t-a}$ where a is uniquely determined by the inequality $r_a \leq m < r_{a+1}$. Since $\psi_L^{r_t-1}(\mathbf{S}) = \psi_R^{r_t-1}(\mathbf{S})$, we have $w_H(\psi^{r_t-1}(\mathbf{S})) \geq 2$. So using properties P1 and P2 we get $w_H(\psi^m(\mathbf{S})) \geq 2^{t-a}$. This implies $w_H(\psi^m(\mathbf{S})) > 2^{t-a}$ since we assumed $w_H(\psi^m(\mathbf{S})) \neq 2^{t-a}$. Again, using P1 and P2 we have $w_H(\psi^{l+1}(\mathbf{S})) > 2^{a-b} \cdot 2^{t-a} = 2^{t-b}$ which contradicts the fact $d_H(\psi_L^l(\mathbf{S}), \psi_R^l(\mathbf{S})) = 2^{t-b}$. Thus $w_H(\psi^m(\mathbf{S})) = 2^{t-a}$ and so equation (22) holds.

To obtain the form of $L_{2^t}(\mathbf{S})$ we consider two cases based on the value of w.

Case 1: $w \leq t$

From the definition of w in the theorem statement it can be shown that $n - r_i = t - i$ for $i = w, \cdots, t$, which implies

$$L = 2^n - (2^{n-r_1} + \cdots + 2^{n-r_{w-1}} + 2^{t-w} + 2^{t-w-1} + \cdots + 2^0).$$ (23)

From equations (18), (23) and Lemma 2 this means that the left and right halves are equal from the $(r_w - 1)$-th step to $(n-1)$-th step of the execution of the Games-Chan algorithm. Using the fact that $n - r_w = t - w$, this implies that the 2^{t-w+1}-vector considered during the $(r_w - 1)$-th step

$$\psi^{r_w-1}(\mathbf{S}) = (\psi^{r_w-1}(\mathbf{S})_0, \cdots, \psi^{r_w-1}(\mathbf{S})_{2^{t-w+1}-1}) = (1, \cdots, 1)$$ (24)

is an all 1 vector. From the definition of w, equation (24) also implies that $w_H(\psi^{r_w-2}(\mathbf{S})) = 2^{t-w+1}$. That is

$$d_H(\psi_L^{r_w-3}(\mathbf{S}), \psi_R^{r_w-3}(\mathbf{S})) = 2^{t-w+1}. \tag{25}$$

By equation (25) and using the definition of l in equation (21) we have $l \le r_w - 3$. We consider two cases based on the value of l.

Case 1a: $r_{w-1} \le l \le r_w - 3$
We first note that this case occurs only when the binary expansion of L as in equation (18) satisfies $r_{w-1} \le r_w - 3$. Throughout this case we use the fact that $n - r_w = t - w$. From the definition of l in equation (21) we have $d_H(\psi_L^l(\mathbf{S}), \psi_R^l(\mathbf{S})) = 2^{t-w+1}$. We already know that making 2^{t-w+1} changes in $\psi^l(\mathbf{S})$ so that $\psi_L^l(\mathbf{S}) = \psi_R^l(\mathbf{S})$ is necessary to achieve the the smallest linear complexity possible by making $k_{min}(L) = 2^t$ changes in $\mathbf{S}^{(2^n)}$. But we have to decide for each of the 2^{t-w+1} positions where $\psi_L^l(\mathbf{S})$ and $\psi_R^l(\mathbf{S})$ differ, whether the change should be made in $\psi_L^l(\mathbf{S})$ or at the corresponding position in $\psi_R^l(\mathbf{S})$. In this case there is a unique of making these 2^{t-w+1} changes so that the linear complexity of the 2^{n-l-1}-periodic binary sequence with period equal to either of the equal halves obtained by forcing $\psi_L^l(\mathbf{S}) = \psi_R^l(\mathbf{S})$ is as small as possible. Next we describe a unique way of making these changes.

Let $\psi^{l+1}(\mathbf{S}') = \psi_L^l(\mathbf{S}) = \psi_R^l(\mathbf{S})$ be the 2^{n-l-1}-vector obtained after forcing $\psi_L^l(\mathbf{S}) = \psi_R^l(\mathbf{S})$ such that the linear complexity of the 2^{n-l-1}-periodic binary sequence with period $\psi^{l+1}(\mathbf{S}')$ is as small as possible. The left and right halves of the vectors considered are not equal from the r_{w-1}-th step to the $(r_w - 2)$-th step of the Games-Chan algorithm when executed with input sequence $\mathbf{S}$. From equation (24) $\psi^{r_w-1}(\mathbf{S})$ is a 2^{t-w+1}-vector with all 1s. Hence for all $v = r_{w-1}, r_{w-1} + 1, \cdots, r_w - 2$ due to the procedure of the Games-Chan algorithm we have

$$\sum_{j=0}^{2^{r_w-v-1}-1} \psi^v(\mathbf{S})_{i+j2^{t-w+1}} = 1 \quad \text{for} \quad i = 0, \cdots, 2^{t-w+1} - 1. \tag{26}$$

Let p_i, $0 \le p_i \le 2^{n-l-1} - 1$, $i = 0, \cdots, 2^{t-w+1} - 1$, be the positions where $\psi_L^l(\mathbf{S})$ and $\psi_R^l(\mathbf{S})$ differ. This means $w_H(\psi^{l+1}) = 2^{t-w+1}$ with 1s at positions p_i, $i = 0, \cdots, 2^{t-w+1} - 1$. As equation (26) is valid for $v = l + 1$, this implies that the mapping $p_i \mapsto p_i \bmod 2^{t-w+1}$ is one-one and onto since otherwise $w_H(\psi^{r_w-1}(\mathbf{S})) < 2^{t-w+1}$. Hence for each p_i, $i = 0, \cdots, 2^{t-w+1} - 1$, only one of the choices, that is, changing $\psi_L^l(\mathbf{S})_{p_i}$ or $\psi_R^l(\mathbf{S})_{p_i}$ results in the 2^{n-l-1}-vector $\psi^{l+1}(\mathbf{S}')$ that satisfies

$$\sum_{j=0}^{2^{r_w-l-2}-1} \psi^{l+1}(\mathbf{S}')_{i+j2^{t-w+1}} = 0 \quad \text{for} \quad i = 0, \cdots, 2^{t-w+1} - 1. \tag{27}$$

The contribution to $L(\mathbf{S})$ during the first $l - 1$ steps of the algorithm is

$$(2^{n-1} + 2^{n-2} + \cdots + 2^{n-l}) - \sum_{i=1}^{w-1} 2^{n-r_i} = 2^n - 2^{n-l} - \sum_{i=1}^{w-1} 2^{n-r_i}.$$

Thus the 2^t-error linear complexity of $\mathbf{S}$ is of the form

$$L_{2^t}(\mathbf{S}) = 2^n - 2^{n-l} - \sum_{i=1}^{w-1} 2^{n-r_i} + C, \tag{28}$$

where C is the linear complexity of the 2^{n-l-1}-periodic binary sequence with period $\psi^{l+1}(\mathbf{S}')$. By equation (27) and Lemma 4 the value C in equation (28) satisfies

$$C = L((\psi^{l+1}(\mathbf{S}'))^\infty) \leq 2^{n-l-1} - 2^{t-w+1}. \tag{29}$$

Also, $\psi^{l+1}(\mathbf{S}')$ is not the all zero vector from the definition of l in equation (21), which implies $C \geq 1$. Thus from equations (28) and (29) $L_{2^t}(\mathbf{S})$ is in the form $L_{w,l,C}$ given in equation (20).

Case 1b: $r_{j-1} \leq l \leq r_j - 2, 1 \leq j \leq w - 1$
From the definition of l in equation (21) we have $d_H(\psi_L^l(\mathbf{S}), \psi_R^l(\mathbf{S})) = 2^{t-j+1}$ Also, by equation (22) we have $w_H(\psi^{r_j-1}(\mathbf{S})) = 2^{t-j+1}$. Since $j \neq w$ we have $n - r_j > t - j$ and so $\psi^{r_j-1}(\mathbf{S})$ is not an all 1 vector. More specifically if

$$G = \{g : \psi^{r_j-1}(\mathbf{S})_g = 0, g = 0, \cdots, 2^{n-r_j+1} - 1\}$$

then

$$|G| = 2^{n-r_j+1} - 2^{t-j+1}. \tag{30}$$

Using a similar argument as that in Case 1a we have

$$L_{2^t}(\mathbf{S}) = 2^n - 2^{n-l} - \sum_{i=1}^{j-1} 2^{n-r_i} + C, \tag{31}$$

where C is the linear complexity of the 2^{n-l-1}-periodic binary sequence with period $\psi^{l+1}(\mathbf{S}')$, which is equal to either of the equal halves obtained by forcing $\psi_L^l(\mathbf{S}) = \psi_R^l(\mathbf{S})$ such that the lowest possible linear complexity is achieved. The left and right halves of the vectors considered from the l-th step to the (r_j-2)-th step are not equal. So by equation (30) due to the procedure of the Games-Chan algorithm we have

$$\sum_{f=0}^{2^{r_j-l-1}-1} \psi^l(\mathbf{S})_{i+f2^{n-r_j+1}} = 0 \quad \text{for} \quad i \in G \tag{32}$$

and

$$\sum_{f=0}^{2^{r_j-l-1}-1} \psi^l(\mathbf{S})_{i+f2^{n-r_j+1}} = 1 \quad \text{for} \quad i \in \{0, \cdots, 2^{n-r_j+1} - 1\} - G. \tag{33}$$

Let p_i, $0 \leq p_i \leq 2^{n-l-1} - 1$, $i = 0, \cdots, 2^{t-j+1} - 1$, be the positions where $\psi_L^l(\mathbf{S})$ and $\psi_R^l(\mathbf{S})$ differ. This means $w_H(\psi^{l+1}(\mathbf{S})) = 2^{t-j+1}$. By equations (32) and (33), this implies that the mapping $p_i \mapsto p_i \bmod 2^{n-r_j+1}$ is one-one since otherwise $w_H(\psi^{r_j-1}(\mathbf{S})) < 2^{t-j+1}$. We can see the mapping is not onto from equation (30). Also, each element in G does not occur as the inverse image of any element of the set $\{p_i : i = 0, \cdots, 2^{t-j+1}\}$. We split the summation in equation (32) into two separate summations involving terms exclusively from $\psi_L^l(\mathbf{S})$ or $\psi_R^l(\mathbf{S})$. For each $i \in G$ we have

$$\Sigma_L(l, i) = \sum_{f=0}^{2^{r_j-l-2}-1} \psi_L^l(\mathbf{S})_{i+f2^{n-r_j+1}}$$

and   (34)

$$\Sigma_R(l, i) = \sum_{f=0}^{2^{r_j-l-2}-1} \psi_R^l(\mathbf{S})_{i+f2^{n-r_j+1}}.$$

For each $i \in G$, from equations (32) and (34) we know that $\Sigma_L(l, i) + \Sigma_R(l, i) = 0$ which implies $\Sigma_L(l, i) = \Sigma_R(l, i) = 0$ or $\Sigma_L(l, i) = \Sigma_R(l, i) = 1$. Note that none of the terms involved in the summations of equation (32) can be altered when forcing $\psi_L^l(\mathbf{S}) = \psi_R^l(\mathbf{S})$. Using these remarks it can be shown that by making appropriate changes at one of the positions p_i or $p_i + 2^{n-l-1}$, for each $i = 0, \cdots, 2^{t-j+1}$ in $\psi^l(\mathbf{S})$, we can only guarantee that $w_H(\psi^{l+1}(\mathbf{S}'))$ is even by forcing $\psi_L^l(\mathbf{S}) = \psi_R^l(\mathbf{S})$. Thus the value C in equation (31) satisfies $1 \le C \le 2^{n-l-1} - 1$. Hence $L_{2^t}(\mathbf{S})$ is in the form $L_{j,l,C}$, $1 \le j \le w-1$, as in equation (19).

Case 2: $w = t + 1$
The proof in this case is similar to that for Case 1 and both forms in equations (19) and (20) are identical.
    This completes the proof of the theorem.   □

4   Counting Functions

In this section we derive expressions for the number of sequences in $\mathcal{A}(L)$ with fixed $k_{min}(L)$-error linear complexity. We need the following generalization of [1, Lemma 3].

Lemma 6. Let $\mathbf{S} \in \mathcal{A}(L)$ such that $1 \le L \le 2^n - 2^r$, $r = 1, \cdots, n-1$. Let $\mathbf{S}'$ be a 2^n-periodic binary sequence corresponding to the polynomial

$$\mathbf{S}'(x) = \mathbf{S}(x) + \sum_{t=0}^{g} x^{i_t},$$

where $0 \le g \le 2^r - 1$ and $i_t \in \{0, \cdots, 2^n - 1\}$, $t = 0, \cdots, g$. If the mapping $i_t \mapsto i_t \bmod 2^r$ is one-one, then we have $L(\mathbf{S}') > L(\mathbf{S})$.

Theorem 2. *Let $\mathcal{N}_{k_{min}(L)}(\mathcal{C})$ be the number of sequences in $\mathcal{A}(L)$, $1 < L < 2^n$, with fixed $k_{min}(L)$-error linear complexity $\mathcal{C}$. Let L be represented as*

$$L = 2^n - (2^{n-r_1} + 2^{n-r_2} + \cdots + 2^{n-r_t}),$$

where $r_0 = 0 < r_1 < r_2 < \cdots < r_t < n+1 = r_{t+1}$ and $1 \leq t \leq n-1$. Let $L_{j,l,C}$ be defined as in equations (19) and (20) and let $w = \min\{i : r_i = n+i-t, 1 \leq i \leq t+1\}$. Then for $1 \leq j \leq w$, if $1 \leq C \leq 2^{n-l-1} - 2^{n-r_j+1}$ then

$$\mathcal{N}_{k_{min}(L)=2^t}(L_{j,l,C}) = 2^{\rho(j,l,C)},$$

where

$$\rho(j,l,C) = 2^n - 2^{n-l} - \sum_{i=1}^{j-1} 2^{n-r_i} + \sum_{i=0}^{w-j-1}(r_{w-i} - r_{w-i-1} - 1)2^{t-w+i+1} \tag{35}$$
$$+ (r_j - l - 1)2^{t-j+1} + C - 1.$$

Also, $\mathcal{N}_{k_{min}(L)=2^t}(0) = 2^{\rho(0)}$, where $\rho(0) = \sum_{i=0}^{w-2}(r_{w-i} - r_{w-i-1} - 1)2^{t-w+i+1} + (r_1 - 1)2^t$ and $\mathcal{N}_{k_{min}(L)=2^t}(\mathcal{C}) = 0$ for all C not in the form $L_{j,l,C}$ as in equations (19) and (20).

Proof. From equations (19) and (20) the $k_{min}(L)$-error linear complexity of $\mathbf{S} \in \mathcal{A}(L)$ is of the form

$$L_{j,l,C} = 2^n - \sum_{i=1}^{j-1} 2^{n-r_i} - 2^{n-l} + C \quad \text{for} \quad 1 \leq j \leq w \tag{36}$$

where $r_{j-1} \leq l \leq r_j - 2$ (For $l = r_w - 2$, there exist no positive values for C in equation (20) and hence no valid values for $L_{w,l,C}$). We determine the counting function for the number of sequences in $\mathcal{A}(L)$ with $k_{min}(L)$-error linear complexity equal to each of the values $L_{j,l,C}$ in equation (36) when $1 \leq C \leq 2^{n-l-1} - 2^{n-r_j+1}$. From the definition of l in equation (21) and by equation (22), for any $\mathbf{S} \in \mathcal{A}(L)$ if $r_{j-1} \leq l \leq r_j - 2$ we know

$$w_H(\psi^{l+1}(\mathbf{S})) = w_H(\psi^{r_j-1}(\mathbf{S})) = 2^{t-j+1}. \tag{37}$$

We consider two cases based on the value of w.

Case 1: $w \leq t$
From equation (24) for any $\mathbf{S} \in \mathcal{A}(L)$ the 2^{t-w+1}-vector $\psi^{r_w-1}(\mathbf{S})$ is an all 1 vector.

Let $\mathcal{D}^1(j,l,C)$ be the number of distinct 2^{n-l-1}-vectors $\psi^{l+1}(\mathbf{S})$ over all $\mathbf{S} \in \mathcal{A}(L)$ such that the 2^{n-r_w+1}-vector $\psi^{r_w-1}(\mathbf{S})$ is an all 1 vector. To determine $\mathcal{D}^1(j,l,C)$ we make the following observations.

(i) By equation (22) it is evident that during the execution of Games-Chan algorithm form the $l+1$-th step to the $(n-1)$-th step the Hamming weight of the vectors considered does not change between two consecutive steps except when going from the $(r_i - 1)$-th step to the r_i-th step for $i = j, \cdots, t$.

(ii) Using (i) the procedure of the Games-Chan algorithm also implies that over all sequences in $\mathcal{A}(L)$ for any integer a such that $l + 1 \leq a < r_j$ or $r_i \leq a < r_{i-1}$ for some $i \in \{j, \cdots, t\}$, the number of distinct vectors in the a-th step that result in a fixed vector $\mathbf{v}$ in the $(a + 1)$-th step is $2^{w_H(\mathbf{v})}$.

(iii) The definition of w implies $n - r_w = t - w$.

From these observations and by using property P1 recursively we obtain

$$\mathcal{D}^1(j, l, C) = \prod_{i=0}^{w-j-1} (2^{r_w - i - r_{w-i-1} - 1})^{2^{t-w+i+1}} (2^{r_j - l - 2})^{2^{t-j+1}}. \tag{38}$$

Recall that $\psi^{l+1}(\mathbf{S}')$ is the 2^{n-l-1}-vector obtained by forcing $\psi_L^l(\mathbf{S}) = \psi_R^l(\mathbf{S})$ so that the least linear complexity is achieved by making $k_{min}(L)$ errors in $\mathbf{S}^{(2^n)}$. Let $\mathcal{D}^2(j, l, C)$, $1 \leq C \leq 2^{n-l-1} - 2^{n-r_j+1}$, be the number of choices for $\psi^{l+1}(\mathbf{S}')$ such that the linear complexity of the 2^{n-l-1}-periodic sequence with period $\psi^{l+1}(\mathbf{S}')$ is C. By equation (5), we have

$$\mathcal{D}^2(j, l, C) = 2^{C-1} \quad \text{for} \quad 1 \leq C \leq 2^{n-l-1} - 2^{n-r_j+1}. \tag{39}$$

Over all $\mathbf{S} \in \mathcal{A}(L)$, for a fixed $\psi^{l+1}(\mathbf{S}) = \mathbf{v}$ with $w_H(\mathbf{v}) = 2^{n-r_w+1}$ and for a fixed choice of $\psi^{l+1}(\mathbf{S}')$ with $L((\psi^{l+1}(\mathbf{S}'))^\infty) = C$, the number of possibilities, denoted by $\mathcal{D}^3(w, l, C)$, for $\psi^l(\mathbf{S})$ such that $\psi_L^l(\mathbf{S}) + \psi_R^l(\mathbf{S}) = \mathbf{v}$ and $d_H(\psi^l(\mathbf{S}), \psi^{l+1}(\mathbf{S}') \mid \psi^{l+1}(\mathbf{S}')) = 2^{n-r_w+1}$ is

$$\mathcal{D}^3(w, l, C) = 2^{2^{t-j+1}}, \tag{40}$$

where $\psi^{l+1}(\mathbf{S}') \mid \psi^{l+1}(\mathbf{S}')$ is the 2^{n-l}-vector formed by concatenating two copies of $\psi^{l+1}(\mathbf{S}')$.

Let p_i, $0 \leq p_i \leq 2^{n-l-1} - 1$, $i = 0, \cdots, 2^{t-j+1} - 1$, be the positions where $\psi_L^l(\mathbf{S})$ and $\psi_R^l(\mathbf{S})$ differ. From Cases 1a and 1b of the proof of Theorem 1 the mapping $p_i \mapsto p_i \bmod 2^{n-r_j+1}$ is one-one. Using this mapping and the condition $1 \leq C \leq 2^{n-l-1} - 2^{n-r_j+1}$, by Lemma 6 for fixed $\psi^{l+1}(\mathbf{S})$ and $\psi^{l+1}(\mathbf{S}')$ each of the $2^{2^{t-j+1}}$ possibilities for $\psi^l(\mathbf{S})$ satisfies

$$L(\psi_L^l(\mathbf{S})) > C \quad \text{and} \quad L(\psi_R^l(\mathbf{S})) > C. \tag{41}$$

By equations (38)-(41), using properties P3 and P4 recursively we obtain

$$\mathcal{N}_{2^t}(L_{j,l,C}) = \mathcal{P}_0 \mathcal{P}_1 \cdots \mathcal{P}_{l-1} \mathcal{D}^1(j, l, C) \mathcal{D}^2(j, l, C) \mathcal{D}^3(j, l, C). \tag{42}$$

We have

$$\mathcal{P}_0 \mathcal{P}_1 \cdots \mathcal{P}_{l-1} = \prod_{i=1}^{j-1} (\mathcal{P}_{r_i-1} \cdots \mathcal{P}_{r_i-2})(\mathcal{P}_{r_j-1} \cdots \mathcal{P}_{l-1})$$

$$= \left(\prod_{i=1}^{j-1} 2^{\sum_{z=1}^{r_i - r_{i-1} - 1} 2^{n-r_i+z}} \right) 2^{\sum_{z=0}^{l-r_j-1} 2^{n-l+z}}. \tag{43}$$

By equations (38)-(41) and (43) a straightforward algebraic simplification of the right hand side of equation (42) gives $\mathcal{N}_{2^t}(L_{j,l,C}) = 2^{\rho(j,l,C)}$ with $\rho(j,l,C)$ as in equation (35). We note that the condition in equation (41) is necessary to avoid double counting in determining the number of distinct possibilities for $\psi^l(\mathbf{S})$ over all $\mathbf{S} \in \mathcal{A}(L)$ such that $\psi^{l+1}(\mathbf{S})$ and $\psi^{l+1}(\mathbf{S}')$ are fixed.

Case 2: $w = t + 1$

In this case we note that the two possibilities for vectors in the $(n-1)$-th step of the Games-Chan algorithm are 01 and 10. Using this it can be shown that the expression for $\mathcal{D}^1(j,l,C)$ in equation (38) holds for $w = t + 1$. The remaining details are similar to those in Case 1.

To obtain $\mathcal{N}_{2^t}(0)$ we only have to count the number of $\mathbf{S} \in \mathcal{A}(L)$ with $w_H(\mathbf{S}) = 2^t$. By equation (22) and property P1 the expression for $\mathcal{N}_{2^t}(0)$ follows using an argument similar to that for finding $\mathcal{D}^1(j,l,C)$ as in equation (38).

This completes the proof of the theorem. $\square$

5   Conclusion

In this paper we studied the k-error linear complexity of 2^n-periodic binary sequences by performing a rigorous analysis of the Games-Chan algorithm. We derived an expression for all the possible values of k-error linear complexities of 2^n-periodic binary sequences with fixed linear complexity when k is the minimum number of changes needed to the lower the linear complexity. For certain sets of these values, we obtained the corresponding number of sequences with fixed linear complexity and k-error linear complexity. Our results further research in analyzing the stability of linear complexity of 2^n-periodic binary sequences. These results, however, have limited importance for practical cryptography in part due to the restriction to 2^n-periodic sequences.

Acknowledgements

The first author thanks Dr. Zongming Fei for providing office space and resources while researching for this paper. The authors also thank anonymous reviewers for their helpful suggestions.

References

1. Fengxiang, Z., Wenfeng, Q.: The 2-error linear complexity of 2^n-periodic binary sequences with linear complexity $2^n - 1$. Journal of Electronics (China) 24(3), 390–395 (2007)
2. Games, R.A., Chan, A.H.: A fast algorithm for determining the complexity of a pseudo-random sequence with period 2^n. IEEE Trans. Inform. Theory 29(1), 144–146 (1983)
3. Kurosawa, K., Sato, F., Sakata, T., Kishimoto, W.: A relationship between linear complexity and k-error linear complexity. IEEE Trans. Inform. Theory 46(2), 694–698 (2000)

4. Meidl, W.: On the stability of 2^n-periodic binary sequences. IEEE Trans. Inform. Theory 51(3), 1151–1155 (2005)
5. Meidl, W., Venkateswarlu, A.: Remarks on the k-error linear complexity of p^n-periodic sequences. Design, Codes and Cryptography 42(2), 181–193 (2007)
6. Rueppel, R.A.: Analysis and Design of Stream Ciphers. Springer, Heidelberg (1986)
7. Stamp, M., Martin, C.F.: An algorithm for the k-error linear complexity of binary sequences with period 2^n. IEEE Trans. Inform. Theory 39(4), 1398–1401 (1993)

On the Exact Success Rate of Side Channel Analysis in the Gaussian Model

Matthieu Rivain

Oberthur Technologies & University of Luxembourg
m.rivain@oberthurcs.com

Abstract. Nowadays, Side Channel Analysis is one of the most power-ful cryptanalytic technique against cryptosystems embedded in portable devices such as smart cards. Faced with this threat, it is of crucial impor-tance to precisely determine what is achievable by a given side channel adversary against a cryptosystem producing a given side channel leak-age. This can be answered by evaluating the success rate of an attack according to the adversary capacities and to the leakage properties.

In this paper, we investigate the issue of evaluating the success rate of side channel analysis in the widely admitted Gaussian leakage model. We introduce a new approach that allows us to efficiently compute the success rate of an attack in this model and we apply it to the two main families of side channel analysis: differential side channel analysis and profiling side channel analysis.

1 Introduction

Side Channel Analysis (SCA) is a cryptanalytic technique that consists in an-alyzing the physical leakage produced during the execution of a cryptographic algorithm embedded on a physical device (*e.g.* execution time [13], power con-sumption [12], electromagnetic emanations [8]). Some kinds of SCA exploit this *side channel leakage* to recover information on the operation flow that may de-pend on the secret key (*e.g.* Simple SCA [12], Timing Attacks [13]). These can be circumvent by ensuring that the operation flow is independent of the secret key. Other kinds of SCA exploit the fact that the side channel leakage is sta-tistically dependent on the intermediate variables of the computation. Some of these variables are themselves related to small parts of the secret key which enables key recovery attacks. These second kinds of SCA are particularly power-ful and securing cryptographic implementation against them constitutes a real challenge.

SCA targeting intermediate variables divides into two main categories: *differ-ential SCA* and *profiling SCA*. Differential SCA relies on correlation techniques [12,4]. Based on several leakage observations, the attacker estimates a correlation between the leakage and different predictions on the value of a key-dependent intermediate variable. According to the obtained correlation values, the attacker is able to (in)validate some hypotheses on the secret key. Profiling SCA [6,19] is based on the maximum likelihood approach. It assumes that the attacker owns a

R. Avanzi, L. Keliher, and F. Sica (Eds.): SAC 2008, LNCS 5381, pp. 165–183, 2009.
© Springer-Verlag Berlin Heidelberg 2009

profile of the leakage according to the values of some key-dependent intermediate variables. This profile is involved to derive the likelihood of some key hypotheses given the observed leakage.

Faced with the threat of side channel analysis, a crucial issue is to quantify the efficiency of the different attacks according to the adversary capacities and the leakage statistical properties. For this purpose, a natural metric is the success rate, namely the probability that an attack succeeds in recovering the correct key (or in isolating it in a restricted set). A straightforward way to evaluate the success rate is to estimate it empirically by performing the attack several times. However such an approach is costly in time and may even become impossible for attacks with medium or high complexity. It is therefore not suitable to efficiently and precisely determine the resistance of an embedded device if this one is not quite weak. To tackle this issue, it is of particular interest to investigate efficient ways to compute (or at least to precisely estimate) the success rate of an attack without requiring to perform it many times.

Previous investigations have been done regarding this issue [7,16,22]. These works investigate differential SCA in a noisy context. They provide an approximation of the required number of leakage measurements for a successful attack [7,16] and an approximation of the success rate [22]. For the sake of generality, these works do not take into account the relationship between the different key candidates (which depends on the target algorithm logical properties and on the leakage statistical properties) and only focus on the good key guess. However, the success rate depends on the joint behavior of the different candidates and this relationship cannot be neglected while looking for a precise estimation of the success rate. Concerning profiling SCA, to the best of our knowledge no solution for the success rate evaluation has been proposed in the literature so far. This is a lack since these attacks are considered as the strongest form of side channel analysis.

In this paper, we address the issue of evaluating the success rate of a side channel key recovery attack. We analyze both differential SCA and profiling SCA under the widely admitted assumption that the noise in the leakage has a Gaussian distribution. We show that the result of these attacks can be expressed as a multivariate Gaussian random variable which leads to an efficient way for determining their success rates.

The rest of the paper is organized as follows. Section 2 introduces some preliminaries. Section 3 presents the side channel theoretical model considered in this paper. In Sections 4 and 5, we respectively analyze differential SCA and profiling SCA. Based on these analyses, Section 6 shows how to efficiently evaluate the success rate of the focused attacks. Finally an empirical validation is provided in Section 7 and concluding remarks are given in Section 8.

2   Preliminaries

The calligraphic letters, like $\mathcal{X}$, are used to denote finite sets (*e.g.* $\mathbb{F}_2^n$). The corresponding large letter X denotes a random variable over $\mathcal{X}$, while the lowercase letter x denotes a particular realization of X. The probability of an event

ev is denoted by P $[ev]$. In case X has a continuous distribution, the notation $x \mapsto P[X = x]$ is further used to denote the probability density function (pdf) of X. The expectation and the variance of a random variable X are respectively denoted by E $[X]$ and Var $[X]$. The covariance between two random variables X and Y is denoted by Cov $[X, Y]$. The Gaussian distribution of dimension T with T-size expectation vector m and $T \times T$ covariance matrix Σ is denoted by $\mathcal{N}(m, \Sigma)$, and the corresponding pdf is denoted by $\phi_{\Sigma, m}$. We recall that this pdf is defined for every $x \in \mathbb{R}^T$ as:

$$\phi_{\Sigma, m}(x) = \frac{1}{\sqrt{(2\pi)^T |\Sigma|}} \exp\left(-\frac{1}{2}(x - m)' \Sigma^{-1} (x - m)\right),$$

where $(x - m)'$ denotes the transpose of the vector $(x - m)$ and $|\Sigma|$ denotes the determinant of the matrix Σ. If the dimension T equals 1, then the Gaussian distribution is said to be *univariate* and the single element of the covariance matrix is the variance that is denoted by σ^2. If $T > 1$, the Gaussian distribution is said to be *multivariate*.

3   Side Channel Model

A formal modeling of side channel key recovery attacks has been initiated by Standaert *et al.* in [21]. The theoretical model introduced hereafter follows the outlines of their work.

3.1   Side Channel Key Recovery Attacks

Let $\mathsf{E_{SK}}$ be a cryptographic algorithm E parameterized by a secret key SK. Let K be a random variable representing a guessable part of SK. Let X be a random variable representing a part of a public value such as an input (resp. output) of $\mathsf{E_{SK}}$. Let S be a random variable representing the result of an intermediate computation of $\mathsf{E_{SK}}$ that satisfies $S = \varphi(X, K)$ for a given function $\varphi : \mathcal{X} \times \mathcal{K} \to \mathcal{S}$. We denote by L the random variable that represents the side channel leakage generated by the computation (and/or the handling) of S on a physical implementation of $\mathsf{E_{SK}}$. We shall further denote by L (s) the random variable $(\mathsf{L}|S = s)$.

A side channel key recovery attack targeting the signal S aims at recovering the value k^* taken by K on a given physical implementation of $\mathsf{E_{SK}}$. For such a purpose, the attacker collects several, say N, leakage measurements $(l_i)_i$ resulting from the computation of $\varphi(x_i, k^*)$ for N inputs $(x_i)_i$. Namely, the l_i's are realizations of the random variables L $(\varphi(K, x_i))$ that are assumed to be mutually independent. Then, the attack makes use of a *distinguisher*, that is a function D which, from the leakage measurements vector $\mathbf{l} = (l_1, \cdots, l_N)$ and the corresponding inputs vector $\mathbf{x} = (x_1, \cdots, x_N)$, outputs a *distinguishing vector* $\mathbf{d} = (d_k)_{k \in \mathcal{K}}$. If the distinguisher is sound and if the leakage brings enough information on S, then $k^* = \mathrm{argmax}_{k \in \mathcal{K}} \, d_k$ holds with a non-negligible probability.

Finally, a side channel adversary can be defined as the composition of a distinguisher with a strategy to select the algorithm inputs *i.e.* the x_i values. These can be randomly drawn (in a known plaintext/ciphertext attack setting) or they can be chosen by the adversary (in a chosen plaintext attack setting). In this paper, we do not assume a specific strategy. Rather, we investigate the success rate of an attack according to the inputs vector $\mathbf{x}$.

3.2   Gaussian Leakage Model

In practice, the leakage measurements are composed of several samples, say T, corresponding to several successive instants in time. The leakage L can hence be modeled by a T-size random vector. In the *Gaussian leakage model*, the leakage $L(s)$ resulting from the computation of any signal $s \in \mathcal{S}$ has a Gaussian distribution: $L(s) \sim \mathcal{N}(m_s, \Sigma_s)$.

Remark 1. The Gaussian model assumption is both very usual in the side channel literature (see for instance [6,15,19,21]) and fairly realistic in practice (see for instance [15, §4]).

For clarity and without ambiguity, we shall respectively denote by m_{x,k^*} and Σ_{x,k^*} the mean vector $m_{\varphi(x,k^*)}$ and the covariance matrix $\Sigma_{\varphi(x,k^*)}$.

3.3   Success Rate

The success rate is a classical metric in side channel analysis. Usually, a key recovery attack is considered successful if the distinguishing vector satisfies $k^* = \operatorname{argmax}_{k \in \mathcal{K}} d_k$. In [21], the authors propose to extend the notion of success rate to different orders. The oth order success rate of a side channel attack using a distinguisher D and a public vector $\mathbf{x}$, and targeting a secret key k^* is defined as:

$$\text{Succ-}o^{\mathsf{D}}_{\mathbf{x},k^*} = \mathrm{P}\left[\left(l_i \leftarrow \mathrm{L}\left(\varphi(k^*, x_i) \right) \right)_i \; ; \; \mathbf{d} \leftarrow \mathsf{D}(\mathbf{x}, \mathbf{l}) \; : \; k^* \in \operatorname*{argmax-}_{k \in \mathcal{K}} o \; d_k \right] ,$$

where $\operatorname{argmax-}o_{k \in \mathcal{K}} d_k$ denotes the set of the o elements $k \in \mathcal{K}$ that maximize d_k. The notion of order is motivated by the fact that an attacker may perform an off-line exhaustive search after the side channel analysis. A oth order success means that the attacker has at the most o key guesses to test after the attack in order to recover the correct one.

Remark 2. In [21], the authors also suggest to use another metric: the so-called *guessing entropy* [17,5]. This one is defined as the expected rank of the good key guess in the distinguishing vector, namely it indicates the average number of key guesses to test after the side channel analysis. This notion is discussed in Appendix A where we show that it can be expressed with respect to the success rates at the different orders.

Our Approach. In order to determine the exact success rate of an attack, we must investigate the multivariate probability distribution of the distinguishing vector. This distribution can be expressed with respect to the inputs vector $\mathbf{x}$, the secret key k^* and the leakage distribution parameters $(m_s, \Sigma_s)_{s \in \mathcal{S}}$. In the rest of the paper, we will investigate the two main families of side channel analysis: differential SCA and profiling SCA. We will show that under the Gaussian assumption, the multivariate distribution of the distinguishing vector is (or at least can be precisely approximated by) a multivariate Gaussian distribution. This will enable us to show how the success rate of such attacks can be efficiently computed.

4 Differential Side Channel Analysis

4.1 Description

Differential side channel analysis uses correlation techniques as distinguisher. Several variants have been proposed in the literature [1,4,3,12,18]. In this paper, we focus on the Pearson correlation coefficient since it is the most widely used and seems to be the most efficient technique in practice [4]. Note that our analysis could be easily extended to other differential distinguishers that rely on correlation computations [1,3,12,18]. The adversary is assumed to own a model of the side channel leakage that is a function $M : \mathcal{X} \times \mathcal{K} \to \mathbb{R}$ such that $M(x, k)$ is linearly related to the expectation of the leakage $L(\varphi(x, k))$. The attack consists in estimating, for every key guess $k \in \mathcal{K}$, the linear correlation between the *prediction* $M(X, k)$ (*i.e.* the predicted value of the leakage for the guess k) and the observable leakage $L(\varphi(X, k^*))$. This correlation is estimated based on the prediction vector $(M(x_1, k), \cdots, M(x_N, k))$ and the leakage measurements vector $\mathbf{l}$ by the following coefficient:

$$\rho_k = \frac{\frac{1}{N} \sum_i \left(M(x_i, k) - \frac{1}{N} \sum_j M(x_j, k) \right) \left(l_i - \frac{1}{N} \sum_j l_j \right)}{\sqrt{\frac{1}{N} \sum_i \left(M(x_i, k) - \frac{1}{N} \sum_j M(x_j, k) \right)^2} \sqrt{\frac{1}{N} \sum_i \left(l_i - \frac{1}{N} \sum_j l_j \right)^2}} . \qquad (1)$$

If the model is sound, the prediction vector for the correct key guess is significantly correlated to the leakage measurements vector. As a result, for N large enough, ρ_k is expected to be maximal for $k = k^*$.

Since the correlation distinguisher takes as input a set of 1-size leakage measurements, we investigate hereafter the distribution of this distinguisher in the univariate Gaussian model.

4.2 Distinguisher Distribution

Let us first denote by τ_x the occurrence ratio of an element $x \in \mathcal{X}$ through the inputs vector $\mathbf{x}$, *i.e.* :

$$\tau_x = \frac{|\{i; x_i = x\}|}{N} . \qquad (2)$$

We shall further denote by $\overline{\mathsf{M}}_k$ and $\widehat{\sigma}_k$ the mean and the standard deviation of the prediction vector $\big(\mathsf{M}(x_i, k)\big)_i$, namely:

$$\overline{\mathsf{M}}_k = \sum_{x \in \mathcal{X}} \tau_x \mathsf{M}(x, k) \quad \text{and} \quad \widehat{\sigma}_k^2 = \sum_{x \in \mathcal{X}} \tau_x \big(\mathsf{M}(x, k) - \overline{\mathsf{M}}_k\big)^2 \,.$$

Instead of focusing on ρ_k, we focus in the sequel on the following coefficient:

$$\dot{\rho}_k = \frac{1}{\widehat{\sigma}_k N} \sum_{i=1}^{N} \big(\mathsf{M}(x_i, k) - \overline{\mathsf{M}}_k\big) \mathsf{l}_i \,. \tag{3}$$

The distribution of $(\dot{\rho}_k)_{k \in \mathcal{K}}$ is indeed more convenient to analyze than the one of $(\rho_k)_{k \in \mathcal{K}}$. Moreover, one can verify that the ratio $\dot{\rho}_k / \rho_k$ equals the standard deviation of the leakage measurement vector l. Consequently, $\dot{\rho}_k / \rho_k$ is positive and constant with respect to the key guess k. As a result, argmax-$o_{k \in \mathcal{K}}$ ρ_k = argmax-$o_{k \in \mathcal{K}}$ $\dot{\rho}_k$ holds for every k and hence, the success rate of the attack is fully determined by the distribution of the vector $(\dot{\rho}_k)_{k \in \mathcal{K}}$. The next proposition provides us with the exact distribution of this vector.

Proposition 1. *The vector* $(\dot{\rho}_k)_{k \in \mathcal{K}}$ *has a multivariate Gaussian distribution whose expectation satisfies for every* $k \in \mathcal{K}$:

$$\mathrm{E}\,[\dot{\rho}_k] = \frac{1}{\widehat{\sigma}_k} \sum_{x \in \mathcal{X}} \tau_x \big(\mathsf{M}(x, k) - \overline{\mathsf{M}}_k\big) m_{x, k^*} \,, \tag{4}$$

and whose covariance satisfies for every $(k_1, k_2) \in \mathcal{K}^2$:

$$\mathrm{Cov}\,[\dot{\rho}_{k_1}, \dot{\rho}_{k_2}] = \frac{1}{N \widehat{\sigma}_{k_1} \widehat{\sigma}_{k_2}} \sum_{x \in \mathcal{X}} \tau_x \big(\mathsf{M}(x, k_1) - \overline{\mathsf{M}}_{k_1}\big) \big(\mathsf{M}(x, k_2) - \overline{\mathsf{M}}_{k_2}\big) \sigma_{x, k^*}^2 \,. \tag{5}$$

Proof. Since the l_i's are drawn from Gaussian distributions $\mathcal{N}\,(m_{x_i, k^*}, \sigma_{x_i, k^*})$ and since the vector $(\dot{\rho}_k)_{k \in \mathcal{K}}$ is a linear transformation of l, one deduces that $(\dot{\rho}_k)_{k \in \mathcal{K}}$ has a multivariate Gaussian distribution.

Now, for every $x \in \mathcal{X}$, we have $N \tau_x$ elements among the x_i's that are equal to x. This, together with (3) immediately leads to (4). Then, the mutual independence of the l_i's and the bilinearity of the covariance imply (5). ◇

Proposition 1 gives the exact distribution of the distinguishing vector $(\dot{\rho}_k)_{k \in \mathcal{K}}$. This makes it possible to precisely compute the success rate of a differential SCA that involves the Pearson correlation coefficient (see Section 6).

If the model is sound, namely if $\mathsf{M}(x, k)$ is linearly related to $m_{x,k}$, then (4) implies that the expectation of $\dot{\rho}_k$ is maximal for the good key guess $k = k^*$ which shows the soundness of the attack.

From (4) we see that the distinguishing vector expectation does not depend on the leakage variance nor on the number of leakage measurements. Conversely, (5) shows that the covariance matrix depends on these parameters. If the leakage variance is multiplied by a factor λ then so does the covariance matrix. And if the number of measurements is multiplied by a factor λ then the covariance matrix is multiplied by $1/\lambda$. As a result, if the leakage variance is increased by a given factor, the number of leakage measurements must also be increased by the same factor to keep unchanged the distinguisher distribution and hence the attack success rate.

Another interesting observation is that the distribution of $(\dot{\rho}_k)_{k\in\mathcal{K}}$ does not fully depend on the inputs vector $\mathbf{x}$ but only on the different ratios τ_x's. A usual choice, for a chosen plaintext differential SCA, is to set these ratios at $\tau_x = 1/|\mathcal{X}|$. For a known plaintext/ciphertext differential SCA, assuming that the x_i's are uniformly drawn, we further have $\tau_x \approx 1/|\mathcal{X}|$ for N large enough. We investigate this setting hereafter.

Uniform Setting. We investigate here the setting where the x_i's are chosen such that $\tau_x = 1/|\mathcal{X}|$ holds for every x. We further assume that the target signal S can be expressed as $S = \psi(X \oplus K)$ where ψ is a balanced function (*i.e.* the cardinal of $\psi^{-1}(s)$ is constant for every $s \in \mathcal{S}$).

In the uniform setting, the previous study can be simplified. In this setting, the mean and the standard deviation of the prediction vector are constant with respect to k^*. Indeed, for every $k \in \mathcal{K}$, we have $\overline{M}_k = \frac{1}{|\mathcal{S}|}\sum_{s\in\mathcal{S}} M(s)$ and $\widehat{\sigma}_k = \sqrt{\frac{1}{|\mathcal{S}|}\sum_{s\in\mathcal{S}}\left(M(s) - \overline{M}\right)^2}$. Hence, we can focus on the following coefficient:

$$\ddot{\rho}_k = \frac{1}{N}\sum_{i=1}^{N} M(x_i, k)l_i . \tag{6}$$

Once again $\ddot{\rho}_k/\rho_k$ is positive and constant with respect to k which implies that focusing on ρ_k instead of $\ddot{\rho}_k$ does not affect the success rate of the attack. The following corollary gives the distribution of $(\ddot{\rho}_k)_{k\in\mathcal{K}}$.

Corollary 1. *The vector $(\ddot{\rho}_k)_{k\in\mathcal{K}}$ has a multivariate Gaussian distribution whose expectation satisfies for every $k \in \mathcal{K}$:*

$$\mathrm{E}\left[\ddot{\rho}_k\right] = \frac{1}{|\mathcal{X}|}\sum_{x\in\mathcal{X}} M(x, k)m_{x,k^*} , \tag{7}$$

and whose covariance satisfies for every $(k_1, k_2) \in \mathcal{K}^2$:

$$\mathrm{Cov}\left[\ddot{\rho}_{k_1}, \ddot{\rho}_{k_2}\right] = \frac{1}{N|\mathcal{X}|}\sum_{x\in\mathcal{X}} M(x, k_1)M(x, k_2)\sigma^2_{x,k^*} . \tag{8}$$

Proof. Corollary 1 straightforwardly holds from Proposition 1 by setting $\overline{M}_k$ to 0 and $\widehat{\sigma}_k$ to 1. ◇

An interesting property of the uniform setting is stressed in the following proposition.

Proposition 2. *Let $(d_k)_{k\in\mathcal{K}}$ and $(d'_k)_{k\in\mathcal{K}}$ be the distributions of the vector $(\ddot{\rho}_k)_{k\in\mathcal{K}}$ for two secret keys $k_1^* \in \mathcal{K}$ and $k_2^* \in \mathcal{K}$ respectively. In the uniform setting, the distributions $(d_{k\oplus k_1^*})_{k\in\mathcal{K}}$ and $(d'_{k\oplus k_2^*})_{k\in\mathcal{K}}$ are indentical.*

Proof. In the uniform setting, we have $M(x, k) = M(\psi(x \oplus k))$ and $m_{x,k^*} = m_{\psi(x \oplus k^*)}$. Hence, from (7) we get $E\left[d_{k \oplus k_1^*}\right] = E[d'_{k \oplus k_2^*}]$ for every $k \in \mathcal{K}$ and from (8) we get $\text{Cov}\left[d_{k_1 \oplus k_1^*}, d_{k_2 \oplus k_1^*}\right] = \text{Cov}[d'_{k_1 \oplus k_2^*}, d'_{k_2 \oplus k_2^*}]$ for every $(k_1, k_2) \in \mathcal{K}^2$. Finally, since $(d_k)_{k \in \mathcal{K}}$ and $(d'_k)_{k \in \mathcal{K}}$ are both Gaussian then they are identical. $\diamond$

Proposition 2 shows that the vector $(\ddot{p}_{k \oplus k^*})_{k \in \mathcal{K}}$ has the same distribution for every k^*. Moreover, the event $k^* \in \text{argmax-}o_{k \in \mathcal{K}} \; \ddot{p}_k$ can be rewritten as $0 \in \text{argmax-}o_{k \in \mathcal{K}} \; \ddot{p}_{k \oplus k^*}$. Since the distribution of $(\ddot{p}_{k \oplus k^*})_{k \in \mathcal{K}}$ is independent of k^*, we get that, in the uniform setting, the success rate is constant with respect to k^*. Therefore, one only needs to analyze the distribution of $(\ddot{p}_{k \oplus k^*})_{k \in \mathcal{K}}$ for a given secret key (*e.g.* for $k^* = 0$) to get the distribution and the success rate of $(\ddot{p}_k)_{k \in \mathcal{K}}$ for any secret key k^*.

5   Profiling Side Channel Analysis

5.1   Description

Profiling Side Channel Analysis assumes an adversary that owns a *profile* of the side channel leakage (also called *template* in the literature from the initial work of Chari *et al.* [6]). More precisely, the adversary owns an estimation of the pdf $1 \mapsto P[L = 1|S = s]$ for every $s \in \mathcal{S}$. In practice, this estimation is obtained through a *profiling phase* on a physical implementation identical to the targeted one (except the secret key) and that is under the attacker control (see for instance [2,6,14,19]).

The attack consists in estimating the likelihood of a key guess k, *i.e.* the probability that K is equal to k, given the leakage measurements vector 1 and the inputs vector $\mathbf{x}$. Assuming that K is uniformly distributed (which is very usual in cryptography), it can be checked that this probability satisfies:

$$P[K = k|(1, \mathbf{x})] = \alpha \prod_{i=1}^{N} P[L = l_i|S = \varphi(x_i, k)] , \qquad (9)$$

where α denotes a value constant with respect to k.

In the Gaussian model, the leakage pdf $1 \mapsto P[L = 1|S = s]$ is the Gaussian pdf ϕ_{Σ_s, m_s}. Estimating such a pdf amounts to estimate the parameters (m_s, Σ_s) for every $s \in \mathcal{S}$. In the sequel, we shall denote these estimations by $\widehat{m}_s$ and $\widehat{\Sigma}_s$. For clarity and without ambiguity, the parameters $\widehat{m}_{\varphi(x,k)}$ and $\widehat{\Sigma}_{\varphi(x,k)}$ are further denoted by $\widehat{m}_{x,k}$ and by $\widehat{\Sigma}_{x,k}$.

For computational reasons, one usually processes the logarithm of the estimated likelihood and averages it on the number of leakage measurements. Moreover, since α is constant with respect to k, it is usually ignored. On the whole, one computes the *log-likelihood* $\mathcal{L}_k = \frac{1}{N} \log(P[K = k|(1, \mathbf{x})]/\alpha)$. In the Gaussian model, $\mathcal{L}_k$ satisfies:

$$\mathcal{L}_k = \frac{1}{2N} \sum_{i=1}^{N} \left(\log\left((2\pi)^T |\widehat{\Sigma}_{x_i,k}|\right) - (1_i - \widehat{m}_{x_i,k})' \, \widehat{\Sigma}_{x_i,k}^{-1} (1_i - \widehat{m}_{x_i,k}) \right) . \qquad (10)$$

In the next section, we investigate the distribution of the log-lekelihood distinguisher under the Gaussian model assumption.

5.2   Distinguisher Distribution

Let us first introduce few notations. The element of the ith row and of the jth column of a matrix A is denoted by $A[i,j]$ while the ith element of a vector V is denoted by $V[i]$. A' denotes the transpose of a matrix (or a vector) A. The notation $\|\cdot\|$ is used to denote the Euclidian norm while the notation $\|\cdot\|_{hs}$ refers to the Hilbert-Schmidt matrix norm defined by $\|A\|_{hs} = \sqrt{\sum_{i,j} A[i,j]^2}$. We shall further denote by A^2 the product $A'A$ and by $A^{-1/2}$ any matrix satisfying $(A^{-1/2})'A^{-1/2} = A$ (*e.g.* the Cholesky decomposition matrix). Finally the trace of A is denoted by $\text{Tr}(A)$.

The next proposition provides a precise approximation of the distribution of the likelihood vector $(\mathcal{L}_k)_{k\in\mathcal{K}}$ (the proof is given in Appendix B).

Proposition 3. *The distribution of the vector $(\mathcal{L}_k)_{k\in\mathcal{K}}$ tends toward a multivariate Gaussian distribution as N grows. Moreover, for every $k \in \mathcal{K}$, the expectation of $\mathcal{L}_k$ satisfies:*

$$
\mathrm{E}\left[\mathcal{L}_k\right] = \frac{1}{2} \sum_{x\in\mathcal{X}} \tau_x \left(\log\left(2\pi|\widehat{\Sigma}_{x,k}|\right) - \left\|\widehat{\Sigma}_{x,k}^{-1/2}\left(m_{x,k^*} - \widehat{m}_{x,k}\right)\right\|^2 \right.
$$
$$
\left. - \text{Tr}\left(\widehat{\Sigma}_{x,k}^{-1/2}\,\Sigma_{x,k^*}\,(\widehat{\Sigma}_{x,k}^{-1/2})'\right)\right), \quad (11)
$$

and for every $(k_1, k_2) \in \mathcal{K}^2$, the covariance between $\mathcal{L}_{k_1}$ and $\mathcal{L}_{k_2}$ satisfies:

$$
\text{Cov}\left[\mathcal{L}_{k_1}, \mathcal{L}_{k_2}\right] = \frac{1}{N} \sum_{x\in\mathcal{X}} \tau_x \left(\frac{1}{2}\left\|\widehat{\Sigma}_{x,k_1}^{-1/2}\,\Sigma_{x,k^*}\,(\widehat{\Sigma}_{x,k_2}^{-1/2})'\right\|_{hs}^2 \right.
$$
$$
\left. + \left(m_{x,k^*} - \widehat{m}_{x,k_1}\right)'\widehat{\Sigma}_{x,k_1}^{-1}\,\Sigma_{x,k^*}\,\widehat{\Sigma}_{x,k_2}^{-1}\left(m_{x,k^*} - \widehat{m}_{x,k_2}\right)\right). \quad (12)
$$

Proposition 3 gives an approximation of the distribution of the log-likelihood vector $(\mathcal{L}_k)_{k\in\mathcal{K}}$ which becomes quickly tight as N grows (see Appendix C). As shown in Section 6, this approximation can be used to estimate the success rate of profiling SCA. The computational cost of (11) and (12) is $O(|\mathcal{X}|T^3)$ where T denotes the leakage dimension. The total cost of computing the distribution parameters is hence $O(|\mathcal{K}|^2|\mathcal{X}|T^3)$. This may be prohibitive if the leakage dimension is high. However, the leakage dimension can be reduced by pre-processing the leakage measurements [2,20]. In practice, $T = 3$ is often sufficient to catch most of the side channel information [2,20].

In order to simplify our analysis, let us make the following assumption.

Assumption 1 (Constant Covariance Assumption). *The covariance matrix Σ_s is the same for all signals $s \in \mathcal{S}$.*

Remark 3. This assumption is quite usual in the literature (see for instance [19,11,21]). The noise in the leakage is indeed often independent of the target signal. This is especially true if most of the noise amount is produced by a noise generator (independent of the target algorithm) as a countermeasure to side channel analysis.

Observing the expectation of $\mathcal{L}_k$ (11), one identifies three sums. The first one and the third one only involve the leakage covariance matrices and/or their estimations. Therefore, under the constant covariance assumption, these sums are constant with respect to k and hence, they provide no discrimination between the different key candidates. Actually, only the second sum in (11) provides some discrimination which depends on the leakage means $m_{x,k}$ (corresponding to the different processed signals $s = \varphi(x,k)$). If these means are clearly dissociated and if their estimations $\widehat{m}_{x,k}$ are precise, then the second sum is around zero for and only for the good key guess k^*. As a result, the expectation of $\mathcal{L}_k$ is maximized for the good key guess $k = k^*$ which illustrates the attack soundness.

From (11) and (12) we also see that, unlike for differential SCA, increasing the number of leakage measurements and increasing the leakage variance do not have a complementary effect on the distinguisher distribution. However, it has a complementary effect on the success rate: if the leakage covariance matrix is multiplied by a factor λ (and assuming that its estimation is also multiplied by λ) then the attacker must multiply the number of measurements by a factor λ in order to keep the success rate constant. This fact is formally demonstrated in Appendix D. We hence remark (according to the analysis in Section 4.2) that Differential SCA and Profiling SCA are affected in the same way by the leakage noise increase. Besides, when the leakage noise is amplified, the ratio between the efficiencies[1] of both attacks remains constant.

As final remark, let us mention that Proposition 2 also applies to the log-likelihood vector $(\mathcal{L}_k)_{k \in \mathcal{K}}$. Besides, in the uniform setting (see Section 4.2), the success rate of the profiling SCA is also constant with respect to k^*.

6   Success Rate Evaluation

In accordance with the analyses of Sections 4.2 and 5.2, we assume that the distribution of the distinguishing vector $\mathbf{d} = (d_k)_{k \in \mathcal{K}}$ is a multivariate Gaussian $\mathcal{N}(m_{\mathbf{d}}, \Sigma_{\mathbf{d}})$. In this section we present two approaches to compute the success rate of a side channel key recovery attack, once the parameter of this distribution have been determined.

In the first approach, we show that the success rate can be expressed as a sum of Gaussian cumulative distribution functions (cdf). It can hence be estimated by numerically computing these cdf. The second approach consists in simulating the multivariate Gaussian vector **d** several times in order to get a precise estimation of the success rate.

---

[1] By efficiency, we mean the required number of leakage measurements to succeed the attack (with high probability).

6.1 Numerical Computation

We show hereafter that the success rate can be expressed as a sum of Gaussian cdf. For this purpose, we need to introduce the *comparison vector* that is the $(|\mathcal{K}| - 1)$-size vector $\mathbf{c} = (c_k)_{k \in \mathcal{K}/\{k^*\}}$ defined for every $k \in \mathcal{K}/\{k^*\}$ by:

$$c_k = d_{k^*} - d_k \ . \tag{13}$$

If all the coordinates of this vector are positive then the attack succeeds in isolating the good key guess as first candidate. If n coordinates are negative then the attack rates the good key guess as the $(n + 1)^{\text{th}}$ candidate; in other words, it succeeds at the $(n + 1)^{\text{th}}$ order. The comparison vector is a linear transformation of the distinguishing vector by a $((|\mathcal{K}| - 1) \times |\mathcal{K}|)$-matrix P whose expression straightforwardly follows from (13). This implies that the comparison vector has a multivariate Gaussian distribution $\mathcal{N}(m_{\mathbf{c}}, \Sigma_{\mathbf{c}})$ where $m_{\mathbf{c}} = P m_{\mathbf{d}}$ and $\Sigma_{\mathbf{c}} = P \Sigma_{\mathbf{d}} P'$.

Let $\alpha \subseteq \{1, \cdots, |\mathcal{K}| - 1\}$ be a set of indices and let I_α and S_α be the $(|\mathcal{K}| - 1)$-size vectors defined by:

$$I_\alpha[i] = \begin{cases} -\infty & \text{if } i \in \alpha \\ 0 & \text{if } i \notin \alpha \end{cases} \quad \text{and} \quad S_\alpha[i] = \begin{cases} 0 & \text{if } i \in \alpha \\ +\infty & \text{if } i \notin \alpha \end{cases} \ .$$

The vector $\mathbf{c}$ has exactly n negative coordinates if and only if there exists a set α of cardinal n s.t. $I_\alpha < \mathbf{c} < S_\alpha$. Since the intervals $([I_\alpha, S_\alpha])_\alpha$ are disjoints, the probability that exactly n coordinates of $\mathbf{c}$ be negative can be written as:

$$p_n = \sum_{\alpha; |\alpha| = n} P\left[I_\alpha \leq \mathbf{c} \leq S_\alpha\right] \ . \tag{14}$$

The oth order success rate equals the sum $p_0 + p_1 + \cdots + p_{o-1}$ which from (14) gives:

$$\text{Succ-}o = \sum_{\alpha; |\alpha| < o} P\left[I_\alpha \leq \mathbf{c} \leq S_\alpha\right] = \sum_{\alpha; |\alpha| < o} \Phi_{m_{\mathbf{c}}, \Sigma_{\mathbf{c}}}(I_\alpha, S_\alpha) \ , \tag{15}$$

where $\Phi_{m, \Sigma}$ denotes the Gaussian cdf that satisfies $\Phi_{m, \Sigma} : (a, b) \mapsto \int_a^b \phi_{m, \Sigma}(x) \, dx$.

Relation (15) shows that the oth order success rate can be computed by performing $\sum_{i < o} \binom{|\mathcal{K}| - 1}{i}$ multivariate Gaussian cdf calculations (on $(|\mathcal{K}| - 1)$-size Gaussian vectors). The numerical computation of multivariate Gaussian cdf is a classical issue in statistics. Some solutions exist (see for instance [9,10]) that can be used to precisely compute the success rate according to (15).

This approach has some drawbacks. Firstly, the numerical computations of Gaussian cdf may be difficult with covariance matrices having particular forms and/or quite high dimensions. For instance it requires that the covariance matrix is not singular, which is not always the case in our context. Yet another drawback of this approach is that the computation of high order success rates requires an important number of Gaussian cdf computations. Regarding these issues, a possible alternative is presented in the next section.

6.2   Gaussian Simulation

Another possibility to compute the success rate is to perform a Gaussian simulation. The principle is to simulate several times the distribution $\mathcal{N}(m_\mathbf{d}, \Sigma_\mathbf{d})$. This amounts to randomly pick up several distinguishing vectors each one corresponding to the result of an attack. The success rate is estimated based on these different results. In other words this approach works as an attack simulation but instead of performing the attack several times, we perform several Gaussian random vectors simulation which is clearly more efficient especially when the number of leakage measurements is high and/or the leakage dimension is high. Another advantage of this approach is that the success rate at the different orders as well as the guessing entropy (see Appendix A) can all be computed using the same simulated distinguishing vectors. Finally Gaussian simulation is sound even when the covariance matrix is singular which may happen in our context.

7   Empirical Validation

In order to empirically validate the theoretical analyses conducted in the previous sections we performed several simulations. We chose $S = X \oplus K$ as target signal where X and K are 8 bits variables. The leakage means $(m_s)_{s \in \mathcal{S}}$ and the leakage covariance matrix Σ (assumed constant for the different signals $s \in \mathcal{S}$) were drawn with random coefficients. Their dimensions were set to 1 for differential SCA, and to 3 for profiling SCA (this is a typical dimension when subspace-based profiling is involved [2,20]). The attacker model/estimations were first assumed to be exact (i.e. $\mathsf{M}(s) = m_s$, $\widehat{m}_s = m_s$ and $\widehat{\Sigma} = \Sigma$) and then assumed to be slightly erroneous (by inserting random errors).

On the one hand, the success rate was estimated empirically by simulating the attack. Namely, the leakage measurements l_i corresponding to random inputs x_i were randomly picked up according to the leakage parameters $(m_{x_i, k^*}, \Sigma_{x_i, k^*})$. The attack was performed several times (few thousands) on such simulated measurements in order to obtain an empirical success rate. On the other hand, the success rate was estimated using our approach. We computed the distinguishing vector expectation and covariance matrix (such as described in Sections 4.2 and 5.2) according to the leakage parameters and assuming $\tau_x = 1/256$ for every x. Then we performed Gaussian simulations (see Section 6.2) to get an estimation of the success rate.

As expected, for differential SCA, the different success rates obtained with our approach always match almost perfectly the success rates obtained by attack simulations. For profiling SCA, the success rates obtained with our approach also match quite well the success rates obtained by attack simulations. The precision of this matching depends on the number of leakage measurements required for the attack to succeed (with a high probability). When this number is quite low (i.e. around few hundreds), our estimation slightly overvalues the real success rate. This overvaluation becomes less marked as the number of required leakage

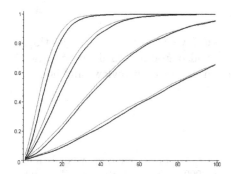

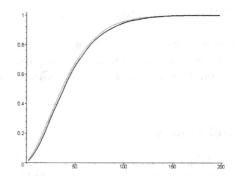

Fig. 1. Success rates of different profiling SCA attacks over an increasing number of leakage measurements

Fig. 2. Success rates of a profiling SCA attack over an increasing number of leakage measurements

measurements increases. As an illustration the success rate of four attacks requiring different amounts of leakage measurements is plotted Figure 1. Success rates obtained by attack simulation are plotted in black while the corresponding ones obtained with our approach are plotted in grey. The convergence can be clearly observed. Figure 2 shows both success rates for an attack requiring around 200 leakage measurements. When moving up to 500 required leakage measurements, the curves completely mix up.

The different empirical results that we obtained have demonstrated the soundness of our theoretical analysis. They also show that the approximation $\tau_x \approx 1/|\mathcal{X}|$ is sound when the x_i's are randomly drawn (*i.e.* in a known plaintext/ciphertext attack setting).

8   Conclusion and Open Issues

In this paper, we have investigated the issue of evaluating the success rate of side channel analysis in the Gaussian leakage model. For the two main families of SCA, namely differential SCA and profiling SCA, we have shown that the distinguishing vector resulting from the attack has (or at least quickly tends towards) a multivariate Gaussian distribution. This allowed us to exhibit an efficient way to compute the success rate of such an attack according to the number of leakage measurements and to the leakage distribution parameters. Finally, our analysis was validated empirically by a large number of attack simulations.

Our analysis stresses several interesting open issues. Future works could focus on chosen plaintext attacks and investigate how the choice of the target inputs may affect the success rate of an attack. Another interesting issue is the tolerance for a distinguisher to the error on the leakage model. How does an error on the attacker model/estimations affect the success rate of the attack ? Finally, extension of our analysis to protected implementations (for instance by masking techniques) would be of great interest to quantify their security.

Acknowledgements

I would like to thank Emmanuel Prouff for his reviews and for his suggestions that substantially improved the paper. I am also grateful to Jean-Sébastien Coron, Emmanuelle Dottax and Christophe Giraud for their reviews and helpful comments.

References

1. Akkar, M.-L., Bévan, R., Dischamp, P., Moyart, D.: Power analysis, what is now possible.. In: Okamoto, T. (ed.) ASIACRYPT 2000. LNCS, vol. 1976, pp. 489–502. Springer, Heidelberg (2000)
2. Archambeau, C., Peeters, E., Standaert, F.-X., Quisquater, J.-J.: Template attacks in principal subspaces. In: Goubin, L., Matsui, M. (eds.) CHES 2006. LNCS, vol. 4249, pp. 1–14. Springer, Heidelberg (2006)
3. Bévan, R., Knudsen, E.W.: Ways to enhance differential power analysis. In: Lee, P.J., Lim, C.H. (eds.) ICISC 2002. LNCS, vol. 2587, pp. 327–342. Springer, Heidelberg (2003)
4. Brier, É., Clavier, C., Olivier, F.: Correlation power analysis with a leakage model. In: Joye, M., Quisquater, J.-J. (eds.) CHES 2004. LNCS, vol. 3156, pp. 16–29. Springer, Heidelberg (2004)
5. Cachin, C.: Entropy Measures and Unconditional Security in Cryptography. PhD thesis (1997)
6. Chari, S., Rao, J., Rohatgi, P.: Template attacks. In: Kaliski Jr., B.S., Koç, Ç.K., Paar, C. (eds.) CHES 2002. LNCS, vol. 2523, pp. 13–28. Springer, Heidelberg (2003)
7. Clavier, C., Coron, J.-S., Dabbous, N.: Differential power analysis in the presence of hardware countermeasures. In: Paar, C., Koç, Ç.K. (eds.) CHES 2000. LNCS, vol. 1965, pp. 252–263. Springer, Heidelberg (2000)
8. Gandolfi, K., Mourtel, C., Olivier, F.: Electromagnetic analysis: Concrete results. In: Koç, Ç.K., Naccache, D., Paar, C. (eds.) CHES 2001. LNCS, vol. 2162, pp. 251–261. Springer, Heidelberg (2001)
9. Genz, A.: Numerical Computation of Multivariate Normal Probabilities. Journal of Computational and Graphical Statistics 1, 141–149 (1992)
10. Genz, A.: Comparison of Methods for the Computation of Multivariate Normal Probabilities. Computing Science and Statistics 25, 400–405 (1993)
11. Gierlichs, B., Lemke-Rust, K., Paar, C.: Templates vs. Stochastic methods. In: Goubin, L., Matsui, M. (eds.) CHES 2006. LNCS, vol. 4249, pp. 15–29. Springer, Heidelberg (2006)
12. Kocher, P.C., Jaffe, J., Jun, B.: Differential power analysis. In: Wiener, M. (ed.) CRYPTO 1999. LNCS, vol. 1666, pp. 388–397. Springer, Heidelberg (1999)
13. Kocher, P.C.: Timing attacks on implementations of diffie-hellman, RSA, DSS, and other systems. In: Koblitz, N. (ed.) CRYPTO 1996. LNCS, vol. 1109, pp. 104–113. Springer, Heidelberg (1996)
14. Lemke-Rust, K., Paar, C.: Gaussian mixture models for higher-order side channel analysis. In: Paillier, P., Verbauwhede, I. (eds.) CHES 2007. LNCS, vol. 4727, pp. 14–27. Springer, Heidelberg (2007)
15. Mangard, S., Oswald, E., Popp, T.: Power Analysis Attacks – Revealing the Secrets of Smartcards. Springer, Heidelberg (2007)

16. Mangard, S.: Hardware countermeasures against DPA – A statistical analysis of their effectiveness. In: Okamoto, T. (ed.) CT-RSA 2004. LNCS, vol. 2964, pp. 222–235. Springer, Heidelberg (2004)
17. Massey, J.: Guessing and Entropy. IEEE ISIT, 204 (1994)
18. Messerges, T., Dabbish, E., Sloan, R.: Investigations of Power Analysis Attacks on Smartcards. In: The USENIX Workshop on Smartcard Technology (Smartcard 1999), pp. 151–161 (1999)
19. Schindler, W., Lemke, K., Paar, C.: A stochastic model for differential side channel cryptanalysis. In: Rao, J.R., Sunar, B. (eds.) CHES 2005. LNCS, vol. 3659, pp. 30–46. Springer, Heidelberg (2005)
20. Standaert, F.-X., Archambeau, C.: Using subspace-based template attacks to compare and combine power and electromagnetic information leakages. In: Oswald, E., Rohatgi, P. (eds.) CHES 2008. LNCS, vol. 5154, pp. 411–425. Springer, Heidelberg (2008)
21. Standaert, F.-X., Malkin, T.G., Yung, M.: A Unified Framework for the Analysis of Side-Channel Key Recovery Attacks. Cryptology ePrint Archive, Report 2006/139 (2006), http://eprint.iacr.org/
22. Standaert, F.-X., Peeters, E., Rouvroy, G., Quisquater, J.-J.: An Overview of Power Analysis Attacks against Field Programmable Gate Arrays. IEEE 94(2), 383–394 (2006)

A   Guessing Entropy

The guessing entropy [17,5] is defined as the expected number of key guesses to test before recovering a target key value. As pointed out in [21], the guessing entropy is relevant in the context of side channel analysis since it indicates the average workload to perform after side channel analysis. Let $\text{rank}_k(\mathbf{d})$ denote the index $i \in \{1, \cdots, |\mathcal{K}|\}$ such that d_k is the ith higher element of $\mathbf{d}$. The guessing entropy of a side channel attack using a distinguisher D and a public vector $\mathbf{x}$, and targeting a secret key k^* is formally defined as:

$$\text{GE}^{\mathsf{D}}_{\mathbf{x},k^*} = \mathrm{E}\left[\left(l_i \leftarrow \mathrm{L}\left(\varphi(k^*, x_i)\right)\right)_i ; \; \mathbf{d} \leftarrow \mathsf{D}(\mathbf{x}, \mathbf{l}) : \text{rank}_{k^*}(\mathbf{d})\right] . \qquad (16)$$

The guessing entropy is related to the success rate of every order. In fact, the correct key guess is rated at the oth rank in the distinguishing vector if and only it is rated among the o first candidates but it is not rated among the $o - 1$ first candidates. As a result, the probability that the correct key guess be rated at the oth rank satisfies for every o: $\mathrm{P}\left[\text{rank}_{k^*}(\mathbf{d}) = o\right] = \text{Succ-}o - \text{Succ-}(o - 1)$, where Succ-0 is naturally defined at zero. This brings to the following relation:

$$\text{GE} = \sum_{o=1}^{|\mathcal{K}|} o \, \mathrm{P}\left[\text{rank}_{k^*}(\mathbf{d}) = o\right] = |\mathcal{K}| - \sum_{o=1}^{|\mathcal{K}|-1} \text{Succ-}o . \qquad (17)$$

B   Proof of Proposition 3

The proof of Proposition 3 makes use of the following lemma.

Lemma 1. *Let X be a T-size random vector having a Gaussian distribution $\mathcal{N}(0, \Sigma)$. Let A_1 and A_2 be two $(T \times T)$-matrices and let m_1 and m_2 be two*

T-size vectors. Let Q_1 and Q_2 be two quadratic forms defined, for $j = 1, 2$, by $Q_j = (X + m_j)' A_j^2 (X + m_j)$. For $j = 1, 2$, the expectation of Q_j satisfies:

$$\mathrm{E}[Q_j] = \|A_j m_j\|^2 + \mathrm{Tr}(A_j \Sigma A_j') . \tag{18}$$

And the covariance of Q_1 and Q_2 satisfies:

$$\mathrm{Cov}[Q_1, Q_2] = 2 \|A_1 \Sigma A_2'\|_{hs}^2 + 4 \, m_1' A_1^2 \Sigma A_2^2 m_2 . \tag{19}$$

Proof. We have $Q_j = \sum_{i=1}^T (A_j (X + m_j))[i]^2$ which leads to:

$$\mathrm{E}[Q_j] = \sum_{i=1}^T \mathrm{E}\left[(A_j (X + m_j))[i]^2\right] \tag{20}$$

$$= \sum_{i=1}^T \mathrm{E}[(A_j (X + m_j))[i]]^2 + \sum_{i=1}^T \mathrm{Var}[(A_j (X + m_j))[i]] , \tag{21}$$

since $\mathrm{E}[Y^2] = \mathrm{Var}[Y] + \mathrm{E}[Y]^2$ holds for every random variable Y. From $(X + m_j) \sim \mathcal{N}(m_j, \Sigma)$ we have $A_j (X + m_j) \sim \mathcal{N}(A_j m_j, A_j \Sigma A_j')$ which directly yields (18).

The quadratic form Q_j can be rewritten as $Q_j = (A_j X)^2 + (A_j m_j)^2 + 2m_j' A_j^2 X$ for $j = 1, 2$. By bilinearity, $\mathrm{Cov}[Q_1, Q_2]$ satisfies:

$$\mathrm{Cov}[Q_1, Q_2] = \mathrm{Cov}\left[(A_1 X)^2, (A_2 X)^2\right]$$
$$+ 2 \, \mathrm{Cov}\left[(A_1 X)^2, m_2' A_2 X\right] + 2 \, \mathrm{Cov}\left[(A_2 X)^2, m_1' A_1 X\right]$$
$$+ 4 \, \mathrm{Cov}[m_1' A X, m_2' A X] . \tag{22}$$

We claim the three following relations:

$$\mathrm{Cov}\left[(A_1 X)^2, (A_2 X)^2\right] = 2 \|A_1 \Sigma A_2'\|_{hs}^2 , \tag{23}$$

$$\mathrm{Cov}\left[(A_1 X)^2, m_2' A_2^2 X\right] = \mathrm{Cov}\left[(A_2 X)^2, m_1' A_1^2 X\right] = 0 , \tag{24}$$

$$\mathrm{Cov}\left[m_1' A_1^2 X, \, m_2' A_2^2 X\right] = m_1' A_1^2 \Sigma A_2^2 m_2 . \tag{25}$$

These relations together with (22) result in (19) and state the correctness of Lemma 1. Relation (25) straightforwardly holds from the bilinearity of the covariance and by symmetry of A_1^2 (*i.e.* $(A_1^2)' = A_1^2$). Relations (23) and (24) are stated hereafter.

First, let us show (23). The covariance between $(A_1 X)^2$ and $(A_2 X)^2$ can be rewritten as:

$$\mathrm{Cov}\left[(A_1 X)^2, (A_2 X)^2\right] = \sum_{i,j} \mathrm{Cov}\left[(A_1 X)[i]^2, (A_2 X)[j]^2\right]$$

$$= \sum_{i,j} \left(\mathrm{E}\left[(A_1 X)[i]^2 (A_2 X)[j]^2\right] - \mathrm{E}\left[(A_1 X)[i]^2\right] \mathrm{E}\left[(A_2 X)[j]^2\right] \right) . \tag{26}$$

Since the expectations of $A_1 X$ and $A_2 X$ equal zero, the expectation of the product $(A_1 X)[i]^2 (A_2 X)[j]^2$ is the Gaussian forth order moment that is known to satisfy:

$$E\left[(A_1 X)[i]^2 (A_2 X)[j]^2\right] = E\left[(A_1 X)[i]^2\right] E\left[(A_2 X)[j]^2\right]$$
$$+ 2 \operatorname{Cov}\left[(A_1 X)[i], (A_2 X)[j]\right]^2 . \quad (27)$$

Hence, (26) gives:

$$\operatorname{Cov}\left[(A_1 X)^2, (A_2 X)^2\right] = 2 \sum_{i,j} \operatorname{Cov}\left[(A_1 X)[i], (A_2 X)[j]\right]^2 . \quad (28)$$

Since we have $\operatorname{Cov}\left[(A_1 X)[i], (A_2 X)[j]\right] = (A_1 \Sigma A_2')[i,j]$, one deduces that (28) finally results in (23).

We now show the correctness of (24). We have:

$$\operatorname{Cov}\left[(A_1 X)^2, m_2' A_2^2 X\right] = \sum_i \operatorname{Cov}\left[(A_1 X)[i]^2, m_2' A_2^2 X\right] . \quad (29)$$

Since X has a zero mean, every term of the previous sum is a Gaussian third order moment and is hence equal to zero. This way, we get (24).   ◇

We give hereafter the proof of Proposition 3.

Proof. (Proposition 3) Since the l_i's are independently drawn from Gaussian distributions $\mathcal{N}(m_{x_i,k^*}, \Sigma_{x_i,k^*})$ and since, for every x, there is a ratio τ_x of the x_i's that equal x, Relation (10) and Lemma 1 directly lead to (11) and (12).

Now, $(\mathcal{L}_k)_{k\in\mathcal{K}}$ can be expressed as a linear transformation of the vector $\sum_{i=1}^N l_i$ and of the vector $\left(\sum_{i=1}^N l_i[j_1] l_i[j_2]\right)_{1\le j_1,j_2\le T}$. The first one has a multivariate Gaussian distribution and, from the multivariate central limit theorem, the second one tends toward a multivariate Gaussian distribution as N grows. Hence $(\mathcal{L}_k)_{k\in\mathcal{K}}$ tends toward a multivariate Gaussian distribution as N grows.   ◇

C   Convergence of the Log-Likelihood Distribution

According to (10), the log-likelihood $\mathcal{L}_k$ can be expressed as the sum of $|\mathcal{X}|$ values $\mathcal{L}_{k,x}$ that are defined by:

$$\mathcal{L}_{k,x} = \frac{\tau_x}{2} \log\left((2\pi)^T |\widehat{\Sigma}_{x,k}|\right) - \frac{1}{2N} \sum_{\substack{i=1 \\ x_i=x}}^N (l_i - \widehat{m}_{x,k})' \widehat{\Sigma}_{x,k}^{-1} (l_i - \widehat{m}_{x,k}) . \quad (30)$$

The first term is constant and the second term is a sum of $N\tau_x$ elements of the form $X' A^2 X$ where A is the matrix $\widehat{\Sigma}_{x,k}^{-1}$ and X is a Gaussian random variable $\mathcal{N}(m_{x,k^*} - \widehat{m}_{x,k}, \Sigma_{x,k^*})$. The distribution of such a sum is given in the following lemma. At first, let us recall that the chi-square distribution with n degrees of freedom $\chi^2(n)$ is the distribution obtained by summing n independent $\mathcal{N}(0,1)$-distributed random variables.

Lemma 2. *Let $(X_j)_j$ be n independent T-size random vectors having a Gaussian distribution $\mathcal{N}(m, \Sigma)$, let A be a $(T \times T)$-matrix and let $(Q_j)_j$ be the quadratic forms defined as $Q_j = X_j' A^2 X_j$. The sum of the Q_j satisfies:*

$$\sum_{j=1}^{n} Q_j = \beta + G + \sum_{i=1}^{T} \alpha_i C_i , \qquad (31)$$

where $\beta = n(A \cdot m)^2$, $\alpha_i = (A \Sigma A')[i, i]$, G is an univariate Gaussian random variable, C_i are T chi-square random variables with n degrees of freedom.

Proof. For $j = 1, 2$, we have $Q_j = (A X_j)^2$. Denoting by $\overline{X}_j$ the centered random variable $X_j - m$, we get $Q_j = (Am)^2 + 2 m' A^2 \overline{X}_j + (A \overline{X}_j)^2$ and hence, $\sum_j Q_j = \beta + 2 \sum_j m' A^2 \overline{X}_j + \sum_j \sum_i (A \overline{X}_j)[i]^2$.

After denoting $2 \sum_j m' A^2 \overline{X}_j$ by G and $\frac{1}{\alpha_i} \sum_j (A \overline{X}_j)[i]^2$ by C_i, we get (31). Now, G is Gaussian since it is defined as a sum of Gaussian random variables. Moreover, the covariance matrix of $A X_j$ being equal to $A \Sigma A'$, we have, for every j: $\alpha_i = \text{Var}\left[(A \overline{X}_j)[i]\right]$. This implies that $\frac{1}{\sqrt{\alpha_i}} (A \overline{X}_j)[i]$ is $\mathcal{N}(0, 1)$-distributed for every j, hence by definition C_i is $\chi^2(n)$-distributed.   ◇

A chi-square distribution with n degrees of freedom quickly tends towards a Gaussian distribution as n grows. A rule of thumb in probability theory is to consider the approximation $\chi^2(n) \approx \mathcal{N}(n, 2n)$ quite reasonable for $n \geq 30$. From Lemma 2, $\mathcal{L}_{k,x}$ is a sum between a constant, a Gaussian random variable and T chi-square random variables with $N\tau_x$ degrees of freedom. Therefore, for $N\tau_x$ large enough, we can consider that $\mathcal{L}_{k,x}$ has a Gaussian distribution. If this holds for every $x \in \mathcal{X}$ then the distribution of $\mathcal{L}_k$ can fairly be approximated by a Gaussian.

D   Profiling SCA – Number of Leakage Measurements *vs.* Leakage Variance

Let us denotes the leakage covariance matrix by Σ and its estimation by $\widehat{\Sigma}$. Under the constant covariance assumption, (11) and (12) can be rewritten as:

$$\text{E}\left[\mathcal{L}_k\right] = C_1 - \frac{1}{2} \sum_{x \in \mathcal{X}} \tau_x \left\| \widehat{\Sigma}^{-1/2} (m_{x,k^*} - \widehat{m}_{x,k}) \right\|^2 , \qquad (32)$$

and

$$\text{Cov}\left[\mathcal{L}_{k_1}, \mathcal{L}_{k_2}\right] = C_2 + \frac{1}{N} \sum_{x \in \mathcal{X}} \tau_x (m_{x,k^*} - \widehat{m}_{x,k_1})' \widehat{\Sigma}^{-1} \Sigma \widehat{\Sigma}^{-1} (m_{x,k^*} - \widehat{m}_{x,k_2}) , \qquad (33)$$

where C_1 and C_2 are some values constant with respect to k that satisfy $C_1 = \log\left(2\pi|\widehat{\Sigma}|\right) + \text{Tr}\left(\widehat{\Sigma}^{-1/2} \Sigma \left(\widehat{\Sigma}^{-1/2}\right)'\right)$ and $C_2 = \frac{1}{2N} \left\| \widehat{\Sigma}^{-1/2} \Sigma (\widehat{\Sigma}^{-1/2})' \right\|_{hs}^2$.

We show in Section 6.1 that the success rate depends of the distribution of the comparison vector $\mathbf{c} = (c_k)_{k \in \mathcal{K}/\{k^*\}}$ that is defined, for Profiling SCA, by $c_k = \mathcal{L}_{k^*} - \mathcal{L}_k$ for every $k \in \mathcal{K}$. Assuming $(\mathcal{L}_k)_{k \in \mathcal{K}}$ Gaussian, $\mathbf{c}$ has a Gaussian distribution whose parameters satisfies:

$$\mathrm{E}\,[c_k] = \mathrm{E}\,[\mathcal{L}_{k^*}] - \mathrm{E}\,[\mathcal{L}_k] \;, \tag{34}$$

and

$$\mathrm{Cov}\,[c_{k_1}, c_{k_2}] = \mathrm{Var}\,[\mathcal{L}_{k^*}] + \mathrm{Cov}\,[\mathcal{L}_{k_1}, \mathcal{L}_{k_2}] - \mathrm{Cov}\,[\mathcal{L}_{k^*}, \mathcal{L}_{k_1}] - \mathrm{Cov}\,[\mathcal{L}_{k^*}, \mathcal{L}_{k_2}] \;. \tag{35}$$

From these expressions, we can see that the constant terms C_1 and C_2 of (32) and (33) cancel each other out in the expectation and the covariance matrix of $\mathbf{c}$. It thus appears that multiplying the leakage covariance matrix by a factor λ (and assuming that its estimation is also multiplied by λ) results in the multiplication of $m_\mathbf{c}$ and $\Sigma_\mathbf{c}$ by $1/\lambda$ while multiplying the number of leakage measurements by λ results in the multiplication of $\Sigma_\mathbf{c}$ by $1/\lambda$.

One can verify that the Gaussian cdf satisfies for every (a, b):

$$\Phi_{m/\lambda, \Sigma/\lambda^2}(a, b) = \Phi_{m, \Sigma}(\lambda a, \lambda b) \;. \tag{36}$$

As shown in Section 6.1, the success rate can be expressed as a sum of cdf $\Phi_{m_\mathbf{c}, \Sigma_\mathbf{c}}$ with inputs in $\{0, +\infty, -\infty\}^{|\mathcal{K}|-1}$. One thus deduces from (36) that multiplying $m_\mathbf{c}$ by $1/\lambda$ and $\Sigma_\mathbf{c}$ by $1/\lambda^2$ keeps the success rate unchanged. Hence we obtain that multiplying the leakage covariance matrix and multiplying the number of leakage measurements have complementary effects on the success rate of Profiling SCA.

Algebraic and Correlation Attacks against Linearly Filtered Non Linear Feedback Shift Registers

Côme Berbain[1], Henri Gilbert[1], and Antoine Joux[2]

[1] Orange Labs
38-40 rue du Général Leclerc, 92794 Issy-les-Moulineaux, France
[2] DGA and Université de Versailles
45 avenue des Etats-Unis, 78035 Versailles Cedex, France
come.berbain@orange-ftgroup.com,
henri.gilbert@orange-ftgroup.com,
antoine.joux@prism.uvsq.fr

Abstract. The filter generator is a well known and extensively studied stream cipher construction. It consists of a Linear Feedback Shift Register (LFSR) filtered by a non linear Boolean function. In this paper we focus on the dual construction, namely a linearly filtered Non linear Feedback Shift Register (NFSR). We show that the existing algebraic and correlation attacks against the filter generator can be transposed to mount algebraic or correlation attacks against this dual construction. We investigate such attacks and extend them to the case where a linearly filtered NFSR is combined linearly with one or more non linearly filtered LFSRs. We apply our algebraic attack to a modified version of Grain-128, resulting in an attack requiring 2^{105} computations and 2^{39} keystream bits. Even though this attack does not apply to the original Grain-128, it shows that the use of a NFSR is not sufficient to avoid all algebraic attacks.

1 Introduction

Stream ciphers represent, together with block ciphers, one of the two main classes of symmetric encryption algorithms. They produce a pseudo-random keystream sequence of numbers over a small alphabet, typically the binary alphabet $\{0, 1\}$. To encrypt a plaintext sequence, each plaintext symbol is combined with the corresponding symbol of the keystream sequence using a group operation, usually the exclusive or operation over $\{0, 1\}$.

A classic way to build a random number generator is to use a Linear Feedback Shift Register (LFSR) and to apply a non-linear Boolean function f to the current LFSR state to produce the keystream. This construction is known as the filter generator. It has been extensively studied over the past years resulting in a large number of criteria for the design of such ciphers. For example, the correlation and fast correlation attacks [23,24,9,19] against this scheme can be avoided if the function f has a high order of correlation immunity or satisfies

R. Avanzi, L. Keliher, and F. Sica (Eds.): SAC 2008, LNCS 5381, pp. 184–198, 2009.

certain criteria [15]. Algebraic attacks [12,2,6,11] led to the notion of algebraic immunity of Boolean function.

Recently new stream ciphers using Non linear Feedback Shift Registers (NFSRs) were proposed as an alternative to LFSR-based stream ciphers. Finalist candidates to the eSTREAM project like Trivium [8] or Grain [17] are using one or several NFSRs combined or not with LFSRs.

In this paper, we analyze the dual of the classical filter generator construction, i.e. a Non linear Feedback Shift Register with a linear output function. We show that it is easy to formally express any internal state variable as a linear combination of the initial state variables and of keystream bits, and this (more surprisingly) allows mounting algebraic or correlation attacks against such a scheme. We extend our analysis to the case where a linearly filtered NFSR is linearly combined with one or several non linearly filtered LFSRs. This allows us to mount an attack against a modified version of Grain-128. Even though this attack does not apply to the original Grain-128, it shows that the use of a NFSR is not sufficient to avoid all algebraic attacks. In particular it contradicts the common idea that the increase in the degree due to the NFSR allows to avoid algebraic attacks.

The paper is organized as follows: in Section 2, we introduce linearly filtered NFSRs and we explain why they might seem to naturally resist algebraic attacks. In Section 3, we introduce a simple formal technique applicable to any linearly filtered NFSR. In Section 4 and 5, we show how to mount algebraic and correlation attacks against these schemes; in Section 6, we extend our attacks to linear combinations of a linearly filtered NFSR and non-linearly filtered LFSRs and we present our attack against a modified version of Grain-128.

2   Linearly Filtered NFSRs

The filter generator, i.e. a LFSR filtered by a nonlinear Boolean function, has been widely studied, and some ciphers are based on this construction like WG [14] or Sfinks[7], two of the candidates to the eSTREAM competition.

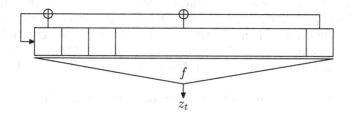

Fig. 1. Filter Generator

A large number of attack techniques applicable to filter generators have been proposed like correlation and fast correlation attacks [13,18,19,20] or algebraic and fast algebraic attacks [12,11,1,2].

For now on, we consider the dual construction, i.e. we swap the update and output functions, i.e. the linear and the non linear functions. The resulting system is a linearly filtered Non Linear Feedback Shift Register (NFSR).

More formally we consider an n-bit NFSR and denote its initial state by $(x_0, \ldots, x_{n-1})$. This NFSR is updated using a Boolean function g:

$$x_{t+n} = g(x_t, \ldots, x_{t+n-1})$$

At each iteration, the function g is applied to the current internal state of the NFSR and a new value x_{t+n} is produced. We denote by d_g the total degree of g, i.e. the number of variables in the ANF (algebraic normal form) representation of g.

The considered output function is linear, i.e. each keystream bit z_t is a linear combination of the internal state of the NFSR at time t.

$$z_t = \bigoplus_{k=0}^{n-1} \alpha_k x_{t+k}$$

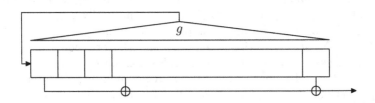

Fig. 2. Linearly filtered NFSR

All the results of this paper can be applied to any linearly filtered NFSR. However in order to render our presentation easier to follow, we will illustrate our results with a first simple example inspired from Grain (a second example inspired from Grain-128 will be introduced in Section 6). In our first example, we took the NFSR from version 1 of Grain stream cipher [17] and slightly modified the output function. In the original Grain, the output of the cipher is composed of a linear combination of the NFSR internal state and of a Boolean function of the LFSR internal state and of a single bit of the NFSR internal state. We removed the LFSR and the associated Boolean function and kept the linear filter of the NFSR.

This NFSR is 80-bit long and it is governed by the recurrence:

$$x_{t+80} = g(x_t, x_{t+1}, \ldots, x_{t+79}),$$

where the expression of nonlinear feedback function g is given by

$$g(x_t, x_{t+1}, \ldots, x_{t+79}) = x_{t+62} \oplus x_{t+60} \oplus x_{t+52} \oplus x_{t+45} \oplus x_{t+37} \oplus x_{t+33} \oplus x_{t+28}$$
$$\oplus x_{t+21} \oplus x_{t+14} \oplus x_{t+9} \oplus x_t \oplus x_{t+63}x_{t+60} \oplus x_{t+37}x_{t+33}$$

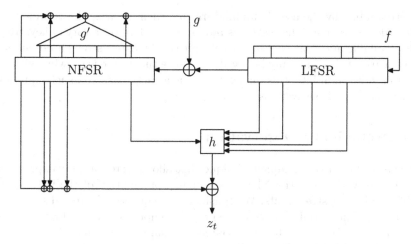

Fig. 3. Grainv1

$$\oplus\, x_{t+15}x_{t+9} \oplus x_{t+60}x_{t+52}x_{t+45} \oplus x_{t+33}x_{t+28}x_{t+21}$$
$$\oplus\, x_{t+63}x_{t+45}x_{t+28}x_{t+9} \oplus x_{t+60}x_{t+52}x_{t+37}x_{t+33}$$
$$\oplus\, x_{t+63}x_{t+60}x_{t+21}x_{t+15} \oplus x_{t+63}x_{t+60}x_{t+52}x_{t+45}x_{t+37}$$
$$\oplus\, x_{t+33}x_{t+28}x_{t+21}x_{t+15}x_{t+9}$$
$$\oplus\, x_{t+52}x_{t+45}x_{t+37}x_{t+33}x_{t+28}x_{t+21}.$$

The cipher output bit z_t is given by the following linear function of the current NFSR internal state:

$$z_t = x_{t+1} \oplus x_{t+2} \oplus x_{t+4} \oplus x_{t+10} \oplus x_{t+31} \oplus x_{t+43} \oplus x_{t+56} \oplus x_{t+63}$$

One of the motivations for NFSR based stream ciphers is that they are generally believed to be naturally immune against algebraic attacks. In fact due to the structure of the NFSR, each internal state variable can be written as a function of degree d_g of the n previous internal state variables. Consequently the degree of the algebraic expression of any internal state variable in the initial variables $(x_0, \ldots, x_{n-1})$ is growing. In the case of a linearly filtered NFSR, the algebraic expression of any keystream bit is a linear combination of the internal state variable. Consequently, the degree of the algebraic expression of each keystream bit is also growing.

In our example, the first 17 keystream bits can be written as linear combinations of the initial state variables. The next 17 keystream bits can be written as polynomials of degree 6. The next 17 polynomials are of degree 10, and the degree keeps growing until it reaches the number of variables. The size of the "blocks" of equations of constant degree is determined by the difference between the position of the feedback (80 in our example) and the position of the tap of highest index in the expression of the update function (63 in our example).

Algebraic attacks as independently discovered by Courtois and Meier [12] and by Ars and Faugère [3] try to reduce the degree of the polynomials corresponding to keystream bits by the use of annihilators, and then linearize the obtained system in order to solve it. These attacks require that a large quantity of keystream bits be available and that each keystream bit can be expressed as a polynomial of fixed degree. It is commonly believed that since the degree is growing with the number of keystream bits, this makes algebraic attacks based on such equations inefficient against these systems.

3   A Preliminary Observation

We now introduce a very simple technique that allows us to formally express any internal state variable of the NFSR as a linear combination of the initial state variables and of keystream bits. We consider the sequence of internal state variables $(x_i)_{i \geq 0}$. The initial state of the NFSR is composed of the n first variables $(x_0, \ldots, x_{n-1})$. We can relate the variables x_i thanks to the non linear update function of the NFSR g. This leads to increasing degrees as stated earlier. We propose instead to use the linear filtering function to derive linear relations between these variables. We recall the expression of z_t for any t:

$$z_t = \bigoplus_{k=0}^{n-1} \alpha_k x_{t+k}$$

We prove the correctness of this technique by induction. We denote by i the highest value $0 \leq k \leq n - 1$ such that α_k is not equal to zero. Let us consider the first bit of keystream which is dependent on x_n. We can write

$$z_{n-i} = x_n \oplus \bigoplus_{k=0}^{i-1} \alpha_k x_{k+n-i}$$

By exchanging the terms z_{n-i} and x_n, we can express x_n as a linear combination of a keystream bit and of a subset of the initial state variables $(x_0, \ldots, x_{n-1})$.

Let us now assume that for all $j \leq t$, all bits x_j can be expressed as a linear combination of the initial state variables and of keystream bits. Let us consider the keystream bit z_{t+1-i}. It results from the definition of i that this is the first keystream bit which depends of x_{t+1}. We can write

$$z_{t+1-i} = x_{t+1} \bigoplus_{k=0}^{i-1} \alpha_k x_{k+t+1-i}$$

By exchanging the terms z_{t+1-i} and x_{t+1}, we can express x_{t+1} as a linear combination of a keystream bit and of variables x_j with $j < t + 1$. By applying the induction assumption, we can replace all these variables x_j by their respective linear combination and we finally express x_{t+1} as a linear combination of keystream bits and of the initial state variables $(x_0, \ldots, x_{n-1})$.

The complexity of building such linear expressions only depends on the number of variables we want to express since the computation can be done in a efficient way by following the induction process we just described. If we want to be able to express N variables, our technique requires $n \cdot N$ computations and $(n+1) \cdot N$ bits of memory.

In our example, the former technique gives:

$$x_{80} = z_{17} \oplus x_{76} \oplus x_{60} \oplus x_{48} \oplus x_{27} \oplus x_{21} \oplus x_{19} \oplus x_{18}$$

$$x_{81} = z_{18} \oplus x_{77} \oplus x_{61} \oplus x_{49} \oplus x_{28} \oplus x_{22} \oplus x_{20} \oplus x_{19}$$

$$x_{82} = z_{19} \oplus x_{78} \oplus x_{62} \oplus x_{50} \oplus x_{29} \oplus x_{23} \oplus x_{21} \oplus x_{20}$$

$$x_{83} = z_{20} \oplus x_{79} \oplus x_{63} \oplus x_{51} \oplus x_{30} \oplus x_{24} \oplus x_{22} \oplus x_{21}$$

$$x_{84} = z_{21} \oplus x_{80} \oplus x_{64} \oplus x_{52} \oplus x_{31} \oplus x_{25} \oplus x_{23} \oplus x_{22}$$

The variable x_{84} depends on x_{80}. By a simple substitution, we get:

$$x_{84} = z_{21} \oplus z_{17} \oplus x_{76} \oplus x_{64} \oplus x_{60} \oplus x_{52} \oplus x_{48} \oplus x_{31} \oplus x_{27} \oplus x_{25} \oplus x_{23} \oplus x_{22} \oplus x_{21} \oplus x_{19} \oplus x_{18}$$

4   Algebraic Attacks

The above observation allows to express each of the variables x_i as a linear combination of the initial state variables and of keystream bits. We denote by L_t the linear expression associated with variable x_t.

Before mounting an algebraic attack against the linearly filtered NFSR, we need to establish a basic property on algebraic immunity of function $g + x_n$, where g is a Boolean function of n variables $(x_0, \ldots, x_{n-1})$ and x_n is an extra Boolean variable.

Theorem 1. *Let g be a Boolean function of n inputs and degree d_g and let h be an annihilator of g, i.e. we have $hg = 0$ or $h(g+1) = 0$. Then $h(x_n + 1)$ is an annihilator of $g + x_n$ (resp. $(g + x_n + 1))$ and we have*

$$AI(g + x_n) \leq AI(g) + 1$$

Let us consider the case where $hg = 0$. We have

$$h(x_n + 1)(g + x_n) = hg(x_n + 1) + h(x_n + 1)x_n$$
$$= 0 \cdot (x_n + 1) + h \cdot 0 = 0$$

The case where h is an annihilator of $(g+1)$ is similar. This shows that $h(x_n+1)$ is an annihilator of $g + x_n$ (resp. $(g + x_n + 1)$).

We can now mount an algebraic attack against the linearly filtered NFSR.

Theorem 2. *Let g be a Boolean function. If the filter generator, i.e. a n-bit LFSR filtered by g is vulnerable to an algebraic attack of complexity T that is using M keystream bits and an annihilator of g of degree d, then we can mount an algebraic attack against a linearly filtered NFSR of n bits updated with g by using an annihilator of $g + x_n$ or $g + x_n + 1$ of degree at most $d+1$ with $M' = M + \binom{n}{d+1}$ keystream bits and of complexity upper bounded by $(M')^\omega + n \cdot M'$.*

In order to mount our attack, we first use the technique introduced at the previous section to express each variable x_t as a linear combination of the initial state variables and of keystream bits. We then use the update function of the NFSR g. We have

$$x_{t+n} = g(x_t, \ldots, x_{t+n-1}).$$

By replacing each variable x_t by its linear expression L_t, we get

$$L_{t+n} = g(L_t, \ldots, L_{t+n-1})$$

which is an algebraic equation of degree d_g.

Since we get an algebraic equation for each keystream bit, we can use an annihilator h of g of degree d to mount an algebraic attack using the annihilator of degree at most $d + 1$

$$h(L_t, \ldots, L_{t+n-1}) \cdot (L_{t+n} + 1).$$

This attack will use at most M' bits of keystream with

$$M' = \sum_{k=0}^{d+1} \binom{n}{k},$$

and will have time complexity at most

$$n \cdot M' + M'^{\omega}$$

where ω is the exponent of the linear solving algorithm (2.8 for the Strassen algorithm).

However looking directly for low degree annihilators of $g + x_{t+n}$ can allow to derive equations on the initial state of the NFSR of degree lower than $d + 1$. Moreover since for all polynomials of n inputs there exists an annihilator of g of degree at most $\lceil \frac{n}{2} \rceil$, there exists an annihilator of $g + x_n$ degree at most $\lceil \frac{n+1}{2} \rceil$. When n is even, $\lceil \frac{n+1}{2} \rceil$ and $\lceil \frac{n}{2} \rceil + 1$ are equal but when n is odd, $\lceil \frac{n+1}{2} \rceil$ is strictly lower than $\lceil \frac{n}{2} \rceil + 1$. This shows that the bound on the algebraic immunity of $g + x_n$ given by Theorem 1. is not an equality and that it is sometimes possible to find annihilators of $g + x_n$ that have a lower degree than the one achieved by deriving an annihilator of $g + x_n$ from an annihilator of g as presented earlier.

For our example, since g is of degree 6, it is straightforward to derive algebraic equations of degree 6. Moreover we can remark that the degree of $(x_{t+28} \oplus 1)(x_{t+60} \oplus 1)g(x_t, \ldots, x_{t+79})$ is only 4. We consequently derive equations of degree 4 in 80 variables, which allow us to recover the initial state of the NFSR by linearization with a complexity of 2^{49} operations using 2^{21} keystream bits and memory.

5   Correlation Attacks

Our preliminary observation of Section 3, which allows us to mount algebraic attacks against linearly filtered NFSRs, can also be used to build correlation attacks against these schemes.

Theorem 3. *Let g be the Boolean function. If the filter generator, i.e. a n-bit LFSR filtered by g is vulnerable to a correlation attack of complexity T that is using M keystream bits, then a linearly filtered NFSR of n bits updated with g is also vulnerable to a correlation attack of complexity $T + n \cdot M$ that is using M keystream bits.*

In order to mount a correlation attack against the linearly filtered NFSR, we first look for a linear approximation of the update function g. Let us denote by $\mathcal{L}_g$ such a linear approximation and by ϵ its bias. With probability $\frac{1}{2} + \epsilon$ we have

$$x_{t+n} = \mathcal{L}_g(x_t, \ldots, x_{t+n-1}),$$

and by replacing each variable by its linear expression as for algebraic attacks, we get with the same probability

$$L_{t+n} = \mathcal{L}_g(L_t, \ldots, L_{t+n-1}).$$

We can mount a correlation attack in order to recover the initial state of the NFSR. Classical techniques of correlation attacks can be applied in order to build parity check equations on a small number of variables, like relations filtering or collision search [20]. In order to improve the efficiency of the correlation attack, a Fast Walsh Transform computation can be used as in [10]. One can also (almost equivalently) notice that the problem of recovering the initial state of the NFSR from the above equations can be viewed as an instance of the Learning Parity with Noise Problem LPN and consequently be solved by the techniques described in [21,22], where the Fast Walsh Transform is also used in an essential way.

In our example, the best linear approximation of the update function g is

$$\mathcal{L}_g = x_{t+62} \oplus x_{t+60} \oplus x_{t+52} \oplus x_{t+45} \oplus x_{t+37} \oplus x_{t+28} \oplus x_{t+21} \oplus x_{t+14} \oplus x_t.$$

It matches the function g with probability $\frac{594}{1024}$. We can mount two correlation attacks against our scheme: in the first one we filter the linear relations in order to retain only those relations involving the $m < n$ variables x_0 to x_{m-1}, while in the second attack we derive new linear approximation equations (of lower bias) involving $m < n$ unknown variables x_0 to x_{m-1} by combining the available approximate equations pairwise, and retaining only those pairs of relations for which the $n - m$ last coefficients collide. Then in both cases we use a Fast Walsh Transform computation in order to compute the correlation and to determine the correct value of the m bits. The first technique above allows us to recover 40 bits with 2^{52} operations and 2^{42} keystream bits, while the second technique allows us to recover 30 bits with 2^{35} operations and 2^{33} keystream bits.

6 Linearly Filtered NFSR Combined with Non Linearly Filtered LFSRs

While having high non linearity and resistance to algebraic equations, NFSRs have the drawback that it is more difficult, contrary to the case of LFSRs, to prove useful statistical properties like period length or linear complexity. A possible solution to this problem is to combine a NFSR with a LFSR: the LFSR brings its good statistical properties while the NFSR is a highly non-linear component. This is the approach of the stream cipher Grain [17].

This led us to consider linear combinations of a linearly filtered NFSR and one or several non linearly filtered LFSRs. Let us consider a m-bit LFSR of initial state $(y_0, \ldots, y_{m-1})$ filtered by a Boolean function h of degree d_h linearly combined with a linearly filtered NFSR of n bits updated by a Boolean function g of degree d_g.

6.1   Algebraic Attacks

The preliminary observation we made on a single linearly filtered NFSR can be easily adapted to this case. Each keystream bit can be written as

$$z_t = \bigoplus_{k=0}^{n-1} \alpha_k x_{t+k} \oplus h(y_t, \ldots, y_{t+m-1})$$

This allows us to write each variable x_t as the sum of a polynomial of degree d_h in the LFSR initial state variables $(y_0, \ldots, y_{m-1})$ and a linear combination of the NFSR initial state variables $(x_0, \ldots, x_{n-1})$ and of keystream bits. The polynomial of degree d_h is a sum of several instances of the function h, and is thus involving LFSR variables y_i. Using the LFSR feedback polynomials, we can express all the instances of h as polynomials of degree d_h in the LFSR initial state variables $(y_0, \ldots, y_{m-1})$.

The number of instances of h in the expression of x_t is equal to the number of keystream bit involved in the expression of x_t and consequently is determined by the difference between the taps of the non linear update function. However this number is growing with t and the complexity of finding the algebraic expression of N variable x_t as a polynomial in the initial state variables of the LFSR and NFSR can be bounded by $n \cdot N^2$.

The extension of our preliminary observation to the considered scheme allows us to derive equations to mount an algebraic attack in the same way as earlier. We replace the expression of each bit x_t in the update function g and we get equations of degree $d_g \cdot d_h$ in $n + m$ variables. Here again classical techniques of algebraic cryptanalysis may allow to reduce the degree of this system of equations. For example finding an annihilator of $g + x_{t+n+1} + 1$ of degree $d < d_g$ allows us to reduce the degree of the equations to $d \cdot d_h$. The total complexity of the attack is $n \cdot M^2 + M^\omega$ where

$$M = \sum_{k=0}^{d} \binom{n+m}{k},$$

and d the final degree of the set of equations.

It is possible to extend algebraic attacks to the case where p non linearly filtered LFSRs are linearly combined with a linearly filtered NFSR. In that case, it is possible to mount an algebraic attack against such a scheme by writing equations of degree at most $d_g \cdot \max_i d_{h_i}$ in $n + \sum_i m_i$ variables, where the $i-th$ of the p LFSRs has size m_i and a filtering function of degree d_{h_i}.

6.2   Application to a Modified Version of Grain-128

In 2006 the eSTREAM project invited the authors of the hardware candidates with a 80-bit key length to present a 128-bit version of their cipher. Grain-128 was introduced by Hell, Johansson, Maximov, and Meier [16] as a response to this invitation. It is built on the same principle as Grain, but uses a 128-bit key and 128-bit IVs: it uses a 128-bit NFSR updated by a function g, a 128-bit LFSR, and an output function. We denote the NFSR internal state at clock t by $X_t = (x_t, \ldots, x_{t+127})$ and the LFSR internal state at clock t by $Y_t = (y_t, \ldots, y_{t+127})$.

In order to achieve a very efficient design for hardware purposes, the authors have chosen a small degree for the function g. The update of the NFSR internal state is governed by the relation

$$x_{t+128} = y_t \oplus x_t \oplus x_{t+26} \oplus x_{t+56} \oplus x_{t+92} \oplus x_{t+96} \oplus x_{t+3}x_{t+67} \oplus x_{t+11}x_{t+13}$$
$$\oplus x_{t+17}x_{t+18} \oplus x_{t+27}x_{t+59} \oplus x_{t+40}x_{t+48} \oplus x_{t+61}x_{t+65} \oplus x_{t+68}x_{t+84}.$$

In order to avoid attacks, two bits of the NFSR internal state instead of one in Grainv1 are input to the non-linear output function h:

$$h(X_t, Y_t) = x_{t+12}y_{t+8} \oplus y_{t+13}y_{t+20} \oplus x_{t+95}y_{t+42} \oplus y_{t+60}y_{t+79} \oplus x_{t+12}x_{t+95}y_{t+95}$$

Each keystream bit can be written as a the XOR of a linear combination of the NFSR internal state, a bit of the LFSR internal state, and the output of function h:

$$z_t = L(X_t) \oplus y_{t+93} \oplus h(X_t, Y_t)$$

with

$$L(X_t) = x_{t+2} \oplus x_{t+15} \oplus x_{t+36} \oplus x_{t+45} \oplus x_{t+64} \oplus x_{t+73} \oplus x_{t+89}$$

In the paper describing Grain-128, the authors discuss the resistance of the algorithm to algebraic attacks: "In Grain-128, an NFSR is used to introduce nonlinearity together with the function h(). Solving equations for the initial 256 bit state is not possible due to the nonlinear update of the NFSR. The algebraic degree of the output bit expressed in initial state bits will be large in general and also varying in time. This will defeat any algebraic attack on the cipher."

We now introduce a modified version of Grain-128. In this version, we replace the two bits of the NFSR internal state that were input into the non-linear output function h by two bits of the LFSR internal state. As stated by the authors of Grain-128, the non-linearity of the algorithm comes from the NFSR and from this function h. We now present an algebraic attack against this modified version, which shows that with the modified version of the function h it is possible to

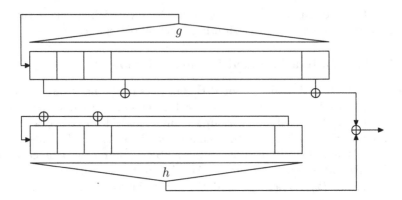

Fig. 4. Combination of a linearly filtered NFSR and a non linearly filtered LFSR

write equations of constant degree even tough the non-linear update of the NFSR is supposed to make the degree vary in time, as claimed by the authors of Grain-128.

For the modified version, the new function h depends only of the LFSR internal state. In order to keep the properties of function h in particular the number of variables and the degree, we replace x_{t+12} and x_{t+95} by y_{t+12} and y_{t+94}. This choice of the two new taps is not significant for our attack.

$$h(Y_t) = y_{t+8}y_{t+12} \oplus y_{t+13}y_{t+20} \oplus y_{t+42}y_{t+94} \oplus y_{t+60}y_{t+79} \oplus y_{t+12}y_{t+94}y_{t+95}$$

In order to keep the keystream dependent on the two bits x_{t+12} and x_{t+95}, we add them to the linear part of the output. Consequently each keystream bit can now be written as:

$$z_t = L(X_t) \oplus x_{t+12} \oplus x_{t+95} \oplus y_{t+93} \oplus h(Y_t)$$

This modified version is almost identical to the case illustrated in Figure 4. The only difference is the influence of the LFSR output on the update of the NFSR. Using the technique described for algebraic attacks against combination of a NFSR and a LFSR, we can express each variable bit of the NFSR internal state x_t as the sum of a polynomial of degree d_h in the LFSR initial state variables $(y_0, \ldots, y_{127})$, of a linear combination of the NFSR initial state variables $(x_0, \ldots, x_{127})$ and of keystream bits. The polynomial of degree d_h is a sum of several instances of the function h and of linear combinations of the LFSR initial state variables stemming from the term y_{t+93} for different values of t. As already explained this allows us, by replacing the expression of each bit x_t and of the bit y_t inside of the update function of the NFSR, to write equations of degree $d_g \cdot d_h = 6$ in the 256 variables of the LFSR and NFSR initial states. By linearizing these equations, we can recover the 256 variables in time $128 \cdot M^2 + M^{2.73}$, where

$$M = \sum_{k=0}^{6} \binom{256}{k}.$$

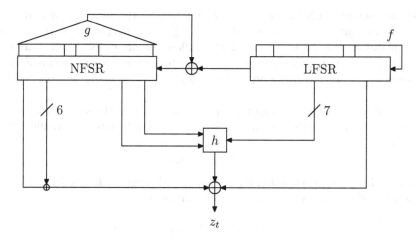

Fig. 5. The original Grain-128

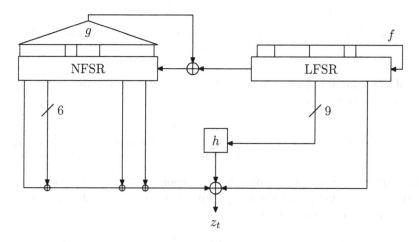

Fig. 6. Modified version of Grain-128

This result in an attack of complexity 2^{105} using 2^{39} keystream bits.

The described attack is not applicable to the original Grain-128 due to the non-linearity in the NFSR state variables. However it shows that the argument used by the authors to justify the immunity against algebraic attack was incomplete.

6.3   Correlation Attacks

Unlike the simple linearly filtered NFSR case, it seems difficult to mount correlation attack against a combination of a linearly filtered NFSR and non linearly filtered LFSRs by using our preliminary observation. This is due to the increasing number of instances of the h function. If the functions g and h are replaced by linear approximations, the resulting bias is decreasing as the number of the instances of h grows. Consequently unless the biases for g and h are very strong and the number of required keystream bits is very low, it seems difficult to mount correlation attacks against these schemes by using our preliminary observation.

It is however possible to mount a correlation attack in the special case where the linear filtering function of the NFSR has one single non-zero coefficient. In that case we have

$$z_t = x_t \oplus h(y_t, \ldots, y_{t+m-1}).$$

By using a linear approximation $\mathcal{L}_g$ of bias ϵ_g and weight w and a linear approximation $\mathcal{L}_h$ of bias ϵ_h, we can derive approximate relations

$$x_{t+n} \simeq \mathcal{L}_g(x_t, \ldots, x_{t+n-1})$$

$$\simeq \bigoplus_{j=0}^{w} x_{t+i_j}$$

$$z_{t+n} \oplus h(y_{t+n}, \ldots, y_{t+n+m-1}) \simeq \bigoplus_{j=0}^{w} z_{t+i_j} \oplus h(y_{t+i_j}, \ldots, y_{t+i_j+m-1})$$

By replacing the non-linear outputs of h with its linear approximation $\mathcal{L}_h$, we get approximate relations

$$z_{t+n} \oplus \mathcal{L}_h(y_{t+n}, \ldots, y_{t+n+m-1}) \simeq \bigoplus_{j=0}^{w} z_{t+i_j} \oplus \mathcal{L}_h(y_{t+i_j}, \ldots, y_{t+i_j+m-1}),$$

which can be re-expressed as approximate relations involving the initial state bits of the LFSR. By using the Piling up Lemma, we can compute the equivalent bias of approximate relations. We get:

$$\epsilon = \epsilon_g (2\epsilon_h)^{w+1}$$

Consequently if the bias ϵ_g and ϵ_h are large enough and if the weight w of the linear approximation of g is small enough, a correlation attack is possible against the construction of Figure 4 in this special case.

7   Conclusion

In this paper, we have shown that the dual case of the filter generator, i.e. a linearly filtered NFSR, is vulnerable to the same kind of attacks than the filter generator. We described how to mount algebraic and correlation attacks against this scheme. These attacks were illustrated on an example NFSR taken from the Grain stream cipher. We then extended these attacks to combinations of a linearly filtered NFSR and one or several non linearly filtered LFSRs. We illustrated the latter extension by an algebraic attack on a modified version of Grain-128, which can be broken in 2^{105} computations with 2^{39} keystream bits. This attack is not applicable to the original Grain-128 but is shows that the use of a NFSR is not sufficient to avoid all algebraic attacks. As far as we know, none of these attacks is directly applicable to stream cipher candidates submitted to the eSTREAM competition or recognized stream ciphers like SNOW 2.0 or MUGI.

The techniques presented in this paper can be easily extended to ciphers in which $t > 1$ bits of the current state are non-linearly updated at each step while t or more linear combinations of the state bits are output as keystream bits. It is an open question whether those attack techniques can be also extended to ciphers in which $t > 1$ state bits are non-linearly updated, while only $t' < t$ linear combinations of the state bits are output.

References

1. Armknecht, F., Krause, M.: Algebraic attacks on combiners with memory. In: Boneh, D. (ed.) CRYPTO 2003. LNCS, vol. 2729, pp. 162–175. Springer, Heidelberg (2003)
2. Ars, G., Faugère, J.: An Algebraic Cryptanalysis of Nonlinear Filter Generators using Gröbner Basis. Technical report, INRIA (2003)
3. Ars, G., Faugère, J.-C.: An algebraic cryptanalysis of nonlinear filter generators using groebner basis. INRIA (2003)
4. Berbain, C.: Analyse et conception d'algorithmes de chiffrement flot. PhD thesis, Université Paris. Diderot, Paris 7 (2007)
5. Berbain, C., Gilbert, H., Maximov, A.: Cryptanalysis of grain. In: Robshaw, M.J.B. (ed.) FSE 2006. LNCS, vol. 4047, pp. 15–29. Springer, Heidelberg (2006)
6. Braeken, A., Lano, J.: On the (Im)Possibility of practical and secure nonlinear filters and combiners. In: Preneel, B., Tavares, S. (eds.) SAC 2005. LNCS, vol. 3897, pp. 159–174. Springer, Heidelberg (2006)
7. Braeken, A., Lano, J., Mentens, N., Preneel, B., Verbauwhede, I.: SFINKS: A Synchonous Stream Cipher for Restricted Hardware Environments. In: eSTREAM, ECRYPT Stream Cipher Project (2005)
8. De Cannière, C., Preneel, B.: Trivium: Specifications. eSTREAM, ECRYPT Stream Cipher Project (2005)
9. Canteaut, A., Trabbia, M.: Improved fast correlation attacks using parity-check equations of weight 4 and 5. In: Preneel, B. (ed.) EUROCRYPT 2000. LNCS, vol. 1807, pp. 573–588. Springer, Heidelberg (2000)
10. Chose, P., Joux, A., Mitton, M.: Fast correlation attacks: An algorithmic point of view. In: Knudsen, L.R. (ed.) EUROCRYPT 2002. LNCS, vol. 2332, pp. 209–221. Springer, Heidelberg (2002)
11. Courtois, N.T.: Fast algebraic attacks on stream ciphers with linear feedback. In: Boneh, D. (ed.) CRYPTO 2003. LNCS, vol. 2729, pp. 176–194. Springer, Heidelberg (2003)
12. Courtois, N., Meier, W.: Algebraic Attacks on Stream Ciphers with Linear Feedback. In: Biham, E. (ed.) EUROCRYPT 2003. LNCS, vol. 2656, pp. 345–359. Springer, Heidelberg (2003)

13. Englund, H., Johansson, T.: A new simple technique to attack filter generators and related ciphers. In: Handschuh, H., Hasan, M.A. (eds.) SAC 2004. LNCS, vol. 3357, pp. 39–53. Springer, Heidelberg (2004)

14. Gong, G., Nawaz, Y.: The WG Stream Cipher. eSTREAM, ECRYPT Stream Cipher Project (2005)

15. Gouget, A., Sibert, H.: Revisiting correlation-immunity in filter generators. In: Adams, C., Miri, A., Wiener, M. (eds.) SAC 2007. LNCS, vol. 4876, pp. 378–395. Springer, Heidelberg (2007)

16. Hell, M., Johansson, T., Maximov, A., Meier, W.: A Stream Cipher Proposal: Grain-128. eSTREAM, ECRYPT Stream Cipher Project (2006)

17. Hell, M., Johansson, T., Meier, W.: Grain - A Stream Cipher for Constrained Environments. eSTREAM, ECRYPT Stream Cipher Project (2005)

18. Johansson, T., Jönsson, F.: Fast correlation attacks based on turbo code techniques. In: Wiener, M. (ed.) CRYPTO 1999. LNCS, vol. 1666, pp. 181–197. Springer, Heidelberg (1999)

19. Johansson, T., Jönsson, F.: Improved fast correlation attacks on stream ciphers via convolutional codes. In: Stern, J. (ed.) EUROCRYPT 1999. LNCS, vol. 1592, pp. 347–362. Springer, Heidelberg (1999)

20. Leveiller, S., Zémor, G., Guillot, P., Boutros, J.: A new cryptanalytic attack for PN-generators filtered by a boolean function. In: Nyberg, K., Heys, H.M. (eds.) SAC 2002. LNCS, vol. 2595, pp. 232–249. Springer, Heidelberg (2003)

21. Levieil, É., Fouque, P.-A.: An improved LPN algorithm. In: De Prisco, R., Yung, M. (eds.) SCN 2006. LNCS, vol. 4116, pp. 348–359. Springer, Heidelberg (2006)

22. Lyubashevsky, V.: The parity problem in the presence of noise, decoding random linear codes, and the subset sum problem. In: Chekuri, C., Jansen, K., Rolim, J.D.P., Trevisan, L. (eds.) APPROX 2005 and RANDOM 2005. LNCS, vol. 3624, pp. 378–389. Springer, Heidelberg (2005)

23. Meier, W., Staffelbach, O.: Fast correlation attacks on stream ciphers. In: Günther, C.G. (ed.) EUROCRYPT 1988. LNCS, vol. 330, pp. 301–314. Springer, Heidelberg (1988)

24. Siegenthaler, T.: Correlation-immunity of Non-linear Combining Functions for Cryptographic Applications. IEEE Transactions on Information Theory 30, 776–780 (1984)

A Cache Timing Analysis of HC-256

Erik Zenner

Technical University of Denmark
Department of Mathematics
e.zenner@mat.dtu.dk

Abstract. In this paper, we describe a cache-timing attack against the stream cipher HC-256, which is the strong version of eStream winner HC-128. The attack is based on an abstract model of cache timing attacks that can also be used for designing stream ciphers. From the observations made in our analysis, we derive a number of design principles for hardening ciphers against cache timing attacks.

Keywords: Cryptanalysis, side-channel attack, cache timing attack, stream cipher, HC-256.

1 Introduction

Cache timing attacks are a new class of side-channel attacks. They received significant attention after being applied to the Advanced Encryption Standard (AES) independently by Bernstein [1] and Osvik, Shamir, and Tromer [12,13] in 2005[1]. The idea is that in some settings, the adversary can obtain information about the cache accesses of a legitimate party by measuring timings. Optimised software implementations of the AES turned out to be particularly vulnerable to this kind of attack.

The discovery was met with great interest. Subsequent research verified the correctness of the findings [11,10,9,15], improved the attack technically [14,3,8] or algorithmically [5], and devised and analysed countermeasures [6,4,16].

However, the focus of the attacks was on the AES, and the countermeasures mainly targeted the *implementation* of cryptographic designs. In this paper, we take a different approach: We discuss how cipher *designers* can make such attacks more difficult. In order to demonstrate our approach, instead of considering a block cipher like AES, we analyse the stream cipher HC-256.

1.1 Cache Timing Attacks

Cache timing attacks exploit that loading data into a CPU register is faster when done from cache than from RAM. By measuring cache timings, an observer can obtain information about the inner state of a cipher. In the following, we give a simplified description of cache timing attacks; for a more complete description, see e.g. [13,9].

---

[1] For prior work on cache timing attacks, see the references contained in [1] and [12].

R. Avanzi, L. Keliher, and F. Sica (Eds.): SAC 2008, LNCS 5381, pp. 199–213, 2009.

Cache workings: The CPU cache of modern processors is organised into blocks of s bytes. Correspondingly, RAM is considered to be (virtually) divided into s-byte blocks. When loading data from RAM into a CPU register, the system first checks whether the corresponding RAM block already lies in cache. If yes, it is loaded directly from cache, which is very fast. If not, it is first loaded from RAM to cache, which takes longer. Mapping from RAM to cache is typically by a simple modulo operation, i.e. if the cache can hold n blocks and if the data lies in RAM block a, then it is loaded into cache block a mod n. This means that neighbouring data in RAM (e.g. tables) stays clustered in cache.

A simple attack: As an example, consider the **prime-then-probe** method presented in [13]. The adversary starts by filling all the cache with his own data. Then the legitimate user U gets the read/write token. U loads the data required for his own computations into cache, where it evicts the adversary's data. When the adversary reobtains the read/write token, he tries to reload his own data from cache. For each cache block, if this takes long, it means that U has evicted the corresponding data.

From this analysis, the adversary obtains a profile of cache blocks that have been used by U. This profile is a noisy version of the cache blocks that have been used for the encryption. By repeating the experiment a number of times, a good approximation of the real cache access profile can be obtained.

Note that the adversary does not learn the *data* that was written in the cache by U – he learns something about the *addresses* of the data that was used. In the case of the AES, this corresponds to the indices of the S-box entries used for encryption, which in turn can be used for an attack.

Practicality: Cache timing attacks require cache timing measurements of sufficient precision. In addition, the experiment has to be repeated sufficiently often. Obviously, these requirements are not always met. However, they are relevant in shared server and in sandbox scenarios, and Bertoni et al. [3] show how to use cache timings if the adversary has physical access to a device, making the attack much more realistic.

Responsibility: Some researchers claim that defending against side-channel attacks should be the responsibility of the implementer, not the cipher designer. However, this view is not shared by everyone. As an example, the AES was chosen partially due to its inherent resistance against side-channel attacks (see e.g. Section 7 of [1]). The reason is that algorithms are designed only once, but implemented many times on many platforms. Thus, if side-channel attacks can be avoided in the design phase, implementation becomes easier, which seems to be preferable. In order to emphasise the designer's responsibility, we use a simplified terminology in this paper: We say that a cipher can be "broken" in a cache timing model if an unprotected implementation is vulnerable to a cache timing attack.

2   The Stream Cipher HC-256

HC-256 was proposed by Wu in [17], and its reduced version HC-128 is part of the eStream portfolio [7]. The cipher is based on two large, key-based tables (i.e., no fixed S-boxes) that change content over time. With each call to the keystream generation function, the cipher updates one table entry and outputs one 32-bit keystream word.

Notation: HC-256 requires a 256-bit key K and a 256-bit IV IV. It uses two tables P and Q, which contain 1024 32-bit words each. Table entries are identified by $P[i]$ and $Q[i]$.

In the following, $\oplus$ denotes xor, $||$ concatenation (most significant bits first), $\ggg$ a circular right shift, $\boxplus$ addition modulo 2^{32}, and $\boxminus$ subtraction modulo 2^{10}.

If X is a word, we denote by $X^{(b..a)}$ the bits $b..a$, where $b > a$. For all notations, the most significant bits are written to the left, while the least significant bits are written to the right. Thus, we can write $X = X^{(31..0)}$.

Auxiliary Functions: The following auxiliary functions on 32-bit variables are used:

$$f_1(x) = (x \ggg 7) \oplus (x \ggg 18) \oplus (x \gg 3)$$
$$f_2(x) = (x \ggg 17) \oplus (x \ggg 19) \oplus (x \gg 10)$$
$$g_1(x,y) = ((x \ggg 10) \oplus (y \ggg 23)) \boxplus Q[(x \oplus y)^{(9..0)}]$$
$$g_2(x,y) = ((x \ggg 10) \oplus (y \ggg 23)) \boxplus P[(x \oplus y)^{(9..0)}]$$
$$h_1(x) = Q[00||x^{(7..0)}] \boxplus Q[01||x^{(15..8)}] \boxplus Q[10||x^{(23..16)}] \boxplus Q[11||x^{(31..24)}]$$
$$h_2(x) = P[00||x^{(7..0)}] \boxplus P[01||x^{(15..8)}] \boxplus P[10||x^{(23..16)}] \boxplus P[11||x^{(31..24)}]$$

Key/IV Setup: For initialisation, the key is split into 32-bit words $K[0], \ldots, K[7]$, and the IV is split into 32-bit words $IV[0], \ldots, IV[7]$. With the help of an auxiliary array $W[0], \ldots, W[2559]$ and a global counter variable r, the algorithm can be described as in Figure 1.

Init(K, IV)
1. For $i = 0, \ldots, 7$:
2. $W[i] = K[i]$
3. For $i = 8, \ldots, 15$:
4. $W[i] = IV[i - 8]$
5. For $i = 16, \ldots, 2559$:
6. $W[i] = f_2(W[i - 2]) \boxplus W[i - 7] \boxplus f_1(W[i - 15]) \boxplus W[i - 16] \boxplus i$
7. For $j = 0, \ldots, 1023$:
8. $P[j] = W[j + 512]$
9. $Q[j] = W[j + 1536]$
10. Set $r = -4096$
10. Repeat 4096 times:
10.   **Next()**   (* Ignore the output *)

Fig. 1. Key/IV setup for HC-256

Next()
1. Set $j = r \bmod 1024$
2. If $((r \bmod 2048) \in \{0, \ldots, 1023\})$:
3. $P[j] = P[j \boxminus 1024] \boxplus P[j \boxminus 10] \boxplus g_1(P[j \boxminus 3], P[j \boxminus 1023])$
4. $z_r = h_1(P[j \boxminus 12]) \oplus P[j]$
5. Else:
6. $Q[j] = Q[j \boxminus 1024] \boxplus Q[j \boxminus 10] \boxplus g_2(Q[j \boxminus 3], Q[j \boxminus 1023])$
7. $z_r = h_2(Q[j \boxminus 12]) \oplus Q[j]$
8. $r = r + 1$

Fig. 2. Keystream generation for HC-256

Keystream Generation: The r-th call to the **Next()** function updates one table entry and produces one 32-bit output word z_r. The function is described in Figure 2. Note that $r = 0$ for the first output word.

3   Mapping Measurements to Inner State

3.1   Preliminaries

State-dependent table lookups: From a cache timing attack, the adversary learns (part of) the table indices that were accessed by the encryption algorithm. However, most table lookups made by HC-256 depend on a public counter which is known to the adversary anyway. The only exceptions are the state-dependent table lookups within the functions g_1 and h_1 (leaking information about table P) as well as g_2 and h_2 (table Q).

Observable index bits: Ideally, the adversary would learn the full table index from each cache observation. In practice, however, the cache is organised in blocks that store several RAM table entries. Thus, all the adversary can learn from his measurements is the cache block containing the table entry.

In the following, we assume that the tables P and Q are perfectly aligned with the cache blocks[2]. Thus, the tables themselves can be considered as being split into blocks that have the same size as the cache blocks.

The cache block size is processor dependent and varies typically between 16 and 128 byte. In the following, we use a cache block size of 64 byte, since it is currently particularly wide-spread (e.g. in Pentium 4 and Athlon). Since tables P and Q have 1024 entries with an entry size of 4 byte each, each table block contains 16 table entries, and each table consists of 64 blocks. Thus, by measuring cache timings, the adversary learns index bits 4..9, but not 0..3.

Note that if the cache block size is smaller (larger) than 64 byte, he will obtain more (less) information about the table entries.

---

[2] If this is not the case, our attack becomes easier, since unaligned table entries leak additional information about the inner state.

3.2 Keystream Generation vs. Key/IV Setup

The functions g_1, g_2, h_1, and h_2 are accessed both during keystream generation and key/IV setup. However, during key/IV setup, *all* entries in the tables W, Q and P are accessed at least once. Thus, the adversary will obtain no additional information compared to the standard model.

Instead, we concentrate on the keystream generation phase, where we make repeated measurements for each output block. This is modelled by giving the adversary access to two oracles:

- KEYSTREAM(i): The adversary requests the cipher to return the i-th keystream block to him. The block length depends on the cipher design.
- SCT_KEYSTREAM(i): The adversary obtains an unordered list of all cache accesses made by KEYSTREAM(i).

A justification and generalisation of this model is given in Section 6.

3.3 Initial State Candidates

If the adversary calls the SCT_KEYSTREAM(R) oracle, this corresponds to a call to the **Next()** function. Consider such a call for a round r with $(r \bmod 2048) \in \{0, \ldots, 1023\}$, i.e. code lines 3 and 4 are executed. From functions g_1 and h_1, he observes either 4 or 5 accesses to table Q, as follows.

Function h_1: In function h_1, table Q is accessed at indices $(00||P[j \boxminus 12]^{(7..0)})$, $(01||P[j \boxminus 12]^{(15..8)})$, $(10||P[j \boxminus 12]^{(23..16)})$, and $(11||P[j \boxminus 12]^{(31..24)})$. While in general, the adversary does not know which table access belongs to which variable, things are more obvious here. Each of the four 10-bit indices starts with a unique 2-bit prefix and can thus be clearly assigned to one of the four variables. Thus, if it were not for code line 3, the adversary could immediately determine the upper half-bytes for $P[j \boxminus 12]$.

Function g_1: However, in the same function call, g_1 accesses table Q at index $(P[j \boxminus 3] \oplus P[j \boxminus 1023])^{(9..0)}$. This index can have any of the prefixes $00, 01, 10$, or 11. Thus, we can not distinguish it from one of the accesses by h_1 which has the same prefix (unless it accidentially uses the same cache block, which happens with probability $1/16$).

Concluding, for three of the four bytes in $P[j \boxminus 12]$, we know precisely their upper half-byte. For the fourth one, we normally have two candidates, which we can not distinguish without additional information. In addition, for $(P[j \boxminus 3] \oplus P[j \boxminus 1023])$, we know exactly what the bits 9 and 8 are, and we have two candidate assignments for bits 7..4.

Functions h_2 and g_2: Exactly the same observations hold for table P for rounds r with $(r \bmod 2048) \in \{1024, \ldots, 2047\}$.

4   Reconstructing the Full Inner State

4.1   Notation

Before considering several calls to the **Next()** function, we have to define a unique notation for the table entries. Since the table is constantly updated, we have to make it clear which of a succession of values in e.g. table cell $P[12]$ we mean.

To this end, for table P, we write P_u when we mean the u-th value that was updated for this table, where P_0 is updated in round $r = 0$. As an example, table cell $P[12]$ has the value P_{-1012} after initialisation, obtains value P_{12} in round $r = 12$ and value P_{1036} in round $r = 2060$.

Similarly, for table Q, we write Q_u when we mean the u-th value that was updated for this table, where Q_0 is updated in round $r = 1024$. As an example, table cell $Q[12]$ has the value Q_{-1012} after initialisation, obtains value Q_{12} in round $r = 1036$ and value P_{1036} in round $r = 3084$.

The following table describes the the relationship between rounds and sequence words that are used for the attack.

Round	Table P	Table Q
$0, \ldots, 1023$	$P_0, \ldots, P_{1023}$	-
$1024, \ldots, 2047$	-	$Q_0, \ldots, Q_{1023}$
$2048, \ldots, 3071$	$P_{1024}, \ldots, P_{2047}$	-
$3072, \ldots, 4095$	-	$Q_{1024}, \ldots, Q_{2047}$
$4096, \ldots, 5119$	$P_{2048}, \ldots, P_{3071}$	-
$5120, \ldots, 6143$	-	$Q_{2048}, \ldots, Q_{3071}$
$6144, \ldots, 7167$	$P_{3072}, \ldots, P_{4095}$	-
$7168, \ldots, 8191$	-	$Q_{3072}, \ldots, Q_{4095}$

4.2   Step 1: Determining the Half-Bytes

The purpose of the first step is to uniquely identify the correct assignments to the upper half-bytes of $P_{1024}, \ldots, P_{3083}$ and $Q_{1024}, \ldots, Q_{3071}$.

Measurement: By using the SCT_KEYSTREAM() oracle for rounds

$$r = 25, \ldots, 1023, \ r = 2048, \ldots, 3071, \ r = 4096, \ldots, 5119, \ r = 6144, \ldots, 6176,$$

the adversary observes partial information about table entries as described in Subsection 3.3. This gives him 2 candidate assignments for each of the following lines:

From h_1				From g_1
$P_{13}^{(7..4)}$	$P_{13}^{(15..12)}$	$P_{13}^{(23..20)}$	$P_{13}^{(31..28)}$	$P_{22}^{(9..4)} \oplus P_{-998}^{(9..4)}$
$P_{14}^{(7..4)}$	$P_{14}^{(15..12)}$	$P_{14}^{(23..20)}$	$P_{14}^{(31..28)}$	$P_{23}^{(9..4)} \oplus P_{-997}^{(9..4)}$
$\ldots$	$\ldots$	$\ldots$	$\ldots$	$\ldots$
$P_{3092}^{(7..4)}$	$P_{3092}^{(15..12)}$	$P_{3092}^{(23..20)}$	$P_{3092}^{(31..28)}$	$P_{3101}^{(9..4)} \oplus P_{2081}^{(9..4)}$

In particular, for the equations $P_{1033} \oplus P_{13}, \ldots, P_{3092} \oplus P_{2072}$, we have 2 candidates for the bits 7..4 from g_1. At the same time, from h_1, we have 1 candidate (with probability $\approx 3/4$) or 2 candidates (with probability $\approx 1/4$) for bits 7..4 of the corresponding values $P_{13}, \ldots, P_{3092}$.

A simple consistency check: We will now try to figure out which of the two candidates for each g_1 equation is the correct one. First note that with probability $1/16$ there is really only one candidate for this equation, namely if bits 7..4 are the same as for h_1. If this is not the case, there are three subcases:

1. For the corresponding h_1 values, there is only 1 candidate each. In this case (which happens with prob. $\approx 9/16$), checking by xoring those h_1 values will always identify the correct candidate for the g_1 value.
2. One of the h_1 values has 1 candidate and one has 2 candidates. In this case (which happens with prob. $\approx 6/16$), there is only one wrong combination of h_1 candidates, and it is identical to the wrong g_1 candidate with probability $1/16$. Thus, the test identifies the wrong g_1 candidate with probability $15/16$.
3. Both h_1 values have 2 candidates. In this case (which happens with prob. $\approx 1/16$), there are 3 wrong combinations of h_1 candidates. They identify the wrong g_1 candidate with probability $\frac{15 \cdot 15 \cdot 14}{16^3}$.

Concluding, the probability of identifying a wrong g_1 candidate by a simple test is

$$\frac{1}{16} + \frac{15}{16} \cdot \left(\frac{9}{16} \cdot 1 + \frac{6}{16} \cdot \left(\frac{15}{16} \right) + \frac{1}{16} \cdot \left(\frac{15 \cdot 15 \cdot 14}{16^3} \right) \right) \approx 0.9646.$$

Consequence: In the following, we will thus assume that the correct candidates for equations $P_{1033} \oplus P_{13}, \ldots, P_{3092} \oplus P_{2072}$ have been identified. In reality, there will be a small number of such equations that have two candidates, but the percentage is small enough not to significantly influence the analysis in the following sections (it will only make an implementation of the attack slightly messier).

Once the correct candidates for equations $P_{1033} \oplus P_{13}, \ldots, P_{3092} \oplus P_{2072}$ are known, we can also identify the correct candidates for the h_1 values of the same lines. Thus, in the following, we can assume that the upper half-bytes are known for the h_1 values under consideration, i.e. the sequence words $P_{1024}, \ldots, P_{3083}$.

By repeating the same procedure for rounds

$$r = 1049, \ldots, 2047, \ r = 3072, \ldots, 4095, \ r = 5120, \ldots, 6143, \ r = 7168, \ldots, 7188,$$

the same bits can be determined for sequence words $Q_{1024}, \ldots, Q_{3071}$.

4.3 Step 2: Reducing the Number of Candidates

In the second step, we will reduce the number of candidates for $Q_{1024}, \ldots, Q_{3059}$ and $P_{2048}, \ldots, P_{3071}$ from 2^{16} to 2^8.

Sequence words $Q_{1024}, \ldots, Q_{2035}$: Let us consider the calls to the function

$$z_r = h_2(Q[j \boxminus 12]) \oplus Q[j]$$

that occur in rounds $r = 3084, \ldots, 4095$. They access the sequence words Q_{1024}, $\ldots, Q_{2047}$ and $P_{1024}, \ldots, P_{2047}$. According to Subsection 4.2, we know all upper half-bytes for these entries. Now we have to try and learn as much as possible about the remaining inner state from this information.

Let $\gamma_0, \ldots, \gamma_3 = (00 \| Q[j \boxminus 12]^{(7..0)}), \ldots, (11 \| Q[j \boxminus 12]^{(31..24)})$. Then we can re-write the above equation as follows:

$$z_r \oplus Q[j] = P[\gamma_0] \boxplus P[\gamma_1] \boxplus P[\gamma_2] \boxplus P[\gamma_3] \tag{1}$$

Remember that the adversary knows the keystream word z_r. Also note that for $Q[j]$, $Q[j \boxminus 12]$ and for all $P[\gamma_i]$ involved, we know the upper half-bytes. We will now proceed by guessing the remaining 16 bits of $Q[j \boxminus 12]$ and then verifying the result by using eq. (1).

If the equation would use $\oplus$ instead of $\boxplus$, verification would be straightforward. We would use the upper half-bytes to obtain 16 linear equations in $GF(2)$. Since we also have to guess 16 bit for $Q[j \boxminus 12]$, only one false guess would pass this test on average.

However, for addition, we have to take carries into account. If we write $A^I, \ldots, A^{IV}$ instead of $A^{(7..4)}, \ldots, A^{(31..28)}$ for the four upper half-bytes of a word A, then we can write 4 verification equations as follows:

$$
\begin{aligned}
z_r^{I} \;\oplus\; Q[j]^{I} &= P[\gamma_0]^{I} \;\boxplus\; P[\gamma_1]^{I} \;\boxplus\; P[\gamma_2]^{I} \;\boxplus\; P[\gamma_3]^{I} \;\boxplus\; c_0 \\
z_r^{II} \;\oplus\; Q[j]^{II} &= P[\gamma_0]^{II} \;\boxplus\; P[\gamma_1]^{II} \;\boxplus\; P[\gamma_2]^{II} \;\boxplus\; P[\gamma_3]^{II} \;\boxplus\; c_1 \\
z_r^{III} \;\oplus\; Q[j]^{III} &= P[\gamma_0]^{III} \;\boxplus\; P[\gamma_1]^{III} \;\boxplus\; P[\gamma_2]^{III} \;\boxplus\; P[\gamma_3]^{III} \;\boxplus\; c_2 \\
z_r^{IV} \;\oplus\; Q[j]^{IV} &= P[\gamma_0]^{IV} \;\boxplus\; P[\gamma_1]^{IV} \;\boxplus\; P[\gamma_2]^{IV} \;\boxplus\; P[\gamma_3]^{IV} \;\boxplus\; c_3
\end{aligned}
$$

Here, $c_0, \ldots, c_3$ are the carry values, taken from $\{0, 1, 2, 3\}$.

Thus, if we want to use the above equations to verify our guess for $Q[j \boxminus 12]$, we have to guess the carry values, too. In total, this gives us $2^{16} \cdot 2^8 = 2^{24}$ possible guesses. On the other hand, we have 16 verification bits. This means that on average, 2^8 guesses for $Q[j \boxminus 12]$ will survive the test. For the table entries $Q_{1024}, \ldots, Q_{2035}$, we write these guesses into a table.

Sequence words $Q_{2036}, \ldots, Q_{3059}$: It remains to reconstruct the remaining words $Q_{2036}, \ldots, Q_{3059}$, which can be done in a similar manner by considering rounds $r = 5120, \ldots, 6143$. These rounds use the sequence words $Q_{2036}, \ldots, Q_{3071}$, as well as some of the sequence words $P_{2048}, \ldots, P_{3071}$. Using the same technique as above, we can reduce the number of candidates for $Q_{2036}, \ldots, Q_{3059}$ to approximately 2^8 candidates each.

Sequence words $P_{2048}, \ldots, P_{3071}$: The same technique can also be applied to reduce the number of candidates for the sequence words $P_{2048}, \ldots, P_{3071}$. We do this by considering the rounds $r = 4108, \ldots, 5119$, which use sequence words

$P_{2048}, \ldots, P_{3071}$ as well as $Q_{1024}, \ldots, Q_{2047}$. From this, we can reduce the number of candidates for $P_{2048}, \ldots, P_{3071}$ to 2^8. Afterwards, we consider rounds $6144, \ldots, 6155$, which use sequence words $P_{3060}, \ldots, P_{3083}$ as well as some of the table entries $Q_{2048}, \ldots, Q_{3071}$.

Resulting table: For $Q_{1024}, \ldots, Q_{3059}$ and $P_{2048}, \ldots, P_{3071}$, the surviving candidate words are written in a table. The total size of this table is $3060 \cdot 2^8 \cdot 4 \approx 3 \cdot 2^{20}$ byte, i.e. 3 MByte.

4.4   Step 3: Backtracking Attack

In the final step, we reduce the number of candidates for $Q_{1024}, \ldots, Q_{2047}$ and $P_{2048}, \ldots, P_{3071}$ to one.

Reconstructing table Q: Consider code line 6 as it is called in round $r = 5120$. It has the following form:

$$Q_{2048} = Q_{1024} \boxplus Q_{2038} \boxplus g_2(Q_{2045}, Q_{1025}).$$

This means that the equation uses the variables $Q_{1024}, Q_{1025}, Q_{2038}, Q_{2045}, Q_{2048}$ and an entry of table P with unknown index. For each of these 6 variables, we have an average of 2^8 possible assignments. If we guess all of these assignments, we obtain 2^{48} possible candidates. Since wrong guesses for the 32-bit values satisfy a linear equation with probability $1/2^{32}$, only $\approx 2^{16}$ of them remain as valid states.

We proceed in the same way for round $r = 5121$, which requires variables $Q_{1025}, Q_{1026}, Q_{2039}, Q_{2046}, Q_{2049}$ and an entry of table P. Note that Q_{1025} is already known from last round, meaning that we only have to guess 5 variables[3]. Our search space increases to $2^{16} \cdot 2^{40} = 2^{56}$, then it collapses to 2^{24} when filtering out the assignments that don't fulfil the equation.

Repeating the same step for round $r = 5122$ increases our search tree to 2^{64}, then collapsing it to 2^{32}. For round $r = 5123$, however, *two* of the required variables are already known. This means that only 4 variables have to be guessed, and the search tree expands to 2^{64} and reduces itself to 2^{32} after verification.

It continues to behave that way until round $r = 5127$. In this round, we need *three* variables that have already been guessed before. This means that the tree only expands to 2^{56} candidates and then collapses back to 2^{24}. From now on, the tree size will reduce itself with every round, until round $r = 5130$ when it has size ≈ 1 after verification, i.e. only valid guesses remain. From now on, every candidate guess can be verified right away.

Concluding, after running through rounds $r = 5120, \ldots, 6143$, we have reconstructed the correct solution for table entries $Q_{1024}, \ldots, Q_{2047}$.

---

[3] Of course, there is also a possibility that the table entry for table P repeats itself, but this probability is not very high in the first rounds. Should this happen by chance, the attack becomes even more efficient.

Reconstructing table P: Note that from the guesses above, a significant number of entries for table P have already been reconstructed. There are numerous possiblities for determining the remaining entries. A very simple one would be to run the same attack as above, using code line 3 instead of line 6. Note that this requires extra cache timings to reduce the number of candidates for $P_{3072}, \ldots, P_{4095}$ to 2^8, each.

A more intelligent approach uses code line 4 for rounds $r = 5008, \ldots, 5119$. This code line requires only two guesses from table P (with high probability at least one of them is known anyway) and allows verification against the full 32-bit keystream word (16 bits of which have not yet been taken into account). This technique should rapidly identify the missing entries for table P.

5   Consequences

5.1   Cost of the Attack

The above attack retrieves the full contents of tables P and Q at the beginning of round $r = 6144$. Given such a snapshot of both tables, we can run the generator forwards to generate previously unknown keystream bits. We can also run it backwards to retrieve the key (the state update function and the key/IV setup are invertible). This shows that an attack is even possible for a synchronous cache adversary (not only for an asynchronous adversary, as suspected by Bernstein [2])[4].

The main computational step is the backtracking attack, which requires less than $5 \cdot 2^{64} < 2^{67}$ computational steps that consist in verifying one equation. Since the key/IV setup of HC-256 has to compute the same equation $4096 = 2^{12}$ times (plus does a number of other computations), the effort is less than trying 2^{55} keys in a brute-force setting. The memory requirements are around 3 MB for the candidate tables, plus a little memory for the search tree (implementing it in a depth-first search fashion keeps the memory consumption low). In addition, we have assumed the availability of precise cache measurements for 6148 chosen rounds, and of 2048 known keystream words. We point out that our attack is not optimised in any way. It is likely that a better attack can be found using less cache measurements and computational effort. Nonetheless, the huge number of necessary cache timing measurements required for this attack indicates how difficult it would be to apply a similar attack in the standard model.

If the attack is run on a processor with a different cache block size, efficiency is influenced. For example, if the cache block size is only 32 byte instead of 64 byte, the adversary learns 7 bit for each table lookup. In this case, no backtracking phase is required at all – the solution can already be determined by the reduction step in Subsection 4.3. On the other hand, if the cache block size is e.g. 128 byte, then only 5 bits for each table lookup are recovered, and the backtracking gets a lot more expensive.

---

[4] For a definition of synchronous and asynchronous cache adversaries, see Section 6.

5.2   Design Recommendations

While trying to break HC-256 (and doing initial analysis of other stream ciphers), we met a number of obstacles that might be possible defense mechanisms against cache timing attacks. Some of them may be known to cipher designers already, but to the best of our knowledge, they have not been documented. Thus, we provide a short list of design recommendations that make cache timing attacks more difficult if use of tables can not be avoided altogether:

1. Make as many table accesses for one function call as possible. This makes things harder for a synchronous adversary, who has to match the observed indices to the inner state. For HC-256, this matching was relatively easy, which made the attack possible in the first place.
2. Make the inner state size large compared to the information obtained from one cache measurement. In the case of HC-256, one call to **Next()** yields 32 bit of keystream information and 52 bit of side-channel information. Because of the large inner state, this means that at least $65,536/84 \approx 780$ precise cache access measurements (or many more noisy ones) have to be made to retrieve the inner state.
3. Exploit that the least significant bits of the table index remain unknown (in our analysis, bits 3..0). This can be achieved by using state update and output generation functions that generate a lot of diffusion without the use of S-boxes. As an example, functions using carry (like addition and multiplication) are suitable for this purpose.
4. Use variable tables instead of (fixed content) S-boxes. This gives the adversary insecurity both about the input and the output of the tables.

6   Attack Model

In the following, we justify and generalise the abstract attack model that was used for our attack, such that it also can be used to analyse other stream ciphers.

6.1   Motivation

Whether or how cache timing attacks can be used against a cipher depends on the details of the system deploying it. This is not helpful for cipher designers who are not allowed to make assumptions about the deployment environment. While it is possible that certain attack options are not available in a fielded system, the cipher designer must not rely on this unavailability.

Thus, he works under worst-case assumptions. As an example, while most practical systems will not give the adversary 2^{40} plaintext/ciphertext pairs, ciphers are nonetheless designed to withstand an attack that has this amount of information available. Unless we want to make very detailed restrictions on how the cipher is to be used, overestimating the adversary's abilities is a key strategy.

A cipher designer who is concerned about cache timing attacks has to proceed in the same way. He has to assume that the adversary gets the maximum amount of information, and then see what damage this would do to the cipher. Ciphers secure under such a model will most likely be secure in practice, too.

6.2 Standard Adversary

In traditional (non-side-channel) stream cipher design, the adversary is assumed to have the following oracles available:

- KEYSETUP(): The adversary requests the cipher to be re-initialised with a new key. No output is returned.
- IVSETUP(N): The adversary requests the cipher to be re-initialised with the initialisation vector N that has not been used before. No output is returned.
- KEYSTREAM(i): The adversary requests the cipher to return the i-th keystream block to him. The block length depends on the cipher design.

An adversary is considered successful if he can distinguish an instance of the stream cipher from a random function producing appropriately formatted (but random) answers to his oracle queries. A cipher is considered secure if for any adversary, the success probability is less than that of the generic adversary using the same computational resources on brute-force key testing.

6.3 Synchronous Cache Adversary

The notion of synchronous cache attacks was introduced by Osvik et al. in [13]. In such an attack, the adversary interacts with the encryption code through some type of interface, and he obtains additional information by making cache measurements before or after execution of this code.

In our model, such an adversary can use the same oracles as the standard adversary. In addition, he also has access to the following cache timing oracles:

- SCT_KEYSETUP(): The adversary obtains a list of all cache accesses made by KEYSETUP().
- SCT_IVSETUP(N): The adversary obtains a list of all cache accesses made by IVSETUP(N).
- SCT_KEYSTREAM(i): The adversary obtains a list of all cache accesses made by KEYSTREAM(i).

In particular, this reflects accurate measurements in a prime-then-probe attack, which seems to be the strongest SCT technique to date; making weaker assumptions would not cover this attack method adequately. Note that the attack described in Sections 3 and 4 use the synchronous attack model.

6.4 Asynchronous Cache Adversary

While a synchronous adversary has to wait until user U has finished the execution of a certain operation, asynchronous adversaries run *in parallel* to U. This is possible e.g. on processors with hyperthreading. In this setting, the adversary can constantly monitor the cache state, which gives him an *ordered* list of all cache accesses made during the observation.

Osvik et al. [13] assume that the adversary obtains no additional information beyond the cache accesses. However, from a designer's point of view, we can not restrict ourselves in this way. It is easy to imagine an adversary who both controls some of the input/output data and observes cache behaviour. Thus, in our model, an asynchronous cache adversary has access to the standard oracles as well as the following side-channel oracles:

- ACT_KeySetup(): The adversary obtains a list of all cache accesses made by KeySetup() in *chronological order.*
- ACT_IVSetup(N): The adversary obtains a list of all cache accesses made by IVSetup(N) in *chronological order.*
- ACT_Keystream(i): The adversary obtains a list of all cache accesses made by Keystream(i) in *chronological order.*

6.5 Discussion

Our attack model abstracts away a number of practical difficulties the adversary might encounter:

- The encryption process is not the only one using the cache. Cache accesses made by other processes generate false positives. Thus, instead of a list of encryption cache accesses, a real-world adversary only obtains a list of cache blocks that have *not* been used by the encryption process.
- Cache timing measurements are subject to timing noise. Thus, the list obtained by the adversary may contain false information that has to be filtered out by statistical or analytical methods.
- The granularity of the measurements may not correspond to the above oracle calls. This depends on how time sharing on the processor is organised.
- The adversary may be unable to choose the IV, or to observe the keystream.

Thus, the model has to be considered as being generous towards the adversary. However, while doing one measurement only creates a noisy version of the cache access list, repeating the measurement and using statistical methods will often eliminate most of the noise.

In order to do this, the function calls have to be repeated under the same key and IV. While at the first glance, this seems to be IV re-use and thus a breach of the security contract, a second look shows that this is not the case at all. All the security contract disallows is re-using the IV for a *different* plaintext, i.e. IV re-use for the *same* plaintext is allowed. In particular, it is easy to imagine scenarios where the rightful user *decrypts* the same ciphertext several times (e.g. an entry in an encrypted database that is accessed repeatedly). Thus, in certain settings, obtaining the necessary measurements might actually be possible.

Note that in addition to analysing cipher resistance against cache timing attacks, the model can also be used to derive security margins for the standard model. If the best cache timing attack against a given cipher requires a large number of cache measurements, then the cipher may be considered as being more robust than one that can be broken by only a few calls to the side-channel oracles. Thus, analysing a cipher in our model achieves a similar effect as analysing modified (e.g. reduced-round) versions of a design: Even though an attack may not constitute a break in the standard model, it indicates how far we are from attacking the full cipher according to specification.

7 Conclusions

In this paper, we have described a cache-timing attack against the stream cipher HC-256, which is the strong version of eStream winner HC-128. The attack was based on a abstract model of cache timing attacks that can also be used for designing stream ciphers. From the observations made in our analysis, we have derived a number of design principles for hardening ciphers against cache timing attacks.

Acknowledgements

The author wishes to thank D.A. Osvik, D.J. Bernstein, G. Bertoni, C. de Cannière, L.R. Knudsen, and S. Lucks for discussions that helped improving this paper.

References

1. Bernstein, D.: Cache timing attacks on AES (2005),
   http://cr.yp.to/papers.html#cachetiming
2. Bernstein, D.: Leaks (February 2007),
   http://cr.yp.to/streamciphers/leaks.html
3. Bertoni, G., Zaccaria, V., Breveglieri, L., Monchiero, M., Palermo, G.: AES power attack based on induced cache miss and countermeasure. In: International Symposium on Information Technology: Coding and Computing (ITCC 2005), vol. 1, pp. 586–591. IEEE Computer Society, Los Alamitos (2005)
4. Blömer, J., Krummel, V.: Analysis of countermeasures against access driven cache attacks on AES. In: Adams, C., Miri, A., Wiener, M. (eds.) SAC 2007. LNCS, vol. 4876, pp. 96–109. Springer, Heidelberg (2007)
5. Bonneau, J., Mironov, I.: Cache-collision timing attacks against AES. In: Goubin, L., Matsui, M. (eds.) CHES 2006. LNCS, vol. 4249, pp. 201–215. Springer, Heidelberg (2006)
6. Brickell, E., Graunke, G., Neve, M., Seifert, S.: Software mitigations to hedge AES against cache-based software side-channel vulnerabilities (2006),
   http://eprint.iacr.org/2006/052.pdf
7. The eStream Portfolio, http://www.ecrypt.eu.org/stream/portfolio.pdf
8. Neve, M., Seifert, J.-P.: Advances on access-driven cache attacks on AES. In: Biham, E., Youssef, A.M. (eds.) SAC 2006. LNCS, vol. 4356, pp. 147–162. Springer, Heidelberg (2007)
9. Neve, M., Seifert, J., Wang, Z.: Cache time-behavior analysis on AES (2006),
   http://www.cryptologie.be/document/Publications/AsiaCSS_full_06.pdf
10. Neve, M., Seifert, J., Wang, Z.: A refined look at bernstein's AES side-channel analysis. In: Proc. AsiaCSS 2006, p. 369. ACM, New York (2006)
11. O'Hanlon, M., Tonge, A.: Investigation of cache-timing attacks on AES (2005),
    http://www.computing.dcu.ie/research/papers/2005/0105.pdf
12. Osvik, D., Shamir, A., Tromer, E.: Cache attacks and countermeasures: The case of AES (2005), http://eprint.iacr.org/2005/271.pdf

13. Osvik, D.A., Shamir, A., Tromer, E.: Cache attacks and countermeasures: The case of AES. In: Pointcheval, D. (ed.) CT-RSA 2006. LNCS, vol. 3860, pp. 1–20. Springer, Heidelberg (2006)

14. Percival, C.: Cache missing for fun and profit. Paper accompanying a talk at BSDCan 2005 (2005), http://www.daemonology.net/papers/htt.pdf

15. Salembier, R.: Analysis of cache timing attacks against AES. Scholarly Paper, ECE Department, George Mason University, Virginia (May 2006), http://ece.gmu.edu/courses/ECE746/project/F06_Project_resources/ Salembier_Cache_Timing_Attack.pdf

16. Wang, Z., Lee, R.: New cache designs for thwarting software cache-based side channel attacks. In: Proc. ISCA 2007, pp. 494–505. ACM, New York (2007)

17. Wu, H.: A new stream cipher HC-256. In: Roy, B., Meier, W. (eds.) FSE 2004, vol. 3017, pp. 226–244. Springer, Heidelberg (2004)

An Improved Fast Correlation Attack
on Stream Ciphers

Bin Zhang[1,2] and Dengguo Feng[2]

[1] Laboratory of Algorithmics, Cryptology and Security,
University of Luxembourg,
6, rue Coudenhove-Kalergi, L-1359, Luxembourg
[2] State Key Laboratory of Information Security,
Institute of Software, Chinese Academy of Sciences, Beijing, 100190, China
bin.zhang@uni.lu

Abstract. At Crypto'2000, Johansson and Jönsson proposed a fast
correlation attack on stream ciphers based on the Goldreich-Rubinfeld-
Sudan algorithm. In this paper we show that a combination of their
approach with techniques for substituting keystream and evaluating
parity-checks gives us the most efficient fast correlation attack known
so far. An application of the new algorithm results in the first-known
near-practical key recovery attack on the shrinking generator with the
parameters suggested by Krawczyk in 1994, which was verified in the 40-
bit data LFSR case for which the only previously known efficient attacks
were distinguishing attacks.

Keywords: Stream ciphers, Correlation attacks, Linear feedback shift
register (LFSR), Shrinking generator.

1   Introduction

Fast correlation attacks are one of the most important attacks against LFSR-
based stream ciphers [18]. The aim is to recover the initial state of the involved
LFSR with complexity as low as possible. The earliest work dates back to [22,17],
followed by a large number of algorithms [1,2,3,9,10,12,14,15,16,19,20,25]. The
basic idea of a fast correlation attack is to regard the truncated keystream as
the noisy version of the underlying LFSR sequence, transmitted through a binary symmetric channel (BSC) with some crossover probability, as shown in
Figure 1. Thus, restoring the initial state of the LFSR is equivalent to decoding
the keystream segment by some method.

Such an attack usually has two phases: in the preprocessing phase, the
attacker constructs a large number of parity-checks according to the linear re-
curring structure of the LFSR; in the realtime phase, the attacker uses these precomputed parity-checks to decode a given keystream segment of certain length.
The preprocessing phase can be done once for all, and usually takes a relatively
longer time than that of the realtime processing phase. The efficiency of a fast
correlation attack highly depends on the cost of the pre-computation for finding

R. Avanzi, L. Keliher, and F. Sica (Eds.): SAC 2008, LNCS 5381, pp. 214–227, 2009.
© Springer-Verlag Berlin Heidelberg 2009

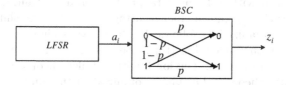

Fig. 1. Model for a fast correlation attack

desirable parity-checks. It is commonly believed that the unrealistically large preprocessing complexity is a significant barrier for decoding in the highly noisy cases. A challenging problem in this field is how to efficiently decode in these highly noisy cases with the processing complexities as low as possible, while not substantially increasing the preprocessing complexity?

In this paper, we propose a new fast correlation attack that allows us to solve this problem for noise up to 0.499 and for a LFSR of arbitrary form used in the BSC model. Our algorithm is a combination of Johansson and Jönsson's algorithm proposed at Crypto'2000 [15] with techniques for substituting keystream and evaluating parity-checks. To our knowledge, such a combination has not been studied before. We give a thorough theoretical analysis of the new algorithm, which is supported by a number of simulation results. Our algorithm can reliably decode in the highly noisy cases with considerably lower data/time/memory complexities than previously known to be possible. Besides, in all the noise and LFSR length cases considered herein, our algorithm can successfully decode without substantially increasing the preprocessing complexities. Therefore, the new algorithm is more efficient than any previously known relevant attacks and largely extends the practical application scope of fast correlation attacks.

To illustrate its power, we use the new algorithm to evaluate the security of the shrinking generator (SG) [4], which is considered as one of the strongest stream ciphers currently available. A shrinking generator consists of two LFSR's, say the data LFSR B and the control LFSR S. LFSR B is irregularly decimated by the regularly clocked LFSR S according to the following rule: the output bit of the data LFSR B is taken if and only if the current output bit of the control LFSR S is 1. For a shrinking generator with the parameters suggested by Krawczyk in [11] (LFSR B of length 61 and LFSR S of similar length), which has shown remarkable resistance against various attacks, the best known cryptanalytic results are those presented in [13] and [24]. More precisely, to restore the initial state of the data LFSR B in such a shrinking generator, the attack in [13] requires 2^{42} operations and 2^{42} keystream bits, while the attack in [24] needs 140000 keystream bits and 2^{57} operations after a pre-computation of 2^{43} operations. Both of the attacks still draw the interest of academics to this day. In contrast, given 10146 keystream bits and at most $2^{33.23}$-byte memory, our new attack against the same shrinking generator works in $2^{35.86}$ operations after a pre-computation of $2^{39.9}$ operations. This is the first-known near-practical key recovery attack against the shrinking generator with the suggested parameters in [11]. We verified our attack on the shrinking generator with a 40-bit data LFSR B on a Pentium 4 PC, in which case the only previously known efficient attacks were distinguishing attacks [5].

This paper is organized as follows. We first give a high level review of Johansson and Jönsson's algorithm [15] in Section 2. Then a description of our new algorithm is presented in Section 3 with detailed theoretical analysis. The simulation results, theoretical estimates of our algorithm and comparisons with the best previously known fast correlation attacks are provided in Section 4. The application of our algorithm to the shrinking generator with the parameters recommended by Krawczyk is described in Section 5 together with comparisons with other known attacks. Finally, some conclusions are given in Section 6.

2   Review of Johansson and Jönsson's Algorithm

Let us first specify the notations used in this paper.

- a_i is the ith output bit of the LFSR.
- z_i is the ith keystream bit.
- $P(z_i = a_i) = p = 0.5 + \varepsilon$ ($\varepsilon > 0$) is the correlation probability.
- L is the length of the LFSR.
- k ($k < L$) is the number of initial state bits to be determined.
- t is the weight of the parity-checks.
- $q = \frac{1}{2} + 2^{t-1}\varepsilon^t$ is the folded noise in a parity-check of weight t.
- N is the length of the available keystream.
- $(\cdot)^T$ is the transpose of a vector or a matrix.
- $\mathbf{b}_f \in \mathbf{F}_2^f$ is a binary column vector $\mathbf{b}_f = (b_{i_1}, b_{i_2}, \ldots, b_{i_f})^T$.
- $\Omega(\mathbf{v}_{L-k})$ is the expected number of parity-checks specified by $\mathbf{v}_{L-k}$.
- n is the number of $\mathbf{v}_{L-k}$s appearing in all the parity-checks.
- $\oplus$ is the bit-wise exclusive or.
- $\cdot$ is the inner product of two binary vectors.
- $\lceil x \rceil$ is the smallest integer greater than or equal to x.

At Crypto'2000, Johansson and Jönsson proposed a fast correlation attack on stream ciphers [15] based on the Goldreich-Rubinfeld-Sudan algorithm [6]. Although they model the decoding problem as the problem of learning a binary linear multivariate polynomial, it can be easily shown that their algorithm can also be interpreted in the BSC model in Figure 1. Hence, we use the BSC model as a unified framework hereafter.

In the preprocessing phase of Johansson and Jönsson's algorithm, the attacker constructs parity-checks of the following form:

$$\mathbf{1}_t^T \cdot \mathbf{a}_t = \mathbf{x}_k^T \cdot \mathbf{a}_k \oplus \mathbf{v}_{L-k}^T \cdot \mathbf{a}_{L-k} , \tag{1}$$

where $\mathbf{1}_t$ denotes the t-dimensional all-one vector, $\mathbf{a}_t = (a_{i_1}, a_{i_2}, \ldots, a_{i_t})$ (here i_j for $1 \leq j \leq t$ are arbitrary indices among the output bits), $\mathbf{a}_k = (a_0, a_1, \ldots, a_{k-1})$ and $\mathbf{a}_{L-k} = (a_k, \ldots, a_{L-1})$. In contrast with other fast correlation attacks that use parity-checks with $\mathbf{v}_{L-k} = \mathbf{0}$, in (1), $\mathbf{v}_{L-k}$ can take non-zero values. Thus, we can use several groups of parity-checks corresponding to different $\mathbf{v}_{L-k}$s here. We need not know the true value of $\mathbf{a}_{L-k}$ when determining $\mathbf{a}_k$ by evaluating the parity-checks with a *fixed* $\mathbf{v}_{L-k}$, as done in (2):

$$1_t^T \cdot \mathbf{z}_t \oplus \mathbf{x}_k^T \cdot \mathbf{a'}_k = \mathbf{x}_k^T \cdot (\mathbf{a}_k \oplus \mathbf{a'}_k) \oplus 1_t^T \cdot \mathbf{e}_t \oplus \mathbf{v}_{L-k}^T \cdot \mathbf{a}_{L-k} \ . \tag{2}$$

In (2), $\mathbf{a'}_k$ is the guessed value of $\mathbf{a}_k$, $\mathbf{z}_t = (z_{i_1}, z_{i_2}, \ldots, z_{i_t})$, $\mathbf{e}_t = (e_{i_1}, e_{i_2}, \ldots, e_{i_t})$ is the random noise vector satisfying $\mathbf{z}_t = \mathbf{a}_t \oplus \mathbf{e}_t$ and $P(e_{i_j} = 0) = P(a_{i_j} = z_{i_j}) = 0.5 + \varepsilon$ for $1 \leq j \leq t$. Note that $\mathbf{v}_{L-k}^T \cdot \mathbf{a}_{L-k}$ is independent of $\mathbf{a}_k$ and takes either 0 or 1 in (2).

Thus, in the realtime processing phase, the attacker can evaluate the left sides of $\Omega(\mathbf{v}_{L-k})$ parity-checks like (2) and record the number of times that $1_t^T \cdot \mathbf{z}_t \oplus \mathbf{x}_k^T \cdot \mathbf{a'}_k = 0$. There will exist a deviation (more or less) from $\frac{1}{2} \cdot \Omega(\mathbf{v}_{L-k})$ in the recorded number when $\mathbf{a'}_k$ is correctly guessed, and such a deviation should not be observed otherwise. To restore $\mathbf{a}_k$, the attacker sums up all the squared values of such deviations and accepts the guess resulting in the highest record as the correct one. Please see the following description of Johansson and Jönsson's algorithm.

Parameters: t, k, n
Pre-computation
    pre-compute n groups of parity-checks like (1)
    with n different $\mathbf{v}_{L-k}$ values
Input: keystream $\mathbf{z}_N = (z_0, z_1, \ldots, z_{N-1})$
Processing
    for all the 2^k possible values $\mathbf{a'}_k$ of $\mathbf{a}_k$ **do**
        let $B_{\mathbf{a'}_k} = 0$
        for each group of parity-checks with a fixed $\mathbf{v}_{L-k}$ **do**
            evaluate the left side of each parity-check like (2),
            and store the total number of times that
                $1_t^T \cdot \mathbf{z}_t \oplus \mathbf{x}_k^T \cdot \mathbf{a'}_k = \mathbf{0}$ as A
            update $B_{\mathbf{a'}_k} = B_{\mathbf{a'}_k} + (2A - \Omega(\mathbf{v}_{L-k}))^2$
        end for
        store $B_{\mathbf{a'}_k}$ in an array U
    end for
    search for the highest value B^* in U and accept the
    corresponding guess $\mathbf{a}^*_k$
Output: $\mathbf{a}_k = (a_0, a_1, \cdots, a_{k-1})$ or a small list of candidates

---

After restoring $\mathbf{a}_k$, other bits of the initial state $\mathbf{a}_L = (a_0, \ldots, a_{L-1})$ can be determined with a much lower complexity, e.g. using the method in [25]. The most time-consuming step in the above algorithm is to substitute the keystream bits into the parity-checks and then to evaluate them. Since the number of the employed parity-checks is often very large and there are 2^k possible values for $\mathbf{a}_k$, this step would take a lot of time, i.e., $2^k k \sum_{\mathbf{v}_{L-k}} \Omega(\mathbf{v}_{L-k})$ operations, when it is done in the straightforward way as in [15]. This is the main bottleneck of Johansson and Jönsson's algorithm.

In the following, we will improve this algorithm by efficiently fulfilling the substitution and evaluation step. To our knowledge, such an improvement has not been studied before.

3   Our Improved Version

3.1   The Main Difference

First note that in (2), we can randomly select a value for $\mathbf{a}_{L-k}$, this does not influence the work of the parity-checks, but makes them easier to follow. Thus we have

$$1_t^T \cdot \mathbf{z}_t \oplus \mathbf{x}_k^T \cdot \mathbf{a}'_k \oplus \mathbf{v}_{L-k}^T \cdot \mathbf{a}''_{L-k} = \mathbf{x}_k^T \cdot (\mathbf{a}_k \oplus \mathbf{a}'_k) \oplus 1_t^T \cdot \mathbf{e}_t \oplus \zeta \,, \qquad (3)$$

where $\mathbf{a}'_k$ is the guessed value of $\mathbf{a}_k$, $\mathbf{a}''_{L-k}$ is the value assigned to $\mathbf{a}_{L-k}$ and $\zeta = 0$ or 1 depending on $\mathbf{a}''_{L-k}$. Other notations are the same as those in (2). In the preprocessing phase, we constructed n groups of such parity-checks, each of which is specified by a fixed $\mathbf{v}_{L-k}$ and has an expected cardinality $\Omega(\mathbf{v}_{L-k})$.

In the processing phase, the attacker evaluates the left sides of (3) and records the number of times that $1_t^T \cdot \mathbf{z}_t \oplus \mathbf{x}_k^T \cdot \mathbf{a}'_k \oplus \mathbf{v}_{L-k}^T \cdot \mathbf{a}''_{L-k} = 0$. To avoid the high time complexity in the substitution and evaluation step of Johansson and Jönsson's algorithm, we proceed as follows.

For a fixed set of parity-checks specified by $\mathbf{v}_{L-k}$, group the parity-checks according to the value of $\mathbf{x}_k$ and define an integer-valued function

$$h_{\mathbf{v}_{L-k}}(\mathbf{x}_k) = \sum_{\mathbf{x}_k}(-1)^{1_t^T \cdot \mathbf{z}_t \oplus \mathbf{v}_{L-k}^T \cdot \mathbf{a}''_{L-k}} \qquad (4)$$

for all the coefficient vectors $\mathbf{x}_k$ appearing in this group of parity-checks. If a value of $\mathbf{x}_k$ does not appear in these parity-checks, we let $h_{\mathbf{v}_{L-k}}(\mathbf{x}_k) = 0$ in (4). Now consider the walsh transform of $h_{\mathbf{v}_{L-k}}(\mathbf{x}_k)$, i.e.,

$$H_{\mathbf{v}_{L-k}}(\omega) = \qquad (5)$$
$$\sum_{\mathbf{x}_k \in \mathbf{F}_2^k} h_{\mathbf{v}_{L-k}}(\mathbf{x}_k)(-1)^{\omega^T \cdot \mathbf{x}_k} = \sum_{\Omega(\mathbf{v}_{L-k})}(-1)^{1_t^T \cdot \mathbf{z}_t \oplus \mathbf{v}_{L-k}^T \cdot \mathbf{a}''_{L-k} \oplus \omega^T \cdot \mathbf{x}_k}.$$

In (5), note that when $\omega = \mathbf{a}'_k$, we have $H_{\mathbf{v}_{L-k}}(\omega) = \Omega(\mathbf{v}_{L-k})_0 - \Omega(\mathbf{v}_{L-k})_1$, where $\Omega(\mathbf{v}_{L-k})_i$ is the number of i for $i = 0$ or 1. Thus given the guess $\mathbf{a}'_k$, the number of times that $1_t^T \cdot \mathbf{z}_t \oplus \mathbf{x}_k^T \cdot \mathbf{a}'_k \oplus \mathbf{v}_{L-k}^T \cdot \mathbf{a}''_{L-k} = 0$ is $u(\mathbf{v}_{L-k}) = \Omega(\mathbf{v}_{L-k})_0 = \frac{\Omega(\mathbf{v}_{L-k}) + H_{\mathbf{v}_{L-k}}(\omega)}{2}$ for this group of parity-checks. For simplicity, we let $u(\mathbf{v}_{L-k}) = \frac{\Omega(\mathbf{v}_{L-k}) + |H_{\mathbf{v}_{L-k}}(\omega)|}{2}$, i.e., $u(\mathbf{v}_{L-k})$ is always greater than or equal to $\frac{\Omega(\mathbf{v}_{L-k})}{2}$.

We can use the fast Walsh transform (FWT) to *simultaneously* compute the 2^k values of $h_{\mathbf{v}_{L-k}}(\mathbf{x}_k)$'s Walsh transform function. Therefore, the total time complexity of the above substitution and evaluation step is $\sum_{\mathbf{v}_{L-k}}(2^k k + \Omega(\mathbf{v}_{L-k})$ $(t+k))$ operations, among which $\Omega(\mathbf{v}_{L-k})(t+k)$ operations are required by the preparation of $h_{\mathbf{v}_{L-k}}(\mathbf{x}_k)$ and $2^k k$ operations are required by the FWT. The memory cost is around $c \cdot 2^k + \sum_{\mathbf{v}_{L-k}}(\ t\lceil \log_2 N \rceil + L)\Omega(\mathbf{v}_{L-k})$-bit, among which $c \cdot 2^k$ bits are the memory consumption in the FWT computation (c is a small constant and is usually ≤ 32) and $\sum_{\mathbf{v}_{L-k}}(\ t\lceil \log_2 N \rceil + L)\Omega(\mathbf{v}_{L-k})$ bits are used for the storage of parity-checks. (see Theorem 1 in Section 3.2 for an explanation). Compared to the straightforward method in [15], the gain in efficiency is obvious. To take the best of the above method and Johansson and Jönsson's idea for constructing parity-checks, we propose the following improved algorithm.

Parameters: t, k, n
Pre-computation
      pre-compute n groups of parity-checks like (1)
      with n different $\mathbf{v}_{L-k}$ values
Input: keystream $\mathbf{z}_N = (z_0, z_1, \ldots, z_{N-1})$
Processing
      let $B_\omega = 0$ for the 2^k possible values of ω
      for each group of parity-checks specified by $\mathbf{v}_{L-k}$ **do**
          let $\mathbf{a}_{L-k}$ take a randomly assigned value
          define a function $h_{\mathbf{v}_{L-k}}(\mathbf{x}_k)$ as in (4)
          apply FWT to compute $H_{\mathbf{v}_{L-k}}(\omega)$ for the 2^k
          possible values of ω
          update $B_\omega = B_\omega + (H_{\mathbf{v}_{L-k}}(\omega))^2/4$ for the 2^k
          possible values of ω
      end for
      search for $B_\omega \geq T$ and accept the corresponding
      ω as a candidate for $\mathbf{a}_k$
Output: $\mathbf{a}_k = (a_0, a_1, \cdots, a_{k-1})$ or a small list of candidates

---

Here T is the threshold determined by the success rate of the whole attack and $B_\omega = B_\omega + (H_{\mathbf{v}_{L-k}}(\omega))^2/4$ accumulates the squared biases for each guess of $\mathbf{a}_k$. The above description illustrates the structure of our improved algorithm. In practical programming, there may be some differences made for optimization.

3.2 Theoretical Analysis

Now we give a theoretical justification of our improved algorithm. First note that the expected number $\Omega(\mathbf{v}_{L-k})$ of the parity-checks with a fixed pattern $\mathbf{v}_{L-k}$ is $\frac{\binom{N}{t}}{2^{L-k}}$. Thus from (3), if $\mathbf{a}'_k$ is correctly guessed, there will exist a deviation $\Omega(\mathbf{v}_{L-k})2^{t-1}\varepsilon^t$ from $\frac{1}{2} \cdot \Omega(\mathbf{v}_{L-k})$ in the number of times that $\mathbf{1}_t^T \cdot \mathbf{z}_t \oplus \mathbf{x}_k^T \cdot \mathbf{a}'_k \oplus \mathbf{v}_{L-k}^T \cdot \mathbf{a}''_{L-k} = 0$. Otherwise, such a bias should not be observed. This is the basis of our algorithm.

In our improved algorithm, the accumulated squared bias is $\sum_{\mathbf{v}_{L-k}} \big(u(\mathbf{v}_{L-k})$
$- \frac{\Omega(\mathbf{v}_{L-k})}{2} \big)^2 = \sum_{\mathbf{v}_{L-k}} \big(\frac{|H_{\mathbf{v}_{L-k}}(\omega)| + \Omega(\mathbf{v}_{L-k})}{2} - \frac{\Omega(\mathbf{v}_{L-k})}{2} \big)^2 = \sum_{\mathbf{v}_{L-k}} \frac{(H_{\mathbf{v}_{L-k}}(\omega))^2}{4}.$
This is just B_ω calculated in the above algorithm. Hence, we can rewrite the judgement condition $B_\omega \geq T$ in our algorithm as

$$B_\omega \geq T \Leftrightarrow \sum_{\mathbf{v}_{L-k}} \big(u(\mathbf{v}_{L-k}) - \frac{\Omega(\mathbf{v}_{L-k})}{2} \big)^2 \geq T. \tag{6}$$

Obviously, when $\mathbf{a}'_k$ is correctly guessed, $u(\mathbf{v}_{L-k})$ follows the binomial distribution $(\Omega(\mathbf{v}_{L-k}), q)$, otherwise it follows the binomial distribution $(\Omega(\mathbf{v}_{L-k}), \frac{1}{2})$. If we use the normal distribution to approximate the binomial distribution, then $\sum_{\mathbf{v}_{L-k}} \big(u(\mathbf{v}_{L-k}) - \frac{\Omega(\mathbf{v}_{L-k})}{2} \big)^2 \geq T$ is equivalent to

$$\frac{\Omega(\mathbf{v}_{L-k})n}{4q(1-q)} \geq \sum_{\mathbf{v}_{L-k}} \frac{(u(\mathbf{v}_{L-k}) - \frac{\Omega(\mathbf{v}_{L-k})}{2})^2}{\Omega(\mathbf{v}_{L-k})q(1-q)} \geq \frac{T}{\Omega(\mathbf{v}_{L-k})q(1-q)} \tag{7}$$

when $\mathbf{a}'_k$ is correctly guessed and to

$$\Omega(\mathbf{v}_{L-k})n \geq \sum_{\mathbf{v}_{L-k}} \frac{(u(\mathbf{v}_{L-k}) - \frac{\Omega(\mathbf{v}_{L-k})}{2})^2}{(\frac{1}{2}\sqrt{\Omega(\mathbf{v}_{L-k})})^2} \geq \frac{4T}{\Omega(\mathbf{v}_{L-k})} \tag{8}$$

when $\mathbf{a}'_k$ is wrongly guessed. (7) can be rewritten as

$$\frac{\Omega(\mathbf{v}_{L-k})n}{4q(1-q)} \geq \sum_{\mathbf{v}_{L-k}} \big(\frac{u(\mathbf{v}_{L-k}) - \Omega(\mathbf{v}_{L-k})q + \Omega(\mathbf{v}_{L-k}) \cdot 2^{t-1}\varepsilon^t}{\sqrt{\Omega(\mathbf{v}_{L-k})q(1-q)}} \big)^2 = \tag{9}$$

$$\sum_{\mathbf{v}_{L-k}} \big(\frac{u(\mathbf{v}_{L-k}) - \Omega(\mathbf{v}_{L-k})q}{\sqrt{\Omega(\mathbf{v}_{L-k})q(1-q)}} + \frac{\Omega(\mathbf{v}_{L-k})2^{t-1}\varepsilon^t}{\sqrt{\Omega(\mathbf{v}_{L-k})q(1-q)}} \big)^2 \geq \frac{T}{\Omega(\mathbf{v}_{L-k})q(1-q)}.$$

Note that when $\mathbf{a}'_k$ is correctly guessed, $\frac{u(\mathbf{v}_{L-k}) - \Omega(\mathbf{v}_{L-k})q}{\sqrt{\Omega(\mathbf{v}_{L-k})q(1-q)}}$ follows the standard normal distribution $\mathcal{N}(0,1)$. When $\mathbf{a}'_k$ is wrongly guessed, the variable $\frac{u(\mathbf{v}_{L-k}) - \frac{\Omega(\mathbf{v}_{L-k})}{2}}{\frac{1}{2}\sqrt{\Omega(\mathbf{v}_{L-k})}}$ also follows the standard normal distribution $\mathcal{N}(0,1)$. (8) means that $\sum_{\mathbf{v}_{L-k}} \frac{(u(\mathbf{v}_{L-k}) - \frac{\Omega(\mathbf{v}_{L-k})}{2})^2}{(\frac{1}{2}\sqrt{\Omega(\mathbf{v}_{L-k})})^2}$ follows the centrally chi-squared distribution when $\mathbf{a}'_k$ is wrongly guessed, while (9) implies that when $\mathbf{a}'_k$ is correctly guessed, $\sum_{\mathbf{v}_{L-k}} \frac{(u(\mathbf{v}_{L-k}) - \frac{\Omega(\mathbf{v}_{L-k})}{2})^2}{\Omega(\mathbf{v}_{L-k})q(1-q)}$ follows the non-central chi-square distribution. This is the criterion used to filter out the wrong guesses.

More precisely, let $\Gamma(y) = \int_0^{+\infty} e^{-x}x^{y-1}dx$ denote the gamma function. Since there are n different $\mathbf{v}_{L-k}$s, the probability density function of a centrally chi-squared distribution is $\phi_1(x) = \frac{x^{\frac{n-2}{2}}e^{-\frac{x}{2}}}{2^{\frac{n}{2}}\Gamma(\frac{n}{2})}$ for $x > 0$ and the probability density function of a non-centrally chi-squared distribution is $\phi_2(x) = \frac{e^{-\frac{(x+\delta^2)}{2}}}{2^{\frac{n}{2}}} \sum_{j=0}^{\infty} \frac{x^{j-1+\frac{n}{2}} \delta^{2j}}{\Gamma(j+\frac{n}{2}) 2^{2j} j!}$ for $x > 0$, where $\delta^2 = \sum_{\mathbf{v}_{L-k}} (\frac{\sqrt{\Omega(\mathbf{v}_{L-k}) 2^{t-1} \varepsilon^t}}{\sqrt{q(1-q)}})^2$. Thus the probability that the right $\mathbf{a}_k$ could result in $B_{\mathbf{a}_k} \geq T$ is

$$P_{right} = \int_{\frac{T}{\Omega(\mathbf{v}_{L-k}) q(1-q)}}^{\frac{\Omega(\mathbf{v}_{L-k}) n}{4q(1-q)} + 0.5} \phi_2(x) dx , \tag{10}$$

while a wrong guess $\mathbf{a}'_k$ would pass the test with the probability

$$P_{wrong} = \int_{\frac{4T}{\Omega(\mathbf{v}_{L-k})}}^{\Omega(\mathbf{v}_{L-k}) n + 0.5} \phi_1(x) dx . \tag{11}$$

In our algorithm, we can control these two probabilities by carefully choosing T. A typical case is that $P_{wrong} < 2^{-k}$ with some P_{right}, i.e., none of the wrong guesses could pass the test, while the right guess could pass with some constant probability.

As a summary of the results in Section 3.1 and 3.2, we have

Theorem 1. *If the success rate of our improved algorithm is set to be P_{right} and $P_{wrong} < 2^{-k}$, then its time complexity is $\sum_{\mathbf{v}_{L-k}} (2^k k + \Omega(\mathbf{v}_{L-k})(t+k))$ operations, its memory complexity is at most $c \cdot 2^k + \sum_{\mathbf{v}_{L-k}} (t \lceil log_2 N \rceil + L) \Omega(\mathbf{v}_{L-k})$-bit and its data complexity is N-bit keystream determined by (10), (11) and $\Omega(\mathbf{v}_{L-k})$.*

Proof. For the time complexity, note that there are n groups of parity-checks constructed, and for each group used in the processing phase, $\Omega(\mathbf{v}_{L-k})(t+k)$ operations are required by the function $h_{\mathbf{v}_{L-k}}(\mathbf{x}_k)$ and $2^k k$ operations are required by the FWT. For the memory complexity, note that we use the following straightforward way to store a parity-check of the form $\mathbf{1}_t^T \cdot \mathbf{z}_t \oplus \mathbf{x}_k^T \cdot \mathbf{a}'_k \oplus \mathbf{v}_{L-k}^T \cdot \mathbf{a}''_{L-k}$,

1. $t \cdot \lceil log_2 N \rceil$ bits to represent the t integers $i_1, i_2, \ldots, i_t$.
2. k bits to represent the coefficient vector $\mathbf{x}_k$.
3. if $\mathbf{v}_{L-k} \neq 0$, then $L - k$ bits to represent it.

Thus, $t \lceil log_2 N \rceil + L$ bits are needed by one parity-check. There are totally $\Omega(\mathbf{v}_{L-k})$ parity-checks in each group, so $(t \lceil log_2 N \rceil + L) \Omega(\mathbf{v}_{L-k})$ bits are required by each group. In addition, $c \cdot 2^k$ bits are required for the computation of the FWT, where c is a small constant determined by k and by the size of a float precision floating-point number. In our analysis, $c \leq 32$ is enough for the current usage. The data complexity is determined by the success rate of the whole attack and we can use (10), (11) and $\Omega(\mathbf{v}_{L-k}) = \frac{\binom{N}{t}}{2^{L-k}}$ to determine it. $\square$

We use the traditional time/memory trade-off to pre-compute the parity-checks with a fixed pattern $\mathbf{v}_{L-k}$. That is, first compute and sort in a table the formal expression of the sum of $\lfloor \frac{t}{2} \rfloor$ a_is in the L initial state bits. Then sort the table and compute the formal sum of the other $\lceil \frac{t}{2} \rceil$ a_is. An exclusive-or collision value equal to $\mathbf{v}_{L-k}$ in the table provides us a parity-check. For the keystream of N-bit length, the time/memory complexities in the preprocessing phase are about $N^{\lceil \frac{t}{2} \rceil} \log_2 N$ operations (even $N^{\lceil \frac{t}{2} \rceil}$ operations by some hashing techniques) and $N^{\lfloor \frac{t}{2} \rfloor}(t\lceil \log_2 N \rceil + L)$-bit, respectively.

4   Simulations, Theoretical Estimates and Comparisons

In general, the simulation results of our algorithm match the theory very well. First we give some experimental results of our algorithm in the same scenarios as those considered in [15]. We only list some typical cases in Table 1[1], other cases are omitted due to limited space.

Table 1. Simulations and comparisons with the basic algorithm in [15]

attack	[15]	ours	[15]	ours	[15]	ours	[15]	ours	[15]	ours
L	40	40	40	40	60	60	60	60	60	60
p	0.64	0.64	0.55	0.55	0.57	0.57	0.68	0.68	0.6	0.6
t	2	3	2	3	2	3	3	5	2	5
N	$4 \cdot 10^5$	$5 \cdot 10^3$	$4 \cdot 10^5$	10^4	$4 \cdot 10^7$	$2 \cdot 10^5$	$1.5 \cdot 10^5$	$8 \cdot 10^3$	$4 \cdot 10^7$	$9 \cdot 10^3$
$time \approx$	3 min	5 sec	3 min	25 sec	106 min	6 min	4.6 min	20 sec	13 min	1 min

We implemented our attack in C on a Pentium 4 PC running under windows XP. To make an accurate comparison, we also implemented one instance ($L = 40$, $p = 0.64$, $N = 4 \cdot 10^5$) of the basic algorithm in [15], the time cost is about 4 minutes (instead of 3 minutes). Although the time results of Johansson and Jönsson's algorithm listed in Table 1 are from [15] and are obtained on a different platform (Sun Ultra-80 running under Solaris), the above instance we implemented shows that the time comparisons are still meaningful. Our algorithm can successfully decode with largely reduced data complexities. For example, in the case that $L = 40$ and $p = 0.55$, our algorithm needs only 10^4 bits, while the attack in [15] needs $4 \cdot 10^5$ bits to decode. The longest pre-computation time of our algorithm in Table 1 are tens of minutes occurring in the two cases that $L = 60$ with $t = 5$. In other cases, the pre-computation costs are all negligible.

Next, we compare our algorithm to the two algorithms in [3,25], which are the best previously known fast correlation attacks. Table 2 and 3 show that our algorithm compares favorably to these two attacks.

---

[1] The feedback polynomial of the 40-bit LFSR used in Table 1 and 2 is $1 + x + x^3 + x^5 + x^9 + x^{11} + x^{12} + x^{17} + x^{19} + x^{21} + x^{25} + x^{27} + x^{29} + x^{32} + x^{33} + x^{38} + x^{40}$. The feedback polynomial of the 60-bit LFSR in Table 1 is not released in [15], we simply choose the polynomial $x^{60} + x + 1$ in our experiments. Since both the attack in [15] and our algorithm are applicable to arbitrary form LFSR, this choice does not influence the experimental results.

Table 2. Experimental results, theoretical predictions of our algorithm and comparisons with the best previously known attacks with success rate close to 1

p	attack	L	N	t	k	time	memory	pre $-$ computation
0.531	[3]	40	80000	3	1	2^{31}	$2^{34.1}$	2^{37}
	[25]	40	2^{22}	2	20	2^{24}	$2^{32.8}$	2^{27}
	ours	40	**40000**	3	12	$\mathbf{2^{20}}$	$\mathbf{2^{25}}$	$\mathbf{2^{30.6}}$
	[3]	89	2^{28}	3	1	2^{44}	$2^{35.81}$	2^{61}
	[25]	89	2^{32}	3	26	2^{32}	$2^{41.12}$	2^{64}
	ours	89	$\mathbf{2^{28}}$	3	22	$\mathbf{2^{29.8}}$	$\mathbf{2^{27.3}}$	$\mathbf{2^{56}}$
0.51	[3]	40	80000	3	1	2^{40}	$2^{40.54}$	2^{37}
	[25]	40	2^{24}	2	22	2^{29}	$2^{35.13}$	2^{29}
	ours	40	**50000**	3	16	$\mathbf{2^{24.13}}$	$\mathbf{2^{26.7}}$	$\mathbf{2^{31.3}}$

Table 3. Theoretical estimates of our algorithm and comparisons with the algorithm in [25] with success rate close to 1 (the result in [25] for the case $L = 61$ and $p = 0.501$ is not correct, here we list the correct result obtained according to the formulas in [25])

p	attack	L	N	t	k	time	memory	pre $-$ computation
0.501	[25]	61	2^{36}	2	22	2^{43}	$2^{49.71}$	2^{42}
	ours	61	$\mathbf{2^{31}}$	2	21	$\mathbf{2^{28.8}}$	$\mathbf{2^{31.3}}$	$\mathbf{2^{39.3}}$
0.531	[25]	103	2^{36}	3	29	2^{34}	$2^{38.51}$	2^{72}
	ours	103	$\mathbf{2^{32}}$	3	24	$\mathbf{2^{31.6}}$	$\mathbf{2^{29.2}}$	$\mathbf{2^{64}}$

In Table 2 and 3, we let $n \leq 12$ in our algorithm, other parameters are explicitly listed in the tables. Here we only give the parameters that result in *uniform* complexities, i.e., the trade-off between data/time/memory and pre-computation complexities is as balanced as possible. Other choices of parameters are also allowed, but they do not have the uniform property.

As in [3,25], we have implemented the case that $L = 40$ and $p = 0.531$ in Table 2 in C on the platform mentioned above. It takes less than 4 seconds to restore the initial state of the involved LFSR after a pre-computation with negligible time. Compared to the attack in [3] which takes a few *days* for pre-computation and that in [25] whose pre-computation lasts for a few *hours*, the gain on efficiency in the preprocessing phase is obvious. This mainly comes from the fact that our algorithm can decode with much lower data complexities. In fact, the keystream requirement in our attack is 2 times smaller than that in [3] and is 2^6 times smaller than that in [25] in this implemented case. Other complexities listed in Table 2 and 3 are theoretical predictions derived according to $(7) - (11)$ and Theorem 1.

From the above simulation results and theoretical estimates, we can see that our algorithm makes real-life fast correlation attacks much more reachable for the noise cases that only theoretical estimates are known by previous methods. This will have an impact on the choice of the secure parameters for the corresponding stream ciphers, as shown in Section 5.

5   Application to the Shrinking Generator

The shrinking generator (SG) was proposed in [4] at Crypto'93. So far, various key recovery and distinguishing attacks on the shrinking generator have been proposed [4,5,8,13,18,23,24], but none of them can practically threaten the security of the shrinking generator with the suggested parameters in [11]. In the following, we will demonstrate a key recovery attack against the same shrinking generator that is near-practical when measured in time/data/memory and preprocessing complexities.

More precisely, let the output sequence of data LFSR B used in a shrinking generator be $b = b_0, b_1, \cdots$. The cryptanalysis of the shrinking generator is usually composed of two phases. First, a new sequence $\hat{b} = \hat{b}_0, \hat{b}_1, \cdots$ associated with b by the relation $P(\hat{b}_i = b_i) = \frac{1}{2} + \varepsilon_i$ $(\varepsilon_i > 0)$ is constructed either by the method in [7] or by the method in [24]. Second, a decoding algorithm is applied to recover the initial state of LFSR B from $\hat{b}$. Here we do not focus on how the sequence $\hat{b}$ is constructed, but on how to efficiently exploit the existing correlations between b and $\hat{b}$. According to [24], the average biases between $\hat{b}$ and b for different keystream lengths N are shown in Table 4.

Table 4. The average biases for different keystream lengths N using the method in [24], which are obtained by a pre-computation of about 4 hours by Mathematica on a Pentium 4 processor

N	240	3000	8000	10000	140000
ε	0.0542	0.021	0.020	0.0195281	0.00982376

For the shrinking generator with the data LFSR B of length 61 and the control LFSR S of similar length, as suggested in [11], we show how to decode $\hat{b}$ using our algorithm in $2^{35.86}$ operations. From Table 4, the correlation between $\hat{b}$ and b is 0.5195281 if $N = 10000$. Note that 10146 keystream bits are needed to get a sequence $\hat{b}$ of 10000 bits. We choose the following parameters in our algorithm: $k = 27$, $n = 12$, $t = 5$, $T = 8.6 \times 10^8$, then $\Omega(\mathbf{v}_{L-k}) = 48457895$, $P_{right} = 97.42\%$ and $P_{wrong} = 2^{-32.16}$. Thus the total time complexity is $12 \cdot (2^{27} \cdot 27 + 48457895 \cdot (27 + 5)) = 2^{35.86}$ operations, the memory complexity is at most $32 \cdot 2^{27} + 48457895 \cdot 12 \cdot (61 + 5 \cdot \lceil \log_2 10000 \rceil) = 2^{36.23}$-bit after a pre-computation of $10000^3 = 2^{39.9}$ operations. We can see that the data and time complexities of our attack are all practical, while the memory and preprocessing complexities are near-practical. The comparisons of our attack with other known attacks on the same target shrinking generator are given in Table 5, which shows that our attack is the best known attack against the shrinking generator with the suggested parameters.

At Eurocrypt'2003, a distinguishing attack on the shrinking generator is proposed in [5]. The best result given in [5] is to distinguish a shrinking generator with the data LFSR B having a weight 4 polynomial of degree 10000 using

Table 5. Comparisons of different attacks on the shrinking generator with the suggested parameters in [11]

	[18]	[8]	[13]	[24]	ours
N	few	$2^{10.23}$	2^{42}	140000	**10146**
$time$	2^{80}	2^{77}	2^{42}	2^{57}	$\mathbf{2^{35.86}}$

2^{32} output bits. Note that an arbitrary weight feedback polynomial of degree r is known to have a weight 4 multiple of degree around $2^{r/3}$ and $10000 = 2^{13.2877} = 2^{r/3}$, the distinguishing attack in [5] is only applicable to arbitrary data LFSR's of length around 40. Please see the following example.

Example 1. Consider the polynomial $x^{40} + x^{38} + x^{35} + x^{32} + x^{28} + x^{26} + x^{22} + x^{20} + x^{17} + x^{16} + x^{14} + x^{13} + x^{11} + x^{10} + x^9 + x^8 + x^6 + x^5 + x^4 + x^3 + 1$, it can be easily checked that its weight 4 multiple is $x^{24275} + x^{6116} + x^{1752} + 1$. Note that the degree 24275 is higher than the degree 10000 given in [5]. □

To further show the advantages of our decoding algorithm, consider a shrinking generator with the data LFSR B of length 40. We launch a key recovery attack on such a shrinking generator with the following attack parameters: $N = 8119$, $t = 3$, $k = 20$, $n = 9$, $p = 0.52$ and $T = 10^6$. Thus, given $2^{26.5}$-bit memory, the time complexity is $2^{27.62}$ operations after a pre-computation of 2^{26} operations. We implemented our attack in C on the same platform as that in Section 4. The realtime decoding lasts for about 10 seconds to output the correct initial state of LFSR B after a pre-computation of negligible time. Since an efficient key recovery attack is usually believed to be stronger than an efficient distinguishing attack on the same cipher, we conclude that our attack is stronger than that in [5].

Remarks on the security of the shrinking generator. For a shrinking generator with the data LFSR having a known connection, our results show that we should use a LFSR of longer length than that recommended by Krawczyk in [11]. If a security level of 2^{80} is needed, the length of the data LFSR should be at least 128-bit and the control LFSR should be of similar length. This comes from the following attack scenario: $L = 128$, $N = 140000$, $p = 0.50982376$, $k = 71$, $t = 6$ and $n = 16$. The corresponding time, memory and preprocessing complexities are $2^{81.15}$ operations, 2^{76}-bit and $2^{59.4}$ operations. An alternative way for strengthening the security of a shrinking generator is to use an unknown connection for the data LFSR at the expense of more hardware complexity. This is originally suggested by the designers in [4], but all the known cryptanalysis results on the shrinking generator so far are achieved under the known connection assumption. Our result on the shrinking generator could be seen as an end of such a research routine if a shrinking generator with the suggested parameters in [11] is employed.

6   Conclusions

In this paper, we proposed an improved fast correlation attack based on the combination of Johansson and Jönsson's algorithm with techniques for substituting keystream and evaluating parity-checks. Both the simulations and theoretical estimates show that the new algorithm is more efficient than all the previously known fast correlation attacks in general. The importance of such an algorithm is that the secure parameters formerly proposed for the corresponding stream ciphers have to be re-evaluated by our decoding method, which is verified by our cryptanalytic result on the shrinking generator with the parameters recommended by Krawczyk in 1994. We believe that our method will be useful in the cryptanalysis of those LFSR-based stream ciphers that other attacks, e.g. algebraic attacks, cannot deal with efficiently.

Acknowledgements. We would like to thank the anonymous reviewers for very helpful comments. This paper is supported by the National Natural Science Foundation of China (Grant No. 60603018) and the National High Technology Research and Development 863 Programm of China (No. 2007AA01Z470).

References

1. Canteaut, A., Trabbia, M.: Improved fast correlation attacks using parity-check equations of weight 4 and 5. In: Preneel, B. (ed.) EUROCRYPT 2000. LNCS, vol. 1807, pp. 573–588. Springer, Heidelberg (2000)
2. Chepyzhov, V.V., Johansson, T., Smeets, B.: A simple algorithm for fast correlation attacks on stream ciphers. In: Schneier, B. (ed.) FSE 2000. LNCS, vol. 1978, pp. 181–195. Springer, Heidelberg (2001)
3. Chose, P., Joux, A., Mitton, M.: Fast correlation attacks: An algorithmic point of view. In: Knudsen, L.R. (ed.) EUROCRYPT 2002. LNCS, vol. 2332, pp. 209–221. Springer, Heidelberg (2002)
4. Coppersmith, D., Krawczyk, H., Mansour, Y.: The shrinking generator. In: Stinson, D.R. (ed.) CRYPTO 1993. LNCS, vol. 773, pp. 22–39. Springer, Heidelberg (1994)
5. Ekdahl, P., Johansson, T.: Predicting the shrinking generator with fixed connections. In: Biham, E. (ed.) EUROCRYPT 2003. LNCS, vol. 2656, pp. 330–344. Springer, Heidelberg (2003)
6. Goldreich, O., Rubinfeld, R., Sudan, M.: Learning polynomials with queries: the highly noisy case. In: 36th Annual Symposium on Foundations of Computer Science, Milwaukee, Wisconsin, pp. 294–303 (1995)
7. Golić, J.D.: Correlation analysis of the shrinking generator. In: Kilian, J. (ed.) CRYPTO 2001. LNCS, vol. 2139, pp. 440–457. Springer, Heidelberg (2001)
8. Golić, J.D., O'Connor, L.: Embedding and probabilistic correlation attacks on clock-controlled shift registers. In: De Santis, A. (ed.) EUROCRYPT 1994. LNCS, vol. 950, pp. 230–243. Springer, Heidelberg (1995)
9. Golić, J.D.: Iterative optimum symbol-by-symbol decoding and fast correlation attack. IEEE Trans. Inform. Theory 47, 3040–3049 (2001)
10. Golić, J.D., Hawkes, P.: Vectorial approach to fast correlation attacks. Designs, Codes and Cryptography 35, 5–19 (2005)

11. Krawczyk, H.: The shrinking generator: some practical considerations. In: Anderson, R. (ed.) FSE 1993. LNCS, vol. 809, pp. 45–46. Springer, Heidelberg (1994)
12. Johansson, T., Jönsson, F.: Fast correlation attacks based on turbo code techniques. In: Wiener, M. (ed.) CRYPTO 1999. LNCS, vol. 1666, pp. 181–197. Springer, Heidelberg (1999)
13. Johansson, T.: Reduced complexity correlation attacks on two clock-controlled generators. In: Ohta, K., Pei, D. (eds.) ASIACRYPT 1998. LNCS, vol. 1514, pp. 342–357. Springer, Heidelberg (1998)
14. Johansson, T., Jönsson, F.: Improved fast correlation attacks on stream ciphers via convolutional codes. In: Stern, J. (ed.) EUROCRYPT 1999. LNCS, vol. 1592, pp. 347–362. Springer, Heidelberg (1999)
15. Johansson, T., Jönsson, F.: Fast correlation attacks through reconstruction of linear polynomials. In: Bellare, M. (ed.) CRYPTO 2000. LNCS, vol. 1880, pp. 300–315. Springer, Heidelberg (2000)
16. Lu, Y., Vaudenay, S.: Faster correlation attack on bluetooth keystream generator E0. In: Franklin, M. (ed.) CRYPTO 2004. LNCS, vol. 3152, pp. 407–425. Springer, Heidelberg (2004)
17. Meier, W., Staffelbach, O.: Fast correlation attacks on certain stream ciphers. Journal of Cryptology, 159–176 (1989)
18. Menezes, A.J., Van Oorschot, P.C., Vanstone, S.A.: Handbook of Applied Cryptography. CRC, Boca Raton (1996)
19. Mihaljević, M.J., Fossorier, M.P.C., Imai, H.: A low-complexity and high-performance algorithm for the fast correlation attack. In: Schneier, B. (ed.) FSE 2000. LNCS, vol. 1978, pp. 196–212. Springer, Heidelberg (2001)
20. Mihaljević, M.J., Fossorier, M.P.C., Imai, H.: Fast correlation attack algorithm with listing decoding and an application. In: Schneier, B. (ed.) FSE 2000. LNCS, vol. 1978, pp. 196–212. Springer, Heidelberg (2001)
21. Shannon, C.E.: A Mathematical theory of communication. Bell Syst. Tech., J. 27 (1948)
22. Siegenthaler, T.: Decrypting a class of stream ciphers using ciphertext only. IEEE Transactions on Computer C-34, 81–85 (1985)
23. Simpson, L.R., Golić, J.D., Dawson, E.: A probabilistic correlation attack on the shrinking generator. In: Boyd, C., Dawson, E. (eds.) ACISP 1998. LNCS, vol. 1438, pp. 147–158. Springer, Heidelberg (1998)
24. Zhang, B., Wu, H., Feng, D., Bao, F.: A fast correlation attack on the shrinking generator. In: Menezes, A. (ed.) CT-RSA 2005. LNCS, vol. 3376, pp. 72–86. Springer, Heidelberg (2005)
25. Zhang, B., Feng, D.: Multi-pass fast correlation attack on stream ciphers. In: Biham, E., Youssef, A.M. (eds.) SAC 2006. LNCS, vol. 4356, pp. 234–248. Springer, Heidelberg (2007)

A Three-Property-Secure Hash Function

Elena Andreeva and Bart Preneel

SCD-COSIC, Dept. of Electrical Engineering, Katholieke Universiteit Leuven,
{Elena.Andreeva,Bart.Preneel}@esat.kuleuven.be

Abstract. This paper proposes a new hash construction based on the widely used Merkle-Damgård (MD) iteration [13,9]. It achieves the three basic properties required from a cryptographic hash function: collision (Coll), second preimage (Sec) and preimage (Pre) security. We show property preservation for the first two properties in the standard security model and the third Pre security property is proved in the random oracle model. Similar to earlier known hash constructions that achieve a form of Sec (eSec [16]) property preservation [4,17], we make use of fixed key material in the iteration. But while these hashes employ keys of size at least logarithmic in the message length (in blocks), we only need a small constant key size. Another advantage of our construction is that the underlying compression function is instantiated as a keyless primitive.

The Sec security of our hash scheme, however, relies heavily on the standard definitional assumption that the target messages are sufficiently random. An example of a practical application that requires Sec security and satisfies this definitional premise on the message inputs is the popular Cramer-Shoup encryption scheme [8]. Still, in practice we have other hashing applications where the target messages are not sampled from spaces with uniform distribution. And while our scheme is Sec preserving for uniform message distributions, we show that this is not always the case for other distributions.

1   Introduction

Hash functions in cryptography are used to compress inputs of arbitrary length to outputs of a fixed size. A typical way to build a hash function is to iteratively apply a fixed-input length compression function. Practical hash functions today are predominantly based on this principle and the most widespread application of an iterative construction is the Merkle-Damgård (MD) hash [13,9]. The main security feature of the MD hash is its collision (Coll) security preservation, which means that if the compression function is collision secure, then the iterated hash function is collision secure as well. But collision security is not the only security property required from hash functions. A good hash function should also be second preimage (Sec) and preimage (Pre) secure.

The recent attacks of Wang et al. [18,19,20], however, have revealed weaknesses in the expected ideal collision strength of the SHA-0 and SHA-1 hash functions. These and earlier MD5 collision attacks suggest that designing a collision secure hash function may turn out to be difficult. With the loss of the

R. Avanzi, L. Keliher, and F. Sica (Eds.): SAC 2008, LNCS 5381, pp. 228–244, 2009.

Coll security guarantee current hash functions fail also to provide apt security for the weaker security properties of Sec and Pre (see the attacks of [11] and the counterexamples of [2]).

The National Institute of Standards and Technology (NIST) of US has in turn addressed the problem by announcing a call for new hash functions [14]. The minimal security requirements stated in the call for proposals are Coll, Sec and Pre security with computational complexity of order $2^{n/2}$, $2^{n-\ell}$ and 2^n, respectively. Here the hash values are of n bits and the Sec security is expressed in terms of the message length in blocks (2^ℓ). We believe that proposals for new hash functions should provide guarantees for security preservation for not only Coll, but also Sec and Pre security. Preservation proofs are important because they allow one to rely on the hash function strength with respect to a concrete security property, independently of the weaknesses of the other properties.

The problem of designing a property-preserving iterations has been earlier investigated by [2,5,6,4,7,10,17]. Although these papers sometimes aim for properties different from the ones mentioned above, showing a property preserving hash function is one of their main goals.

In this paper we propose a new iterative hash function, that is based on the MD hash principle, and provably preserves the notions of Coll and Sec security in the standard security model and achieves Pre security when the compression function is instantiated as a random oracle. Our reduction for Coll is tight and we lose a factor of the message length (in blocks) in the Sec preservation. In the estimated Sec gap we are also able to mount the Sec attacks of [11,1]. Still, as we show, these attacks are only possible for target messages of a very specific structure. Finding a preimage message takes approximately 2^n evaluations of the compression function when it is modeled as a random oracle.

Our hash design benefits from a keyless compression function and makes use of keys in the iteration. We call this the *keyless compression function – keyed iteration* setting. Compared to the dedicated key setting of [6] (keyed compression function – keyed iteration), we achieve the three basic properties with a more practically understood and employed primitive, namely a keyless compression function. In the iterative portion of our design, we have reasons to believe that achieving security guarantees for Sec security is hard without the use of some form of randomization (provided by the keys in our case). A publicly known key selects a single function from a family of hash functions and once chosen at random it remains fixed for the hash algorithm. Note however, that any security claims for keyed hash functions hold only as long as the keys are generated honestly. If the keys are maliciously chosen, then they become exploitable constants and could potentially give rise to future attacks. A possible way to employ fixed keys in practice is to make the key selection process open and fixed in standard.

Achieving Coll and Sec preservation in the standard security and Pre security in the random oracle model partially attempts to answer the question from [2] if a multi-property preserving hash transform is realizable in the standard model. To achieve a seven-property-preserving hash function the authors of [2] benefit from the use of a random oracle for the mask (key) generation. Also, compared to [4,17], which show the eSec property preservation, we use the key material sparsely. While the latter hash constructions use keys of length at least logarithmic in the message size (in blocks), we only need keys of constant length, $b + 2n$ bits, where b and n are the block and hash sizes, respectively.

Together with the basic hash function, we present some generalized versions of it. These vary according to the order in which the input values are processed by the compression function F. Still, the optimal input ordering for F in the iteration heavily depends on the specifications of the concrete compression function.

In the line of this work, another interesting problem has come to our notice. While our hash scheme offers a theoretically sound Sec preservation proof, in practical scenarios this result may lack the claimed strength. Why does this discrepancy occur? The standard Sec security definition assumes a uniform target message space distribution. We use this fact in our hash design to extract randomness from the message and mix it with the chaining portion of the iteration. A prominent example application that requires Sec security and where the hash inputs are chosen uniformly at random is the Cramer-Shoup cryptosystem [8][1]. However, in some applications, the target message space may not have the uniform distribution. By building a Sec secure compression function, we are able to demonstrate a Sec attack on our hash only for such biased distributions.

On the other hand, working with non-uniform target message distributions allows for better message visibility. Some messages are hashed with higher probability and thus are more predictable. This interpretation deviates from the Sec definition and is a shift towards the notion of target collision security, or eSec from [16], where the messages are fully predictable (chosen) by the adversary. This observation can be interpreted in two ways. One way to think about the problem is to work with variants of the Sec definitions that take into account the target message distribution. Another solution may be to provide appropriate message input randomization to guarantee the randomness of the message inputs. In the final part of our paper we provide a short discussion on the issue.

2   Security Definitions

NOTATION. Let ε be the empty string. $x \| y$ denotes the concatenation of strings x and y. If x is a string, then $x|_z^{\mathtt{msb}}$ and $x|_y^{\mathtt{lsb}}$ specify the most z and least y, respectively, significant bits of x. $|x|$ is the length in bits of the string x and $x|_i^j$ is the substring of x containing the i-th through j-th bit of x, inclusive.

If S is a set, then $x \xleftarrow{\$} S$ denotes the uniformly random selection of an element from S. We let $y \leftarrow \mathsf{A}(x)$ and $y \xleftarrow{\$} \mathsf{A}(x)$ be the assignment to y of the output of a deterministic and randomized algorithm A, respectively, when run on input x.

---

[1]   "For this purpose, we will use a family of hash functions, such that given a randomly chosen tuple of group elements and randomly chosen hash function key, it is computationally infeasible to find a different tuple of group elements that hashes to the same value using the given hash key."

Definition of target collision resistance from [8] matching the standard Sec security one.

An *adversary* is an algorithm with polynomial running time, possibly with access to some oracles. To avoid trivial lookup attacks, it will be our convention to include in the time complexity of an adversary A its running time and its code size (relative to some fixed model of computation).

In our *keyless compression function-keyed iteration* setting we model the fixed-input-size function to be a keyless compression function. The iterative arbitrary-input-size hash function on the other hand is a family of functions indexed by a fixed random key.

SECURITY OF COMPRESSION FUNCTIONS. Let $F : \{0,1\}^{b+n} \rightarrow \{0,1\}^n$ be a compression function that takes inputs of fixed size $(n+b)$ bits and maps them to outputs of size n. First we define the following advantage measures for Coll and Sec security for a fixed adversary A and message length $\lambda \in \mathbb{N}$:

$$\mathbf{Adv}_F^{\mathrm{Coll}}(A) = \Pr\left[\; M', M \xleftarrow{\$} A(\varepsilon) \;:\; M \neq M' \text{ and } F(M) = F(M') \;\right]$$

$$\mathbf{Adv}_F^{\mathrm{Sec}[\lambda]}(A) = \Pr\left[\; M \xleftarrow{\$} \{0,1\}^\lambda \;;\; M' \xleftarrow{\$} A(M) \;:\; M \neq M' \text{ and } F(M) = F(M') \;\right]$$

$$\mathbf{Adv}_F^{\mathrm{Pre}[\lambda]}(A) = \Pr\left[\; M \xleftarrow{\$} \{0,1\}^\lambda \;;\; Y \leftarrow F(M) \;;\; M' \xleftarrow{\$} A(Y) \;:\; F(M') = Y \;\right]$$

We say that F is (t, ϵ) *atk* secure for atk $\in \{\text{Sec}, \text{Pre}\}$ if $\mathbf{Adv}_F^{\mathrm{atk}[\lambda]}(A) < \epsilon$ for all adversaries A running in time at most t and $\lambda = b+n$. Note that it is impossible to define security for the case of Coll in an analogous way. Indeed, if collisions on F exist, then an adversary A that simply prints out a collision that is hardcoded into it always has advantage 1. Rather than defining Coll security through the non-existence of an algorithm A, we follow Rogaway's human-ignorance approach [15] and use the above advantage function as a metric to relate the advantage of an adversary A against the hash function to that of an adversary B against the compression function.

SECURITY OF HASH FUNCTIONS. A hash function family is a function $H : \mathcal{K} \times \mathcal{M} \rightarrow \mathcal{Y}$ where the key space $\mathcal{K}$ and the target space $\mathcal{Y}$ are finite sets of bit strings. The message space $\mathcal{M}$ could be infinitely large; we assume that there exists at least one $\lambda \in \mathbb{N}$ such that $\{0,1\}^\lambda \subseteq \mathcal{M}$. The key K is an index that selects a instance from the function family. Following [16], we use the following advantage measures:

$$\mathbf{Adv}_H^{\mathrm{Coll}}(A) = \Pr\left[\; K \xleftarrow{\$} \mathcal{K} \;;\; (M, M') \xleftarrow{\$} A(K) \;:\; \begin{array}{c} M \neq M' \text{ and} \\ H(K, M) = H(K, M') \end{array} \right]$$

$$\mathbf{Adv}_H^{\mathrm{Sec}[\lambda]}(A) = \Pr\left[\; \begin{array}{c} K \xleftarrow{\$} \mathcal{K} \;;\; M \xleftarrow{\$} \{0,1\}^\lambda \\ M' \xleftarrow{\$} A(K, M) \end{array} \;:\; \begin{array}{c} M \neq M' \text{ and} \\ H(K, M) = H(K, M') \end{array} \right]$$

$$\mathbf{Adv}_H^{\mathrm{Pre}[\lambda]}(A) = \Pr\left[\; \begin{array}{c} K \xleftarrow{\$} \mathcal{K} \;;\; M \xleftarrow{\$} \{0,1\}^\lambda \\ Y \leftarrow H(K, M) \;;\; M' \xleftarrow{\$} A(K, Y) \end{array} \;:\; H(K, M') = Y \;\right]$$

For atk $=$ Coll, we say that H is (t, ϵ) atk secure if $\mathbf{Adv}_H^{\mathrm{atk}}(A) < \epsilon$ for all adversaries A running in time at most t. For atk $\in \{\text{Sec}, \text{Pre}\}$, we say that H is

(t, ϵ) atk secure if $\mathbf{Adv}_{H}^{atk[\lambda]}(A) < \epsilon$ for all adversaries A running in time at most t and for all $\lambda \in \mathbb{N}$ such that $\{0, 1\}^{\lambda} \subseteq \mathcal{M}$.

Our security claims in the random oracle model consider (q_{RO}, ϵ) atk-security, where q_{RO} is the total number of queries that the adversary A makes to the a random oracle. In the same model, we assume that the compression function $F : \{0, 1\}^{b+n} \rightarrow \{0, 1\}^n$ behaves as a random oracle. That means F is chosen uniformly at random from the set of all functions with the respective domain and range space and is publicly computable function.

SECURITY PRESERVATION. Our goal is to build an infinite-domain hash function family H out of a limited-domain compression function F so that the hash function "inherits" its Coll and Sec security from the natural analogues of these properties for F. For atk $=$ Sec, we say that H *preserves* atk security if H is (t, ϵ) atk secure whenever F is (t', ϵ') atk secure, for some well-specified relation between $t, t', \epsilon, \epsilon'$. For the case of Coll, we have to be more careful because, as pointed out before, (t, ϵ)-Coll security cannot be defined for the keyless compression function F. Rather, we follow Rogaway [15] by saying that collision resistance is preserved if, for an explicitly given Coll adversary A against H, there exists a corresponding, explicitly specified Coll adversary B, as efficient as A, that finds collisions for F.

3   The Basic Construction

3.1   The $\mathcal{BCM}$ Hash Function

In this section we present our hash mode. We refer to it as the backwards chaining mode, or the $\mathcal{BCM}$ hash (see Fig. 1).

THE HASH FUNCTION. The $\mathcal{BCM}_F$ hash uses a fixed-input-length compression function $F : \{0, 1\}^{b+n} \rightarrow \{0, 1\}^n$ where $b \geq n$ and takes as inputs a message M of arbitrary length and a key $K = K_1 \| K_2 \| K_3$ of fixed length $(b + 2n)$ bits, where $|K_2| = b$ and $|K_1| = |K_3| = n$. For security and practical reasons we set a bound on the minimal and maximal message length λ, or $n < \lambda < 2^c$ where typically $c = 64$ and $c < n$. The message is preprocessed with a standard MD

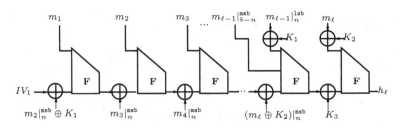

Fig. 1. The $\mathcal{BCM}$ Construction. The message M is MD strengthened. K_1, K_2 and K_3 are randomly chosen and fixed keys of length n, b and n bits, respectively.

strengthening [12]. That is, a single 1 bit is appended to the message M followed by as many zeros as needed and the binary encoding of $|M|$ in 64 bits. We denote the MD padding and strengthening function by pad and $m_1 \| \ldots \| m_\ell \leftarrow \mathsf{pad}(M)$, such that $|m_i| = b$ for $i = 1$ to ℓ.

The $\mathcal{BCM}_\mathrm{F}$ hash function can be described as follows. It XORs the key K_1 and the most significant n bits of block m_2 with the fixed initial chaining variable IV_1 (e.g. $IV_1 = 0^n$). The message block m_1 together with the resulting value from the XOR computation form the input to the first application of F. The current-in-line message block m_i and the chaining variable h_{i-1} XORed with the most significant n bits of the next-in-line message block m_{i+1} are the following inputs to the compression function F in the iteration for $i = 1$ to $\ell - 2$.

The one but last block is interpreted differently than the rest of the message blocks. Here the difference is that the least significant n bits of $m_{\ell-1}$ are XORed with the key K_1, while the chaining variable $h_{\ell-2}$ is XORed with $K_2|_n^{\mathrm{msb}}$ and $m_\ell|_n^{\mathrm{msb}}$. The order of processing the inputs is preserved also in the $(\ell-1)$st block. The final input to the last compression function is provided by the last message block m_ℓ and the chaining variable $h_{\ell-1}$ XORed with keys K_2 and K_3, respectively.

We describe our construction in pseudocode below (Alg. 1) and give a graphical representation in Fig. 1.

Algorithm 1. $\mathcal{BCM}_\mathrm{F}(K, M)$:

> $m_1 \| \ldots \| m_\ell \leftarrow \mathsf{pad}(M)$
> $h_0 = IV_1,\ g_1 = h_0 \oplus K_1 \oplus m_2|_n^{\mathrm{msb}}$
> $h_1 = \mathrm{F}(m_1, g_1)$
> **for** $i = 2$ to $\ell - 2$ **do**
> $\quad g_{i-1} = m_{i+1}|_n^{\mathrm{msb}} \oplus h_{i-1}$
> $\quad h_i = \mathrm{F}(m_i, g_{i-1})$
> **end for**
> $g_{\ell-2} = (K_2 \oplus m_\ell)|_n^{\mathrm{msb}} \oplus h_{\ell-2}$
> $h_{\ell-1} = \mathrm{F}(m_{\ell-1}|_{b-n}^{\mathrm{msb}} \| (m_{\ell-1}|_n^{\mathrm{lsb}} \oplus K_1), g_{\ell-2})$
> $h_\ell = \mathrm{F}(m_\ell \oplus K_2, h_{\ell-1} \oplus K_3)$
> **return** h_ℓ

The $\mathcal{BCM}$ hash of a single strengthened message block m_1 is computed as $h_1 = \mathrm{F}(m_1 \oplus K_2, IV_1 \oplus K_1 \oplus K_3)$. And when the message is two blocks long, then $h_1 = \mathrm{F}(m_1|_{b-n}^{\mathrm{msb}} \| m_1|_n^{\mathrm{lsb}} \oplus K_1, IV_1 \oplus K_1 \oplus (m_2 \oplus K_2)|_n^{\mathrm{msb}})$ and the final output hash is computed as $h_2 = \mathrm{F}(m_2 \oplus K_2, h_1 \oplus K_3)$.

EFFICIENCY. The $\mathcal{BCM}_\mathrm{F}$ hash mode is a streaming hashing mode that compared to the known MD mode delays the processing with n bits in start-up time. Also as in the MD hash function a single message block is processed per call to the compression function F. Although we lack any concrete efficiency measurements, we expect a small loss in efficiency (compared to MD) due to the constant storage of extra n bits in memory.

DISCUSSION ON THE DESIGN CHOICES. We choose to XOR IV_1 with the key K_1 to provide additional randomization on the initialization value. The rest of the XOR choices, namely XORing the chaining variables with the most significant n bits of the incoming message blocks provide the randomization on the chaining values necessary for the Sec security preservation. Also, to achieve the Sec security we have to disallow any fixed inputs introduced by the message padding and strengthening. Hence, we XOR $m_{\ell-1}|_n^{\text{lsb}}$ and m_ℓ with the keys K_1 and K_2, respectively, and we additionally use the key K_3 to randomize the final chaining hash value $h_{\ell-1}$. When F is modeled as a random oracle, the Pre property of $\mathcal{BCM}_{\text{F}}$ is easily satisfiable as long as the message has a minimal length of n bits. An interesting observation is that none of the applied randomization techniques contributes for a Pre preservation in the standard security model.

3.2   Possible Variants

We discuss modifications on the basic $\mathcal{BCM}$ construction with respect to the order of input values to the compression function F. The variants include the different chaining iterations on F, such that F takes as inputs any ordering of: $(A_1 = m_1|_n^{\text{msb}}, B_1 = m_1|_{b-n}^{\text{lsb}}, C_1 = IV_1 \oplus K_1 \oplus m_2|_n^{\text{msb}}), (A_i = m_i|_n^{\text{msb}}, B_i = m_i|_{b-n}^{\text{lsb}}, C_i = h_{i-1} \oplus m_{i+1}|_n^{\text{msb}})$ for $i = 2$ to $\ell - 2$, $(A_{\ell-1} = m_{\ell-1}|_n^{\text{msb}}, B_{\ell-1} = m_{\ell-1}|_{n+1}^{b-2n} \| m_{\ell-1}|_{b-2n+1}^{b-n} \oplus K_1, C_{\ell-1} = h_{\ell-2} \oplus (m_\ell \oplus K_2)|_n^{\text{msb}})$ and finally $(A_\ell = (m_\ell \oplus K_2)|_n^{\text{msb}}, B_\ell = (m_\ell \oplus K_2)|_{b-n}^{\text{lsb}}, C_\ell = h_{\ell-1} \oplus K_3)$. The indices denote the position of the input values in the iteration, e.g. (A_1, B_1, C_1) forms the set of input values to the first application of F. There are at most six permuted input sets to F (per call to F). As long as the inputs in the final call to F are ordered identically for messages of any arbitrary length, then the security properties of the basic $\mathcal{BCM}$ carry through to any chaining iteration that switches the input wires to F in any chosen, but specified order.

Let $S_i^1 = \{A_i, B_i, C_i\}$ for $i = 1$ to ℓ be the sets containing the input values to F of the same index i. We then define the sets S_i^j for $j = 2$ to 6 and $i = 1$ to ℓ to be the rest of the possible orderings of the base set S_j^1, or these are $S_i^2 = \{A_i, C_i, B_i\}, S_i^3 = \{C_i, A_i, B_i\}, S_i^4 = \{C_i, B_i, A_i\}, S_i^5 = \{B_i, C_i, A_i\}$ and $S_i^6 = \{B_i, A_i, C_i\}$. Let $P_i^j : S_i^1 \to S_i^j$ where $j = 1$ to 6 and $i = 1$ to ℓ. With P_i^a we then denote any arbitrarily chosen mapping from S_i^1 to S_i^j for any $j = 1$ to 6 (i is a fixed input parameter), while P^f stands for the final mapping from S_ℓ^1 to S_ℓ^j for some randomly chosen and fixed j.

The $\mathcal{GBCM}$ hash of a 1-block message m_1 is $h_1 = \text{F}(P_1^a(A_1, B_1, C_1))$ with $(A_1 = (m_1 \oplus K_2)|_n^{\text{msb}}, B_1 = (m_1 \oplus K_2)|_{b-n}^{\text{lsb}}, C_1 = IV_1 \oplus K_1 \oplus K_3)$.

The $\mathcal{GBCM}$ hash of a 2-block strengthened message is $h_2 = \text{F}(P^f(A_2, B_2, C_2))$ for $(A_2 = (m_2 \oplus K_2)|_n^{\text{msb}}, B_2 = (m_2 \oplus K_2)|_{b-n}^{\text{lsb}}, C_2 = h_1 \oplus K_3)$ where $h_1 = \text{F}(P_1^a(A_1, B_1, C_1))$ and $(A_1 = m_1|_n^{\text{msb}}, B_1 = m_1|_{n+1}^{b-n} \| m_1|_n^{\text{lsb}} \oplus K_1, C_1 = IV_1 \oplus K_1 \oplus (m_2 \oplus K_2)|_n^{\text{msb}})$.

We then summarize the variants of $\mathcal{BCM}$ by exhibiting a generalized $\mathcal{GBCM}$ construction and describe it in pseudocode in Algorithm 2.

Algorithm 2. $\mathcal{GBCM}_F(K, M)$:

$m_1 \| \dots \| m_\ell \leftarrow \mathsf{pad}(M)$
$h_0 = IV_1,$
for $i = 1$ to $\ell - 1$ **do**
$\quad h_i = F(P_i^a(A_i, B_i, C_i))$
end for
$h_\ell = F(P^f(A_\ell, B_\ell, C_\ell))$
return h_ℓ

---

4 Property Preservation of the $\mathcal{BCM}$ ($\mathcal{GBCM}$) Construction

In this section we provide the full proofs for Coll and Sec security preservation in the standard security model and we show Pre security of the $\mathcal{BCM}_F$ construction when the compression function is instantiated as a random oracle. We provide the proofs of $\mathcal{GBCM}$ in the Appendix.

Theorem 1. *If there exists an explicitly given adversary A that (t, ϵ)-breaks the Coll security of $\mathcal{BCM}_F$ ($\mathcal{GBCM}_F$), then there exists an explicitly given adversary B that (t', ϵ')-breaks the Coll security of F for $\epsilon' \geq \epsilon$ and $t' \leq t + 2\ell \cdot \tau_F$. Here, τ_F is the time required for the evaluation of F and $\ell = \lceil (\lambda + 65)/b \rceil$ where λ is the maximum message length of the two messages output by A.*

Proof. Given a Coll adversary A against the iterated hash $\mathcal{BCM}_F$, we construct a Coll adversary B against the compression function F. B generates at random a key $K \xleftarrow{\$} \{0,1\}^{b+2n}$ with $K = K_1 \| K_2 \| K_3$ where $|K_1| = |K_3| = n$ and $|K_2| = b$. B runs A on input K. Finally, A outputs a colliding pair of messages M and M', such that $\mathcal{BCM}_F(K, M) = \mathcal{BCM}_F(K, M')$. We investigate the following two cases:

1. If $|M| \neq |M'|$, then the inputs to the last compression function differ (due to the present message length encoding in m_ℓ) and therefore a collision on the final F occurs, or $m_\ell \oplus K_2 \neq m'_{\ell'} \oplus K_2$ where $F(m_\ell \oplus K_2, h_{\ell-1} \oplus K_3) = F(m'_{\ell'} \oplus K_2, h'_{\ell'-1} \oplus K_3)$. B then outputs $(m_\ell \oplus K_2, h_{\ell-1} \oplus K_3)$ and $(m'_{\ell'} \oplus K_2, h'_{\ell'-1} \oplus K_3)$ as a valid colliding pair.
2. Else if $|M| = |M'|$, then $\ell = \ell'$. If $m_\ell \oplus K_2 \| h_{\ell-1} \oplus K_3 \neq m'_\ell \oplus K_2 \| h'_{\ell-1} \oplus K_3$, then a collision occurs again in the last application of F. Else B proceeds in the following way.
   B parses the inputs to the $(\ell-1)$st application of F as $(m_{\ell-1}|_{b-n}^{\mathtt{msb}} \| (m_{\ell-1}|_n^{\mathtt{lsb}} \oplus K_1), g_{\ell-2})$ and $(m'_{\ell-1}|_{b-n}^{\mathtt{msb}} \| (m'_{\ell-1}|_n^{\mathtt{lsb}} \oplus K_1), g'_{\ell-2})$. If these inputs differ, then they constitute a valid collision pair for B, else $g_{\ell-2} = g'_{\ell-2}$ and hence $h_{\ell-2} = h'_{\ell-2}$ because of the previous equality for $m_\ell \oplus K_2|_n^{\mathtt{msb}} = m'_\ell \oplus K_2|_n^{\mathtt{msb}}$. Following the iteration principle B parses the previous inputs as $m_{\ell-2} \| g_{\ell-3}$ and $m'_{\ell-2} \| g'_{\ell-3}$ and proceeds in the same manner backwards.

The inequality of the message inputs M and M' guarantees the existence of an index $i > 0$, such that $m_i\|g_{i-1} \neq m'_i\|g'_{i-1}$ where $F(m_i\|g_{i-1}) = F(m'_i\|g'_{i-1})$. B outputs then the colliding pair $(m_i\|g_{i-1}, m'_i\|g'_{i-1})$ for the $\max(i)$ satisfying the former statement.

Whenever A succeeds, then B also succeeds with the same advantage. The time complexity of B is at most the time complexity of A plus two evaluations of $\mathcal{BCM}_F$ over messages M and M' taking time $2\ell \cdot \tau_F$. □

Theorem 2. *For* atk $=$ Sec, *if the compression function* F *is* (t', ϵ') atk *secure, then the iterated function* $\mathcal{BCM}_F$ *(*$\mathcal{GBCM}_F$*) is* (t, ϵ) atk *secure for* $\epsilon \leq \ell \cdot \epsilon'$ *and* $t \geq t' - 2\ell \cdot \tau_F$. *Here,* τ_F *is the time required for the evaluation of* F *and* $\ell = \lceil (\lambda + 65)/b \rceil$ *where* λ *is the maximum message length of the two messages output by* A.

Proof. Given a Sec[λ] adversary A against $\mathcal{BCM}_F$, we construct a Sec adversary B against the compression function F. B receives a random challenge message $m\|h$. For a randomly chosen index i the goal of the adversary B is to construct a challenge message M and a key K, which have B's challenge message $m\|h$ embedded, such that when A outputs its second preimage message M', then a collision for M and M' can be found at the ith block of M. To simulate A's view correctly, however, B generates M and K, such that they are uniformly distributed. The proof goes as follows.

B chooses a random index $i \xleftarrow{\$} \{1, \ldots, \ell = \lceil(\lambda+65)/b\rceil\}$ and a random message M of length λ. B has now to successfully embed his challenge $m\|h$ at position i in the target strengthened message $m_1\| \ldots \|m_\ell \leftarrow \mathsf{pad}(M)$ and in the chaining iterative portion of $\mathcal{BCM}_F(K, M)$. Let $\hat{M} = m_1\| \ldots \|m_\ell$ be the strengthened message M. Depending on the outcomes for i, B takes its decisions as follows:

1. If $i = 1$, then B sets $m_1 \leftarrow m$ and $K_1 \leftarrow IV_1 \oplus h \oplus m_2|^{\mathsf{msb}}_n$. Except block m_1, the rest of $\hat{M}$ is unaltered. B chooses $K_2\|K_3 \xleftarrow{\$} \{0,1\}^{b+n}$. Two special cases arise in the case when $\lambda < b - 65$ or $\lambda < 2b - 65$. In the former case, B proceeds as described below for $i = \ell$, and in the latter case as for $i = \ell - 1$.
2. If $i \in \{2, \ldots, \ell - 2\}$, then B continues as follows. B sets $m_i \leftarrow m$ and computes the intermediate chaining value h_{i-1} with $K_1 \xleftarrow{\$} \{0,1\}^n$. B sets $m_{i+1}|^{\mathsf{msb}}_n \leftarrow h_{i-1} \oplus h$. Then B chooses the keys $K_2\|K_3$ at random. Except modifying blocks m_i and $m_{i+1}|^{\mathsf{msb}}_n$, B leaves the rest of $\hat{M}$ unaltered.
3. If $i = \ell - 1$, then B sets $m_{\ell-1}|^{\mathsf{msb}}_{b-n} \leftarrow m|^{\mathsf{msb}}_{b-n}$ and $K_1 \leftarrow (m_{\ell-1} \oplus m)|^{\mathsf{lsb}}_n$, and computes the intermediate chaining value $h_{\ell-2}$. $h_{\ell-2}$ is set to $IV_1 \oplus K_1$ when $b - 65 \leq \lambda < 2b - 65$. B sets $K_2|^{\mathsf{msb}}_n \leftarrow h_{\ell-2} \oplus h \oplus m_\ell|^{\mathsf{msb}}_n$. Then B chooses at random $K_2|^{\mathsf{lsb}}_{b-n}\|K_3 \xleftarrow{\$} \{0,1\}^b$. With the exception of $m_{\ell-1}|^{\mathsf{msb}}_{b-n}$, the rest of $\hat{M}$ remains unaltered.
4. If $i = \ell$, then B chooses at random $K_1 \xleftarrow{\$} \{0,1\}^n$ and computes the intermediate hash value $h_{\ell-1}$. Note that if $\lambda < b - 65$, then the chaining value is computed as $IV_1 \oplus K_1$. B then sets $K_3 \leftarrow h \oplus h_{\ell-1}$ and $K_2 \leftarrow m \oplus m_\ell$. $\hat{M}$ remains unchanged.

After $m\|h$ is successfully embedded, B proceeds by running A on inputs message M where $M = \mathsf{pad}^{-1}(\hat{M})$(the non-strengthened version of $\hat{M}$ with the applied modifications) and key $K = K_1\|K_2\|K_3$. Note that both M and K are uniformly distributed. Initially B chooses uniformly at random M and then modifies some of its blocks as prescribed in the former cases. However, the modified blocks of M are assigned only independent random values. Hence, the resulting final challenge message M is also uniformly distributed. The key K is constructed in a similar way.

On inputs M and K A returns a second preimage message M', such that $\mathcal{BCM}_{\mathrm{F}}(K, M) = \mathcal{BCM}_{\mathrm{F}}(K, M')$. For the rest of the proof B acts identically as in the Coll proof. With probability $1/\ell$ B finds the colliding pair at the correct position i (at which B embedded $m\|h$) and outputs the colliding inputs for F as its valid second preimage. If A succeeds with advantage ϵ, then B also succeeds with advantage ϵ/ℓ. The time complexity of B is at most the time complexity of A plus two evaluations of $\mathcal{BCM}_{\mathrm{F}}$. This completes the proof. □

Theorem 3. *If the compression function F is instantiated as a random oracle, then the iterated function $\mathcal{BCM}_{\mathrm{F}}$ ($\mathcal{GBCM}_{\mathrm{F}}$) is $(q_{\mathrm{RO}}, \epsilon)$ Pre$[\lambda]$ secure where $\epsilon \leq q_{\mathrm{RO}}/2^n$ and q_{RO} is the number of queries to the random oracle.*

Proof. Let A be a Pre$[\lambda]$ adversary on the iterated hash function $\mathcal{BCM}_{\mathrm{F}}$. Given a challenge hash value Y and key $K = K_1\|K_2\|K_3$, the goal of A is to invert Y, which is computed as $Y = \mathcal{BCM}_{\mathrm{F}}(K, M)$ for a randomly chosen message $M \xleftarrow{\$} \{0,1\}^\lambda$ and a key $K \xleftarrow{\$} \mathcal{K}$.

We investigate the following two cases: 1. $n < \lambda \leq b - 65$ and 2. $\lambda > b - 65$. In the first case $Y = \mathrm{F}(m_1 \oplus K_2, IV_1 \oplus K_1 \oplus K_3)$. The adversary A knows the message length, respectively the applied strengthening bits, the fixed IV_1 value and the random key values $K_1\|K_2\|K_3$. However, A has no information of at least n bits of the message input m_1. Thus, the only way to find a valid preimage message for $\mathcal{BCM}_{\mathrm{F}}$ is to exhaustively query the random oracle F on chosen inputs for the missing part of m_1. The probability to find the correct preimage message per single query is $1/2^n$ and after q_{RO} queries to the random oracle A succeeds to invert Y with probability $q_{\mathrm{RO}}/2^n$.

In the second case $Y = \mathrm{F}(m_\ell \oplus K_2, h_{\ell-1} \oplus K_3)$. Here A knows the keys K_2 and K_3, the message length λ and the respective strengthening bits used in the last message block m_ℓ. Still, B does not know the intermediate chaining value $h_{\ell-1}$. Again as in the former case, A needs to invert Y and its success ϵ is bound by $q_{\mathrm{RO}}/2^n$. □

A Pre COUNTEREXAMPLE IN THE STANDARD MODEL. Surprisingly, the presented $\mathcal{BCM}_{\mathrm{F}}$ does not provide Pre property preservation when the compression function is a Pre secure hash and not modeled as a random oracle. Here we provide a counterexample compression function, which is Pre secure as long as the underlying compressing function is also Pre secure.

Let F be defined as $\mathrm{F}(m\|h) = \mathrm{CE}_1(m)$. If $\mathrm{CE}_1 : \{0,1\}^b \to \{0,1\}^n$ is (t',ϵ') Pre secure compressing function, then F is also (t,ϵ) Pre secure function with $\epsilon \leq \epsilon'$ and $t \geq t'$.

A Pre[λ] adversary A on the iterated hash function $\mathcal{BCM}_F$ succeeds in constant time with probability one in breaking the Pre[λ] security. A is given a challenge hash value $Y = \mathcal{BCM}_F(M, K)$ for a random $M \xleftarrow{\$} \{0,1\}^\lambda$ and a random key $K \leftarrow \mathcal{K}$ with $K = K_1 \| K_2 \| K_3$. A succeeds by outputting any message M' of length λ, because all these messages result in $Y = F(m_\ell \oplus K_2, h_{\ell-1} \oplus K_3) = CE_1(m_\ell \oplus K_2)$. Here we made an assumption that λ is such that m_ℓ consists only of padding and strengthening bits.

4.1   Second Preimage Attacks Beyond $2^{n-\ell}$

From [11] and [1] we know that earlier Merkle-Damgård based constructions are prone to Sec attacks in a bit more that $2^{n-\ell}$ compression function calls when the target messages are of size 2^ℓ blocks. The latter attacks apply to our scheme given that the target messages are of a specific format. Let the challenge messages be parsed as a sequence of b-bit blocks. When these contain fixed and predictable message chunks in their n most significant bits (e.g. $m_i|_n^{\mathtt{msb}} = 0^n$), the mentioned attacks can be mounted on our hash construction. But even then the attacks are in no contradiction with our claimed Sec security result. In our Sec security proof we loose a factor of ℓ (number of message blocks), while the attacks are valid in the estimated security gap (between the exhibited $2^{n-\ell}$ and the ideal 2^n Sec security).

To build either an expandable message or a diamond structure used in the attacks, an adversary searches for collisions on F. These are possible by going over different values only in the least significant $(b-n)$ bits of the message blocks chosen by the adversary. The adversary then commits to an intermediate hash value h_i. Next, in both the expandable message attack [11] and the diamond structure [1] second preimage attacks, the adversary has to connect from h_i to a chaining value in the target message M. Here the requirement for a specific message format comes into play. If the message blocks differ in their most significant n bits, the adversary's probability to connect correctly is small. That is because he has to have predicted in advance $m_j|_n^{\mathtt{msb}}$, given that he successfully connects to a chaining value h_j. The best adversarial strategy here is to exploit message blocks repetitions in their most significant n bits. Then if all these are equal, the attacks become feasible in approximately $2^{n-\ell}$ steps (compression function calls).

One way to fully block this type of attacks is to XOR the chaining values with the output of a function f that takes as input the complete forth-coming message block of b bits. In the iteration we would replace the $m_i|_n^{\mathtt{msb}}$ with $f(m_i)$ for all $i = 1$ to ℓ. To achieve the property preservation the function f has to be instantiated as a random oracle, which turns the suggested scheme into a less efficient variant (linear number of calls to RO) of the ROX [2] hash.

5   Security Discussion or Where Theory Meets Practice

Our scheme preserves the Coll security of the compression function F due to the MD strengthening and achieves the $2^{n/2}$ security level if F is an ideal function.

The Sec security property is preserved through the randomization of the final inputs of F with keys K_2 and K_3, and the intermediate hash values with parts of the message blocks.

Notice that the standard Sec definition we use for the latter result assumes the uniform distribution U on the message space $\mathcal{M}$. This allows us to extract randomness from the challenge messages, rather than adding extra key material (e.g. log number of keys in Shoup's hash). Where is the caveat here? When the messages are not sufficiently random, the $\mathcal{BCM}$ hash scheme does not reach the claimed Sec security level. More precisely, we do not need the randomness from the whole message source but we extract it only from the n most significant bits of every b-bit chunk of message. Then messages without sufficient entropy in those most significant bits can introduce Sec weaknesses in our scheme, as we show. Non-uniform distributions that allow for such attacks are some concrete distributions of messages with low entropy. Next we exhibit such a counterexample.

THE LOW ENTROPY MESSAGES COUNTEREXAMPLE. In our counterexample we construct a contrived Sec secure compression function and specify the format of the messages that occur with the highest probability according the challenge messages distribution.

Let the challenge message space $\mathcal{M}^{D_l}$ be assigned the distribution D_l. Here we define $\mathcal{M}^{D_l} = \{0,1\}^\lambda$ where $\lambda > 2b - 65$. According to D_l for any $m_i|_n^{\text{lsb}}$ with $i = 2$ to ℓ ($\ell = \lceil (\lambda)/b \rceil$) the messages $m_1 \| 0^n \| m_2|_{b-n}^{\text{lsb}} \| \dots \| 0^n \| m_\ell|_{b-n}^{\text{lsb}}$ appear with high probability $(1 - \epsilon')$ while all the rest of the messages occur with negligible probability ϵ'. The most frequent challenge messages contain n bits of 0s in the most significant bits of their b-bit blocks. The counterexample compression function we use is similar to the one from Theorem 3.2 [2].

Theorem 4. *If there exists a (t, ϵ) Sec secure function* G $: \{0,1\}^{b+n} \to \{0,1\}^{n-1}$, *then there exists a $(t, \epsilon - 1/2^{n-1})$ Sec secure compression function* $CE_2 : \{0,1\}^{b+n} \to \{0,1\}^n$ *and an adversary* A *running in constant time with* $\text{Sec}[\lambda, D_l]$*-advantage* $(1 - \epsilon')$ *in breaking* $\mathcal{BCM}_{CE_2}$ *for any challenge message* $M \in \mathcal{M}^{D_l}$ *chosen according to the distribution* D_l.

Note that we additionally parameterize the adversarial advantage by the message space distribution D_l and that ϵ' is the probability for messages different from the specified format to be chosen from $\mathcal{M}^{D_l}$.

Proof. Our CE_2 is given by

$$CE_2(m\|h) = 0^n \qquad \text{if } h = 0^n \text{ or } m|_n^{\text{msb}} = 0^n$$
$$= G(m\|h) \| 1 \text{ otherwise .}$$

If G is (t, ϵ) Sec secure, then CE_2 is $(t, \epsilon - 1/2^{n-1})$ Sec secure; we refer to the full version [3] for the proof.

According to the distribution D_l the messages of the specified format are chosen with high probability $(1 - \epsilon')$ as target messages. Then for any random key $K \xleftarrow{\$} \{0,1\}^{b+2n}$ and target message $M = m_1 \| 0^n \| m_2|_{b-n}^{\text{lsb}} \| \dots \| 0^n \| m_\ell|_{b-n}^{\text{lsb}}$, a $\text{Sec}[\lambda, D_l]$ adversary A finds a second preimage message M', such that $|M| = |M'|$

and M' is of the same format as M. Let M' differ from M only in the least significant $(b - n)$ bits of their second blocks. Then the chaining values in the computation of M and M' are equal immediately after the second application of CE_2. For any $\lambda > 2b - 65$ the chaining value 0^n is propagated further in the chain. Finally $\mathcal{BCM}_{CE_2}(K, M) = \mathcal{BCM}_{CE_2}(K, M')$. $\qquad\Box$

We admit that this is a particularly contrived counterexample for a low entropy message distribution. The minimal requirement on the message structure with this type of counterexample compression functions is a repetition in the most significant n bits of two adjacent b-bit message blocks. Such counterexamples are especially problematic for low entropy message distributions and are also valid for high entropy message distributions. However, in the latter case, the probability for such challenge messages to be chosen is not high on average and we cannot exhibit an efficient Sec adversary.

6   Concluding Discussion

In our opinion the Sec security preservation is one of the hardest security notions to satisfy in an iterative hash mode. At the cost of a logarithmic number of keys to randomize the chaining values and an additional constant b-bit key to randomize the message blocks, Shoup's hash [17] could be modified to also achieve Sec preservation in the standard model. Notice, however, that once the keys are fixed, we can always identify non-uniform distributions for which contrived counterexamples are possible (even if we increase the fixed keys for the randomization of the message blocks from a constant to linear in the message length). One way to avoid this problem is to introduce randomization per message, known also as salting. It is therefore an interesting question to identify the conditions that such a message randomization transform needs to satisfy in order to provide Sec preservation for any target message distribution.

On the other hand, the question of correctly formalizing and satisfying Sec security properties that take into account biased challenge message distributions may be practically relevant. Practical message distributions that deviate from uniform allow for predictability of certain target messages and in our view are a shift from the Sec to the known target collision resistance (TCR/UOWH/eSec) property. Another interesting problem may then be to find ways to achieve Sec security for any message distribution with an efficient hash construction that uses a minimal amount of key material. In our view, one possible way to go around this problem is to correctly identify new assumptions on the compression function.

Acknowledgements

We would like to thank Gregory Neven, Sebastiaan Indesteege and Markulf Kohlweiss for the helpful discussions. Also, we would like to thank the anonymous referees for their useful feedback. This work was supported in part by the European Commission through the IST Programme under Contract IST-2002-507932 ECRYPT, and in part by the IAP Programme P6/26 BCRYPT of the Belgian State (Belgian Science Policy). The first author is supported by a Ph.D. Fellowship from the Flemish Research Foundation (FWO - Vlaanderen).

References

1. Andreeva, E., Bouillaguet, C., Fouque, P.-A., Hoch, J.J., Kelsey, J., Shamir, A., Zimmer, S.: Second preimage attacks on dithered hash functions. In: Smart, N.P. (ed.) EUROCRYPT 2008. LNCS, vol. 4965, pp. 270–288. Springer, Heidelberg (2008)
2. Andreeva, E., Neven, G., Preneel, B., Shrimpton, T.: Seven-property-preserving iterated hashing: ROX. In: Kurosawa, K. (ed.) ASIACRYPT 2007. LNCS, vol. 4833, pp. 130–146. Springer, Heidelberg (2007)
3. Andreeva, E., Neven, G., Preneel, B., Shrimpton, T.: Seven-property-preserving iterated hashing: ROX. Cryptology ePrint Archive, Report 2007/176 (2007)
4. Bellare, M., Rogaway, P.: Collision-resistant hashing: Towards making uOWHFs practical. In: Kaliski Jr., B.S. (ed.) CRYPTO 1997. LNCS, vol. 1294, pp. 470–484. Springer, Heidelberg (1997)
5. Bellare, M., Ristenpart, T.: Multi-property-preserving hash domain extension and the EMD transform. In: Lai, X., Chen, K. (eds.) ASIACRYPT 2006. LNCS, vol. 4284, pp. 299–314. Springer, Heidelberg (2006)
6. Bellare, M., Ristenpart, T.: Hash functions in the dedicated-key setting: Design choices and MPP transforms. In: Arge, L., Cachin, C., Jurdziński, T., Tarlecki, A. (eds.) ICALP 2007. LNCS, vol. 4596, pp. 399–410. Springer, Heidelberg (2007)
7. Coron, J.-S., Dodis, Y., Malinaud, C., Puniya, P.: Merkle-damgård revisited: How to construct a hash function. In: Shoup, V. (ed.) CRYPTO 2005. LNCS, vol. 3621, pp. 430–448. Springer, Heidelberg (2005)
8. Cramer, R., Shoup, V.: Design and analysis of practical public-key encryption schemes secure against adaptive chosen ciphertext attack. SIAM Journal on Computing 33(1), 167–226 (2003)
9. Damgård, I.B.: A design principle for hash functions. In: Brassard, G. (ed.) CRYPTO 1989. LNCS, vol. 435, pp. 416–427. Springer, Heidelberg (1990)
10. Halevi, S., Krawczyk, H.: Strengthening digital signatures via randomized hashing. In: Dwork, C. (ed.) CRYPTO 2006. LNCS, vol. 4117, pp. 41–59. Springer, Heidelberg (2006)
11. Kelsey, J., Schneier, B.: Second preimages on n-bit hash functions for much less than 2^n work. In: Cramer, R. (ed.) EUROCRYPT 2005. LNCS, vol. 3494, pp. 474–490. Springer, Heidelberg (2005)
12. Lai, X., Massey, J.L.: Hash functions based on block ciphers. In: Rueppel, R.A. (ed.) EUROCRYPT 1992. LNCS, vol. 658, pp. 55–70. Springer, Heidelberg (1993)
13. Merkle, R.C.: One way hash functions and DES. In: Brassard, G. (ed.) CRYPTO 1989. LNCS, vol. 435, pp. 428–446. Springer, Heidelberg (1990)
14. NIST. Announcing request for candidate algorithm nominations for a new cryptographic hash algorithm (sha-3) family (2007),
    http://csrc.nist.gov/groups/ST/hash
15. Rogaway, P.: Formalizing human ignorance. In: Nguyên, P.Q. (ed.) VIETCRYPT 2006. LNCS, vol. 4341, pp. 211–228. Springer, Heidelberg (2006)

16. Rogaway, P., Shrimpton, T.: Cryptographic hash-function basics: Definitions, implications, and separations for preimage resistance, second-preimage resistance, and collision resistance. In: Roy, B., Meier, W. (eds.) FSE 2004. LNCS, vol. 3017, pp. 371–388. Springer, Heidelberg (2004)
17. Shoup, V.: A composition theorem for universal one-way hash functions. In: Preneel, B. (ed.) EUROCRYPT 2000. LNCS, vol. 1807, pp. 445–452. Springer, Heidelberg (2000)
18. Wang, X., Yu, H.: How to break MD5 and other hash functions. In: Cramer, R. (ed.) EUROCRYPT 2005, vol. 3494, pp. 19–35. Springer, Heidelberg (2005)
19. Wang, X., Yin, Y.L., Yu, H.: Finding collisions in the full SHA-1. In: Shoup, V. (ed.) CRYPTO 2005. LNCS, vol. 3621, pp. 17–36. Springer, Heidelberg (2005)
20. Wang, X., Yu, H., Yin, Y.L.: Efficient collision search attacks on SHA-0. In: Shoup, V. (ed.) CRYPTO 2005. LNCS, vol. 3621, pp. 1–16. Springer, Heidelberg (2005)

A $\mathcal{GBCM}$ Proofs

A.1   Coll Proof of Theorem 1 for $\mathcal{GBCM}$

Given a Coll adversary A against $\mathcal{GBCM}_F$, we construct a Coll adversary B against the compression function F. B generates at random a key $K \xleftarrow{\$} \{0,1\}^{b+2n}$ with $K = K_1\|K_2\|K_3$ where $|K_1| = |K_3| = n$ and $|K_2| = b$. B runs A on input K. Finally, A outputs a colliding pair of messages M and M', such that $\mathcal{BCM}_F(K, M) = \mathcal{BCM}_F(K, M')$. We investigate the following two cases:

1. If $|M| \neq |M'|$, then the inputs to the last compression function differ (due to the present message length encoding in m_ℓ) and therefore a collision on the final F occurs, or $P^f(A_\ell, B_\ell, C_\ell) \neq P^f(A'_\ell, B'_\ell, C'_\ell)$. Note that the transformation P^f is fixed and identical for messages of arbitrary length. Therefore, a difference in blocks B_ℓ and $B_{\ell'}$ (induced by the applied strengthening) results in difference in the outputs of P^f on the same input wires for F. B outputs $P^f(A_\ell, B_\ell, C_\ell), P^f(A'_\ell, B'_\ell, C'_\ell)$ as a valid colliding pair.
2. Else if $|M| = |M'|$, then $\ell = \ell'$ and the processing of M and M' is symmetric with respect to the inputs of F (the same arbitrary P^a_j for $j = 1$ to $\ell - 1$ is applied at all positions j for both M and M'). Here B proceeds by searching backwards (block-by-block) for distinct F inputs $P^a_j(A_j, B_j, C_j)$ and $P^a_j(A'_j, B'_j, C'_j)$, which result in equal output hash values h_j an h'_j under F. Since $M \neq M'$, then there exists an index $j > 0$, such that $P^a_j(A_j, B_j, C_j) \neq P^a_j(A'_j, B'_j, C'_j)$. Then for the max($j$) that satisfies the inequality, B outputs the corresponding colliding pair $(P^a_j(A_j, B_j, C_j), P^a_j(A'_j, B'_j, C'_j))$.   □

A.2   Sec Proof of Theorem 2 for $\mathcal{GBCM}$

Given a Sec[λ] adversary A against $\mathcal{GBCM}_F$, we construct a Sec adversary B against the compression function F. B receives a random challenge message $m\|h$. Then B chooses a random index $i \xleftarrow{\$} \{1, \ldots, \ell = \lceil(\lambda + 65)/b\rceil\}$ and a random message $M \xleftarrow{\$} \{0,1\}^{\lambda}$. B has to successfully embed his challenge $m\|h$ in the target strengthened message $m_1\|\ldots\|m_\ell \leftarrow \mathsf{pad}(M)$ and in the chaining iterative portion of $\mathcal{GBCM}_{\mathrm{F}}(K,M)$. Let $\hat{M} = m_1\|\ldots\|m_\ell$ be the strengthened message.

Let $(X_i, Y_i, Z_i) \leftarrow P_i^a(A_i, B_i, C_i)$ for $i = 1$ to $\ell - 1$ and $(X_\ell, Y_\ell, Z_\ell) \leftarrow P^f(A_\ell, B_\ell, C_\ell)$. Then B identifies the type of mapping applied in the respective ith position in the iteration for $i = 1$ to $\ell - 1$

1. if $P_i^a = P_i^1$, then $X_i\|Y_i = m$ and $Z_i = h$.
2. if $P_i^a = P_i^2$, then $X_i\|Z_i = m$ and $Y_i = h$.
3. if $P_i^a = P_i^3$, then $Y_i\|Z_i = m$ and $X_i = h$.
4. if $P_i^a = P_i^4$, then $Z_i\|Y_i = m$ and $X_i = h$.
5. if $P_i^a = P_i^5$, then $Z_i\|X_i = m$ and $Y_i = h$.
6. if $P_i^a = P_i^6$, then $Y_i\|X_i = m$ and $Z_i = h$.

Now if $i = 1$, then a value that equals to h translates to B setting $K_1 \leftarrow IV_1 \oplus h \oplus m_2|_n^{\mathsf{msb}}$. If $\lambda < b - 65$, then B proceeds as in case $i = \ell$.

If $i \in \{2 \ldots \ell-2\}$, then equality to h translates to B setting $m_{i+1}|_n^{\mathsf{msb}} \leftarrow h_{i-1} \oplus h$ for a randomly chosen $K_1 \xleftarrow{\$} \{0,1\}^n$ (h_{i-1} is the $(i-1)$st intermediate chaining value computed by B). If $\lambda < 2b - 65$, then B proceeds as in case $i = \ell - 1$.

In both these cases B modifies either block m_1, or blocks m_i and $m_{i+1}|_n^{\mathsf{msb}}$ from the message $\hat{M}$. B also chooses at random $K_2\|K_3 \xleftarrow{\$} \{0,1\}^{b+n}$.

If $i = \ell - 1$, then equality of a value to m is equivalent to B setting the most significant $b - n$ bits of it to $m|_{b-n}^{\mathsf{msb}}$ and $K_1 \leftarrow (m \oplus m_{\ell-1})|_n^{\mathsf{lsb}}$. Equality to h here means that B sets $K_2|_n^{\mathsf{msb}} \leftarrow h \oplus h_{\ell-2} \oplus m_\ell|_n^{\mathsf{msb}}$ ($h_{\ell-2}$ is the $(\ell - 2)$nd intermediate chaining value computed by B when $\lambda \geq 2b - 65$ and $IV_1 \oplus K_1$ when $b - 65 \leq \lambda < 2b - 65$). A chooses $K_2|_{b-n}^{\mathsf{lsb}}\|K_3 \xleftarrow{\$} \{0,1\}^n$. Only the first $b - n$ bits of block $m_{\ell-1}$ are modified from the originally generated message $\hat{M}$.

If $i = \ell$, then

1. if $P^f = P^1$, then $X_\ell\|Y_\ell = m$ and $Z_\ell = h$.
2. if $P^f = P^2$, then $X_\ell\|Z_\ell = m$ and $Y_\ell = h$.
3. if $P^f = P^3$, then $Y_\ell\|Z_\ell = m$ and $X_\ell = h$.
4. if $P^f = P^4$, then $Z_\ell\|Y_\ell = m$ and $X_\ell = h$.
5. if $P^f = P^5$, then $Z_\ell\|X_\ell = m$ and $Y_\ell = h$.
6. if $P^f = P^6$, then $Y_\ell\|X_\ell = m$ and $Z_\ell = h$.

B chooses at random $K_1 \xleftarrow{\$} \{0,1\}^n$ and computes the intermediate hash value $h_{\ell-1}$. An equality to m means that B then sets $K_2 \leftarrow m \oplus m_\ell$ and equality to h that $K_3 \leftarrow h \oplus h_{\ell-1}$ ($h_{\ell-1}$ is the $(\ell - 1)$st intermediate chaining value computed by B when $\lambda \geq b - 65$ and $IV_1 \oplus K_1$ when $\lambda < b - 65$). Except m_ℓ, the rest of $\hat{M}$ remains unchanged.

Now $m\|h$ is successfully embedded and B proceeds by running A on inputs message M where $M = \mathsf{pad}^{-1}(\hat{M})$(the non-strengthened version of $\hat{M}$) and key $K = K_1\|K_2\|K_3$. As in the case of $\mathcal{BCM}$ hash, the message M and key K are uniformly distributed. A returns a second preimage message M', such that $\mathcal{BCM}_{\mathrm{F}}(K,M) = \mathcal{BCM}_{\mathrm{F}}(K,M')$. For the rest of the proof B acts identically as in the Coll proof. With probability $1/\ell$ B finds the colliding pair at the correct position i (at which B embedded $m\|h$) and outputs the colliding inputs for F as its valid second preimage. This completes the proof. □

A.3   Pre Proof of Theorem 3 for $\mathcal{GBCM}$

Proof. Let A be a Pre[λ] adversary on the iterated hash function $\mathcal{GBCM}_{\mathrm{F}}$. Given a challenge hash value Y and key $K = K_1\|K_2\|K_3$, the goal of A is to invert Y, which is computed as $Y = \mathcal{GBCM}_{\mathrm{F}}(K, M)$ for a randomly chosen message $M \xleftarrow{\$} \{0,1\}^\lambda$ and a random key $K \xleftarrow{\$} \mathcal{K}$.

We investigate the following two cases: 1. $n < \lambda \le b - 65$ and 2. $\lambda > b - 65$. In the first case $Y = \mathrm{F}(P^f(A_1, B_1, C_1))$. A knows λ, respectively the applied strengthening bits, IV_1 and $K_1\|K_2\|K_3$. Thus, A knows at most the input B_1 and C_1 and can derive at most $b - n$ bits from the output of P^f . However, A has no information on at least n bits of the message input $A_1 = m_1$. Thus, the only way to find a valid preimage message for $\mathcal{BCM}_{\mathrm{F}}$ is to exhaustively query the random oracle F on chosen inputs for the missing part of m_1. The probability to find the correct preimage message per single query is $1/2^n$ and therefore after q_{RO} queries to the random oracle A succeeds to invert Y with probability $q_{\mathrm{RO}}/2^n$.

In the second case $Y = \mathrm{F}(P^f(A_\ell, B_\ell, C_\ell))$. Here A knows the keys K_2 and K_3, the message length λ and the respective strengthening bits used in the last message block m_ℓ. A can compute at most $b-n$ bits of $P^f(A_\ell, B_\ell, C_\ell)$. But again B does not know the intermediate chaining value $h_{\ell-1}$ and also $C_\ell = h_{\ell-1} \oplus K_3$, then again as in case one, A can at best try to invert Y. A'a success ϵ is bound by $q_{\mathrm{RO}}/2^n$. □

Analysis of the Collision Resistance of RadioGatún Using Algebraic Techniques

Charles Bouillaguet and Pierre-Alain Fouque

Ecole normale supérieure, CNRS, INRIA

Abstract. In this paper, we present some preliminary results on the security of the RadioGatún hash function. RadioGatún has an internal state of 58 words, and is parameterized by the word size, from one to 64 bits. We mostly study the one-bit version of RadioGatún since according to the authors, attacks on this version also affect the reasonably-sized versions. On this toy version, we revisit the claims of the designers and first improve some results. Secondly, given a differential path, we show how to find a message pair colliding more efficiently than the strategy proposed by the authors using algebraic techniques. We experimented this strategy on the one-bit version since we can efficiently find differential path by brute force. Even though the complexity of this collision attack is higher than the general security claim on RadioGatún⟨1⟩, it is still less than the birthday paradox on the size of the internal state.

1 Introduction

RadioGatún is a new hash function, proposed in 2006 by Bertoni, Daemen, Peeters and Van Assche at the Second NIST Hash Workshop. This hash function is very interesting to study since its design is not similar to traditional hash functions. It is not a blockcipher-based hash function such as the Davies-Meyer construction of compression function and it does not use the Merkle-Damgård paradigm to transform a compression function into a hash function. This hash function improves a previous design used in the Panama hash function [12]. RadioGatún has an internal state of 58 words; the size of those words, from one to the recommended 64 bits, define the actual size of the internal state.

1.1 Related Work

The Sponge construction. RadioGatún is the current hash function whose design resemble the sponge construction most. This construction differ significantly from the SHA family : the internal building block transform the internal state bijectively, and there is no message expansion: the input blocks are simply injected into the state. The size of the input block is smaller than the internal state, which is also much bigger than the security parameter. The output can be of arbitrarily length.

Sponge functions were introduced in [2,4], to serve as a reference model for the security of hash functions. Random sponges are an abstraction of a random

R. Avanzi, L. Keliher, and F. Sica (Eds.): SAC 2008, LNCS 5381, pp. 245–261, 2009.

function with a finite internal state. In [2], random sponges were shown to be indifferentiable from a random oracle up to a number of query which depends on the "capacity" of the sponge, a part of its internal state.

The Backtracking Attack. The security of the RadioGatún hash function against differential attacks has been initially studied by Bertoni *et al.* in [3]. In fact, the main security analysis has been done on the one-bit version since for this version, differential paths can be found efficiently. If the one-bit version was structurally broken, the the bigger version would be likely to be broken as well. In [3], the authors describe a strategy to find two colliding messages given a differential path and named it the *trail backtracking* attack. This kind of attack improves the statistical attack which tries as many messages as the inverse of the probability of the differential trail. Such an attack has also been mounted on Panama [19,11] and on Grindahl in [18].

Message Modification and Algebraic Techniques. Expressing the problem of finding a collision as the problem of solving a set of equations is an old technique, that was used to break MD4 first [13]. Message modification was used with great success to attack hash functions from the MD and SHA families [24,22,25,23]. More advanced algebraic techniques, such as Gröbner bases, were used by [21] to improve the message modification part of existing attack against SHA-1.

1.2 Our Results

Our main object of study is the one bit version, RadioGatún⟨1⟩. We show that the backtracking attack can be performed more efficiently and without any backtracking. We use Gröbner basis algorithms to compute the set of all states from which colliding messages can be found using a given trail. The main drawback of this attack is that once we have this set, we need other techniques to go from the initial vector to these states. Actually, our method uses statistical trials until one satisfies the equations characterizing the set. The techniques we use heavily rely on the fact that the non-linear function is quadratic and so the differential of such function gives linear conditions on the states.

RadioGatún⟨1⟩ has an internal state of 58 bits, and it is conjectured in [3] that differential attacks would cost at least 2^{46}. A first technique using only linear algebra yields collisions in 2^{27} evaluations of the round function. We present a second technique using more sophisticated algebraic tools, most notably Gröbner Basis computations, that produces collisions in less than $2^{24.5}$ evaluations of the round function, for any fixed IV. This is more than one million times faster than what the authors of RadioGatún expected. Both attacks are faster than the birthday paradox on the size of the internal state, but the do not break the security claim of the designers of RadioGatún, since they took a high security margin.

The first attack is trivially applicable to RadioGatún⟨ℓ_w⟩ for any value of ℓ_w. The status of the second attack is less clear, but we give some arguments supporting the idea that it will still be applicable when $\ell_w > 1$.

1.3   Organization of the Paper

In section 2, we describe the RadioGatún hash function. Then, we recall the original backtracking attack on RadioGatún presented by its authors and present some surprising experimental results. In section 4, we show how we can improve the backtracking attack using only simple linear algebra. In section 5 and that we can also remove the backtracking by propagating the linear conditions using Gröbner basis computations. Consequently, we derive precise conditions on the states from which colliding pairs can be found.

2   A Brief Description of RadioGatún

RadioGatún is parametrized according to the word length ℓ_w ranging from one to the recommended 64 bits; RadioGatún$\langle \ell_w \rangle$ denote the ℓ_w-bit version of the hash function.

RadioGatún is a hash function based on the sponge paradigm. During the "absorbing" phase, it absorbs an arbitrary number of $3\ell_W$-bit input blocks, and during the "squeezing" phase it produces an arbitrary long output. The input message p is padded, so that its size is a multiple of the input block size. Radio-Gatún absorbs the input message by alternatively XORing 3 message words into the internal state and applying a bijective round function R until the padded message is entirely read. In each round, $\ell_i = 3 \cdot \ell_W$ bits are absorbed. Note that 6 rounds are required in order that $n/2$ bits are hashed, where n is the size of the internal state. This suggests that colliding message will span over at least 6 rounds. Then, the internal state is mixed using 16 blank iterations of R, and finally the outputs is produced by alternatively extracting 2 words of the internal state and applying the round function until enough bits are extracted. The security of a sponge function is not defined in terms of the digest length (since it can be arbitrarily big), but rather according to another parameter called the *capacity*, which is connected to the size of the internal state. For RadioGatún, the authors made a "flat sponge claim" : More precisely, the authors claim that (truncated) RadioGatún$\langle \ell_W \rangle$ is as strong as a random sponge of capacity $19 \cdot \ell_W$. This mean in particular that it should not be possible to find collisions in less than $2^{9.5 \cdot \ell_W}$, while RadioGatún$\langle \ell_W \rangle$ has an internal state of 58 words ($58 \cdot \ell_W$ bits).

From this description, it is easy to see that a collision into the state at the end of the absorbing phase leads to a collision on the output bits. Consequently, the authors of RadioGatún worry about such collisions and named them *internal collisions*. However, in order to analyze such attacks, it seems that the important parameter is not really the capacity, but rather the half of the size of the internal state. We will see in the following, that if we take this security parameter, we have attacks on the one-bit version of RadioGatún.

In RadioGatún, the state is split into two parts: the *Mill* and the *Belt*. The role of the Belt is to have good long-term diffusion property and uses a simple invertible linear update function, while the goal of the mill is to create confusion and uses an invertible non-linear update function. The Belt and the Mill interact with each other in each application of the round function.

The Mill a consists of 19 words $a[i]$, and the Belt b is a matrix of 3 rows and 13 columns. An input block x consists of three words $x[i]$. All indices start from 0. We defer the reader to [3] for a complete description of the hash function. Schematically a round of RadioGatún can be described in the following way:

$$b' \leftarrow L_1(b) \oplus L_3(a) \oplus L_2(x)$$
$$a' \leftarrow L_4 \circ \gamma\Big(a \oplus L_5(x)\Big) \oplus L_6(b')$$

where the L_i are bijective linear mappings, and γ is a word-wise bijective quadratic mapping defined by: $\gamma(a)[i] = a[i] \oplus \overline{a[i+1]} \wedge a[i+2]$, where & denotes bitwise AND, and indices are taken modulo 19.

3   The Trail Backtracking Attack

3.1   Differential Trails

It seems natural to try a differential attack, considering the successes obtained against the MD and SHA families. We call a differential over the round function a *round differential*; it is a pair (Δ_i, Δ_o). Its *differential probability* (DP) is the proportion of states $\mathbf{s}$ such that $R(\mathbf{s}) \oplus R(\mathbf{s} \oplus \Delta_i) = \Delta_o$. A round differential is *possible* if DP > 0. We may want to take into account not only the internal state, but also the message block entering a round. In this case, a round differential is a triple $(\Delta_i, \Delta_x, \Delta_o)$, and it is satisfied by a state $\mathbf{s}$ and a message block $\mathbf{x}$ if the internal state after the injection of $\mathbf{x}$ satisfies the differential $\big(\Delta_i \oplus F_i(\Delta_x), \Delta_o\big)$. The *(restriction) weight* of a differential is defined by: $W_r(\Delta_i, \Delta_o) = -\log_2 DP(\Delta_i, \Delta_o)$.

Since we will have to track the difference between two parallel hashing processes amongst several iterations of the round function, we are lead naturally to the definition of a *collision trail*; it describes the propagation of the difference on the internal state, when given differences on the input block are applied. Such a trail is a sequence of round differentials: $(\Delta_i^0, \Delta_x^0, \Delta_o^0), \dots, (\Delta_i^r, \Delta_x^r, \Delta_o^r)$, where $\Delta_i^0 = 0, \Delta_o^r = 0$, for all $1 \leq k \leq r, \Delta_{k-1}^o = \Delta_k^i$. For each round k, the trail enforces that if the internal states satisfy $\mathbf{s}^k \oplus \mathbf{s}'^k = \Delta_i^k$, and the input message blocks satisfy $\mathbf{x}^k \oplus \mathbf{x}'^k = \Delta_x^k$, then after R the output state pair has difference Δ_o^k. If one finds an input state $\mathbf{s}^0$ and a sequence of message blocks $\mathbf{x}^0, \dots, \mathbf{x}^r$ satisfying all the conditions imposed by the trail, then one has found a collision. The probability that a random message follows the trail is the differential probability (DP) of the trail, and the *differential weight* of $\mathcal{T}$ is defined by $W_r(\mathcal{T}) = -\log_2 DP(\mathcal{T})$.

3.2   The Trail Backtracking Attack

Given a r-round differential trail and an initial state, a naive way to look for a collision would be to try random sequences of r message blocks satisfying the differences specified by the trail until a collision is found. The expected workload of this attack is $r/DP(T)$ evaluations of R. It may very well happen that the input message passes some of the first rounds with the right difference, but then diverges from the trail. This message has an interesting *prefix*, but in the naive attack it is simply thrown away. Additionally, it is useless to hash the end of the message, since we could know in the middle that it would not follow the trail.

In the backtracking attack, however, a right prefix is reused as much as possible. It can be seen as an analogous of Wang's message modification on Davies-Meyer-type compression functions, but adapted to the alternating-input framework. Suppose we have a message that passes the first k rounds, but not the $(k + 1)$-th. Either the choice of $\mathbf{x^k}$ was bad and by choosing another block we can pass the $(k + 1)$-th round, or the previous choices of $\mathbf{x^0}, \ldots, \mathbf{x^{k-1}}$ were bad, and we have to reconsider them (this is what we will call "backtracking").

More precisely, if a right pair enters round k, the difference at the input of the round function will be the same regardless of the value of the input block $\mathbf{x^k}$, as long as it satisfies the specified difference Δ_x^k. Therefore, this right pair can be turned into 2^{ℓ_i} right pairs by simply enumerating all possible values of $\mathbf{x}$. If this results in a right outgoing pair, we can proceed to the next round, and otherwise, we have to backtrack to the previous round. This can be seen as the depth-first exploration of a big 2^{ℓ_i}-ary tree in which nodes are labeled with internal state values and edges are labeled with message blocks (the root being labeled by $\mathbf{s^0}$).

BT_Attack(s, k) :
Given a right pair entering the k-th round, try to go further along a given trail T or backtrack.

- If $k = |T|$, then a collision has been found
- For all possible input block $\mathbf{x^k}$ do
  - if the state $\mathbf{s}$ along with the input block $\mathbf{x}$ pass the k-th round differential of T, *i.e.* if

$$R\left(\mathbf{s} \oplus F_i\left(\mathbf{x}\right)\right) \oplus R\left(\left(\mathbf{s} \oplus \Delta_i^k\right) \oplus F_i\left(\mathbf{x} \oplus \Delta_x^k\right)\right) = \Delta_o^k$$

    then invoke **BT_Attack**$(R\left(\mathbf{s} \oplus F_i\left(\mathbf{x}\right)\right), k+1)$

Fig. 1. Pseudocode of the trail backtracking attack

It may very well happen that the input state $\mathbf{s^0}$, which can be chosen at random by hashing a random message for example, cannot possibly lead to a collision along T. In that case, we just have to generate a new one.

3.3   The Original Complexity Analysis of [3]

The authors of [3] give a generic complexity analysis of the trail backtracking attack. They always assume that the conditions imposed by the round differentials are independent from each other, which means that:

$$W_r(\mathcal{T}) = \sum_{k=0}^{r-1} W_r(\Delta_i^k, \Delta_x^k, \Delta_o^k)$$

Following [3], we assume that we will try N pairs of (random) input state $\mathbf{s}^0$ before finding a collision. We count the number of right pairs entering and going out of each round ; the round with the most incoming pairs is called the *crowded round*, and the round with the less outgoing pairs is called the *lonesome round*. If q pairs enter round k, then we can expect $q \cdot 2^{\ell_i - W_r(\Delta_i^k, \Delta_x^k, \Delta_o^k)}$ pairs to go out. We therefore define the *excess weight* in round k to be:

$$W_e(k) = \sum_{j=0}^{k-1} \left(W_r(\Delta_i^j, \Delta_x^j, \Delta_o^j) - \ell_i \right)$$

The total expected number of pairs entering round k is $N \cdot 2^{-W_e(k)}$, and the expected number of pairs going out of round k is $N \cdot 2^{-\ell_i - W_e(k+1)}$. The analysis now proceeds in two steps:

Evaluate N. We assume that the attack succeeds as long as at least one pair goes out of each round. This imposes $N \geq 2^{\ell_i + W_e(k+1)}$ for all $1 \leq k < r$. This condition is satisfied by setting $N = 2^{\ell_i + \max_k W_e(k+1)}$.

Evaluate the Workload. According to [3], the workload can be approximated by the number of pairs entering the crowded round : $L(\mathcal{T}) \simeq \max_k N \cdot 2^{-W_e(k)} = N \cdot 2^{-\min_k W_e(k)}$. Therefore, by using the previous result, we define the *backtracking cost*:

$$C_b(\mathcal{T}) = \ell_i + \max_{0 \leq j < k \leq r} W_e(k) - W_e(j)$$

The workload of the trail backtracking attack is then: $L(\mathcal{T}) \simeq 2^{C_b(\mathcal{T})}$.

The authors of [3] present arguments that RadioGatún resists the trail backtracking attack, using this complexity analysis. In particular, on RadioGatún⟨1⟩, where the internal state is 58-bit long, they performed an extensive search and did not find collision trail with backtracking cost smaller than 46. If there were no better trail, this would imply that the trail backtracking attack could not possibly be faster than exhaustive search on the one-bit version. Because the description of RadioGatún makes use of intra-word rotation, an operation that has no effect on the one-bit version, it is likely that the many-bit version have better diffusion, and therefore are stronger.

We emphasize that the differential weight is not a relevant indicator, because the backtracking attack may dramatically reduce the cost of finding a collision. A similar phenomenon occurs in the backtracking attack against Grindalh, or in the differential attacks based on message-modification against MD5.

3.4   Experimentation with the Backtracking Attack

We implemented the two required steps of the trail backtracking attack to find collisions : finding differential trails, and actually finding colliding messages using a given trail. Note that [3] only present experimental results about the former. The C++ programs that we developed are available on the webpage of the first author.

Finding Differential Trails. In RadioGatún$\langle 1 \rangle$, it is possible to find collisions by brute force, and a collision describes a collision trail. We therefore looked for collisions extensively, and collected the corresponding trails. The authors of [3] communicated to us their best collision trail on RadioGatún$\langle 1 \rangle$, that we will note T_1. It is completely defined by two colliding messages sharing a 7-block prefix followed by a 8-block colliding part. In octal notations, the two messages are : 0364220 64172767 and 0364220 20435061. The differential weight of this trail is 63, and its backtracking cost is 46.

We eventually found a 7-round trail (called T_2) with backtracking cost 31 and differential weight 45. This surprising result was obtained while looking for 7-round trails, by initializing the internal state with a 9-block random prefix. T_2 is defined by the following colliding messages, again in octal notation: 476356301 6336565 and 476356301 4250471. With the trail backtracking attack, this gives a collision in an expected 2^{31} effort, which is still above the birthday bound.

Searching for a Collision. We used these two trails to find collisions on RadioGatún$\langle 1 \rangle$. It may be argued that we needed to find collisions (to get the trails) before actually being able to find collisions, but there may very well be other methods of finding trails, and we did not consider this problem. Moreover, once a good collision trail is found, it can be used to find collision from any value of the internal state. At the very least, if a technique were found to efficiently find chosen-IV collisions, it could generate collision trails, and therefore be used as a preprocessing step in our attack.

On average, the trail backtracking attack succeeds with 2^{29} evaluations of the round function (when using T_2), which is exactly the complexity of the birthday bound. With T_1, the attack succeeds in $2^{34.5}$ (in average), when the announced complexity was 2^{46}.

In order to get some insight to why the collision search procedure succeeds faster than expected, we observed the number of right pairs going in and out of each round. Figure 2 shows these numbers when the first collision is found (using T_1). The results are similar if we average them on 100 collisions, or to what is obtained with other trails. After round 2 or 3, the pairs pass the next round with abnormally high probability. Apparently, the round differentials are not independent. This would explain why the first rounds have a tendency to "filter" good pairs that pass the subsequent rounds more easily. This may be very specific to the one-bit version of RadioGatún, though. It may also be specific to the way the collision trails were obtained (by actually computing a collision).

In itself, the trail backtracking attack does not break RadioGatún.

round	in. pairs	out. pairs	$W_r(\Delta_i^k, \Delta_x^k, \Delta_o^k)$	experimental weight
0	32.01	30.01	2	2.0
1	32.82	25.82	7	7.0
2	28.62	16.66	11	12.0
3	19.46	12.88	11	6.6
4	15.69	7.61	10	8.1
5	10.42	6.02	11	4.4
6	8.82	0.00	11	8.8
7	0.00	0.00	0	0.0

Fig. 2. (*log₂* of the) Number of pairs going in and out of each round in T_1 ; comparison with the weight of each round

4   Improving the Backtracking Attack on RadioGatún$\langle \ell_W \rangle$

In the next sections, we focus on improving the efficiency of the trail backtracking attack whose complexity is above the birthday bound, using algebraic techniques, as suggested by the authors of RadioGatún themselves.

The only non-linear part in RadioGatún is the "mill function", and more specifically its first component γ. The specific properties of γ are extensively studied in [10, chapter 6]. For now, let us notice that γ has algebraic degree 2 over $\mathbb{F}_2$ (each bit of $\gamma(a)$ can be expressed as a quadratic form in the input bits). Because the rest of the round function is linear, the whole round function R can be expressed as a tuple of $58 \cdot \ell_W$ polynomials of degree 2 in $58 \cdot \ell_W$ variables over $\mathbb{F}_2$ (we denote by $\mathbb{F}_2[\mathbf{s}]$ the set of all polynomials over $58 \cdot \ell_W$ variables corresponding to the bits of the internal state).

It is well-known that if a function is quadratic, then its differential is *linear*. This is the key idea in this preliminary algebraic analysis of RadioGatún. Let us consider the set of internal states $\hat{\mathbf{s}}$ after input injection satisfying the round differential (Δ_i, Δ_o). These state satisfy the following equation: $R(\hat{\mathbf{s}}) \oplus R(\hat{\mathbf{s}} \oplus \Delta_i) = \Delta_o$. Even though R is quadratic, this equation is only *linear* in $\hat{\mathbf{s}}$. Therefore, we know that all the values of $\hat{\mathbf{s}}$ satisfying it lie in an affine space, and thus can be characterized by linear conditions on $\hat{\mathbf{s}}$. These conditions depend on Δ_i and Δ_o, and can be computed efficiently using linear algebra. We denote by $\mathcal{C}(\Delta_i, \Delta_o)$ (or $\mathcal{C}^k$) these conditions. The state entering the round function is given by $\hat{\mathbf{s}} = \mathbf{s} \oplus F_i(\mathbf{x})$. Therefore, conditions on $\hat{\mathbf{s}}$ give two kinds of information:

1. linear conditions on the bits of $\mathbf{s}$.
2. linear conditions between bits of $\mathbf{s}$ and bits of $\mathbf{x}$.

The former can be used to detect incoming pair that will never give rise to an outgoing pair, for any value of $\mathbf{x}$. This allows to stop the exploration of dead branches of the tree earlier. The latter directly gives us some bits of $\mathbf{x}$, as linear combinations of bits of $\mathbf{s}$, and thus allow us to filter the values of $\mathbf{x}$ that do not yield a right outgoing pair. Using these conditions, we can decrease the amount of useless trials in the backtracking attack. Figure 3 shows which bits of the input

k	$\mathbf{x}^k[0]$	$\mathbf{x}^k[1]$	$\mathbf{x}^k[2]$
0	$\mathbf{a}[16]$		$\mathbf{a}[18]+1$
1	$\mathbf{a}[16]+1$		
2			
3		$\mathbf{a}[17]+\mathbf{a}[16]+\mathbf{x}[0]$	$\mathbf{a}[18]$
4	$\mathbf{a}[15]+\mathbf{a}[16]+1$	$\mathbf{a}[17]+1$	$\mathbf{a}[18]$
5		$\mathbf{a}[17]+\mathbf{a}[16]+\mathbf{x}[0]+1$	$\mathbf{a}[18]$
6			

Fig. 3. When using T_2, some bits of the input blocks $\mathbf{x}^k$ are determined by some bits of the incoming state. An empty cell means that the corresponding bit has to be chosen by the attacker.

k	$\mathcal{C}\left(\Delta_i^k, \Delta_x^k, \Delta_o^k\right)$	
0	$x[0]+a[16]$	$x[2]+a[18]+1$
1	$x[0]+a[16]+1$ $a[2]$ $a[5]+a[4]$ $a[8]$ $a[12]$	$a[0]$ $a[3]+1$ $a[6]$ $a[10]+1$ $a[14]+1$
2	$a[4]+1$ $a[7]+a[6]+1$ $a[13]$	$a[7]+a[5]+1$ $a[8]+1$ $a[15]$
3	$x[0]+a[17]+a[16]+x[1]$ $a[1]+1$ $a[13]+1$	$x[2]+a[18]$ $a[3]$ $a[15]$
4	$a[15]+x[0]+a[16]+1$ $x[2]+a[18]$ $a[3]+a[1]$ $a[4]$ $a[7]+a[8]+1$ $a[14]+1$	$a[17]+x[1]+1$ $a[3]+a[0]+1$ $a[3]+a[2]$ $a[6]$ $a[9]$
5	$x[0]+a[17]+a[16]+x[1]+1$ $a[0]$ $a[4]$ $a[8]+a[10]$ $a[11]+1$	$x[2]+a[18]$ $a[2]+1$ $a[7]$ $a[9]+a[10]+1$ $a[15]+1$

Fig. 4. Conditions imposed at the beginning of each round by T_2

message are determined by the internal state, for the trail T_2. The complete set of conditions is given in fig. 4. It must be noted that these conditions can be computed efficiently for all values of the word size ℓ_W (the linear algebra involved is cubic in the word size).

4.1   Experimental Results

We implemented the improved backtracking attack on RadioGatún$\langle 1 \rangle$, using the same two trails T_1 and T_2, so that the result can be compared with the regular attack. For T_2, collisions are found on an average of 2^{27} evaluations of R, which means a speedup of 4 compared to the regular attack. Note that this is below the birthday bound. Annex 4 shows the local conditions imposed by T_2. For T_1, 2^{32}, which is more than 5.5 times faster than the original attack.

5   The Backtrackingless Backtracking Attack

While the technique described in the previous section reduces the amount of backtracking by allowing an earlier filtering of pairs that will not lead to a collision, it does not prevent all backtracking. The reason for this is that this filtering is only *local*: at round k, we cannot yet filter pairs that will not pass round $k + 1$. In this section, we address this issue. We show that it is possible to avoid *all* backtracking by propagating equations backwards from the last round to the first round of the trail. We get a set of equations on the internal states entering the first round ; if a state $\mathbf{s}^0$ satisfies these equations, then we can generate a few collisions at a negligible cost. We achieve some kind of *global* filtering, because we filter at the first round all the pairs that will not pass any of the subsequent rounds (we "push" all the conditions at the root of the tree). We propose to name this attack the *Backtrackingless Backtracking attack*[1].

   In the previous section, we showed how to generate a set of conditions $\mathcal{C}^k$ such that if a state $\mathbf{s}$ satisfies these conditions, then the pair $\left(\mathbf{s}, \mathbf{s} \oplus \Delta_i^k\right)$ will pass round k. In order to pass round $k + 1$, the states going out from round k must also satisfy $\mathcal{C}^{k+1}$. Our objective is to express a new set of conditions on $\mathbf{s}$ such that if these conditions are satisfied, then $\mathbf{s}$ satisfies $\mathcal{C}^k$ *and* $\mathbf{s}'$ satisfies $\mathcal{C}^{k+1}$. We achieve our objective of propagating all the conditions backwards to the first round by recursively applying this process.

5.1   Description

We use standard notions and notations for commutative algebra, that can be found for example in [8]. Formally, we say that a (polynomial) condition (or constraint, or equation) on $\mathbf{s}$ is a polynomial of the ring $\mathbb{F}_2[\mathbf{s}]$ (*i.e.*, a polynomial in which the variables are bits of $\mathbf{s}$). A condition P is satisfied by $\mathbf{s}$ if P vanishes when the variables are substituted with the actual values of bits in $\mathbf{s}$. We can then write $P(\mathbf{s}) = 0$, or, using the notation from the area of logics, $\mathbf{s} \models P$. The set of states satisfying P is then the set of zeroes of P. We will also have to consider the conjunction $\mathcal{C}$ of several such conditions (*i.e.*, systems of polynomial equations). A convenient way to represent such a system is to consider the polynomial *ideal* I generated by the polynomials in $\mathcal{C}$. It contains all the polynomial combinations of

---

[1] Its name is reminiscent of the *inductionless induction* of [7] or of the *splittingless splitting* of [16].

its generators, that is, all the polynomial "consequences" of the original equations. The set of states satisfying C is the set of all common zeroes of all the polynomials in I, which is called the *affine variety* $\mathbf{V}(I)$ associated to I. We say that a set of conditions $\mathcal{D}$ is a consequence of another system C if $I_\mathcal{D} \subseteq I_C$. We note $C \Rightarrow \mathcal{D}$ to describe this situation.

When expressing conditions about the output of the round function, we introduce $58 \cdot \ell_W$ more variables $\mathbf{s'}$, corresponding to the bits of the output. The equations of R are actually equations in $\mathbb{F}_2[\mathbf{s}, \mathbf{s'}]$. We will note I_R the ideal of $\mathbb{F}_2[\mathbf{s}, \mathbf{s'}]$ generated by the equations of R; its affine variety contains all the tuples $(\mathbf{s}, \mathbf{s'})$ such that $\mathbf{s'} = R(\mathbf{s})$. From a geometric point of view, these equations describe the *graph* of the function R (in the same fashion that $y - x^2 = 0$ describes a parabola). Later on, we will use a different representation of these equations, that still describe the same graph.

We need a last tool before defining formally the objects we wish to compute. We need to express conditions on the input of the $(j+1)$-th round as conditions on the output of the j-th round. This is simply done by renaming variables. We define the renaming function $\rho : \mathbb{F}_2[\mathbf{s}] \to \mathbb{F}_2[\mathbf{s'}]$ as $\rho(\mathbf{s}_j) = \mathbf{s'}_j$. This renaming can be extended to operate on ideals : $\rho(I) = \{\rho(P) \mid P \in I\}$. It is straightforward to check that $\rho(I)$ is still an ideal.

New Sets of Conditions. Given a r-round trail $\mathcal{T}$, a sequence of r input blocks $\left(\mathbf{x}^k\right)_{0 \leq k < r}$ and an internal state $\mathbf{s}^0$, we note $\mathbf{s}^{i+1} = R\left(\mathbf{s}^i \oplus F_i\left(\mathbf{x}^i\right)\right)$. Our objective is to build r sets of conditions $\mathcal{D}^k, 0 \leq k < r$ such that if $\mathbf{x}^k \models \mathcal{D}^k$, then for all $j \geq k$, $\mathbf{x}^k \models C^j$ (if the internal state at the input of round k satisfies the conditions $\mathcal{D}^k$, then we know for sure that it will lead to a collision because it satisfies all the subsequent sufficient conditions C^j, for $k \leq j$). In particular, if we are able to find an internal state satisfying $\mathcal{D}^0$, then we get a collision nearly for free. Intuitively, our objective is to transfer simultaneously all the conditions C^i at the beginning of each round to conditions on the internal state $\mathbf{s}^0$ entering the first round. It must be noted that the authors of [3] mentioned the possibility to propagate conditions on the input of the lonesome round to the input of the preceding rounds. Here, we propagate conditions on the internal state.

From the definition of $\mathcal{D}^k$, we can first deduce that $\mathcal{D}^k \Rightarrow C^k$, and then that if $\mathbf{s}^k \models \mathcal{D}^k$, then $\mathbf{s}^{k+1} \models \mathcal{D}^{k+1}$. We also know that there are no conditions on the output of round r (because a collision is already obtained), and therefore: $\mathcal{D}^{r-1} = C^{r-1}$. For $0 \leq j < r - 1$, we can now define $\mathcal{D}^j$ by:

$$\mathcal{D}^j = \left(C^j + I_R + \rho\left(\mathcal{D}^{j+1}\right)\right) \bigcap \mathbb{F}_2[\mathbf{s}]$$

Informally, $\mathcal{D}^j$ is the ideal obtained by writing together the constraints $\mathcal{D}^{j+1}$ on $\mathbf{s'}$, the equations of the round function and the constraints C^j on $\mathbf{s}$. By taking its intersection with $\mathbb{F}_2[\mathbf{s}]$, we eliminate all the polynomials containing a variable from $\mathbf{s'}$. This amounts to considering the consequences of these equations that can be expressed using only the variables of $\mathbf{s}$ – a process known as *eliminating* the variables $\mathbf{s'}$.

Computing the $\mathcal{D}^j$'s. The Hilbert Basis theorem tells us that, like all ideals of a polynomial ring, $\mathcal{D}^j$ admits a finite number of generators; moreover, they can be computed using a computer algebra system: compute a Gröbner basis G of $\mathcal{C}^j + I_R + \rho(\mathcal{D}^{j+1})$ for the lexicographic ordering (or a suitable elimination ordering). The basis G generates $\mathcal{C}^j + I_R + \rho(\mathcal{D}^{j+1})$, but the *elimination theorem* (see [8]) additionally tells us that $G \cap \mathbb{F}_2[\mathbf{s}]$ generates $\mathcal{D}^j$. Now, we claim that $\mathbf{V}(\mathcal{D}^j)$ is exactly the set of all the states that will pass the end of the trail. This is in fact a consequence of the *extension theorem*: if $\mathbf{s} \in \mathbf{V}(\mathcal{D}^j)$, then there exists an "extension" value $\mathbf{s}'$ such that $(\mathbf{s}, \mathbf{s}') \in \mathbf{V}(\mathcal{C}^j + I_R + \rho(\mathcal{D}^{j+1}))$. Because we included the equations of R, this value is necessarily $R(\mathbf{s})$. The conclusion follows by induction on the number of rounds.

To complete the attack, we need to find a message yielding an internal state $\mathbf{s}$ satisfying $\mathcal{D}^0$, starting from the IV (which is the null state); then we would automatically get a collision without any backtracking. Note that being able to just determine a "standalone" state in $\mathbf{V}(\mathcal{D}^0)$ would give a chosen-IV collision attack.

Finding points in an affine variety is difficult in the general case, but becomes easier when a Gröbner basis of the corresponding ideal is known (and it is *very* easy when a Gröbner basis is known for the *lexicographic* ordering). Here, as it result from the process of elimination, $\mathcal{D}^0$ form a Gröbner basis for a certain ordering, which depends on the ordering used for the elimination process. It could be chosen so that $\mathcal{D}^0$ form a lexicographic Gröbner basis (using a block order where the non-eliminated variables are ordered lexicographically), but this may make the elimination process slower. In any case, order change algorithms could be used, such as the Gröbner Walk [6] or FGLM [15].

Reaching $\mathcal{D}^0$. To find real collisions, we need to be able to reach $\mathbf{V}(\mathcal{D}^0)$ starting from the null state. The problem of finding a collision thus reduces to the problem of reaching a state satisfying a set of polynomial conditions. This formulation of the problem is again reminiscent of message modification techniques. This suggest that such powerful techniques could be used here. We did not investigate this problem in detail, and we only tried to hash random messages until all the conditions are satisfied. In this case, the complexity of finding a collision is related to the cardinality of $\mathbf{V}(\mathcal{D}^j)$.

The representation of the condition set $\mathcal{D}^0$ on the initial conditions is not unique. In our case, it forms a Gröbner basis, which is certainly interesting. We have some freedom in the choice of the ordering. The Graded Reverse Lexicographic order produces the system of lowest possible degree, but usually returns a system with more equations than when using the lexicographic ordering (which yields equations of higher degree).

5.2   Implementation and Experimental Results

We implemented the backtrackingless backtracking attack, using an off-the-shelf computer algebra system to perform the algebraic computations, and then we adapted our collision-finding program to use these conditions.

Propagating Conditions. Back to our two trails T_1 and T_2, the process of computing the conditions $\mathcal{D}_0$ involves nontrivial algebraic computation. We used the implementation of the F4 [14] algorithm in the MAGMA computer algebra system to obtain the Gröbner bases. We expected these computations to be very hard (the systems have 100+ variables, and contains the equations of R). It is usually not possible to compute a Gröbner basis of the equations describing directly a cryptographic primitive – MAGMA ate 8Gb of memory and crashed when we tried to compute a Gröbner basis of I_R. However, the computations of our sets of conditions were not only possible, but also unexpectedly fast (less than a second). Computing $\mathcal{D}^0$ for a given trail is usually a matter of less than five seconds on a desktop computer. Even more surprisingly, the conditions $\mathcal{D}^j$ are almost always *linear*, for all trails, except when $j = 0$ on some trails.

- For T_2 in particular, the conditions $\mathcal{D}^5$, $\mathcal{D}^4$, $\mathcal{D}^3$ and $\mathcal{D}^1$ are linear. $\mathcal{D}^2$ contains a few equations of degree 2, and $\mathcal{D}^0$ contains one equation of degree 3, along with 97 quadratic and 15 linear equations.
- For T_1, only $\mathcal{D}^0$ is non-linear ; it contains 26 quadratic and 26 linear equations.

This means that the size of conditions propagated through the round function does not blow up exponentially with the number of round passed. This was unexpected, because the size of the equations describing $R^{(k)}$ grows exponentially with k. In fact, the local conditions computed in section 4 play a *crucial* role here: the Gröbner basis computation are *much* faster and more tractable if there are a few linear conditions on the internal state entering the round. In particular, $\mathcal{C}^0$ is usually almost empty for many trails, and the conditions $\mathcal{D}^0$ are usually bigger and of higher degree than the others.

Actually Finding Collisions. Our collision-finding program just hashes random messages and checks if the resulting internal state satisfies $\mathcal{D}_0$. If it is the case, the previous version of the backtracking attack is run, and succeeds without backtracking. The performance of our straightforward implementation is the following: for T_2, a collision is found with about $2^{24.5}$ evaluations of the round function ($2^{29.5}$ for T_1, which means a speedup of 32 compared to the original attack).

In addition, we estimated the size of the set of states leading to a collision by Monte-Carlo sampling : the probability that a random message yields a state from which a collision is possible for T_1 (resp. T_2) is $2^{-28.42}$ (resp. $2^{-23.4}$). This means that for T_1 (resp. T_2) we have $|T_1| \simeq 2^{29.6}$ (resp. $2^{34.6}$). It is worth noting that the trail that has the best backtracking cost still yields the biggest affine variety. We also note that the running time of our simple implementation is relatively well-correlated to the cardinalities of the affine varieties.

5.3   About the Structure of the Equations

In this short section, we give a few elements in order to explain why the algebraic attack is successful. We discuss the case of the one-bit version, but the discussion also apply to the bigger versions, as we may notice. There are 19 mill equations that we denote by $f_1, \ldots, f_{19}$.

Peeling Off the Diffusion Layer. Compared to algebraic attacks on block ciphers, the situation is quite easier here. First of all, we are not facing a monolithic cipher where all the internal state is unknown but we can attack each round independently of the others (the conditions $\mathcal{S}^j$ bridge the gap between the isolated rounds). The equations we are manipulating therefore only represent a single round, which is much weaker than the whole construction.

Second, the hardness of solving the equations associated with a block cipher come from the alternation of a simple non-linear part (the S-Boxes, or the γ function here) with a linear diffusion layer. Since we are considering a single round here, it is possible to "peel off" the diffusion layer, and to expose the non-linear core directly, by considering a linear combination of the original equations:

$$\begin{pmatrix} g_1 \\ \vdots \\ g_{19} \end{pmatrix} = L_4^{-1} \times \begin{pmatrix} f_1 \\ \vdots \\ f_{19} \end{pmatrix} = \gamma\Big(a \oplus L_5(x)\Big) \oplus \text{linear terms}$$

Thus, it is relatively equivalent to perform our analysis on the equations of γ and on the equations of the mill function. Recall that $\gamma(a)[i]$ is given by:

$$\gamma(a)[i] = a[i+1]a[i+2] \oplus a[i+2] \oplus a[i] \oplus 1$$

Sparsity of the Equations. Computing a Gröbner base of the equations of γ is not easy (MAGMA takes about 10 minutes on a fast machine and requires 2.8 Gbytes of memory to do so, for the degree reverse lexicographic order). However, these equations have a specific structure that can be exploited. They are extremely sparse, each containing only one quadratic term. Moreover, each variable appear in exactly two quadratic terms. This means that if the value of a variable is fixed, two equations become linear. To illustrate how bad a property this is, let us consider a random quadratic form in n variables. It is shown in [17] that on fields of characteristic 2, any quadratic form becomes a special standard form $f = \sum_{i=0}^{n/2} x_{2i} x_{2i+1}$ under the right change of variables (some details omitted for the sake of simplicity). This means that we may have to fix about $n/2$ variables in the new basis before the form becomes linear.

Let us go back to our main computational problem, namely the computation of a Gröbner basis of the ideal generated by $\mathcal{C}^j + I_R + \rho\left(\mathcal{D}^{j+1}\right)$, for a suitable elimination ordering. Along with the equations of R are the linear conditions $\mathcal{C}^j$ imposed on the input bits of each round. These conditions, shown in fig. 4, often fix the value of one bit. Therefore, in conjunction with the removal of the diffusion layer, they can be used to *dramatically* simplify the equations of I_R. In fact, these simplifications are able performed *automatically* by most computer algebra systems. This explains why the elimination process results in mostly linear equations, and terminates so fast.

The REDUCTION Algorithm. Let $\mathcal{B}$ be a set of polynomials. A polynomial P is said to be reduced for $\mathcal{B}$ if no monomial of P lies in the ideal generated by the head terms of $\mathcal{B} - P$. Intuitively, this means that P cannot be "simplified" by a polynomial combination of elements in $\mathcal{P}$. A Basis $\mathcal{B}$ is said to be reduced if each $P \in \mathcal{B}$ is reduced for $\mathcal{B} - P$ (the polynomials of $\mathcal{B}$ cannot simplify each other). The REDUCTION algorithm, which gives a reduced basis from an arbitrary basis, is described in [8, chapter 2, paragraph 7], and in [1, figure 5.2]. Note than when applied to linear polynomials only, it is actually (a version of) the Gaussian reduction algorithm.

REDUCTION is often invoked automatically in computer algebra systems before and after the computation of a Gröbner basis. When the graded-reverse lexicographic ordering is used, it removes some of the quadratic terms in the equations of γ by substituting the linear equations of $\mathcal{C}^j$ in them. Additional tuning of variable order does not seem to be necessary to obtain satisfactory results. However, ordering the variables in the following way: $\mathbf{a}' < \mathbf{a} < \mathbf{x}$ peels off the diffusion layer very nicely, by keeping the number of quadratic term close to the minimum, and making the 19 quadratic terms of γ the head terms of the 19 equations.

6 Extension to $\ell_w > 1$

The main interest in studying RadioGatún$\langle 1 \rangle$, according to [3], is that a collision trail for the one-bit version could be transformed into a collision trail for any n-bit version, with an increased differential weight. In this section, we briefly survey how the result presented in this paper apply to the case where $\ell_W > 1$.

The backtracking attack with local filtering presented in section 4 can be mounted for any value of ℓ_W without any difficulty, as its complexity is polynomial in ℓ_w.

The backtrackingless backtracking attack may be more difficult to implement, as we have no upper-bound on the complexity of the Gröbner basis computations involved in the attack. However, all the arguments given in section 5.3 still apply to the multi-bit case ; the diffusion layer can be gotten rid of as efficiently as in the one-bit case. Then, the multi-bit version of γ is actually a collection of ℓ_w copies of the one-bit version of γ operating independently (the diffusion layer is supposed to connect them).

Unfortunately, we did not implement the attack in the multi-bit case, because we were not able to find any possible differential trail for any value of $\ell_w > 1$. The heuristic argument of [3] regarding the extension of trails from 1-bit to n-bit assumes that the conditions imposed by the round differentials are independent. As we have seen earlier, this is not the case. All the possible trails we knew for the 1-bit version turned out to be impossible to extend to n-bit versions (the round differentials seem to impose contradictory conditions).

In any case, we believe that our technique may come in handy when collision trail will be found for RadioGatún$\langle \ell_w \rangle$ with $\ell_w > 1$ though.

7   Conclusion

We presented an improvement to the trail backtracking attack introduced by the authors of RadioGatún, and which is reminiscent of the well-known message modification applied against the MD and SHA family. We are able to give an algebraic characterization of the internal states that can lead to a collision along a given trail. Finding a message mapping the IV to a state satisfying all these conditions remains an open problem, which is also reminiscent of message modification.

These preliminary remarks on RadioGatún invite some comments : the fact that the round function is only quadratic seems to be exploitable in unpredictable ways. It would be safe to consider functions of higher degree, but the hashing speed would probably be affected. Alternatively, increasing the diffusion effect of the belt in order to exploit the non-linearity of the mill function further seems to be a potential solution to make the backtracking cost of collision trails higher.

Acknowledgement. We thank Guido Bertoni, Joan Daemen, Michaël Peeters and Gilles Van Assche for many useful discussion and comments. The authors are indebted to Christophe de Cannière who helped us with the trail-finding program.

References

1. Becker, T., Weispfenning, V., Kredel, H.: Gröbner bases: a computational approach to commutative algebra. Springer, London (1993)
2. Bertoni, G., Daemen, J., Peeters, M., Van Assche, G.: On the indifferentiability of the sponge construction. In: Smart, N.P. (ed.) EUROCRYPT 2008. LNCS, vol. 4965, pp. 181–197. Springer, Heidelberg (2008)
3. Bertoni, G., Daemen, J., Peeters, M., Assche, G.V.: RadioGatún, a Belt-and-Mill Hash Function. In: Presented at Second Cryptographic Hash Function Workshop, Santa Barbara, California, August 24-25 (2006), http://radiogatun.noekeon.org/
4. Bertoni, G., Daemen, J., Peeters, M., Assche, G.V.: Sponge functions. In: Presented at ECrypt Hash Function Workshop, Barcelona, Spain, May 24 (2007)
5. Biryukov, A. (ed.): FSE 2007. LNCS, vol. 4593. Springer, Heidelberg (2007)
6. Collart, S., Kalkbrener, M., Mall, D.: Converting bases with the gröbner walk. J. Symb. Comput. 24(3/4), 465–469 (1997)
7. Comon, H.: Inductionless induction. In: Robinson, J.A., Voronkov, A. (eds.) Handbook of Automated Reasoning, pp. 913–962. Elsevier and MIT Press (2001)
8. Cox, D., Little, J., O'Shea, D.: Ideals, Varieties, and Algorithms: An Introduction to Computational Algebraic Geometry and Commutative Algebra (Undergraduate Texts in Mathematics), February 2007. Springer, Heidelberg (2007)
9. Cramer, R. (ed.): EUROCRYPT 2005, vol. 3494. Springer, Heidelberg (2005)
10. Daemen, J.: Cipher and hash function design. Strategies based on linear and differential cryptanalysis. PhD thesis, Katholieke Universiteit Leuven (March 1995)
11. Daemen, J., Assche, G.V.: Producing collisions for panama, instantaneously. In: Biryukov [5], pp. 1–18 (2007)

12. Daemen, J., Clapp, C.S.K.: Fast hashing and stream encryption with PANAMA. In: Vaudenay, S. (ed.) FSE 1998. LNCS, vol. 1372, pp. 60–74. Springer, Heidelberg (1998)
13. Dobbertin, H.: Cryptanalysis of md4. J. Cryptology 11(4), 253–271 (1998)
14. Faugère, J.-C.: A new efficient algorithm for computing grobner bases (f4). Journal of Pure and Applied Algebra 139(1-3), 61–68 (1999)
15. Faugère, J.-C., Gianni, P.M., Lazard, D., Mora, T.: Efficient computation of zero-dimensional gröbner bases by change of ordering. J. Symb. Comput. 16(4), 329–344 (1993)
16. Goubault-Larrecq, J., Roger, M., Verma, K.N.: Abstraction and resolution modulo ac: How to verify diffie-hellman-like protocols automatically. J. Log. Algebr. Program. 64(2), 219–251 (2005)
17. Lidl, R., Niederreiter, H.: Finite Fields (Encyclopedia of Mathematics and its Applications), October 1996. Cambridge University Press, Cambridge (1996)
18. Peyrin, T.: Cryptanalysis of GRINDAHL. In: Kurosawa, K. (ed.) ASIACRYPT 2007. LNCS, vol. 4833, pp. 551–567. Springer, Heidelberg (2007)
19. Rijmen, V., Van Rompay, B., Preneel, B., Vandewalle, J.: Producing collisions for PANAMA. In: Matsui, M. (ed.) FSE 2001. LNCS, vol. 2355, pp. 37–51. Springer, Heidelberg (2002)
20. Shoup, V. (ed.): CRYPTO 2005. LNCS, vol. 3621. Springer, Heidelberg (2005)
21. Sugita, M., Kawazoe, M., Perret, L., Imai, H.: Algebraic cryptanalysis of 58-round sha-1. In: Biryukov[5], pp. 349–365 (2007)
22. Wang, X., Lai, X., Feng, D., Chen, H., Yu, X.: Cryptanalysis of the Hash Functions MD4 and RIPEMD. In: Cramer [9], pp. 1–18 (2005)
23. Wang, X., Yin, Y.L., Yu, H.: Finding Collisions in the Full SHA-1. In: Shoup [20], pp. 17–36 (2005)
24. Wang, X., Yu, H.: How to Break MD5 and Other Hash Functions. In: Cramer [9], pp. 19–35 (2005)
25. Wang, X., Yu, H., Yin, Y.L.: Efficient Collision Search Attacks on SHA-0. In: Shoup [20], pp. 1–16 (2005)

A Scheme to Base a Hash Function on a Block Cipher

Shoichi Hirose[1] and Hidenori Kuwakado[2]

[1] Graduate School of Engineering, University of Fukui
hrs_shch@u-fukui.ac.jp
[2] Graduate School of Engineering, Kobe University
kuwakado@kobe-u.ac.jp

Abstract. This article discusses the provable security of an iterated hash function using a block cipher. It assumes the construction using the Matyas-Meyer-Oseas (MMO) scheme for the compression function and the Merkle-Damgård with a permutation (MDP) for the domain extension transform. It is shown that this kind of hash function, MDP-MMO, is indifferentiable from the variable-input-length random oracle in the ideal cipher model. It is also shown that HMAC using MDP-MMO is a pseudorandom function if the underlying block cipher is a pseudorandom permutation under the related-key attack with respect to the permutation used in MDP. Actually, the latter result also assumes that the following function is a pseudorandom bit generator:

$$(E_{IV}(K \oplus \mathtt{opad}) \oplus K \oplus \mathtt{opad}) \| (E_{IV}(K \oplus \mathtt{ipad}) \oplus K \oplus \mathtt{ipad}) \ ,$$

where E is the underlying block cipher, IV is the fixed initial value of MDP-MMO, and $\mathtt{opad}$ and $\mathtt{ipad}$ are the binary strings used in HMAC. This assumption still seems reasonable for actual block ciphers, though it cannot be implied by the pseudorandomness of E as a block cipher. The results of this article imply that the security of a hash function may be reduced to the security of the underlying block cipher to more extent with the MMO compression function than with the Davies-Meyer (DM) compression function, though the DM scheme is implicitly used by the widely used hash functions such as SHA-1 and MD5.

1 Introduction

Background. A hash function is one of the most important primitives in cryptography. It normally consists of a function with fixed input length. This component function is called a compression function. A domain-extension transform is also specified which describes how to apply the compression function to a given input of variable length.

The methods to construct a compression function are classified in two classes: dedicated methods and those using block ciphers. Compression functions of well-known hash functions such as SHA-1/256 are constructed with the dedicated methods. However, they are also regarded as Davies-Meyer functions using dedicated block ciphers known as SHACAL-1/2.

R. Avanzi, L. Keliher, and F. Sica (Eds.): SAC 2008, LNCS 5381, pp. 262–275, 2009.
© Springer-Verlag Berlin Heidelberg 2009

Contribution. The topic of this article is to reduce the security of a hash function to the security of the underlying block cipher. It assumes the construction using the Matyas-Meyer-Oseas (MMO) scheme [14] for the compression function and the Merkle-Damgård with a permutation (MDP) [10] for the domain extension transform. This kind of hash function is called MDP-MMO in this article. A message padding scheme with the MD-strengthening is also assumed for MDP-MMO.

This article mainly discusses two security properties of MDP-MMO: indifferentiability from the variable-input-length (VIL) random oracle and pseudorandomness of HMAC [2,12] using MDP-MMO. Collision-resistance is also mentioned briefly. These results imply that the security of an iterated hash function may be reduced to the security of the underlying block cipher to more extent with the MMO compression function than with the Davies-Meyer (DM) compression function.

It is shown that MDP-MMO is indifferentiable from the VIL random oracle in the ideal cipher model. This work is motivated by the recent work of Gong et al. [9]. They claimed that hash functions indifferentiable from the VIL random oracle in the ideal cipher model can be constructed using the MMO compression function and the domain extension transforms in [8]. The contribution of the current article is to reconstruct the proof using the game playing technique [5]. Also, notice that they did not consider MDP for domain extension.

Indifferentiability of an iterated hash function is often discussed on the assumption that the underlying compression function is a random oracle with fixed input length. Taking the structure of compression functions of widely used hash functions into consideration, it is not satisfactory. For example, DM and MMO compression functions are not indifferentiable from the fixed-input-length (FIL) random oracle [8,13].

It is also shown that HMAC using MDP-MMO is a pseudorandom function (PRF) if the underlying block cipher is a pseudorandom permutation (PRP) under the related-key attack with respect to the permutation used in MDP. Actually, this result also requires that the following function is a pseudorandom bit generator (PRBG):

$$(E_{IV}(K \oplus \mathtt{opad}) \oplus K \oplus \mathtt{opad}) \| (E_{IV}(K \oplus \mathtt{ipad}) \oplus K \oplus \mathtt{ipad}) ,$$

where E is the underlying block cipher, IV is the fixed initial value of MDP-MMO, and $\mathtt{opad}$ and $\mathtt{ipad}$ are the binary strings used in HMAC. It does not seem difficult to design a block cipher with which the function shown above is PRBG, though it cannot be implied by the pseudorandomness of E as a block cipher. It is because any adversary has no control over IV, $\mathtt{ipad}$ and $\mathtt{opad}$.

It can be said that the pseudorandomness of HMAC using MDP-MMO is almost reduced to the pseudorandomness of the underlying block cipher. Intuitively, it is because the chaining variables are fed into the block cipher via the key input and they are not disclosed to the adversary. On the other hand, if the Davies-Meyer compression function is used, then it is difficult to obtain a similar result. For this type of compression function, instead of the chaining variables, the message blocks are fed into the block cipher via the key input. They are selected and controlled fully by the adversary.

Related Work. Coron et al. [8] first discussed the indifferentiability of hash functions from the VIL random oracle. They presented four domain extension transforms: the Merkle-Damgård (MD) transform with prefix-free encoding, the MD transform dropping some output bits, and NMAC/HMAC-like transforms. Then, they showed that hash functions using them are indifferentiable from the VIL random oracle if the underlying compression functions are FIL random oracles. Moreover, they showed that hash functions using them and the DM compression function are indifferentiable from the VIL random oracle in the ideal cipher model.

Chang et al. [7] discussed the indifferentiability of hash functions from the VIL random oracle in the ideal cipher model. They assumed the compression functions using a block cipher in the PGV model [17] and the MD transform with prefix-free encoding for domain extension. They showed that the hash functions using 16 compression functions in the PGV model are indifferentiable from the VIL random oracle in the ideal cipher model. They also showed that the hash function using the MMO compression function is differentiable from the VIL random oracle. On the other hand, as mentioned before, Gong, Lai and Chen claimed that it is possible to construct hash functions indifferentiable from the VIL random oracle in the ideal cipher model even with the MMO compression function [9].

Bellare and Ristenpart gave a new notion called multi-property preservation (MPP) for domain extension [4]. A domain extension transform is called MPP if it preserves multiple security properties of a compression function such as collision-resistance, pseudorandomness, indifferentiability from a random oracle, etc. They also presented the EMD domain extension transform with the MPP property.

Hirose, Park and Yun [10] proposed a MPP domain extension transform called MDP. They also showed that a hash function using MDP and the DM compression function is indifferentiable from the VIL random oracle in the ideal cipher model. Ferguson had originally suggested an example of the MDP transform [11].

HMAC was first proposed by Bellare, Canetti and Krawczyk [2]. It was also shown in the same paper that HMAC is a PRF if the underlying compression function is a PRF with two keying strategies and the iterated hash function is weakly collision-resistant. Bellare proved that HMAC is a PRF under the sole assumption that the underlying compression function is a PRF with two keying strategies [1].

Organization. This article is organized as follows. Some notations and definitions are given in Section 2. The definition of MDP-MMO is given in Section 3. Section 4 is devoted to the indifferentiability of MDP-MMO from the VIL random oracle in the ideal cipher model. The security of HMAC using MDP-MMO as a PRF is discussed in Section 5.

2   Definitions

Let $\mathsf{Func}(D, R)$ be the set of all functions from D to R, and $\mathsf{Perm}(D)$ be the set of all permutations on D. Let $s \xleftarrow{\$} S$ represent that an element s is selected from the set S under the uniform distribution.

2.1   Pseudorandom Bit Generator

Let g be a function such that $g : \{0,1\}^n \to \{0,1\}^l$, where $n < l$. Let A be a probabilistic algorithm which outputs 0 or 1 for a given input in $\{0,1\}^l$. The prbg-advantage of A against g is defined as follows:

$$\mathrm{Adv}_g^{\mathrm{prbg}}(A) =$$

$$\left| \Pr[A(g(k)) = 1 \mid k \xleftarrow{\$} \{0,1\}^n] - \Pr[A(s) = 1 \mid s \xleftarrow{\$} \{0,1\}^l] \right| ,$$

where the probabilities are taken over the coin tosses by A and the uniform distributions on $\{0,1\}^n$ and $\{0,1\}^l$. g is called a pseudorandom bit generator (PRBG) if $\mathrm{Adv}_g^{\mathrm{prbg}}(A)$ is negligible for any efficient A.

2.2   Pseudorandom Function

Let $f : K \times D \to R$ be a keyed function or a function family. $f(k, \cdot)$ is often denoted by $f_k(\cdot)$. Let A be a probabilistic algorithm which has oracle access to a function from D to R. A first asks elements in D and obtains the corresponding elements in R with respect to the function, and then outputs 0 or 1. The prf-advantage of A against f is defined as follows:

$$\mathrm{Adv}_f^{\mathrm{prf}}(A) = \left| \Pr[A^{f_k} = 1 \mid k \xleftarrow{\$} K] - \Pr[A^\rho = 1 \mid \rho \xleftarrow{\$} \mathsf{Func}(D, R)] \right| ,$$

where the probabilities are taken over the coin tosses by A and the uniform distributions on K and $\mathsf{Func}(D, R)$. f is called a pseudorandom function (PRF) if $\mathrm{Adv}_f^{\mathrm{prf}}(A)$ is negligible for any efficient A.

Let $p : K \times D \to D$ be a keyed permutation or a permutation family. The prp-advantage of A against p is defined similarly:

$$\mathrm{Adv}_p^{\mathrm{prp}}(A) = \left| \Pr[A^{p_k} = 1 \mid k \xleftarrow{\$} K] - \Pr[A^\rho = 1 \mid \rho \xleftarrow{\$} \mathsf{Perm}(D)] \right| .$$

p is called a pseudorandom permutation (PRP) if $\mathrm{Adv}_p^{\mathrm{prp}}(A)$ is negligible for any efficient A.

2.3   Pseudorandom Function under Related-Key Attack

Pseudorandom functions under related-key attacks are first formalized by Bellare and Kohno [3]. In this article, we only consider a related-key attack with respect to a permutation π as in [10]. We will refer to this type of related-key attack as the π-related-key attack. Let A be a probabilistic algorithm which has oracle access to a pair of functions from D to R. The prf-rka-advantage of A against f under the π-related-key attack is given by

$$
\mathrm{Adv}_{\pi,f}^{\mathrm{prf\text{-}rka}}(A) =
$$
$$
\left| \Pr[A^{f_k, f_{\pi(k)}} = 1 \mid k \xleftarrow{\$} K] - \Pr[A^{\rho, \rho'} = 1 \mid \rho, \rho' \xleftarrow{\$} \mathsf{Func}(D, R)] \right| ,
$$

where the probabilities are taken over the coin tosses by A and the uniform distributions on K and $\mathsf{Func}(D, R)$. f is called a π-RKA-secure PRF if $\mathrm{Adv}_{\pi,f}^{\mathrm{prf\text{-}rka}}(A)$ is negligible for any efficient A.

For a permutation, the prp-rka-advantage and the π-RKA-secure PRP can also be defined similarly.

2.4 Computationally almost Universal Function Family

Computationally almost universal function families are formalized by Bellare in [1]. Let $f : K \times D \to R$ be a function family. Let A be a probabilistic algorithm which takes no inputs and produces a pair of elements in D. The au-advantage of A against f is defined as follows:

$$
\mathrm{Adv}_f^{\mathrm{au}}(A) = \Pr[f_k(M_1) = f_k(M_2) \wedge M_1 \neq M_2 \mid (M_1, M_2) \leftarrow A \wedge k \xleftarrow{\$} K] ,
$$

where the probabilities are taken over the coin tosses by A and the uniform distribution on K. f is called a computationally almost universal function family if $\mathrm{Adv}_f^{\mathrm{au}}(A)$ is negligible for any efficient A.

2.5 Indifferentiability from Random Oracle

The notion of indifferentiability is introduced by Maurer et al. [15] as a generalized notion of indistinguishability. Then, it is tailored to security analysis of hash functions by Coron et al. [8].

Let C be an algorithm with oracle access to an ideal primitive $\mathcal{F}$. In the setting of this article, C is an algorithm to construct a hash function using $\mathcal{F}$ with fixed input length. Let $\mathcal{H}$ be the VIL random oracle and S be a simulator which has oracle access to $\mathcal{H}$. $S^{\mathcal{H}}$ tries to behave like $\mathcal{F}$ in order to convince an adversary that $\mathcal{H}$ is $C^{\mathcal{F}}$. Let A be an adversary with access to two oracles. The indiff-advantage of A against C with respect to S is given by

$$
\mathrm{Adv}_{C,S}^{\mathrm{indiff}}(A) = \left| \Pr[A^{C^{\mathcal{F}}, \mathcal{F}} = 1] - \Pr[A^{\mathcal{H}, S^{\mathcal{H}}} = 1] \right| ,
$$

where the probabilities are taken over the coin tosses by A, C and S and the distributions of ideal primitives. $C^{\mathcal{F}}$ is said to be indifferentiable from the random oracle $\mathcal{H}$ if there exists a simulator $S^{\mathcal{H}}$ such that $\mathrm{Adv}_{C,S}^{\mathrm{indiff}}(A)$ is negligible for any efficient A.

2.6   Ideal Cipher Model

A block cipher with block length n and key length κ is called an (n, κ) block cipher. Let $E : \{0,1\}^\kappa \times \{0,1\}^n \to \{0,1\}^n$ be an (n, κ) block cipher. Then, $E(k, \cdot) = E_k(\cdot)$ is a permutation for every $k \in \{0,1\}^\kappa$. An (n, κ) block cipher E is called an ideal cipher if E_k is a truly random permutation for every k.

The lazy evaluation of an ideal cipher is described as follows. The encryption oracle receives a pair of a key and a plaintext as a query, and returns a randomly selected ciphertext. On the other hand, the decryption oracle receives a pair of a key and a ciphertext as a query, and returns a randomly selected plaintext. The oracles share a table of triplets of keys, plaintexts and ciphertexts, which are produced by the queries and the corresponding replies. Referring to the table, they select a reply to a new query under the restriction that E_k is a permutation for every k.

3   MDP with MMO Compression Function

We denote concatenation of sequences by $\|$. For sequences $M_1, M_2, \ldots, M_k$, we often denote $M_1 \| M_2 \| \cdots \| M_k$ simply by $M_1 M_2 \cdots M_k$. Let $\mathcal{B} = \{0,1\}^n$ and $\mathcal{B}^+ = \cup_{i=1}^\infty \mathcal{B}^i$.

Let $E : \mathcal{B} \times \mathcal{B} \to \mathcal{B}$ be an (n, n) block cipher. The Matyas-Meyer-Oseas (MMO) compression function [16] $F : \mathcal{B} \times \mathcal{B} \to \mathcal{B}$ with E is defined as follows: $F(s, x) = E_s(x) \oplus x$, where s is a chaining variable and x is a message block.

The MDP transform [10] of F with a permutation π is denoted by $F_\pi^\circ : \mathcal{B} \times \mathcal{B}^+ \to \mathcal{B}$ and defined as follows: For $s \in \mathcal{B}$ and $M_1 M_2 \cdots M_k$ $(M_i \in \mathcal{B})$,

1. $s_0 = s$,
2. $s_i = F(s_{i-1}, M_i)$ for $1 \le i \le k - 1$,
3. $s_k = F(\pi(s_{k-1}), M_k)$,
4. $F_\pi^\circ(s, M_1 M_2 \cdots M_k) \stackrel{\text{def}}{=} s_k$.

The following padding function $\mathsf{pad} : \{0,1\}^* \to \cup_{i=2}^\infty \mathcal{B}^i$ is also prepared:

$$\mathsf{pad}(M) = M \| 10^\ell \| bin(|M|) ,$$

where ℓ is the minimum non-negative integer such that $|M| + \ell \equiv 0 \pmod{n}$, and $bin(|M|)$ is the $(n-1)$-bit binary representation of $|M|$. Thus, the input length of pad is precisely at most $2^{n-1} - 1$.

Now, MDP-MMO is a scheme to construct a hash function using a block cipher $E : \mathcal{B} \times \mathcal{B} \to \mathcal{B}$, a permutation $\pi : \mathcal{B} \to \mathcal{B}$ and an initial value $IV \in \mathcal{B}$ defined as follows:

$$\mathsf{MDP\text{-}MMO}[E, \pi, IV](M) \stackrel{\text{def}}{=} F_\pi^\circ(IV, \mathsf{pad}(M)) .$$

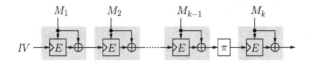

Fig. 1. MDP-MMO$[E, \pi, IV](M)$. $\mathsf{pad}(M) = M_1 M_2 \cdots M_k$

A diagram of MDP-MMO is shown in Figure 1.

4 Security of MDP-MMO

4.1 Collision Resistance

It is easy to see that MDP-MMO$[E, \pi, IV]$ is collision-resistant (CR) if its compression function is CR, that is, it is difficult to compute a pair of distinct (S, X) and (S', X') such that $E_S(X) \oplus X = E_{S'}(X') \oplus X'$. The pseudorandomness of E as a block cipher cannot imply the property. It is easy to find a counterexample. However, it seems still reasonable to assume that a well-designed block cipher such as AES has this property.

The CR of MDP-MMO can also be proved in the ideal cipher model using the technique by Black et al. in [6].

4.2 Indifferentiability from Random Oracle

In this section, we show that MDP-MMO$[E, \pi, IV]$ is indifferentiable from the VIL random oracle in the ideal cipher model. The following theorem states the indifferentiability of MDP-MMO in the ideal cipher model.

Theorem 1. *Let E be an (n, n) block cipher. Let π be a permutation and P_π be the set of its fixed points. Let A be an adversary that asks at most $q_\mathcal{H}$ queries to the VIL oracle, $q_\mathcal{E}$ queries to the FIL encryption oracle and $q_\mathcal{D}$ queries to the FIL decryption oracle. Let l be the maximum number of message blocks in a VIL query. Suppose that $l q_\mathcal{H} + q_\mathcal{E} + q_\mathcal{D} \leq 2^{n-1}$. Then, in the ideal cipher model,*

$$\mathrm{Adv}^{\mathrm{indiff}}_{\mathsf{MDP\text{-}MMO}, S_E, S_D}(A)$$

$$\leq \frac{6 \left(l q_\mathcal{H} + q_\mathcal{E} + q_\mathcal{D}\right)^2 + 14 \left(l q_\mathcal{H} + q_\mathcal{E}\right) q_\mathcal{D} + \left(l q_\mathcal{H} + q_\mathcal{E}\right)^2}{2^{n+1}}$$

$$+ \frac{2 \, l q_\mathcal{H}(q_\mathcal{E} + q_\mathcal{D})}{2^{n-1} - 3 \left(l q_\mathcal{H} + q_\mathcal{E} + q_\mathcal{D}\right) - |P_\pi|}$$

$$+ \frac{(4 \, |P_\pi| + 5) \left(l q_\mathcal{H} + q_\mathcal{E} + q_\mathcal{D}\right) + 21 \, q_\mathcal{D}}{2^{n+1}},$$

where the simulators S_E and S_D are given in Figure 2. S_E is a simulator for the encryption oracle, and S_D for the decryption oracle. S_E makes at most $q_\mathcal{E}$ queries and runs in time $O(q_\mathcal{E}(q_\mathcal{E} + q_\mathcal{D}))$. S_D makes at most $q_\mathcal{D}$ queries and runs in time $O(q_\mathcal{D}(q_\mathcal{E} + q_\mathcal{D}))$.

For Theorem 1, suppose that π has no fixed points. Also suppose that $lq_{\mathcal{H}} + q_{\mathcal{E}} + q_{\mathcal{D}} \leq 2^{n-3}$, and $lq_{\mathcal{H}} \geq 1$, $q_{\mathcal{E}} \geq 1$, $q_{\mathcal{D}} \geq 1$. Then, a looser but simpler bound is obtained:

$$\mathrm{Adv}^{\mathrm{indiff}}_{\mathrm{MDP\text{-}MMO},S_E,S_D}(A) \leq \frac{5\,(lq_{\mathcal{H}} + q_{\mathcal{E}} + q_{\mathcal{D}})^2}{2^{n-1}} \ .$$

Instead of the proof omitted due to the page limit, a brief intuitive idea of the proof is given below.

S_E and S_D simulate the ideal cipher using lazy evaluation. In Figure 2, $\mathcal{P}(s)$ and $\mathcal{C}(s)$ represent the set of plaintexts and that of ciphertexts, respectively, which are available for the reply to the current query with the key s. Both of them are initially $\{0,1\}^n$, and their elements are deleted one by one as the simulation proceeds.

Let (s_i, x_i, y_i) be the triplet determined by the i-th query of the adversary and the corresponding answer, where $E_{s_i}(x_i) = y_i$. For the MMO compression function, s_i is a chaining variable, and x_i is a message block. The triplets naturally defines a graph which initially consists of a single node labeled by the initial value IV and grows as the simulation proceeds. (s_i, x_i, y_i) adds two nodes labeled by s_i and $z_i = x_i \oplus y_i$, and an edge labeled by x_i from s_i to z_i. The additions avoid duplication of nodes with the same labels.

The simulators use two sets $\mathcal{V}$ and $\mathcal{T}$. $\mathcal{V}$ keeps all the labels of the nodes with outgoing edge(s) in the graph. $\mathcal{T}$ keeps all the labels of the nodes reachable from the node labeled by IV following the paths. The procedure getnode(s) returns the sequence of labels of the edges on the path from the node labeled by IV to the node labeled by s.

The simulators select a reply not simply from $\mathcal{C}(s)$ or $\mathcal{P}(s)$ but from $\mathcal{C}(s) \setminus \boldsymbol{S}_{\mathrm{bad}}$ or $\mathcal{P}(s) \setminus \boldsymbol{S}_{\mathrm{bad}}$. It prevents most of the events which make the simulators fail. For example, since $\{y \mid x \oplus y \in \mathcal{T}\} \subset \boldsymbol{S}_{\mathrm{bad}}$, every node in $\mathcal{T}$ has a unique path from the node labeled by IV. Thus, $\tilde{M}$ is uniquely identified at the lines 204 and 304. The most critical work of the simulators is to reply to a decryption query related to the final invocation of the compression function in MDP-MMO$[E, \pi, IV](M)$ for some M. Let (s, x) be such a query to S_D. In order to reply to it properly, the simulator S_D has to ask M to the VIL random oracle H and return $H(M) \oplus x$. Owing to the padding scheme pad, there exist only two possibilities for M, $M^{(0)}$ and $M^{(1)}$, which correspond to the message blocks $\tilde{M}$ fed to the compression functions before the permutation π. Thus, S_D can accomplish the work.

5   Security of HMAC Using MDP-MMO

In this section, we discuss the pseudorandomness of HMAC using the MDP-MMO hash function (HMAC-MDP-MMO). This function is defined as follows:

$$\mathrm{HMAC}[E, \pi, IV](K, M) = H((K \oplus \mathrm{opad})\|H((K \oplus \mathrm{ipad})\|M)) \ ,$$

where H is MDP-MMO$[E, \pi, IV]$ and K is a secret key. A diagram of HMAC-MDP-MMO is given in Figure 3. Let us call $H((K \oplus \mathrm{ipad})\|\cdot)$ inner hashing and $H((K \oplus \mathrm{opad})\|\cdot)$ outer hashing.

Initialize:	Interface $\mathcal{E}(s,x)$:
100: $\mathcal{V} \leftarrow \emptyset$	200: **if** $s \in \mathcal{T}$ **then**
101: $\mathcal{T} \leftarrow \{IV\}$	201: $E_s(x) \xleftarrow{\$} \mathcal{C}(s) \setminus S_{\text{bad}}$
102: $\mathcal{P}(s) \leftarrow \{0,1\}^n$	202: $\mathcal{T} \leftarrow \mathcal{T} \cup \{E_s(x) \oplus x\}$
103: $\mathcal{C}(s) \leftarrow \{0,1\}^n$	203: **else if** $\pi^{-1}(s) \in \mathcal{T}$ **then**

Interface $\mathcal{E}(s,x)$:

200: **if** $s \in \mathcal{T}$ **then**
201:     $E_s(x) \xleftarrow{\$} \mathcal{C}(s) \setminus S_{\text{bad}}$
202:     $\mathcal{T} \leftarrow \mathcal{T} \cup \{E_s(x) \oplus x\}$
203: **else if** $\pi^{-1}(s) \in \mathcal{T}$ **then**
204:     $\tilde{M} \leftarrow \mathsf{getnode}(\pi^{-1}(s))$
205:     **if** $x \in \{lb(M^{(0)}), lb(M^{(1)})\}$ **then**
206:         **if** $x = lb(M^{(0)})$ **then**
207:             $E_s(x) \leftarrow H(M^{(0)}) \oplus lb(M^{(0)})$
208:         **else**
209:             $E_s(x) \leftarrow H(M^{(1)}) \oplus lb(M^{(1)})$
210:         **if** $E_s(x) \notin \mathcal{C}(s)$ **then**
211:             **return fail**
212:     **else**
213:         $E_s(x) \xleftarrow{\$} \mathcal{C}(s)$
214: **else**
215:     $E_s(x) \xleftarrow{\$} \mathcal{C}(s)$
216: $\mathcal{V} \leftarrow \mathcal{V} \cup \{s\}$
217: $\mathcal{P}(s) \leftarrow \mathcal{P}(s) \setminus \{x\}$
218: $\mathcal{C}(s) \leftarrow \mathcal{C}(s) \setminus \{E_s(x)\}$
219: **return** $E_s(x)$

Interface $\mathcal{D}(s,x)$:

300: **if** $s \in \mathcal{T}$ **then**
301:     $D_s(x) \xleftarrow{\$} \mathcal{P}(s) \setminus S_{\text{bad}}$
302:     $\mathcal{T} \leftarrow \mathcal{T} \cup \{D_s(x) \oplus x\}$
303: **else if** $\pi^{-1}(s) \in \mathcal{T}$ **then**
304:     $\tilde{M} \leftarrow \mathsf{getnode}(\pi^{-1}(s))$
305:     **if** $x = H(M^{(0)}) \oplus lb(M^{(0)})$ **then**
306:         $D_s(x) \leftarrow lb(M^{(0)})$
307:     **else if** $x = H(M^{(1)}) \oplus lb(M^{(1)})$ **then**
308:         $D_s(x) \leftarrow lb(M^{(1)})$
309:     **else**
310:         $D_s(x) \xleftarrow{\$} \mathcal{P}(s) \setminus \{lb(M^{(0)}), lb(M^{(1)})\}$
311: **else**
312:     $D_s(x) \xleftarrow{\$} \mathcal{P}(s)$
313: $\mathcal{V} \leftarrow \mathcal{V} \cup \{s\}$
314: $\mathcal{P}(s) \leftarrow \mathcal{P}(s) \setminus \{D_s(x)\}$
315: $\mathcal{C}(s) \leftarrow \mathcal{C}(s) \setminus \{x\}$
316: **return** $D_s(x)$

Fig. 2. Pseudocode for the simulators S_E and S_D. $S_{\text{bad}} = \{y \mid y \in \{0,1\}^n \wedge x \oplus y \in \mathcal{V} \cup \mathcal{T} \cup \pi^{-1}(\mathcal{V} \cup \mathcal{T}) \cup \pi(\mathcal{T}) \cup P_\pi\}$. $\mathsf{pad}(M^{(0)}) = \tilde{M} \| lb(M^{(0)})$, and $\mathsf{pad}(M^{(1)}) = \tilde{M} \| lb(M^{(1)})$. $\tilde{M} = M^{(0)} \| 10^\ell$ $(0 \le \ell \le n - 2)$ and $lb(M^{(0)}) = 0 \| bin(|M^{(0)}|)$. $\tilde{M} = M^{(1)}$ and $lb(M^{(1)}) = 1 \| bin(|M^{(1)}|)$.

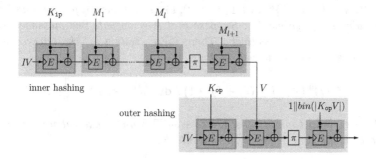

Fig. 3. HMAC$[E, \pi, IV](K, M)$. E is an (n, n) block cipher. $K_{\text{ip}} = K \oplus \text{ipad}$ and $K_{\text{op}} = K \oplus \text{opad}$. $\text{pad}(K_{\text{ip}} \| M) = K_{\text{ip}} M_1 \cdots M_{l+1}$.

We use the technique given by Bellare [1] in the analysis. We can also obtain a similar result based on the pseudorandomness of Prefix-MDP [10] in a more straightforward way. However, to the best of our analysis, the upper bound on the prf-advantage against HMAC-MDP-MMO obtained with this approach is worse than the one given below.

First, the compression function construction is considered. The following lemma says that the MMO compression function is a (π-RKA-secure) PRF when keyed via the chaining variable if the underlying block cipher is a (π-RKA-secure) PRP under the chosen plaintext attack up to the birthday bound. The proof is easy and omitted.

Lemma 1. *Let E be an (n, n) block cipher and F be a function such that $F_K(x) = E_K(x) \oplus x$.*

- *Let A_F be a prf-adversary against F which runs in time at most t and asks at most q queries. Then, there exists a prp-adversary A_E against E such that*

$$\text{Adv}_F^{\text{prf}}(A_F) \leq \text{Adv}_E^{\text{prp}}(A_E) + \frac{q(q-1)}{2^{n+1}} \ ,$$

 where A_E runs in time at most $t + O(q)$ and asks at most q queries.
- *Let π be a permutation. Let $A_{\pi, F}$ be a prf-rka-adversary against F with respect to π which runs in time at most t and asks at most q queries. Then, there exists a prp-rka-adversary $A_{\pi, E}$ against E with respect to π such that*

$$\text{Adv}_{\pi, F}^{\text{prf-rka}}(A_{\pi, F}) \leq \text{Adv}_{\pi, E}^{\text{prp-rka}}(A_{\pi, E}) + \frac{q(q-1)}{2^{n+1}} \ ,$$

 where $A_{\pi, E}$ runs in time at most $t + O(q)$ and asks at most q queries.

The following lemma is on the inner hashing. It says that, if the compression function F is a π-RKA-secure PRF, then the MDP composition of F and π is computationally almost universal. The proof is omitted due to the page limit.

Lemma 2. *Let* $F : \{0,1\}^\kappa \times \{0,1\}^n \to \{0,1\}^\kappa$ *be a function family, and let* $A_{F_\pi^\circ}$ *be an au-adversary against* F_π°. *Suppose that* $A_{F_\pi^\circ}$ *outputs two messages with at most* ℓ_1 *and* ℓ_2 *blocks, respectively. Then, there exists a prf-rka-adversary* $A_{\pi,F}$ *against* F *with respect to* π *such that*

$$\mathrm{Adv}_{F_\pi^\circ}^{\mathrm{au}}(A_{F_\pi^\circ}) \leq (\ell_1 + \ell_2 - 1)\,\mathrm{Adv}_{\pi,F}^{\mathrm{prf\text{-}rka}}(A_{\pi,F}) + \frac{1}{2^\kappa} \ ,$$

where $A_{\pi,F}$ *runs in time at most* $O((\ell_1 + \ell_2)T_F)$ *and makes at most 2 queries.* T_F *represents the time required to compute* F.

Lemma 2 requires a π-RKA-secure compression function. However, the assumption does not seem severe since adversaries are allowed to make only at most 2 queries to the oracles.

The following lemma is on the outer hashing. It says that, if the compression function is a PRF, then the outer-hashing function is also a PRF. The proof is omitted.

Lemma 3. *Let* $F : \{0,1\}^\kappa \times \{0,1\}^n \to \{0,1\}^\kappa$ *be a function family. Let* $\hat{F}^2 : \{0,1\}^\kappa \times \{0,1\}^n \to \{0,1\}^\kappa$ *be a function family defined by*

$$\hat{F}^2(K, X) = F(\pi(F(K, X)), 1\|bin(\kappa + n)) \ ,$$

where $K \in \{0,1\}^\kappa$ *and* $X \in \{0,1\}^n$. *Let* $A_{\hat{F}^2}$ *be a prf-adversary against* $\hat{F}^2$ *that runs in time at most* t *and makes at most* q *queries. Then, there exist prf-adversaries* A_F *and* A_F' *against* F *such that*

$$\mathrm{Adv}_{\hat{F}^2}^{\mathrm{prf}}(A_{\hat{F}^2}) \leq \mathrm{Adv}_F^{\mathrm{prf}}(A_F) + q\,\mathrm{Adv}_F^{\mathrm{prf}}(A_F') \ ,$$

where A_F *runs in time at most* $t + O(q\,T_F)$ *and makes at most* q *queries, and* A_F' *runs in time* $t + O(q\,T_F)$ *and makes at most 1 query.* T_F *represents the time required to compute* F.

The following lemma is Lemma 3.2 in [1]. It says that $h(K_\circ, G(K_i, \cdot))$ is a PRF if $h(K_\circ, \cdot)$ is a PRF and $G(K_i, \cdot)$ is computationally almost universal, where $K_\circ$ and K_i are secret keys chosen uniformly and independently of each other.

Lemma 4 (Lemma 3.2 in [1]). *Let* $h : \{0,1\}^\mu \times \{0,1\}^n \to \{0,1\}^\mu$ *and* $G : \{0,1\}^\kappa \times \mathcal{D} \to \{0,1\}^n$ *be function families. Let* $hG : \{0,1\}^{\mu+\kappa} \times \mathcal{D} \to \{0,1\}^\mu$ *be defined by* $hG(K_\circ\|K_i, M) = h(K_\circ, G(K_i, M))$ *for* $K_\circ \in \{0,1\}^\mu$, $K_i \in \{0,1\}^\kappa$ *and* $M \in \mathcal{D}$. *Let* A_{hG} *be a prf-adversary against* hG *that runs in time at most* t *and makes at most* $q\,(\geq 2)$ *queries each of whose lengths is at most* d. *Then, there exist a prf-adversary* A_h *against* h *and an au-adversary* A_G *against* G *such that*

$$\mathrm{Adv}_{hG}^{\mathrm{prf}}(A_{hG}) \leq \mathrm{Adv}_h^{\mathrm{prf}}(A_h) + \frac{q(q-1)}{2}\mathrm{Adv}_G^{\mathrm{au}}(A_G) \ ,$$

where A_h *runs in time at most* t *and makes at most* q *queries, and* A_G *runs in time* $O(T_G(d))$ *and the two messages it outputs have length at most* d. $T_G(d)$ *is the time to compute* G *on a* d-bit input.

The following theorem is on the pseudorandomness of the NMAC-like function made from $\mathsf{HMAC}[E, \pi, IV](K, \cdot)$ by replacing the first calls of the compression function in inner and outer hashing with two secret keys chosen uniformly and independently of each other. The theorem states that the security of the function as a PRF is reduced to the security of the underlying block cipher as a PRP under the π-related-key attack. It directly follows from Lemmas 1 through 4.

Theorem 2. *Let $\mathcal{B} = \{0,1\}^n$ and E be an (n,n) block cipher. Let $F : \mathcal{B} \times \mathcal{B} \to \mathcal{B}$ be a function such that $F_K(x) = E_K(x) \oplus x$. Let $\hat{F}^2 F_\pi^\circ : \mathcal{B}^2 \times \mathcal{B}^+ \to \mathcal{B}$ be defined by $\hat{F}^2 F_\pi^\circ(K_\circ \| K_i, M) = \hat{F}^2(K_\circ, F_\pi^\circ(K_i, M))$ for $K_\circ, K_i \in \mathcal{B}$ and $M \in \mathcal{B}^+$. Let $A_{\hat{F}^2 F_\pi^\circ}$ be a prf-adversary against $\hat{F}^2 F_\pi^\circ$ that runs in time at most t and makes at most $q \, (\geq 2)$ queries each of which has at most ℓ blocks. Then, there exist prp-adversaries A_E and A_E' against E and a prp-rka-adversary $A_{\pi,E}$ against E with respect to π such that*

$$
\mathrm{Adv}^{\mathrm{prf}}_{\hat{F}^2 F_\pi^\circ}(A_{\hat{F}^2 F_\pi^\circ}) \leq
$$

$$
\mathrm{Adv}^{\mathrm{prp}}_E(A_E) + q \, \mathrm{Adv}^{\mathrm{prp}}_E(A_E') + \ell \, q^2 \mathrm{Adv}^{\mathrm{prp\text{-}rka}}_{\pi,E}(A_{\pi,E}) + \frac{(2\ell+3)q^2}{2^{n+1}} ,
$$

where A_E runs in time at most $t + O(q T_E)$ and makes at most q queries, A_E' runs in time at most $t + O(q T_E)$ and makes at most 1 query, and $A_{\pi,E}$ runs in time $O(\ell T_E)$ and makes at most 2 queries. T_E represents the time required to compute E.

The following lemma says that, even if the secret key of a PRF is replaced by the output of a PRBG, the resulting function remains a PRF. The proof is easy and omitted.

Lemma 5. *Let $g : \{0,1\}^\kappa \to \{0,1\}^{\kappa'}$ be a function and $G : \{0,1\}^{\kappa'} \times \mathcal{D} \to \{0,1\}^n$ be a function family. Let $Gg : \{0,1\}^\kappa \times \mathcal{D} \to \{0,1\}^n$ be a function family defined by $Gg(K, M) = G(g(K), M)$ for $K \in \{0,1\}^\kappa$ and $M \in \mathcal{D}$. Let A_{Gg} be a prf-adversary against Gg that runs in time at most t and makes at most q queries of length at most d. Then, there exist a prbg-adversary A_g against g and a prf-adversary A_G against G such that*

$$
\mathrm{Adv}^{\mathrm{prf}}_{Gg}(A_{Gg}) \leq \mathrm{Adv}^{\mathrm{prbg}}_g(A_g) + \mathrm{Adv}^{\mathrm{prf}}_G(A_G) ,
$$

where A_g runs in time at most $t + O(q T_G(d))$, and A_G runs in time t and makes at most q queries of length at most d.

Now, we can obtain the result on the pseudorandomness of HMAC-MDP-MMO simply by combining Theorem 2 and Lemma 5.

Corollary 1. *Let E be an (n,n) block cipher. Let $g_E : \{0,1\}^n \to \{0,1\}^{2n}$ be a function such that $g_E(K) = (E_{IV}(K_{\mathsf{op}}) \oplus K_{\mathsf{op}}) \| (E_{IV}(K_{\mathsf{ip}}) \oplus K_{\mathsf{ip}})$, where $K_{\mathsf{op}} = K \oplus \mathsf{opad}$ and $K_{\mathsf{ip}} = K \oplus \mathsf{ipad}$. Let A be a prf-adversary against $\mathsf{HMAC}[E, \pi, IV]$ that runs in time at most t and makes at most $q \, (\geq 2)$ queries each of which has* *at most ℓ blocks. Then, there exist prp-adversaries A_E and A'_E against E, a prp-rka-adversary $A_{\pi,E}$ against E with respect to π and a prbg-adversary A_{g_E} such that*

$$\mathrm{Adv}^{\mathrm{prf}}_{\mathrm{HMAC}[E,\pi,IV]}(A) \leq \mathrm{Adv}^{\mathrm{prbg}}_{g_E}(A_{g_E}) + \mathrm{Adv}^{\mathrm{prp}}_E(A_E) + q\,\mathrm{Adv}^{\mathrm{prp}}_E(A'_E)$$

$$+ \ell\,q^2 \mathrm{Adv}^{\mathrm{prp\text{-}rka}}_{\pi,E}(A_{\pi,E}) + \frac{(2\ell+3)q^2}{2^{n+1}} \;,$$

where A_{g_E} runs in time at most $t+O(q\,\ell\,T_E)$, A_E runs in time at most $t+O(q\,T_E)$ and makes at most q queries, A'_E runs in time at most $t + O(q\,T_E)$ and makes at most 1 query, and $A_{\pi,E}$ runs in time $O(\ell\,T_E)$ and makes at most 2 queries.

Actually, we have not completely reduced the security of HMAC-MDP-MMO as a PRF to the security of the underlying block cipher as a PRP under the π-related-key attack. It is easy to see that the function g_E in Corollary 1 may not be a PRBG in general even if E is a PRP. However, it does not seem so difficult to design a block cipher E such that g_E is a PRBG. This is because IV is a fixed initial value chosen by the designer of the hash function and the block cipher. Furthermore, ipad and opad are fixed sequences given by HMAC. Any adversary has no control over them.

We can say that the security of HMAC as a PRF is reduced to the security of the underlying block cipher as a PRP using the MMO scheme to more extent than using the Davies-Meyer scheme.

Acknowledgements

The authors would like to thank Dr. Yoshida and Dr. Ideguchi at Hitachi, Ltd. and Prof. Ohta and Dr. Wang at The University of Electro-Communications for their valuable discussions and comments on this research. The authors would also like to thank anonymous reviewers for their valuable comments. This research was supported by the National Institute of Information and Communications Technology, Japan.

References

1. Bellare, M.: New proofs for NMAC and HMAC: Security without collision-resistance. In: Dwork, C. (ed.) CRYPTO 2006. LNCS, vol. 4117, pp. 602–619. Springer, Heidelberg (2006); The full version is Cryptology ePrint Archive: Report 2006/043, http://eprint.iacr.org/
2. Bellare, M., Canetti, R., Krawczyk, H.: Keying hash functions for message authentication. In: Koblitz, N. (ed.) CRYPTO 1996. LNCS, vol. 1109, pp. 1–15. Springer, Heidelberg (1996)
3. Bellare, M., Kohno, T.: A theoretical treatment of related-key attacks: RKA-PRPs, RKA-PRFs. In: Biham, E. (ed.) EUROCRYPT 2003. LNCS, vol. 2656, pp. 491–506. Springer, Heidelberg (2003)

4. Bellare, M., Ristenpart, T.: Multi-property-preserving hash domain extension and the EMD transform. In: Lai, X., Chen, K. (eds.) ASIACRYPT 2006. LNCS, vol. 4284, pp. 299–314. Springer, Heidelberg (2006); The full version is Cryptology ePrint Archive: Report 2006/399, http://eprint.iacr.org/

5. Bellare, M., Rogaway, P.: Code-based game-playing proofs and the security of triple encryption. Cryptology ePrint Archive, Report 2004/331 (2006), http://eprint.iacr.org/

6. Black, J., Rogaway, P., Shrimpton, T.: Black-box analysis of the block-cipher-based hash-function constructions from PGV. In: Yung, M. (ed.) CRYPTO 2002. LNCS, vol. 2442, pp. 320–335. Springer, Heidelberg (2002)

7. Chang, D., Lee, S.-J., Nandi, M., Yung, M.: Indifferentiable security analysis of popular hash functions with prefix-free padding. In: Lai, X., Chen, K. (eds.) ASIACRYPT 2006. LNCS, vol. 4284, pp. 283–298. Springer, Heidelberg (2006)

8. Coron, J.-S., Dodis, Y., Malinaud, C., Puniya, P.: Merkle-Damgård revisited: How to construct a hash function. In: Shoup, V. (ed.) CRYPTO 2005. LNCS, vol. 3621, pp. 430–448. Springer, Heidelberg (2005)

9. Gong, Z., Lai, X., Chen, K.: A synthetic indifferentiability analysis of some block-cipher-based hash functions. Cryptology ePrint Archive, Report 2007/465 (2007), http://eprint.iacr.org/

10. Hirose, S., Park, J.H., Yun, A.: A simple variant of the Merkle-Damgård scheme with a permutation. In: Kurosawa, K. (ed.) ASIACRYPT 2007. LNCS, vol. 4833, pp. 113–129. Springer, Heidelberg (2007)

11. Kelsey, J.: A comment on draft FIPS 180-2. Public Comments on the Draft Federal Information Processing Standard (FIPS) Draft FIPS 180-2, Secure Hash Standard, SHS (2001)

12. Krawczyk, H., Bellare, M., Canetti, R.: HMAC: Keyed-hashing for message authentication. Network Working Group RFC 2104 (1997)

13. Kuwakado, H., Morii, M.: Compression functions suitable for the multi-property-preserving transform. Cryptology ePrint Archive, Report 2007/302 (2007), http://eprint.iacr.org/

14. Matyas, S.M., Meyer, C.H., Oseas, J.: Generating strong one-way functions with cryptographic algorithm. IBM Technical Disclosure Bulletin 27, 5658–5659 (1985)

15. Maurer, U.M., Renner, R.S., Holenstein, C.: Indifferentiability, impossibility results on reductions, and applications to the random oracle methodology. In: Naor, M. (ed.) TCC 2004. LNCS, vol. 2951, pp. 21–39. Springer, Heidelberg (2004)

16. Menezes, A.J., van Oorschot, P.C., Vanstone, S.A.: Handbook of Applied Cryptography. CRC Press, Boca Raton (1996)

17. Preneel, B., Govaerts, R., Vandewalle, J.: Hash functions based on block ciphers: A synthetic approach. In: Stinson, D.R. (ed.) CRYPTO 1993. LNCS, vol. 773, pp. 368–378. Springer, Heidelberg (1994)

Collisions and Other Non-random Properties for Step-Reduced SHA-256*

Sebastiaan Indesteege[1,2,**], Florian Mendel[3], Bart Preneel[1,2], and Christian Rechberger[3]

[1] Department of Electrical Engineering ESAT/SCD-COSIC, Katholieke Universiteit Leuven. Kasteelpark Arenberg 10, B–3001 Heverlee, Belgium
sebastiaan.indesteege@esat.kuleuven.be
[2] Interdisciplinary Institute for BroadBand Technology (IBBT), Belgium
[3] Institute for Applied Information Processing and Communications
Inffeldgasse 16a, A–8010 Graz, Austria

Abstract. We study the security of step-reduced but otherwise unmodified SHA-256. We show the first collision attacks on SHA-256 reduced to 23 and 24 steps with complexities 2^{18} and $2^{28.5}$, respectively. We give example colliding message pairs for 23-step and 24-step SHA-256. The best previous, recently obtained result was a collision attack for up to 22 steps. We extend our attacks to 23 and 24-step reduced SHA-512 with respective complexities of $2^{44.9}$ and $2^{53.0}$. Additionally, we show non-random behaviour of the SHA-256 compression function in the form of free-start near-collisions for up to 31 steps, which is 6 more steps than the recently obtained non-random behaviour in the form of a semi-free-start near-collision. Even though this represents a step forwards in terms of cryptanalytic techniques, the results do not threaten the security of applications using SHA-256.

Keywords: SHA-256, SHA-512, hash functions, collisions, semi-free-start collisions, free-start collisions, free-start near-collisions.

1 Introduction

In the light of previous break-through results on hash functions such as MD5 and SHA-1, the security of their successors, SHA-256 and sisters, against all kinds of cryptanalytic attacks deserves special attention. This is even more important as many products and services that used to rely on SHA-1 are now migrating to SHA-256.

* This work was supported in part by the IAP Programme P6/26 BCRYPT of the Belgian State (Belgian Science Policy), in part by the European Commission through the IST Programme under Contract IST-2002-507932 ECRYPT, and in part by the Austrian Science Fund (FWF), project P19863. This work was done during a visit of the first author to the Graz University of Technology.
** F.W.O. Research Assistant, Fund for Scientific Research — Flanders (Belgium).

R. Avanzi, L. Keliher, and F. Sica (Eds.): SAC 2008, LNCS 5381, pp. 276–293, 2009.
© Springer-Verlag Berlin Heidelberg 2009

1.1 Previous Work on Members of the SHA-2 Family

Below, we briefly discuss existing work. Results on older variants of the larger MD4 related hash function family, including SHA-1, suggest that the concept of local collisions might also be important for the SHA-2 family. The first published analysis on members of the SHA-2 family, by Gilbert and Handschuh [2], goes in this direction. They show that there exists a 9-step local collision with probability 2^{-66}. Later on, the result was improved by Hawkes *et al.* [3]. By considering modular differences, they increased the probability to 2^{-39}. Using XOR differences, local collisions with probability as high as 2^{-38} where used by Hölbl *et al.* [4]. Local collisions with lower probability but with other properties were studied by Sanadhya and Sarkar in [13].

Now we turn our attention to the analysis of simplified variants of SHA-256. In [17], Yoshida and Biryukov replace all modular additions by XOR. For this variant, a search for pseudo-collisions is described, which is faster than brute force search for up to 34 steps. Matusiewicz *et al.* [8] analysed a variant of SHA-256 where all Σ- and σ-functions are removed. The conclusion is that for this variant, collisions can be found much faster than by brute force search. The work shows that the approach used by Chabaud and Joux [1] in their analysis of SHA-0 is extensible to that particular variant of SHA-256. The message expansion as a building block on its own was studied by Matusiewicz *et al.* [8] and Pramstaller *et al.* [12].

Finally, we discuss previous work that focuses on step-reduced but otherwise unmodified SHA-256. The first study was done by Mendel *et al.* [9]. The results obtained are a practical 18-step collision and a differential characteristic for 19-step SHA-224 collision. Also, an example of a pseudo-near-collision for 22-step SHA-256 is given. Similar techniques have been studied by Matusiewicz *et al.* [8] and recently also by Sanadhya and Sarkar [15]. Using a different technique, Nikolić and Biryukov [10] obtained collisions for up to 21 steps and non-random behaviour in the form of semi-free-start near-collisions for up to 25 steps. Very recently, Sanadhya and Sarkar [16] extended this, and showed a collision example for 22 steps of SHA-256 in [14].

1.2 Our Contribution

We extend the work of Nikolić and Biryukov [10] to collisions for 23- and 24-step SHA-256 with respective time complexities of 2^{18} and $2^{28.5}$ reduced SHA-256 compression function evaluations. These 23- and 24-step attacks are also applied to SHA-512, with complexities of $2^{44.9}$ and $2^{53.0}$ for 23-step SHA-512 and 24-step SHA-512, respectively. Example collision pairs for 23-step SHA-256 and SHA-512, and for 24-step SHA-256 are given. The collision attacks presented in this work do not extend beyond 24 steps, but we investigate several weaker collision style attacks on a larger number of rounds. Our results are summarised in Table 1.

We use the terminology introduced by Lai and Massey [5] for different types of attacks on (iterated) hash functions. A collision attack aims to find two distinct

Table 1. Comparison of our results with the known results in the literature for each type. Effort is expressed in (equivalent) calls to the respective reduced compression functions.

function	steps	type	effort	source	example
SHA-256	18	collision	2^0	[9]	yes
SHA-256	20	collision	$2^{1.58}$	[10]	no
SHA-256	21	collision	2^{15}	[10]	yes
SHA-256	22	collision	2^9	[14]	yes
SHA-256	**23**	collision	2^{18}	this work	yes
SHA-256	**24**	collision	$2^{28.5}$	this work	yes
SHA-512	**23**	collision	$2^{43.9}$	this work	yes
SHA-512	**24**	collision	$2^{53.0}$	this work	no
SHA-256	23	semi-free-start collision	2^{17}	[10]	yes
SHA-256	**24**	semi-free-start collision	2^{17}	this work	no
SHA-224	**25**	free-start collision	2^{17}	this work	no
SHA-256	22	free-start near-collision	2^0	[9]	yes
SHA-256	25	semi-free-start near-collision	2^{34}	[10]	yes
SHA-256	**31**	free-start near-collision	2^{32}, Table 6	this work	no

messages that hash to the same result. In a semi-free-start collision attack, the attacker is additionally allowed to choose the initial chaining value, but the same value should be used for both messages. In a free-start collision attack, a (small) difference may appear in the initial chaining value. Near-collision attacks relax the requirement that the hash results should be equal and allow for small differences.

The structure of this paper is as follows. We give a short description of SHA-256 in Sect. 2. Section 3 gives an alternative description of the semi-free-start collision attack by Nikolić and Biryukov [10], which will make the subsequent description of the new attacks easier to understand. We then discuss our collision attacks on 23- and 24-step SHA-256 in Sect. 4. In Sect. 5, we apply our results to step-reduced SHA-512. Finally, Sect. 6 concludes.

2   Description of SHA-256

This section gives a short description of the SHA-256 hash function, using the notation from Table 2. For a detailed specification, we refer to [11].

The compression function of SHA-256 consists of a message expansion, which transforms a 512-bit message block into 64 expanded message words W_i of 32 bits each, and a state update transformation. The latter updates eight 32-bit state variables $A, \ldots, H$ in 64 identical steps, each using one expanded message word. The message expansion can be defined recursively as follows.

$$W_i = \begin{cases} M_i & 0 \le i < 16 \\ \sigma_1(W_{i-2}) + W_{i-7} + \sigma_0(W_{i-15}) + W_{i-16} & 16 \le i < 64 \end{cases} . \quad (1)$$

Table 2. The notation used in this paper

$X \ggg s$	X rotated over s bits to the right
$X \gg s$	X shifted over s bits to the right
$\overline{X}$	One's complement of X
$X \oplus Y$	Bitwise exclusive OR of X and Y
$X + Y$	Addition of X and Y modulo 2^{32}
$X - Y$	Subtraction of X and Y modulo 2^{32}
$A_i, \cdots, H_i$	State variables at step i, for the first message
$A'_i, \cdots, H'_i$	Idem, for the second message
W_i	i-th expanded message word of the first message
W'_i	Idem, for the second message
δX	Additive difference in X, i.e., $X' - X$
$\delta\sigma_0(X)$	Additive difference in $\sigma_0(X)$, i.e., $\sigma_0(X') - \sigma_0(X)$

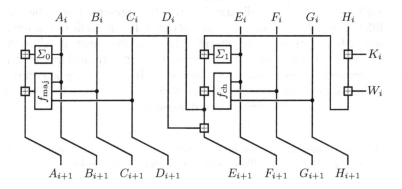

Fig. 1. The state update transformation of SHA-256

The functions $\sigma_0(X)$ and $\sigma_1(X)$ are given by

$$\sigma_0(X) = (X \ggg 7) \oplus (X \ggg 18) \oplus (X \gg 3) \ ,$$
$$\sigma_1(X) = (X \ggg 17) \oplus (X \ggg 19) \oplus (X \gg 10) \ . \tag{2}$$

The state update transformation updates two of the state variables in every step. It uses the bitwise Boolean functions f_{ch} and f_{maj} as well as the GF(2)-linear functions Σ_0 and Σ_1.

$$f_{\mathrm{ch}}(X,Y,Z) = XY \oplus \overline{X}Z \ ,$$
$$f_{\mathrm{maj}}(X,Y,Z) = XY \oplus YZ \oplus XZ \ ,$$
$$\Sigma_0(X) = (X \ggg 2) \oplus (X \ggg 13) \oplus (X \ggg 22) \ ,$$
$$\Sigma_1(X) = (X \ggg 6) \oplus (X \ggg 11) \oplus (X \ggg 25) \ . \tag{3}$$

Figure 1 describes the state update transformation, where K_i is a step constant. Equivalently, it is described by the following equations.

$$T_1 = H_i + \Sigma_1(E_i) + f_{\mathrm{ch}}(E_i, F_i, G_i) + K_i + W_i \ ,$$
$$T_2 = \Sigma_0(A_i) + f_{\mathrm{maj}}(A_i, B_i, C_i) \ ,$$
$$A_{i+1} = T_1 + T_2 \ , \quad B_{i+1} = A_i \ , \quad C_{i+1} = B_i \ , \quad D_{i+1} = C_i \ , \tag{4}$$
$$E_{i+1} = D_i + T_1 \ , \quad F_{i+1} = E_i \ , \quad G_{i+1} = F_i \ , \quad H_{i+1} = G_i \ .$$

After 64 steps, the initial state variables are fed forward using word-wise addition modulo 2^{32}.

3   Review of the Nikolić-Biryukov Semi-Free-Start Collision Attack

In this section, we review the 23-step semi-free-start collision attack by Nikolić and Biryukov [10]. The new results presented in this paper are extensions of this attack. The notations we use are given in Table 2.

The attack uses a nine step differential, which is presented in Table 3. All additive differences are fixed, as well as the actual values of some of the internal state variables. Fixing these values ensures that the differential is followed, as will be explained later. The constants α, β, γ and ϵ are determined by the attack. The first difference is inserted via the message word W_9. There are no differences in expanded message words other than those indicated in Table 3, i.e., only W_9, W_{10}, W_{11}, W_{12}, W_{16} and W_{17} can have a difference.

Table 3. A 9 step differential, using additive differences (left) and conditions on the value (right). Blanks denote zero differences resp. unconstrained values.

step	δA	δB	δC	δD	δE	δF	δG	δH	δW	A	B	C	D	E	F	G	H
8										α				γ			
9									1	α	α			$\gamma+1$	γ		
10	1			1					-1	-1	α	α		-1	$\gamma+1$	γ	
11		1		-1	1				δ_1	α	-1	α	α	ϵ	-1	$\gamma+1$	γ
12			1		-1	1			δ_2	α	α	-1	α	β	ϵ	-1	$\gamma+1$
13				1		-1	1			α	α	α	-1	β	β	ϵ	-1
14					1		-1				α	α	α	-1	β	β	ϵ
15						1						α	α	0	-1	β	β
16							1		1				α	-2	0	-1	β
17								1	-1					-2	0	-1	
18															-2	0	

The attack algorithm consists of two phases. The first phase finds suitable values for the constants α, β, γ and ϵ as well as two expanded message words, W_{16} and W_{17}. A detailed description of this phase of the attack will be given in Sect. 3.2, as it is more instructive to describe the second phase first.

3.1   The Second Phase of the Attack

The second phase of the attack finds, when given suitable values for α, β, γ, ϵ, W_{16} and W_{17}, a pair of messages and a set of initial values that lead to a semi-free-start collision for 23 steps of SHA-256. It works by carefully fixing the internal state at step 11 as indicated in Table 3, and then computing forward and backward. At each step, the expanded message word W_i is computed such that the differential from Table 3 is followed. During this, four extra conditions appear, involving only the constants determined by the first phase of the attack.

$$\sigma_1 \left(W_{16} + 1\right) - \sigma_1 \left(W_{16}\right) - \Sigma_1 \left(\epsilon - 1\right) + \Sigma_1 \left(\epsilon\right)$$
$$- f_{ch} \left(\epsilon - 1, 0, \gamma + 1\right) + f_{ch} \left(\epsilon, -1, \gamma + 1\right) = 0 . \quad (5)$$

$$\sigma_1 \left(W_{17} - 1\right) - \sigma_1 \left(W_{17}\right) - f_{ch} \left(\beta, \epsilon - 1, 0\right) + f_{ch} \left(\beta, \epsilon, -1\right) = 0 . \quad (6)$$

$$\beta = \alpha - \Sigma_0 \left(\alpha\right) . \quad (7)$$

$$f_{ch} \left(\beta, \beta, \epsilon - 1\right) - f_{ch} \left(\beta, \beta, \epsilon\right) = -1 . \quad (8)$$

The first phase guarantees that the constants are such that these conditions are satisfied. The second phase of the attack has a negligible complexity and is guaranteed to succeed. Since there is still a lot of freedom left, many 23-step semi-free-start collisions can be found, with only a negligible additional effort, by repeating this second phase several times. A detailed description of this phase, including the origins of (5)–(8), is given in Appendix A.

3.2   The First Phase of the Attack

The goal of the first phase of the attack is to determine suitable values for the constants α, β, γ and ϵ, as well as two expanded message words, W_{16} and W_{17}. Suitable values imply that the four conditions (5)–(8) are satisfied. Nikolić and Biryukov [10] do not give much detail on this procedure, hence we clarify it below.

1. Make a random choice for γ and ϵ and search for a value of W_{16} such that condition (5) is satisfied. This condition is of the form $\sigma_1 \left(x + 1\right) - \sigma_1 \left(x\right) = \delta$. There exists a simple, generic method to solve equations of this form, which is described in Appendix B. We note however that for this particular case, a faster method exists. An exhaustive search over every possible value of x resulted in the observation that only 6 181 additive differences δ can ever be achieved. These can be stored in a lookup table, together with one or more solutions for each difference. Hence, solving an equation of this form can be done with a simple table lookup.

    If no solution exists, simply retry with different choices for γ and/or ϵ. If the right hand side difference δ is selected uniformly at random, the probability that the equation has a solution is $2^{-19.5}$, so we expect to have to repeat this step about $2^{19.5}$ times.

2. Make a random choice for α, and compute β using (7). Now check condition (8). As described in [10], this equation is satisfied if the bits of β are zero in the positions where the bits of $\epsilon - 1$ and ϵ differ. This occurs with a probability of approximately $1/3$, so this condition is fairly easy to satisfy.

3. The last condition, (6), is of the same form as the first condition, so it can be solved in exactly the same way. The expected probability that a solution exists is again $2^{-19.5}$.

Note that, because not all conditions depend on all of the constants determined in this phase of the attack, the first condition can be treated independently of the last three. Thus, the first and last step of this phase of the attack are executed about $2^{19.5}$ times and the second step about 2^{21} times. One of these steps requires much less work than an evaluation of the compression function of (reduced) SHA-256 — a bit less than one step. Hence, the overall time complexity of the entire attack, when expressed in SHA-256 compression function evaluations, is below 2^{17}.

4   Our Collision Attacks on Step-Reduced SHA-256

In this section we describe a novel, practical collision attack on SHA-256, reduced to 23 steps. It has a time complexity of about 2^{18} evaluations of the reduced SHA-256 compression function. We also extend this to 24 steps of SHA-256, with an expected time complexity of $2^{28.5}$ compression function evaluations.

4.1   23-Step Collision

Our collision attack for SHA-256, reduced to 23 steps, consists of two parts. First, we construct a semi-free-start collision for 23 steps, based on the attack from Sect. 3. Then we transform this semi-free-start collision into a real collision.

Finding "Good" Constants. Finding a 23 step semi-free-start collision is done using the same attack as described in Sect. 3, with a slight change to the first phase. In Sect. 3.2, it was described how to find constants α, β, γ and ϵ such that there exist values for W_{16} and W_{17} ensuring that the conditions (5) and (6) are satisfied. There are still some degrees of freedom left in this process. Indeed, it is possible to determine the constants α, β, γ and ϵ such that there are *many* values for W_{16} and W_{17} satisfying (5) and (6).

We performed an exhaustive search for such good constants. Condition (5) depends only on ϵ and γ. An exhaustive search for this condition can be performed with approximately 2^{37} evaluations of (5), because for each value of ϵ, only some of the bits in γ can have an influence. We found several values for ϵ and γ for which more than 2^{29} choices for W_{16} ensure that (5) is satisfied, for instance

$$\gamma = \texttt{0000017c}_x \ , \quad \epsilon = \texttt{7f5f7200}_x \ . \tag{9}$$

Conditions (6) and (8) depend on ϵ and β, which in turn depends on α through (7). An interesting property is that condition (6) becomes independent of ϵ if we assume that condition (8) is satisfied. Indeed, since this assumption implies that the bits of β are zero where ϵ and $\epsilon - 1$ differ, (6) reduces to

$$\sigma_1 \left(W'_{17} + 1 \right) - \sigma_1 \left(W'_{17} \right) = \overline{\beta} \ . \tag{10}$$

Because of this, an exhaustive search for good values of α and β is feasible. There are many of the optimal values for α and β which are consistent with (several of) the optimal values for ϵ, thus yielding a global optimum. For instance, with γ and ϵ as in (9), the following values for α and β are one of many optimal choices:

$$\alpha = \mathtt{00b321e3}_x \ , \quad \beta = \mathtt{fcffe000}_x \ . \tag{11}$$

There are 2^{16} possible choices for W_{17} which satisfy (6) with these constants. Thus, these values for α, β, γ and ϵ give us an additional freedom of 2^{45} in the choice of W_{16} and W_{17}. This phase can be considered a precomputation, or alternatively, one can reduce the effort spent in this phase by only searching a smaller part of the available search space, which likely leads to less optimal results. It may however be a worthwhile trade-off in practice.

Transforming into a Collision. Note that only 7 expanded message words, W_{11} until W_{17}, are actually fixed to a certain value when constructing a semi-free-start collision, ignoring the freedom left in W_{16} and W_{17} for now. The others are chosen arbitrarily or computed from the message expansion when necessary. Using this freedom, it is possible to construct many semi-free-start collisions with only a negligible additional effort. But it is also possible to use this freedom in a controlled manner to transform the semi-free-start collision into a real collision.

To this end, we first introduce an alternative description of SHA-256. In older variants of the same design strategy, like MD5 or SHA-1, only a single state variable is updated in every step. This naturally leads to a description where only the first state variable is considered. Something similar can be done with the SHA-2 hash functions, even though in the standard description, two state variables are updated in every step.

From the state update equations (4), we derive a series of equations expressing the inputs of the i-th state update transformation, $A_i, \ldots, H_i$, as a function of only A_i through A_{i-7}.

$$
\begin{aligned}
A_i &= A_i \ , \quad B_i = A_{i-1} \ , \quad C_i = A_{i-2} \ , \quad D_i = A_{i-3} \ , \\
E_i &= A_{i-4} + A_i \quad -\Sigma_0(A_{i-1}) - f_{\mathrm{maj}}(A_{i-1}, A_{i-2}, A_{i-3}) \ , \\
F_i &= A_{i-5} + A_{i-1} -\Sigma_0(A_{i-2}) - f_{\mathrm{maj}}(A_{i-2}, A_{i-3}, A_{i-4}) \ , \\
G_i &= A_{i-6} + A_{i-2} -\Sigma_0(A_{i-3}) - f_{\mathrm{maj}}(A_{i-3}, A_{i-4}, A_{i-5}) \ , \\
H_i &= A_{i-7} + A_{i-3} -\Sigma_0(A_{i-4}) - f_{\mathrm{maj}}(A_{i-4}, A_{i-5}, A_{i-6}) \ .
\end{aligned}
\tag{12}
$$

Substituting these into (4) yields an alternative description requiring only a single state variable. This description can be written concisely as

$$A_{i+1} = F \left(A_i, A_{i-1}, A_{i-2}, A_{i-3}, A_{i-4}, A_{i-5}, A_{i-6} \right) + A_{i-7} + W_i \ . \tag{13}$$

The function $F(\cdot)$ encapsulates (4) and (12), except for the addition of the expanded message word W_i and the state variable A_{i-7}. From (12), it is clear that one can easily transform an internal state in the standard description, $\langle A_i, \cdots, H_i \rangle$, to the corresponding internal state in the alternative description, $\langle A_i, \cdots, A_{i-7} \rangle$, and vice versa. Analogous to what is done for MD5 and SHA-1, the initial values can be redefined as $A_{-7}, \cdots, A_0$.

This alternative description of SHA-256 can be used to transform a 23 step semi-free-start collision for SHA-256 into a real collision. Since control over one expanded message word W_i gives full control over one state variable A_{i+1}, control over eight consecutive expanded message words gives full control over the entire internal state.

1. Start from a 23-step semi-free-start collision pair. Set $\langle A_0, \cdots, A_{-7} \rangle$ to the SHA-256 initial values, in the alternative description. Make arbitrary choices for W_0, W_1 and W_2, and recompute the first three steps.
2. The eight message words W_3 until W_{10} are now modified such that A_4 until A_{11} remain unchanged. This implies that the internal state at step 11, $\langle A_{11}, \cdots, H_{11} \rangle$ does not change, and thus we connect to the rest of the semi-free-start collision. More specifically, for every step i, $3 \le i \le 10$, the new value of the i-th message word is computed as

$$W_i = A_{i+1} - F\left(A_i, A_{i-1}, A_{i-2}, A_{i-3}, A_{i-4}, A_{i-5}, A_{i-6}\right) - A_{i-7} \ . \qquad (14)$$

   In the message words W_9 and W_{10} there is an additive difference of 1 and -1, respectively. This does not pose a problem since the construction of the semi-free-start condition guarantees that these will have the intended effect, regardless of the values of W_9 and W_{10}, see Appendix A.
3. Now we need to verify again if conditions (5) and (6) are still satisfied, since they depend on W_{16} and W_{17}, which may have changed. If the conditions are not satisfied, simply restart and make different choices for W_0, W_1 and/or W_2.

   Recall however that we have spent extra effort in the first phase of the attack to choose the constants α, β, γ and ϵ such that there are *many* values for W_{16} and W_{17} that satisfy the conditions. For the constants given in (9) and (11), there are 2^{45} allowed values for these two expanded message words. This translates into a probability of 2^{-19} that the conditions (5) and (6) are indeed still satisfied. We hence expect to have to repeat this procedure about 2^{19} times. Every trial requires an effort equivalent to about 10 steps of SHA-256.
4. After a successful modification of the first message words, the expanded message words W_{18} until W_{22} need to be recomputed, and also the corresponding steps need to be redone. The construction of the semi-free-start collision still guarantees that no differences will be introduced.

If we consider the first phase to be a precomputation, the overall attack complexity is about 2^{18} evaluations of the compression function of SHA-256 reduced to 23 steps. An example collision pair for 23-step reduced SHA-256 is given in Table 4.

Table 4. Example colliding message pair for 23-step reduced SHA-256

M	29f1ebfb 4468041a 1e6565b6 4cc17e75 4ea4f993 33a77104 864a828d 1dcec3d2
	d33d7b02 bcd4a2d7 3b10201d 39953548 8e127f2b 0304fc01 e7118577 43b12ca7
M'	29f1ebfb 4468041a 1e6565b6 4cc17e75 4ea4f993 33a77104 864a828d 1dcec3d2
	d33d7b02 bcd4a2d8 3b10201c 3995d548 91129f2a 0304fc01 e7118577 43b12ca7
H	c77405ea 8bfe2016 ff0531b6 a89b81f6 e98cf052 491a6c62 fd009a40 3969dc83

4.2   24-Step Collision

The same approach can be extended to 24 steps of SHA-256, using the 24-step semi-free-start collision attack given in detail in Sect. 4.3. Simply put, the 23-step attack is simply shifted down by a single step, and no difference is introduced into W_0 by the message expansion in the backward direction.

When turning the semi-free-start collision into a collision, however, the value of the expanded message word W_{16} (which was the non-expanded message word W_{15} in the 23-step attack) should not change. In a straightforward extension of the 23-step collision attack to 24 steps, this extra condition would only be satisfied with a probability of 2^{-32}. Using the available freedom in a better way, this can be improved substantially.

1. Start from a 24-step semi-free-start collision pair. Set $\langle A_0, \cdots, A_{-7} \rangle$ to the SHA-256 initial values. Make an arbitrary choice for W_0 and recompute the first step. Now, it follows from (4) that $(A_2 - W_1)$ is a constant:

$$c_1 = A_2 - W_1 . \tag{15}$$

2. The new value of W_9 is determined from (14), $i.e.$, it depends on A_2 through A_{10}. The state variables A_5 through A_{10} have already been fixed in the semi-free-start collision. If we additionally fix A_4 and A_3 to arbitrary values, it is possible to compute the sum of W_9 and A_2,

$$c_2 = W_9 + A_2 = A_{10} - F(A_9, \cdots, A_3) . \tag{16}$$

3. Combining (1) and (15)–(16), results in

$$W_{16} - \sigma_1(W_{14}) - c_2 + c_1 - W_0 = \sigma_0(W_1) - W_1 . \tag{17}$$

It is easy to find a suitable value for W_1 that ensures that W_{16} has the proper value, if it exists. It suffices to guess the 15 least significant bits of W_1 to compute all 32 bits of W_1, satisfying the above condition with probability 2^{-14}. A conservative estimate is that each trial requires an effort equivalent to one step update of SHA-256.

4. Now all the internal state variables have been fixed. The corresponding message words can be found from (14) and the message expansion. Just as in the 23-step collision attack, however, there are still some conditions left. As explained in Sect. 4.1, these are satisfied with a probability of 2^{-19}.

Table 5. Example colliding message pair for 24-step reduced SHA-256

M	0187e08e 865cedaf 5b69e21a e0f7485e 50b98993 217e4650 51e3cf65 c2997c68
	2c267e16 82ffa4e9 37b5af09 5b28721d 1be35597 7ff22aa1 e807a758 c1519aaa
M'	0187e08e 865cedaf 5b69e21a e0f7485e 50b98993 217e4650 51e3cf65 c2997c68
	2c267e16 82ffa4e9 37b5af0a 5b28721c 1be3f597 82f24aa0 e807a758 c1519aaa
H	1584074c 8b810a94 01ea31b1 81bffd02 d29c817d e4e04b51 b9f5ac4f 6b34d1f8

Hence, the overall expected time complexity is equivalent to about $2^{19} \cdot (2^{14} + 10)$ SHA-256 step computations, or about $2^{28.5}$ evaluations of the SHA-256 compression function reduced to 24 steps. An example collision pair for 24-step reduced SHA-256 is given in Table 5. An extension of this attack method beyond 24 steps fails, because then a difference in the first or in the last message word becomes unavoidable. In [14], another differential than the one shown in Table 3 is used to find 22-step collisions for SHA-256. We tried to use this differential in our extended attacks, but even for 23 steps, using this differential fails.

4.3   Further Extensions

This section discusses further extensions using weaker attack models. The starting point is the 23-step semi-free-start collision attack of Nikolić and Biryukov [10], which was described in Sect. 3.

Semi-Free-Start Collisions for 24 Steps of SHA-256. We keep the entire attack algorithm from Sect. 3 unchanged, but shift everything down by a single step. Because of this, one more message word, W_0, needs to be computed from the message expansion in the reverse direction. From (1), it follows that the additive difference in this word is

$$\delta W_0 = \delta W_{16} - \delta \sigma_1 (W_{14}) - \delta W_9 - \delta \sigma_0 (W_1) \ . \tag{18}$$

None of these expanded message words has a difference, so also $\delta W_0 = 0$. This yields 24-step semi-free-start collisions of SHA-256 with the same complexity of 2^{17} compression function evaluations.

Free-Start Collisions for 25 Steps of SHA-224. SHA-224 differs from SHA-256 in two ways. First, it has different initial values, and second, the output is truncated to the leftmost 224 bits. We can thus extend the 24-step semi-free-start collision of SHA-256 to a 25-step free-start collision of SHA-224 by simply shifting the same attack down one more step. Now a difference will inevitably appear in W_0, which propagates to the initial value H_0. The other initial values, A_0 through G_0 still have a zero difference. Because the word H is truncated away in SHA-224, this results in free-start collisions for 25 steps of SHA-224, with the same complexity. Note that this attack would not apply if a different method of truncation would have been chosen in the design of SHA-224.

Table 6. Experimental results of the free-start near-collision attack on SHA-256. For each number of steps, only the combination of forward/backward steps that gave the best results is shown. For comparison, the expected numbers of solutions for a generic birthday attack with an equal effort are also given.

steps	fwd.	bwd.	k_{min}	2-logarithm of the number of solutions with k							
				≤ 8	≤ 16	≤ 24	≤ 32	≤ 40	≤ 48	≤ 56	≤ 64
25	1	1	2	31.95	32.00	32.00	32.00	32.00	32.00	32.00	32.00
26	1	2	8	24.17	31.55	31.99	32.00	32.00	32.00	32.00	32.00
27	1	3	11	$-\infty$	15.41	26.20	30.65	31.89	32.00	32.00	32.00
28	1	4	18	$-\infty$	$-\infty$	8.77	20.41	27.24	30.63	31.80	31.99
29	1	5	32	$-\infty$	$-\infty$	$-\infty$	1.58	14.31	22.86	28.19	30.93
30	1	6	43	$-\infty$	$-\infty$	$-\infty$	$-\infty$	$-\infty$	10.73	19.58	25.68
31	2	6	53	$-\infty$	$-\infty$	$-\infty$	$-\infty$	$-\infty$	$-\infty$	6.34	15.50
Birthday Attack			57	-143.41	-108.84	-80.49	-56.36	-35.51	-17.37	-1.57	12.14

Free-Start Near-Collisions of SHA-256. Extending the attack to more steps is possible, provided that some differences are allowed both in the initial value and in the hash result, *i.e.*, when considering free-start near-collisions. The starting point is again the 23-step semi-free-start collision attack from Sect. 3. It is extended by adding a number of extra forward and backward steps.

As explained above, no difference is introduced in the first backward step. Note that, in general, the diffusion of differences is slower in the backward direction than in the forward direction. A difference introduced in an expanded message word W_i affects both A_{i+1} and E_{i+1} in the forward direction, as opposed to only H_i in the backward direction. Thus, in the forward direction, all state words can be affected by a single difference in an expanded message word after only four rounds. In the backward direction, this takes eight rounds.

We have done several experiments, each equivalent to an effort of 2^{32} reduced SHA-256 compression function evaluations. The results of our experiments are summarised in Table 6. The first three columns give the total number of steps, the number of extra forward and extra backward steps, respectively. The fourth column gives k_{min}, the smallest Hamming distance found. The last eight columns contain the 2-logarithm of the number of solutions with a Hamming distance k of at most 8, 16, ..., 64 bits.

For comparison, also the expected values for a generic birthday attack with an equal effort of 2^{32} is given. For a generic (free-start) near-collision attack on an ideal n-bit hash function, using the birthday paradox with an effort of 2^w compression function evaluations, the lowest expected Hamming distance is the lowest k for which

$$2^{2w} \cdot \sum_{i=0}^{k} 2^{-n} \binom{n}{i} \geq 1 . \tag{19}$$

For instance, with $w = 32$ and for SHA-256 (*i.e.*, $n = 256$), this gives $k = 57$ bits. Our attack performs significantly better for up to 30 steps of SHA-256.

For 31 steps, we still found 208 free-start near-collisions with a Hamming distance of at most 57 bits, whereas a birthday attack is only expected to find one with the same effort.

5   Collision Attacks on Step-Reduced SHA-512

SHA-512 is a 512-bit hash function from the SHA-2 family. Its structure is very similar to SHA-256. The sizes of all words are increased to 64 bits and the number of rounds is increased to 80. It uses a different initial chaining value, and different step constants. Finally, the GF(2)-linear functions are redefined. Refer to [11] for details on SHA-512. In this section, we extend the collision attacks on SHA-256 that were described in Sect. 4.1 and 4.2 to SHA-512. The first phase of the attacks needs to be adapted, since an exhaustive search as in Sect 3.2 is no longer feasible.

Finding "Good" Constants for SHA-512. Recall from Sect. 3.2 that the goal of the first phase of the attack is to find values for the constants α, β, γ, ϵ such that the conditions (5)–(8) are satisfied for many values of the expanded message words W_{16}, W_{17}. Since an exhaustive search for good constants is infeasible, we suggest the following approach.

1. First, make a list L of additive differences δ for which the equation

$$\sigma_1 (x + 1) - \sigma_1 (x) = \delta \qquad (20)$$

   has many solutions x. This can be accomplished by picking several values for x at random and computing the corresponding δ's. This procedure is likely to quickly find the "good" values for δ, since the more x's correspond to a δ, the more likely we are to find it. Using Appendix B, the number of solutions x for a given δ can be counted efficiently.

2. Since all conditions (5)–(8) will need to be satisfied, we can use (10) instead of (6). Hence, $\overline{\beta}$ should preferably be one of the "good" δ's from the list L. Knowing the value of β, we need to invert (7) to find α. This can, for instance, be done by guessing the 36 most significant bits of α and determining the other bits using (7). A guess succeeds with a probability of about 2^{-36}. Note that (7) cannot necessarily be inverted for all β's.

3. Now we make an arbitrary choice for ϵ which satisfies (8). Denote by l_β the length of the run of least significant "0"-bits in β. Then, (8) is satisfied if and only if the least significant "1"-bit of ϵ lies within the l_β least significant bits. Unfortunately, for SHA-512, this condition eliminates the best values for β.

4. If we choose a "good" value for $\sigma_1 (W_{16} + 1) - \sigma_1 (W_{16})$ from the list L, and since ϵ has already been chosen, (5) can be rewritten as

$$C - f_{\text{ch}} (\epsilon - 1, 0, \gamma + 1) + f_{\text{ch}} (\epsilon, -1, \gamma + 1) = 0 , \qquad (21)$$

   where C is a known constant. The bits in which ϵ and $(\epsilon - 1)$ differ can be corrected by a proper choice of γ. Hence it is advantageous to choose ϵ with

Table 7. Example colliding message pair for 23-step reduced SHA-512

M	0000000017daf2ec 000000004b7adc8e 000000000d01f49d 54cce0ac731eb4c9
	5caf52c6f3e941cd 0224e6b804216305 95bbdc5df5b491c8 9f7f1453e39ee6c0
	3e345efecc818058 93dfcee7a268ce69 90561054da994c54 7262751c31b5bdd0
	54b1d56610b9e802 7f201dfcfce968c0 2b90cc3824ee5f13 05cfd16a7b4c4ab1
M'	0000000017daf2ec 000000004b7adc8e 000000000d01f49d 54cce0ac731eb4c9
	5caf52c6f3e941cd 0224e6b804216305 95bbdc5df5b491c8 9f7f1453e39ee6c0
	3e345efecc818058 93dfcee7a268ce6a 90561054da994c53 7266551c31b5bd18
	54b0b56610b9e801 7f201dfcfce968c0 2b90cc3824ee5f13 05cfd16a7b4c4ab1
H	dd44d89f178803f5 136802b223c880ba bbb80917dda6a3e7 be1f118889bd5415
	98adc37a0f32d151 83d35099922ee2c6 670ac37463f224da e0835506fb66503d

a long run of least significant "0"-bits. This again constrains β, as explained above. If no choice for γ can satisfy (21), retry with a different choice for ϵ and/or β.

Unlike the exhaustive search in Sect. 3.2, this procedure does not guarantee to find the optimal solution. However, experiments show that we can quickly find many good solutions. We found many values for the constants α, β, γ and ϵ for which the conditions (5) and (8) are satisfied for $2^{49.1}$ and 2^{34} values for W_{16} and W_{17}, respectively. Example values are

$$\alpha = 3891fd20b54a8eb9_x \ , \quad \beta = 0001200000000000_x \ , $$
$$\gamma = 00000fff7f7fff46_x \ , \quad \epsilon = 0000100000000000_x \ . \tag{22}$$

23-step Collision. The second phase of the 23-step attack from Sect. 4.1 can directly be applied to SHA-512. With the constants from (22), a single attempt to turn a 23-step semi-free-start collision into a 23-step collision will succeed with an expected probability of $2^{-44.9}$ and costs about half of a reduced SHA-512 compression function evaluation. Hence, this results in a collision attack on 23-step SHA-512 with an expected time complexity of $2^{43.9}$ reduced compression function evaluations. An example collision pair for 23-step reduced SHA-512 is given in Table 7.

24-Step Collision. Also the second phase of the 24-step attack from Sect. 4.2 can be applied to SHA-512. One slight modification is required when determining a suitable value for W_0, due to the redefinition of the σ_0-function in SHA-512. Guessing the 8 least significant bits of W_0 allows to compute all of W_0, satisfying (17) with probability 2^{-8}. This results in a collision attack on 24-step SHA-512 with an expected time complexity of $2^{53.0}$ reduced compression function evaluations.

Further Extensions. The attacks on SHA-512 can also be extended, much like the extensions described for the SHA-256 attacks in Sect. 4.3. Adding more rounds trivially leads to several (semi-) free-start (near-) collision attacks. One noteworthy case is a free-start collision attack on 26 steps of SHA-384. It is analogous to the 25-step free-start collision attack on SHA-224 from Sect. 4.3, but as two words are truncated away in the case of SHA-384, the attack extends to 26 steps.

6   Conclusion

Our results push the limit for cryptanalysis of step reduced but otherwise unmodified SHA-256; we found practical collisions for up to 24 steps. For almost half of the steps (31 out of 64) non-random properties of the compression function are detectable in practice. The results also apply to SHA-512, albeit with higher time complexities.

Acknowledgements

The authors would like to thank Vincent Rijmen for his advice. We acknowledge the use of the VIC computer cluster of K.U.Leuven, which was used to obtain most of the experimental results presented in this paper. Finally, thanks to Ralph Wernsdorf for bringing [14] to our attention.

References

1. Chabaud, F., Joux, A.: Differential collisions in SHA-0. In: Krawczyk, H. (ed.) CRYPTO 1998. LNCS, vol. 1462, pp. 56–71. Springer, Heidelberg (1998)
2. Gilbert, H., Handschuh, H.: Security Analysis of SHA-256 and Sisters. In: Matsui, M., Zuccherato, R.J. (eds.) SAC 2003. LNCS, vol. 3006, pp. 175–193. Springer, Heidelberg (2004)
3. Hawkes, P., Paddon, M., Rose, G.G.: On corrective patterns for the SHA-2 family. Cryptology ePrint Archive, Report 2004/2007 (August 2004) http://eprint.iacr.org/
4. Hölbl, M., Rechberger, C., Welzer, T.: Searching for messages conforming to arbitrary sets of conditions in SHA-256. In: Lucks, S., Sadeghi, A.-R., Wolf, C. (eds.) WEWoRC 2007. LNCS, vol. 4945, pp. 28–38. Springer, Heidelberg (2008)
5. Lai, X., Massey, J.L.: Hash functions based on block ciphers. In: Rueppel, R.A. (ed.) EUROCRYPT 1992. LNCS, vol. 658, pp. 55–70. Springer, Heidelberg (1993)
6. Lipmaa, H., Moriai, S.: Efficient algorithms for computing differential properties of addition. In: Matsui, M. (ed.) FSE 2001. LNCS, vol. 2355, pp. 336–350. Springer, Heidelberg (2002)
7. Lipmaa, H., Wallén, J., Dumas, P.: On the additive differential probability of exclusive-or. In: Roy, B., Meier, W. (eds.) FSE 2004. LNCS, vol. 3017, pp. 317–331. Springer, Heidelberg (2004)
8. Matusiewicz, K., Pieprzyk, J., Pramstaller, N., Rechberger, C., Rijmen, V.: Analysis of simplified variants of SHA-256. In: Proceedings of WEWoRC 2005, LNI P-74, pp. 123–134 (2005)
9. Mendel, F., Pramstaller, N., Rechberger, C., Rijmen, V.: Analysis of step-reduced SHA-256. In: Robshaw, M.J.B. (ed.) FSE 2006. LNCS, vol. 4047, pp. 126–143. Springer, Heidelberg (2006)

10. Nikolić, I., Biryukov, A.: Collisions for step-reduced SHA-256. In: Nyberg, K. (ed.) FSE 2008. LNCS, vol. 5086, pp. 1–15. Springer, Heidelberg (2008)
11. National Institute of Standards and Technology (NIST). FIPS-180-2: Secure Hash Standard (August 2002), http://www.itl.nist.gov/fipspubs/
12. Pramstaller, N., Rechberger, C., Rijmen, V.: Preliminary Analysis of the SHA-256 Message Expansion. In: NIST - First Cryptographic Hash Workshop, October 31-November 1 (2005)
13. Sanadhya, S.K., Sarkar, P.: New local collisions for the SHA-2 hash family. In: Nam, K.-H., Rhee, G. (eds.) ICISC 2007. LNCS, vol. 4817, pp. 193–205. Springer, Heidelberg (2007)
14. Sanadhya, S.K., Sarkar, P.: 22-step collisions for SHA-2. arXiv e-print archive, arXiv:0803.1220v1 (March 2008), http://de.arxiv.org/abs/0803.1220
15. Sanadhya, S.K., Sarkar, P.: Attacking reduced round SHA-256. In: Bellovin, S.M., Gennaro, R., Keromytis, A.D., Yung, M. (eds.) ACNS 2008. LNCS, vol. 5037, pp. 130–143. Springer, Heidelberg (2008)
16. Sanadhya, S.K., Sarkar, P.: Non-linear reduced round attacks against SHA-2 hash family. In: Mu, Y., Susilo, W., Seberry, J. (eds.) ACISP 2008. LNCS, vol. 5107, pp. 254–266. Springer, Heidelberg (2008)
17. Yoshida, H., Biryukov, A.: Analysis of a SHA-256 variant. In: Preneel, B., Tavares, S. (eds.) SAC 2005. LNCS, vol. 3897, pp. 245–260. Springer, Heidelberg (2006)

A   Detailed Description of the Second Phase of the Nikolić-Biryukov Attack

This appendix gives a detailed description of the second phase of the Nikolić-Biryukov attack [10]. When given suitable values for α, β, γ, ϵ, W_{16} and W_{17} by the first phase, as described in Sect. 3.2, it constructs a pair of messages and a set of initial values that lead to a semi-free-start collision for 23 steps of SHA-256.

1. Start at step 11 by fixing the state variables in this step, $A_{11}, \cdots, H_{11}$ as indicated in Table 3. The constants α, β, γ and ϵ are given by the first phase of the attack.
2. Calculate W_{11} such that $A_{12} = \alpha$ and W'_{11} such that $A'_{12} = \alpha$. Now $E_{12} = \beta$ only depends on α, and we find condition (7) from Sect. 3.1.

$$E_{12} = \alpha - \Sigma_0(\alpha) = \beta \ . \tag{23}$$

3. In a similar way, calculate W_{12} such that $E_{13} = \beta$ and W'_{12} such that $E'_{13} = \beta$. This also guarantees that $A_{13} = A'_{13}$ because the majority function absorbs the difference in C_{12}.
4. Calculate W_{13} such that $E_{14} = -1$ and set $W'_{13} = W_{13}$. Now, see Table 3, δE_{14} should be equal to 1. This yields the condition

$$\delta E_{14} = f_{\mathrm{ch}}(\beta, \beta, \epsilon - 1) - f_{\mathrm{ch}}(\beta, \beta, \epsilon) + 2 = 1 \ . \tag{24}$$

It was given before as (8), and is satisfied by the first phase of the attack. Note that this also ensures that $\delta A_{14} = 0$.

5. Calculate W_{14} such that $E_{15} = 0$ and set $W'_{14} = W_{14}$. Since the values of E_{14} and E'_{14} were chosen in the previous step to be fixed points of the function Σ_1, $\delta\Sigma_1 (E_{14}) = \delta E_{14} = 1$ cancels with $\delta H_{14} = -1$. Also, f_{ch} absorbs the difference in E_{14}, so no new differences are introduced.
6. Calculate W_{15} such that $E_{16} = -2$ and set $W'_{15} = W_{15}$. The difference in F_{15} is absorbed by f_{ch}.
7. The value for W_{16} is computed in phase one of the attack. The difference $\delta W_{16} = 1$ is cancelled by the output of f_{ch}. Indeed, since the binary representation of $E_{16} = -2$ is $111 \cdots 10_b$, the f_{ch} function passes only the difference in the least significant bit.
8. Also the value for W_{17} is computed in phase one of the attack. The difference $\delta W_{17} = -1$ cancels with $\delta H_{17} = 1$, thereby eliminating the final difference in the state variables. Thus, a collision is reached.
9. Now, go back to step 11 and proceed in the backward direction. Make an arbitrary choice for W_{10}. The differential from Table 3 is followed because of the careful choice of the state variables in step 11.
10. Make an arbitrary choice for W_9, and proceed one step backward. The difference $\delta W_9 = 1$ cancels with δA_{10} and with δE_{10} such that there is a zero difference in the state variables A_9 through H_9. Now randomly choose W_8 down to W_2 and calculate backward. Because no new differences appear in these expanded message words, there is also a zero difference in the state variables A_2 through H_2.
11. It is not possible to freely choose W_0 or W_1 as 16 expanded message words have already been chosen, i.e., W_2 until W_{17}. Hence, these are computed using the message expansion in the backward direction. Although some of the message words used to compute W_0 and W_1 have differences, these differences always cancel out.
12. Continuing forward from step 18 again, note that the collision is preserved as long as no new differences are introduced via the expanded message words. From the message expansion, it follows that

$$\delta W_{18} = \sigma_1 (W_{16} + 1) - \sigma_1 (W_{16}) - \Sigma_1 (\epsilon - 1) + \Sigma_1 (\epsilon)$$
$$- f_{ch} (\epsilon - 1, 0, \gamma + 1) + f_{ch} (\epsilon, -1, \gamma + 1) = 0 . \quad (25)$$

This is condition (5), which is satisfied by the first phase of the attack.
13. Similarly, in step 19, we require that $\delta W_{19} = 0$, which results in

$$\sigma_1 (W_{17} - 1) - \sigma_1 (W_{17}) - f_{ch} (\beta, \epsilon - 1, 0) + f_{ch} (\beta, \epsilon, -1) = 0 . \quad (26)$$

This condition was given in (6), and is also satisfied by the first phase of the attack.
14. In steps 20–22, the message expansion guarantees that no new differences are introduced. In step 23, however, a difference of 1 is impossible to avoid, hence the attack stops after 23 steps.

Every step in this procedure is guaranteed to succeed, provided that the first phase of the attack supplied suitable constants. Thus, the complexity of the second phase of the attack is negligible. Since there is still a lot of freedom left, many 23-step semi-free-start collisions can be found, with only a negligible additional effort, by repeating this second phase several times.

B   Solving $\mathcal{L}(x + \delta) = \mathcal{L}(x) + \delta'$

This appendix describes a generic method to solve equations of the form $\mathcal{L}(x + \delta) = \mathcal{L}(x) + \delta'$ where δ and δ' are given n-bit additive differences, and $\mathcal{L}$ is an n-bit to n-bit GF(2)-linear transformation. This is similar to the problems studied by Lipmaa and Moriai [6] and Lipmaa et $al.$ [7].

Consider the modular addition $x + \delta$ and let $\Delta = (x + \delta) \oplus x$. This addition is described by the following equations, where x_i is the i-th bit of x and the c_i's are the carry bits:

$$
\begin{array}{llll}
(x + \delta)_i = x_i \oplus \delta_i \oplus c_i & & c_i = \delta_i \oplus \Delta_i & \\
c_{i+1} = f_{\mathrm{maj}}(x_i, \delta_i, c_i) & \Leftrightarrow & c_{i+1} = f_{\mathrm{maj}}(x_i, \delta_i, \delta_i \oplus \Delta_i) & . \quad (27) \\
c_0 = 0 & & c_0 = 0 &
\end{array}
$$

Hence, once we fix both the additive difference δ and the XOR difference Δ, all the carries c_i are fixed. Some of the x_i's are also fixed: when $\Delta_i = 1$ and $i < n - 1$, it must hold that $x_i = c_{i+1} = \delta_{i+1} \oplus \Delta_{i+1}$. The other x_i's can be chosen arbitrarily. Thus, the allowed values for x lie in an affine space. Note that not all additive differences are consistent with all XOR differences, $i.e.$, the following conditions must be satisfied

$$
\begin{cases}
c_0 = \delta_0 \oplus \Delta_0 = 0 \\
\delta_i = \delta_{i+1} \oplus \Delta_{i+1} & \text{when } \Delta_i = 0 \text{ and } i < n - 1
\end{cases} \quad (28)
$$

Solving an equation of the form $\mathcal{L}(x + \delta) = \mathcal{L}(x) + \delta'$ can be done as follows. Let $\Delta' = (\mathcal{L}(x) + \delta') \oplus \mathcal{L}(x)$, $i.e.$, the XOR-difference associated with the modular addition $\mathcal{L}(x) + \delta'$. Since $\mathcal{L}(x + \delta) = \mathcal{L}(x) + \delta'$ and $\mathcal{L}$ is GF(2)-linear, it follows that $\Delta' = \mathcal{L}(\Delta)$. We can thus simply enumerate all the XOR-differences Δ consistent with the given additive difference δ, compute $\Delta' = \mathcal{L}(\Delta)$ and check if this is consistent with the other additive difference δ'. If it is, both additions restrict x to a (different) affine space. The intersection of these spaces, which can be computed by solving a system of linear equations over GF(2), gives the solutions x for the chosen XOR-difference Δ. Note that this intersection may be empty. If no solutions are found for any value of the XOR-difference Δ, the equation $\mathcal{L}(x + \delta) = \mathcal{L}(x) + \delta'$ has no solutions. Note that the number of solutions of the equation can be counted efficiently using this method, as the number of solutions of a linear system over GF(2) is straightforward to compute.

The time complexity of this method is proportional to the minimum of the number of XOR differences consistent with the given additive differences δ or δ'. This follows from the fact that one can easily modify the method to choose Δ' instead of Δ.

Public Verifiability from Pairings in Secret Sharing Schemes⋆

Somayeh Heidarvand and Jorge L. Villar

Universitat Politècnica de Catalunya, Spain
{somayeh,jvillar}@ma4.upc.edu

Abstract. In this paper we propose a new publicly verifiable secret sharing scheme using pairings with close relations to Shoenmakers' scheme. This scheme is efficient, multiplicatively homomorphic and with unconditional verifiability in the standard model. We formalize the notion of Indistinguishability of Secrets and prove that out scheme achieves it under the Decisional Bilinear Square (DBS) Assumption that is a natural variant of the Decisional Bilinear Diffie Hellman Assumption. Moreover, our scheme tolerates active and adaptive adversaries.

Keywords. Public Verification, Secret Sharing, Pairings, Semantic Security, Threshold Cryptography.

1 Introduction

Most of the work on secret sharing dates from the eighties and the nineties, before the invention of Paillier's cryptosystem [10] and the first positive use of pairings in cryptography [8]. As a consequence, little attention has been paid to the potential use of recently invented cryptographic tools in the design of secret sharing schemes. However, pairings have been successfully used in the design of some distributed cryptographic protocols like threshold encryption and threshold signatures, in the last years.

Background. In a secret sharing (SS) scheme, a dealer D wants to share a secret among a set of participants in such a way that only special (qualified) subsets are able to recover the secret. Here we are interested in (t, n)-threshold secret sharing schemes, in which the qualified subsets are those with at least t participants. Since the publication of the seminal paper by Shamir [15], secret sharing has found innumerable applications and is nowadays considered as a fundamental tool for the design of distributed cryptographic protocols.

The first constructions of secret sharing schemes achieved a high level of security (secrecy): The Shamir scheme provides secrecy even in the presence of a passive adversary (i.e., an eavesdropper who controls the secret information of at most $t - 1$ participants) with unlimited computational power. However, these schemes do not provide enough protection against dishonest participants or a dishonest dealer.

⋆ This research was partially supported by the Centre de Recerca Matemàtica (CRM).

R. Avanzi, L. Keliher, and F. Sica (Eds.): SAC 2008, LNCS 5381, pp. 294–308, 2009.

Verifiable secret sharing (VSS) schemes have been introduced in [2] to solve the problem of dishonest dealers and dishonest participants who try to deceive other participants. Feldman's VSS scheme [3] is a very practical scheme in which secrecy is based on a computational assumption related to the Discrete Logarithm problem. However, since a deterministic function of the secret is published, only a weak notion of secrecy (one-wayness) is guaranteed. Pedersen [11], proposed a VSS scheme in which secrecy is guaranteed for an unbounded passive adversary, but verifiability relies on a computational assumption.

In [16], Stadler proposed a publicly verifiable secret sharing (PVSS) scheme. In this scheme, the validity of the shares can be verified by anyone only from the public information. Typically, in a PVSS scheme, the dealer only broadcasts some information to the participants, and no private channels are needed for the distribution of the shares. Shoenmakers' PVSS scheme [14] works in a group in which the Discrete Logarithm problem is intractable. His scheme is quite simple, but to make it publicly verifiable some non-interactive zero-knowledge proofs have been used.

In most publicly verifiable secret sharing schemes [5,14], the verification procedure involves interactive proofs of knowledge. These proofs are made non-interactive by means of the Fiat-Shamir technique [4]. This implies that verifiability relies on the properties of some hash function. Actually, known security proofs for verifiability work only in the Random Oracle Model (ROM), and there is a known negative result about the universal validity of Fiat-Shamir heuristics [6].

There are other known ways to obtain non-interactive zero knowledge proofs without using Fiat-Shamir. For instance, a recent work by Groth et al. [7] shows a generic non-interactive zero knowledge proof for any language in NP in the common reference string model (CRS), that takes advantage of pairings. However, these zero knowledge proofs are still quite inefficient.

Based on a PVSS scheme by Fujisaki and Okamoto [5], Ruiz and Villar in [12] overcame some of the above problems: a new PVSS scheme is proposed which makes use of the additive homomorphic property of Paillier's encryption. The dealer commits to the coefficients of the polynomial of the underlying Shamir SS scheme by broadcasting their encryptions. The resulting PVSS scheme is unconditionally verifiable (in the Standard Model) and the verification protocol is intrinsically non-interactive (i.e., does not make use of Fiat-Shamir heuristics). The main drawback of this scheme is that it requires an additional step of interaction in the sharing phase: the dealer holds a secret/public key pair and every participant sends an encrypted random value to him in order to establish a secure channel through which the corresponding share is sent.

Moreover, compared to Feldman's scheme, the Ruiz-Villar scheme provides a higher level of secrecy called indistinguishability (IND) based on the Decisional Composite Residuosity (DCR) assumption. Due to the unconditional verifiability, this secrecy is guaranteed even in the presence of an active and adaptive adversary.

Another consequence of not using Fiat-Shamir non-interactive zero-knowledge proofs is that the Ruiz-Villar scheme is an additively homomorphic PVSS scheme (i.e., anyone can compute the public information of a sharing of $s_1 + s_2$ from the individual sharing of s_1 and s_2, including the verification information, in such a way that nobody can distinguish this sharing of $s_1 + s_2$ from a direct sharing of the same value).

Boldyreva [1] proposed a threshold signature scheme based on gap Diffie-Hellman groups (that can be naturally instantiated with pairings). The signing key is distributed by using Feldman's VSS and the signature verification procedure takes advantage of the DDH oracle. The techniques used in Boldyreva's paper are somewhat similar to ours but in the different context of threshold signatures.

Contributions. On the one hand, we give two formal definitions of secrecy in publicly verifiable secret sharing, which capture the notion of indistinguishability of shared secrets. We also discuss their relationship.

On the other hand, we propose a new PVSS scheme that overcomes the use of Fiat-Shamir zero-knowledge proofs in a different way than in [12], and which does not require any additional interaction in the sharing phase: we basically replace zero-knowledge proofs in Shoenmakers' scheme by equalities involving bilinear map computations. The resulting scheme has the following features:

- Public Verifiability: a misbehaving dealer or participant is unconditionally detected.
- Secrecy: indistinguishability of secrets is based on the Decisional Bilinear Square Assumption, which is a variant of the Decisional Bilinear Diffie-Hellman Assumption.
- Active adversaries are tolerated (even adaptive ones), whenever there is a majority of honest participants.
- Multiplicatively homomorphic property (compatible with public verifiability).
- Efficiency comparable to Shoenmakers' scheme, with a more efficient dealer but a less efficient verifier.

Reducing the computational cost of the dealer can be desirable in applications in which there exist many potential dealers but a limited amount of participants, as in electronic voting. A variant of the basic scheme is also presented, which allows secret reconstruction via open channels in an efficient way. We also show how the scheme generalizes to linear access structures other than threshold ones.

Organization. The paper is organized as follows: In Section 2 we recall the characteristics of PVSS schemes. Computational secrecy for PVSS schemes is revisited in Section 3. The proposed PVSS scheme is presented in Section 4 and its security is analyzed in Section 5. In Section 6 a variant of the scheme which allows the reconstruction of the secret on public channels is presented. Finally, the multiplicatively homomorphic property of the scheme is discussed in Section 7.

Notation. As usual in cryptography papers, we use the convenient notation $x \xleftarrow{\$} X$ to denote that x is a uniformly distributed random element of a set X.

2   Publicly Verifiable Secret Sharing Schemes

Let $\mathcal{P} = \{P_1, \ldots, P_n\}$ be a set of n participants. We only refer to a (t,n)-threshold access structure, although the schemes proposed in this paper can easily be generalized to any vector space access structure. The dealer D wants to share a secret s between the participants of $\mathcal{P}$ in such a way that every set of at least t participants can recover the secret, and no set of at most $t-1$ participants can get any information about the secret. V is any (external) verifier who wants to check any phase of the scheme.

In a basic secret sharing scheme, three subprotocols are needed: *setup*, *distribution* of the shares and *reconstruction* of the secret. In a PVSS scheme, the dealer is supposed to communicate with participants via open channels. In spite of simplicity, we assume the existence of an authenticated broadcast channel. Furthermore, we can assume the existence of private channels between participants during the secret reconstruction. However, some basic PVSS schemes can be modified in order to remove this last assumption. An additional public *verification* subprotocol is also considered.

Setup. All the parameters of the scheme are generated and published by the dealer D. Also every participant publishes his public key and withholds the corresponding secret key.[1]

Distribution. For a secret s the dealer creates $s_1, s_2, \ldots, s_n$ as the shares of $P_1, P_2, \ldots, P_n$ respectively. The dealer computes and publishes the encrypted shares $E_i(s_i)$ for each participant $P_i \in \mathcal{P}$. He also must publish $PROOF_D$ to ensure the verifier that the published values are encryptions of correct shares.

Verification. From all the public information generated during setup and distribution phases, V verifies non-interactively that the published information is consistent and that every authorized subset of (honest) participants will recover the same secret. If verification fails, the whole protocol is aborted (i.e., honest participants exit the protocol).

Reconstruction. Using his secret key, every participant P_i in a qualified subset $A \subset \mathcal{P}$ decrypts his encrypted share and gets s_i. Then, all participants in A exchange their shares s_i together with a proof $PROOF_{P_i}$ via private channels. Every participant in A locally reconstructs the secret from a subset of t correct

---

[1] This public key could be the encryption with the public key of the dealer of a random value chosen by the participant as a one-time key. However, different instances of the protocol need independent one-time keys, and this adds a new interaction step in the distribution subprotocol, as in [12].

shares (i.e., those with a valid proof). If no private channels are available to participants, they then can send encryptions of the shares instead of the shares themselves.

As usual, adversaries can be classified into *passive* and *active*, depending on the behavior of corrupted participants: a passive adversary cannot change the behavior of a corrupted participant, while an active adversary can change it in any possible way, but in any case the adversary learns all the participant's secret information. Also adversaries can be *static* or *adaptive*. A static adversary decides the participants whom will be corrupted at the very beginning of the protocol, while an adaptive one can decide to corrupt a new participant at any time, as a function of his knowledge. We always consider a *rushing* adversary, who makes corrupted participants wait for honest participants' messages before sending theirs in each communication round.

The three properties required for a PVSS scheme: *correctness*, *verifiability* and *secrecy* are defined below.

Correctness. If the dealer and the participants act honestly, every qualified subset of participants reconstructs the secret s in the reconstruction phase. This obviously implies that the dealer passes the verification subprotocol.

Verifiability. If a dishonest dealer passes the verification subprotocol, then there exists a unique value s such that the honest participants in any qualified subset with at least t honest participants recover s as the secret. We can consider weaker notions of verifiability by tolerating a negligible error probability (statistical verifiability) or by considering a computationally bounded adversary (computational verifiability).

Secrecy. For an honest dealer, the adversary cannot learn any information about the secret, even after the execution of the reconstruction subprotocol by all honest participants. We can also consider weaker notions of secrecy, depending on the type of adversary and the tolerated amount of information he can learn about the secret.

Unconditional secrecy is not possible in PVSS schemes, since the encrypted shares are sent by public channels, so an unbounded adversary can decrypt them and then compute the secret. In the following section we review some notions of computational secrecy (i.e., secrecy against a computationally bounded adversary).

3   Computational Secrecy

PVSS schemes can be related to threshold decryption schemes, in which only qualified subsets can decrypt ciphertext encrypted with a certain public key. In this analogy, the shared secret is the encrypted message and the information published by the dealer is the ciphertext.

Hence, one-way secrecy in PVSS means that the adversary wants to know the whole secret. However, achieving only this secrecy level (as in Feldman's scheme) does not appear to be enough in the real world.

A formalization of the intuitive notion of semantic security for a PVSS scheme was first introduced in [12]. We refine that secrecy notion for the worst case active and adaptive adversary and give two secrecy levels that we call IND1 and IND2. The weaker notion (IND1) informally means that the adversary cannot tell apart the shared secret from a random value. This is a natural definition if the PVSS scheme is seen as a Key Encapsulation Mechanism (KEM).

Definition 1 (Indistinguishability of secrets (IND1)). *We say that a (t, n)-threshold PVSS scheme is IND1-secret if any probabilistic polynomial time $\mathcal{A}$ has a negligible advantage in the following game played against a challenger $\mathcal{C}$. During the game, $\mathcal{A}$ can corrupt a new participant at any time, but up to $t - 1$ participants in total. When $\mathcal{A}$ corrupts a participant, he receives his secret key (only after step 1 in the game). A list of corrupted participants is maintained during the game.*

1. *$\mathcal{C}$ runs the setup subprotocol and sends the public parameters to $\mathcal{A}$ along with the public keys of still uncorrupted participants. $\mathcal{C}$ stores the secret keys of those participants.*
2. *$\mathcal{A}$ sends the public keys of already corrupted participants.*
3. *$\mathcal{C}$ picks two random secrets x_0, x_1 and a random bit $b \in \{0, 1\}$. Then he runs the distribution subprotocol for secret x_0 and sends all the resulting information to $\mathcal{A}$, along with x_b.*
4. *$\mathcal{C}$ runs the reconstruction subprotocol for the set of all uncorrupted participants and sends all the messages exchanged via public channels (if any) to $\mathcal{A}$. No new corruptions are allowed from this point.*
5. *$\mathcal{A}$ outputs a guess bit b'.*

The advantage of $\mathcal{A}$ in that game is defined as $\left| \mathsf{Prob}[b' = b] - \frac{1}{2} \right|$.

The stronger notion (IND2) is similar to (IND1) but now x_0, x_1 are chosen by the adversary.

Definition 2 (Indistinguishability of secrets (IND2)). *We define IND2-secrecy of a (t, n)-threshold PVSS scheme exactly as in the definition of IND1-secrecy but replacing item 3 by*

3' $\mathcal{A}$ selects to secrets x_0, x_1 and sends them to $\mathcal{C}$. Then $\mathcal{C}$ picks a random bit $b \in \{0, 1\}$ and runs the distribution subprotocol for the secret x_b. Finally, $\mathcal{C}$ sends all the resulting information to $\mathcal{A}$.

Clearly, IND2-secrecy implies IND1-secrecy but the converse is not true. However, one can upgrade an IND1-secret PVSS scheme to achieve IND2-secrecy by using the original PVSS scheme to share a random session key K, and then the dealer publishes $K \oplus s$, where s is the secret and $\oplus$ is a suitable group operation. We refer to this PVSS scheme as a *hybrid* PVSS scheme. See Appendix B for a detailed proof of this fact.

4   The Proposed PVSS Scheme

Assume that G is a group of order q, q a prime number, and g and h are two independent generators of this group. Let e be a non-degenerated bilinear map $e : G \times G \to G_1$. This means that the map $e : G \times G \to G_1$ fulfils the following properties:

1. $e(g^\alpha, g^\beta) = e(g, g)^{\alpha\beta}$ for all $\alpha, \beta \in \mathbb{F}_q$.
2. $e(g, g) \neq 1$.
3. $e(x, y)$ is efficiently computable given x and y in G.

We wish to use Shoenmakers' protocol together with the bilinear map to build a (t, n)-threshold PVSS scheme to share a secret in G_1 among the participants $P_1, \ldots, P_n$, where $n \geq 2t - 1$, in such a way that the public verifiability does not require the use of Fiat-Shamir non-interactive zero-knowledge proofs. To do this, the dealer chooses $z_0 \in \mathbb{F}_q^*$ randomly and distributes the secret $S = e(h, h)^{z_0}$ in the following way:

Setup. Every participant P_i chooses a random secret value $d_i \in \mathbb{F}_q^*$ and publishes $h_i = h^{d_i}$ as his public key.

Distribution. The dealer chooses a random polynomial $P(x) = \sum_{j=0}^{t-1} \alpha_j x^j$ of degree at most $t - 1$ with coefficients in $\mathbb{F}_q$ and $\alpha_0 = z_0$. The dealer publishes the commitments $C_j = g^{\alpha_j}$, for $0 \leq j < t$. He also publishes the encryptions of the shares $Y_i = h_i{}^{P(i)}$ for $1 \leq i \leq n$.

Verification. Every (external) verifier can compute the value $X_i = \prod_{j=0}^{t-1} C_j^{i^j}$ for every participant P_i by himself and check the correctness of the shares by checking the equation $e(X_i, h_i) = e(g, Y_i)$. If the verification fails, all participants exit the protocol (i.e., they refuse to take part in the reconstruction subprotocol).

Reconstruction. Let A be a qualified subset of participants. Each participant in A gets the encrypted share $S_i = h^{P(i)}$ by using its private key and computing $S_i = Y_i^{1/d_i}$. Then all participants in A pool their shares. All shares can be verified easily by other participants of A by checking the equation $e(S_i, h_i) = e(Y_i, h)$. After the verification, if there are at least t correct shares, then for an arbitrary set $B \subseteq A$ of t participants which have pooled correct shares, every participant in A can get h^{z_0} by Lagrange interpolation: $\prod_{P_i \in B} S_i{}^{\lambda_i} = \prod_{P_i \in B} (h^{P(i)})^{\lambda_i} = h^{\sum_{P_i \in B} \lambda_i P(i)} = h^{P(0)} = h^{z_0}$, where $\lambda_i = \prod_{P_j \in B \setminus \{P_i\}} \frac{j}{j-i}$ is a Lagrange coefficient. The secret S will be recovered by computing $e(h^{z_0}, h)$. The protocol is summarized in Figure 1.

5   Analysis of the Scheme

5.1   Correctness

Correctness of the scheme means that: (i) an honest D always passes the verification procedure, and (ii) any subset of at least $2t - 1$ participants (which

---

Setup:
- ▷ Each participant P_i computes $(pk, sk) = (h_i, d_i) = (h^{d_i}, d_i)$, where $d_i \xleftarrow{\$} \mathbb{F}_q^*$, and broadcasts pk.

Distribution:
Let $z_0 \xleftarrow{\$} \mathbb{F}_q^*$ and $S = e(h, h)^{z_0}$ be the secret to be shared.
- ▷ D picks a random polynomial $P(x) = \sum_{j=0}^{t-1} \alpha_j x^j$ where $\alpha_0 = z_0$ and $\alpha_j \xleftarrow{\$} \mathbb{F}_q$.
- ▷ D publishes $C_j = g^{\alpha_j}$ and the encrypted shares $Y_i = h_i^{P(i)}$.

Verification:
- ▷ V computes $X_i = \prod_{j=0}^{t-1} C_j^{i^j}$ and checks if $e(X_i, h_i) = e(g, Y_i)$.

Reconstruction:
Let A be a subset of participants running the reconstruction subprotocol.
- ▷ Every $P_i \in A$, sends $S_i = Y_i^{1/d_i}$ to the other participants in A.
- ▷ Every $P_i \in A$, checks if $e(S_j, h_j) = e(Y_j, h)$ for all $P_j \in A \setminus \{P_i\}$ and defines B_i as a subset of t participants that pass the test.
- ▷ Every $P_i \in A$, computes $\prod_{P_j \in B_i} S_j^{\lambda_j} = h^{z_0}$ where $\lambda_j = \prod_{P_k \in B_i \setminus \{P_j\}} \frac{k}{k-j}$ and gets $S = e(h^{z_0}, h)$.

Fig. 1. PVSS scheme with reconstruction via private channels

ensures us that there are at least t honest participants) is always able to recover the secret shared by an honest D. Checking these requirements for the above protocol is straightforward.

5.2 Verifiability

Now we show that if D passes the verification, then all participants in the protocol must behave honestly or will be detected. More precisely, on the one hand, the dealer must be honest in the distribution subprotocol and, on the other hand, participants must be honest in the reconstruction subprotocol.

Verifiability of the Distribution. In the following, we prove that a dishonest D cannot cheat the participants without being detected in the verification. More precisely, if D passes the verification, then all qualified subsets of honest participants will reconstruct the same secret.

Lemma 1. *If V accepts, then there exists a unique polynomial $P(x)$ such that the encrypted share of participant P_i is $Y_i = h_i^{P(i)}$ for $1 \leq i \leq n$.*

Proof. Assume that the share of participant P_i is equal to $Y_i = h_i^s$. Let us write $C_j = g^{\alpha_j}$ for suitable α_j and consider the polynomial $P(x) = \alpha_0 + \alpha_1 x + \ldots + \alpha_{t-1} x^{t-1}$. If V accepts, then for every $1 \leq i \leq n$ the dealer passes the equation $e(X_i, h_i) = e(g, Y_i)$, where $X_i = \prod_{j=0}^{t-1} C_j^{i^j}$. By the definition of $P(x)$, we have $e(g, h_i)^{P(i)} = e(g, h_i)^s$, which leads to $s = P(i)$. The uniqueness of $P(x)$ implies that all sets of t correct shares get the same secret in the reconstruction subprotocol. $\qquad\square$

In the actual protocol, all participants act as verifiers after the secret distribution stage. Then, if D broadcasts corrupt information, then all honest participants drop out of the protocol. Hence, nobody can successfully run the reconstruction subprotocol. It is worth noticing that verifiability of D is unconditional (i.e., does not depend on any computational assumption).

Verifiability of the Shares in the Reconstruction Subprotocol. Consider now that D behaves honestly. Let P_i and P_j be two participants taking part in the reconstruction subprotocol. Suppose P_i opens his secret value $S_i = Y_i{}^s$ to P_j, and P_j behaves honestly.

Lemma 2. *If P_j accepts P_i's value, then $S_i = Y_i^{1/d_i}$, where d_i is the secret key of P_i, that is $h_i = h^{d_i}$.*

Proof. Since P_j accepts P_i's value, then $e(S_i, h_i) = e(Y_i, h)$, and so $e(Y_i{}^s, h^{d_i}) = e(Y_i, h)$. By using the properties of the bilinear map we get $e(Y_i, h)^{sd_i} = e(Y_i, h)$, which results in $sd_i = 1$. So if P_j accepts the secret share of P_i, then $S_i = Y_i^{1/d_i}$. $\qquad\square$

Thus, by the two previous lemmas, all honest participants involved in the reconstruction subprotocol accept only correct shares $S_i = Y_i^{1/d_i} = h^{P(i)}$ (whether the shares come from honest or dishonest participants). If there are at least t honest participants in the subset A running the reconstruction subprotocol, then every honest participant in A accepts at least t correct shares, which lead to the secret $S = h^{P(0)}$ by Lagrange interpolation in the exponent. Notice that this property does not depend on any computational assumption.

The results in this section are summarized in the following theorem.

Theorem 1. *The proposed PVSS scheme is publicly verifiable even in the presence of an unbounded adversary.*

5.3   Secrecy

Our goal now is to show that an active and adaptive probabilistic polynomial time adversary corrupting at most $t - 1$ participants cannot obtain any information about the shared secret S, assuming an honest D. To show this, we first define the following assumption:

Assumption 1 (Decisional Bilinear Square (DBS)). *Let G and G_1 be two groups of prime order q, g be a random generator of G and $e : G \times G \to G_1$ be a non-degenerated bilinear map. For random values μ, ν and s chosen uniformly and independently from $\mathbb{F}_q^*$ and given $h = g^\mu$, $u = g^\nu$, the following probability distributions are polynomially indistinguishable: $D_0 = (g, h, u, T_0 = e(h, h)^\nu)$ and $D_1 = (g, h, u, T_1 = e(h, h)^s)$.*

This assumption is equivalent to the Decisional Bilinear Quotient (DBQ) Assumption, recently introduced in [9], and it is a natural variant of the standard Decisional Bilinear Diffie-Hellman Assumption, in which informally, an adversary aims to tell apart $e(g, g)^{xyz}$ from a random value, given (g, g^x, g^y, g^z). DBS Assumption corresponds to the case $x = y$. See Appendix C for more details about the relations of these assumptions.

Theorem 2. *If the DBS Assumption holds, then the proposed scheme is IND1- secret.*

Proof. Assume that there is an active and adaptive probabilistic polynomial time adversary, $\mathcal{A}$, playing the game in Definition 1 with a non-negligible advantage $\varepsilon_{\mathcal{A}}$. Then we describe a simulator $\mathcal{F}$ that using $\mathcal{A}$ as a subroutine can break the DBS Assumption with a non-negligible advantage $\varepsilon_{\mathcal{F}}$.

1. Once $\mathcal{F}$ receives the description of a random instance of the DBS Problem $(q, G, G_1, e, g, h = g^\mu, u = g^\nu, T_b)$, as described in Assumption 1, he simulates the (t, n)-threshold PVSS scheme as a challenger for $\mathcal{A}$. So $\mathcal{F}$ sends $(n, t, \mathcal{P}, q, G, G_1, e, g, h)$ to $\mathcal{A}$ as the public parameters of the scheme. $\mathcal{A}$ chooses a subset $B_0 \subset \mathcal{P}$ of initially corrupted players and gives it to $\mathcal{F}$. Now $\mathcal{F}$ guesses the set of all players corrupted by $\mathcal{A}$ at the end of the game by choosing at random B such that $B_0 \subset B \subset \mathcal{P}$ and $|B| = t - 1$. Then $\mathcal{F}$ computes the public keys of the players as follows: $\forall P_i \in B \setminus B_0$; $h_i = h^{d_i}$, $d_i \xleftarrow{\$} \mathbb{F}_q^*$ and $\forall P_i \in \mathcal{P} \setminus B$; $h_i = g^{r_i}$, $r_i \xleftarrow{\$} \mathbb{F}_q^*$, and sends them to the adversary.
2. $\mathcal{A}$ sends the public keys of the corrupted players which have been arbitrary chosen by himself.
3. $\mathcal{F}$ chooses $s_i \xleftarrow{\$} \mathbb{F}_q$ and sets $Y_i = h_i^{s_i}$ for all players $P_i \in B$. There exists a unique interpolating polynomial $P(x) = \alpha_0 + \alpha_1 x + ... + \alpha_{t-1}x^{t-1} \in \mathbb{F}_q[x]$, such that $\forall P_i \in B$; $P(i) = s_i$ and $g^{P(0)} = u$. Thus all the coefficients can be uniquely determined for some efficiently computable constants μ_{ij} (that only depend on B) as $\alpha_j = \sum_{P_i \in B} \mu_{ij} s_j + \mu_{0j} \nu$. Now $\mathcal{F}$ computes $C_j = g^{\sum_{P_i \in B} \mu_{ij} s_i} u^{\mu_{0j}}$, $1 \le j \le t - 1$ and sets $C_0 = u$. Then $\forall P_i \in \mathcal{P} \setminus B$, $\mathcal{F}$ sets $Y_i = h_i^{P(i)} = g^{P(i)r_i} = [u \prod_{j=1}^{t-1} C_j^{i^j}]^{r_i}$, and sends all Y_i, all C_j and T_b to $\mathcal{A}$.
4. $\mathcal{A}$ chooses $B_1 \in \mathcal{P} \setminus B_0$ such that $|B_0 \cup B_1| \le t - 1$, and corrupts the participants in B_1. If $B_1 \not\subseteq B \setminus B_0$, then $\mathcal{F}$ exits the game giving a random bit b' as output. Otherwise $\mathcal{F}$ sends the secret key d_i of every participant $P_i \in B_1$ to $\mathcal{A}$.
5. Eventually $\mathcal{A}$ ends by outputting a bit b' which is forwarded by $\mathcal{F}$.

Let Fail denote the event that $\mathcal{F}$ exits the game at step 4. Notice that, if Fail occurs then the probability of success of $\mathcal{F}$ (i.e., $b' = b$) is exactly $1/2$. Otherwise, $\mathcal{F}$ perfectly simulates the challenger for $\mathcal{A}$. On the other hand, the choice of B is independent of all the variables of the secret sharing scheme, and then Fail is independent of the success of $\mathcal{A}$. Thus, the probability of success of $\mathcal{F}$ is $\mathsf{Succ}_{\mathcal{F}} = \frac{1}{2}\mathsf{Prob}[\mathsf{Fail}] + \mathsf{Succ}_{\mathcal{A}} \mathsf{Prob}[\neg\mathsf{Fail}]$ and $\varepsilon_{\mathcal{F}} = \varepsilon_{\mathcal{A}} \mathsf{Prob}[\neg\mathsf{Fail}]$. The (conditional) probability of $\neg\mathsf{Fail}$ can be easily computed as $\mathsf{Prob}[\neg\mathsf{Fail}] = \binom{t-1-|B_0|}{|B_1|}/\binom{n-|B_0|}{|B_1|}$, for any possible choice of B_1, which ranges from 1 if $B_1 = \emptyset$ (that is, in the case of an active but static adversary $\varepsilon_{\mathcal{F}} = \varepsilon_{\mathcal{A}}$) to $\binom{n}{t-1}^{-1}$ if $B_0 = \emptyset$ and $|B_1| = t - 1$ (that is, the worst case adversary). $\qquad\square$

As seen in Section 3, we can modify the basic PVSS scheme to achieve IND2-secrecy by letting the dealer share a random value $K = e(h,h)^{z_0} \in G_1$, and then publish the product $T = KS$, where $S \in G_1$ is the actual secret he wants to share.

6   Secret Reconstruction on Public Channels

In the basic scheme the secret reconstruction supposes the existence of private channels between participants. In this section we remove this requirement without losing any good property of the scheme.

Assume that participant P_i wants to sends his encrypted share, $S_i = Y_i^{1/d_i}$, to P_j. To do that publicly, he chooses a random value ρ and sends $(r, z, w) = (h_i{}^\rho, Y_i{}^\rho, h_j{}^{1/(d_i\rho)})$, where h_i, h_j are the public keys of P_i, P_j respectively. Now everybody can verify the correctness by checking the equations $e(r, Y_i) = e(z, h_i)$ and $e(r, w) = e(h_j, h)$, since Y_i is publicly available from the sharing information broadcast by the dealer. Notice that this verification is unconditional.

Then P_j computes the share of P_i as $e(h,h)^{P(i)} = e(z,w)^{1/d_j}$. From t correct shares, P_i can locally compute the secret $S = e(h,h)^{P(0)}$ as usual, by means of Lagrange interpolation in the exponent. The secrecy of the modified protocol is also based on the DBS Assumption.

Theorem 3. *The protocol is IND1-secret under the DBS Assumption.*

Proof. We only have to modify step 5 of the simulation in the proof of Theorem 2 to provide $\mathcal{A}$ with all the messages exchanged by the uncorrupted participants during the secret reconstruction.

5' For every ordered pair (P_i, P_j) of different uncorrupted participants, $\mathcal{F}$ chooses $\rho \xleftarrow{\$} \mathbb{F}_q^*$ and sends $(r = h_i{}^\rho, z = Y_i{}^\rho, w = h^{r_j/(r_i\rho)})$ to $\mathcal{A}$. Eventually $\mathcal{A}$ ends by outputting a bit b' which is forwarded by $\mathcal{F}$.

Notice that the simulation of the secret reconstruction subprotocol is perfect. Therefore, the advantage of $\mathcal{F}$ fulfils exactly the same equation as in Theorem 2. $\qquad\square$

7   Homomorphic Properties

It is well known that some basic secret sharing schemes have nice homomorphic properties. For instance, in Shamir's scheme, if every participant P_i locally computes a linear combination of his shares s_i and t_i for the secrets s and t, respectively, then he obtains a new share corresponding to the same linear combination of the secrets. This interesting property has found a lot of applications in electronic voting or multiparty computation, for example.

However, if the same idea is applied to a publicly verifiable secret sharing scheme, then new difficulties arise: one wants to compute the sharing information (including verification information) of the operation of two secrets from the information of the individual sharing processes. This seems very hard to achieve if the public verifiability depends on non-interactive zero-knowledge proofs, but it is straightforward in our scheme (as it was in [12]). Our basic scheme has the following multiplicatively homomorphic property. We assume that public keys of the participants are reused for multiple secret sharing.

Proposition 1. *Let* $(C_0, \ldots, C_{t-1}, Y_1, \ldots, Y_n)$ *and* $(\tilde{C}_0, \ldots, \tilde{C}_{t-1}, \tilde{Y}_1, \ldots, \tilde{Y}_n)$ *be the sharing information broadcast by the dealer for secrets S and $\tilde{S}$, respectively. Then, for any $\alpha, \beta \in \mathbb{F}_q^*$ the tuple* $(C_0^\alpha \tilde{C}_0^\beta, \ldots, C_{t-1}^\alpha \tilde{C}_{t-1}^\beta, Y_1^\alpha \tilde{Y}_1^\beta, \ldots, Y_n^\alpha \tilde{Y}_n^\beta)$ *has the same probability distribution as a direct sharing of the secret $S^\alpha \tilde{S}^\beta$.*

The same property applies to the IND2-secret improved scheme. Indeed, it suffices to do the same operation $T^\alpha \tilde{T}^\beta$ with the additional public elements T and $\tilde{T}$.

8   Final Remarks

As in Shoenmakers' scheme, the PVSS scheme proposed in this paper can be easily extended to linear access structures other than the (t, n)-threshold ones by following a standard procedure. Firstly, assign to every participant P_i a vector $\mathbf{v}_i = (v_{i,0}, \ldots, v_{i,t-1}) \in \mathbb{F}_q^t$ for a suitable dimension t, and let $\mathbf{v}_0 = (1, 0, \ldots, 0)$ be the vector associated to the dealer. Then replace the sharing polynomial $P(x)$ by a (dual) vector $\alpha = (\alpha_0, \ldots, \alpha_{t-1})$, and $P(i)$ by the dot product $\mathbf{v}_i \cdot \alpha$. Hence, X_i is computed as $X_i = \prod_{j=0}^{t-1} C_j^{v_{i,j}}$. All the remaining equations are maintained except for Lagrange interpolation coefficients, which are replaced by the coefficients of the expression of $\mathbf{v}_0$ as a linear combination of the vectors associated to a qualified subset of participants.

On the other hand, the proposed PVSS scheme has a performance comparable to Shoenmakers' scheme. Indeed, the dealer's computational effort of computing the non-interactive zero-knowledge proofs ($2n$ exponentiations) and the verification of them by the verifier ($2n$ multi exponentiations) have been replaced by the computation of $2n$ pairings by the verifier. Hence, the dealer's computation complexity is reduced in about a 50%. If we tolerate a positive error probability in the verification procedure, then the verifier can check a random combination of the n equations, reducing the number of pairing computations to only $n + 1$. Moreover, every participant taking part of our scheme's reconstruction subprotocol must compute some extra pairings (typically $2t - 1$), but he does not have to compute and check the non-interactive zero-knowledge proofs (saving 2 exponentiations and $2t - 2$ multi exponentiations).

References

1. Boldyreva, A.: Threshold signatures, multisignatures and blind signatures based on the gap-diffie-hellman-group signature scheme. In: Desmedt, Y.G. (ed.) PKC 2003. LNCS, vol. 2567, pp. 31–46. Springer, Heidelberg (2002)
2. Chor, B., Goldwasser, S., Micali, S., Awerbuch, B.: Verifiable secret sharing and achieving simultaneity in the presence of faults. In: Proc. 26th IEEE Symp. on Found. of Comp. Sci., pp. 383–395 (1985)

3. Feldman, P.: A Practical Scheme for Non-interactive Verifiable Secret Sharing. In: Proceedings 28th IEEE Symp. on Found. of Comp. Sci., pp. 427–437 (1987)

4. Fiat, A., Shamir, A.: How to prove yourself: Practical solutions to identification and signature problems. In: Odlyzko, A.M. (ed.) CRYPTO 1986. LNCS, vol. 263, pp. 186–194. Springer, Heidelberg (1987)

5. Fujisaki, E., Okamoto, T.: A practical and provably secure scheme for publicly verifiable secret sharing and its applications. In: Nyberg, K. (ed.) EUROCRYPT 1998. LNCS, vol. 1403, pp. 32–46. Springer, Heidelberg (1998)

6. Goldwasser, S., Tauman, Y.: On the (In)security of the Fiat-Shamir Paradigm. In: Proc. 44th Annual IEEE Symp. on Found. of Comp. Sci., pp. 102–115 (2003)

7. Groth, J., Ostrovsky, R., Sahai, A.: Perfect non-interactive zero knowledge for NP. In: Vaudenay, S. (ed.) EUROCRYPT 2006. LNCS, vol. 4004, pp. 339–358. Springer, Heidelberg (2006)

8. Joux, A.: A One Round Protocol for Tripartite Diffie-Hellman. In: Bosma, W. (ed.) ANTS 2000. LNCS, vol. 1838, pp. 385–394. Springer, Heidelberg (2000)

9. Libert, B., Vergnaud, D.: Unidirectional chosen-ciphertext secure proxy re-encryption. In: Cramer, R. (ed.) PKC 2008. LNCS, vol. 4939, pp. 360–379. Springer, Heidelberg (2008)

10. Paillier, P.: Public-key cryptosystems based on composite degree residuosity classes. In: Stern, J. (ed.) EUROCRYPT 1999. LNCS, vol. 1592, pp. 223–238. Springer, Heidelberg (1999)

11. Pedersen, T.P.: Non-interactive and information-theoretic secure verifiable secret sharing. In: Feigenbaum, J. (ed.) CRYPTO 1991. LNCS, vol. 576, pp. 129–140. Springer, Heidelberg (1992)

12. Ruiz, A., Villar, J.L.: Publicly Verifiable Secret Sharing from Paillier's Cryptosystem. In: WEWoRC 2005. LNI P-74, pp. 98–108 (2005)

13. Sadeghi, A.-R., Steiner, M.: Assumptions related to discrete logarithms: Why subtleties make a real difference. In: Pfitzmann, B. (ed.) EUROCRYPT 2001. LNCS, vol. 2045, pp. 243–260. Springer, Heidelberg (2001)

14. Schoenmakers, B.: A simple publicly verifiable secret sharing scheme and its application to electronic voting. In: Wiener, M. (ed.) CRYPTO 1999. LNCS, vol. 1666, pp. 148–164. Springer, Heidelberg (1999)

15. Shamir, A.: How to share a secret. Commun. of the ACM 22, 612–613 (1979)

16. Stadler, M.A.: Publicly verifiable secret sharing. In: Maurer, U.M. (ed.) EUROCRYPT 1996. LNCS, vol. 1070, pp. 190–199. Springer, Heidelberg (1996)

A   A Short Description of the Ruiz-Villar PVSS Scheme

Ruiz-Villar PVSS uses the additively homomorphic Paillier cryptosystem to add public verifiability to Shamir's secret sharing scheme over the ring $\mathbb{Z}_N$, where $N = pq$ is an RSA modulus. Let g be an element with multiplicative order N in $\mathbb{Z}_{N^2}^*$ (e.g., $g = 1 + N$) and suppose that only the dealer knows p and q. The distribution subprotocol for a secret $s \in \mathbb{Z}_N$ works as follows:

1. P_i picks $(m_i, r_i) \xleftarrow{\$} \mathbb{Z}_N \times \mathbb{Z}_N^*$ and broadcasts $c_i = g^{m_i} r_i{}^N \bmod N^2$.

2. D picks a random polynomial $P(x) = \sum_{j=0}^{t-1} \alpha_j x^j$ where $\alpha_0 = s$ and $\alpha_j \xleftarrow{\$} \mathbb{Z}_N$. Then D sets $s_i = P(i) \bmod N$.

3. D decrypts all ciphertexts c_i, thus obtaining the pairs (m_i, r_i), and broadcasts $d_i = s_i + m_i \bmod N$.

4. D picks $R_j \xleftarrow{\$} \mathbb{Z}_N^*$ and broadcasts $A_j = g^{\alpha_j} R_j^N \bmod N^2$, for $0 \leq j < t$.

5. D also broadcasts $t_i = R_0 R_1^i \cdots R_{t-1}^{i^{t-1}} r_i \bmod N$, for every $i = 1, \ldots, n$.

For each $1 \leq i \leq n$, a verifier can check $A_0 A_1^i \cdots A_{t-1}^{i^{t-1}} = \frac{g^{d_i}}{c_i} t_i^N \bmod N^2$. Finally, the secret reconstruction subprotocol (on private channels) for a subset A with at least t honest participants, works as follows:

1. Every $P_i \in A$ sends the secret pair (m_i, r_i) to the other participants in A, who check that c_i is the corresponding Pallier's ciphertext.
2. P_i computes the valid shares $s_j = d_j - m_j \bmod N$ for the other participants in A who passed the verification in the previous step.
3. P_i computes s by Lagrange interpolation in $\mathbb{Z}_N$ from a set of t valid shares, as in Shamir's secret sharing scheme.

The above PVSS scheme is unconditionally verifiable and it is IND2-secret under the Decisional Composite Residuosity (DCR) Assumption, and it is also additively homomorphic. The scheme does not make use of Fiat-Shamir non-interactive zero-knowledge proofs: instead it uses the homomorphic property of Paillier's encryption at the cost of an additional communication round in the distribution subprotocol.

B   Generic Transformation from IND1 to IND2-Secrecy

Let us consider an IND1-secret PVSS scheme. Let sharing(S) be the information published by the dealer during the distribution subprotocol for a secret S. Let us assume that the set of possible secrets is a group $\mathcal{G}$, and let $\oplus$ denote the group operation.

A new *hybrid* PVSS scheme can be defined from the original one by letting the dealer choose and share a random secret $K \in \mathcal{G}$ and then publish $T = K \oplus S$ along with sharing(K). Obviously, this modification has no effect on the correctness and the public verifiability properties of the scheme. The reconstruction subprotocol is slightly modified by just adding a last step in which every participant computes $S = K^{-1} \oplus T$ after the computation of K. Let us show that if the basic scheme is IND1-secret, then the hybrid scheme is IND1-secret. Let $\mathcal{A}_2$ be an adversary playing the IND2 game in Definition 2 for the hybrid PVSS scheme, with a non-negligible advantage ε_2. We show an adversary $\mathcal{A}_1$ playing the IND1 game in Definition 1 for the basic scheme, also with a non-negligible advantage ε_1. Let $\mathcal{C}$ be the challenger for $\mathcal{A}_1$ in that game. $\mathcal{A}_1$ will act as a challenger for $\mathcal{A}_2$. In particular, $\mathcal{A}_1$ will forward all the corruption queries and responses of $\mathcal{A}_2$ to and from $\mathcal{C}$ during the game. The only nontrivial part of $\mathcal{A}_1$ is in step 3.

1. $\mathcal{A}_1$ forwards the distribution information from $\mathcal{C}$ to $\mathcal{A}_2$.
2. $\mathcal{A}_1$ forwards the corrupted participants' public keys from $\mathcal{A}_2$ to $\mathcal{C}$.
3. $\mathcal{A}_1$ receives (sharing(K_0), K_b) from $\mathcal{C}$, where $K_0, K_1 \xleftarrow{\$} \mathcal{G}$ and $b \xleftarrow{\$} \{0, 1\}$ are chosen by $\mathcal{C}$. $\mathcal{A}_1$ also receives $S_0, S_1 \in \mathcal{G}$ from $\mathcal{A}_2$. Then, $\mathcal{A}_1$ picks $\beta \xleftarrow{\$} \{0, 1\}$ and sends sharing(K_0) and $T_{b,\beta} = K_b \oplus S_\beta$ to $\mathcal{A}_2$.

4. $\mathcal{A}_1$ forwards the reconstruction information from $\mathcal{C}$ to $\mathcal{A}_2$.
5. If $\mathcal{A}_2$'s output β' equals β, then $\mathcal{A}_1$ outputs $b' = 0$. Otherwise $\mathcal{A}_1$ outputs $b' = 1$.

Notice that if $b = 0$, then $\mathcal{A}_1$ perfectly simulates a challenger for $\mathcal{A}_2$ since $T_{0,\beta} = K_0 \oplus S_\beta$ and then $\mathcal{A}_1$ sent a correct sharing of S_β for a random β. Otherwise $b = 1$, and then the view of $\mathcal{A}_2$ is independent of β. Indeed $T_{1,\beta} = K_1 \oplus S_\beta$, which is independent of sharing(K_0) and S_β. Hence, the probability that $\beta' = \beta$ is exactly $\frac{1}{2}$. So $\varepsilon_1 = \varepsilon_2/2$. On the other hand, $\mathcal{A}_1$ runs within the same time as $\mathcal{A}_2$ plus a small number of group operations.

This hybrid construction can be generalized to an arbitrary symmetric encryption scheme, $T = E_K(S)$, such that for any possible value of S, $E_K(S)$ is pseudorandom. Obviously, the above reduction should be modified to take into account the maximum advantage of an attacker against the pseudorandomness of the encryption scheme.

C   Decisional Bilinear Square and Related Assumptions

We show here that the DBS Assumption is equivalent to the DBQ Assumption, which is defined below.

Assumption 2 (Decisional Bilinear Quotient (DBQ)). *Let G and G_1 be two groups of prime order q, g be a random generator of G and $e : G \times G \to G_1$ be a non-degenerated bilinear map. For $\mu, \nu, s \xleftarrow{\$} \mathbb{F}_q^*$, the probability distributions $D_0 = (g, g^\nu, g^\mu, T_0 = e(g,g)^{\nu/\mu})$ and $D_1 = (g, g^\nu, g^\mu, T_1 = e(g,g)^s)$ are polynomially indistinguishable.*

Lemma 3. *DBQ Assumption implies the DBS Assumption.*

Proof. We can solve the DBQ problem by using a solver for the DBS problem as follows. On input of a DBQ tuple $(g, u = g^\nu, v = g^\mu, T_b)$ we construct a correct DBS tuple $(v, g, u = v^{\nu/\mu}, T_b)$. Indeed, $T_0 = e(g,g)^{\nu/\mu}$ and T_1 is a random value independent of the rest of the tuple. $\square$

Lemma 4. *DBS Assumption implies the DBQ Assumption.*

Proof. Similarly, on input of a DBS tuple $(g, u = g^\nu, v = g^\mu, T_b)$ we construct a correct DBQ tuple $(u, v = u^{\mu/\nu}, g = u^{1/\nu}, T_b)$. Indeed, $T_0 = e(u,u)^\mu = e(u,u)^{(\mu/\nu)(1/\nu)^{-1}}$ and T_1 is random and independent of the rest of the tuple. $\square$

Lemma 5. *DBS Assumption implies the DBDH Assumption.*

Proof. On input of a DBS tuple $(g, u = g^\nu, v = g^\mu, T_b)$ we construct a correct DBDH tuple $(g, u, u^\gamma, v, T_b^\gamma)$ where $\gamma \xleftarrow{\$} \mathbb{F}_q^*$. Indeed, $T_0^\gamma = e(u,u)^{\mu\gamma} = e(g,g)^{\nu(\nu\gamma)\mu}$ and T_1^γ is random and independent of (g, u, u^γ, v). $\square$

These relations are very similar to the relations between the Decisional Diffie Hellman (DDH), the Decisional Square Exponent (DSE) and the Decisional Inverse Exponent (DIE) Assumptions (see [13]).

The Elliptic Curve Discrete Logarithm Problem and Equivalent Hard Problems for Elliptic Divisibility Sequences

Kristin E. Lauter[1] and Katherine E. Stange[2],[*]

[1] Microsoft Research, One Microsoft Way, Redmond, WA 98052
klauter@microsoft.com
[2] Department of Mathematics, Harvard University, Cambridge, MA 02138
stange@math.harvard.edu

Abstract. We define three hard problems in the theory of elliptic divisibility sequences (*EDS Association*, *EDS Residue* and *EDS Discrete Log*), each of which is solvable in sub-exponential time if and only if the elliptic curve discrete logarithm problem is solvable in sub-exponential time. We also relate the problem of EDS Association to the Tate pairing and the MOV, Frey-Rück and Shipsey EDS attacks on the elliptic curve discrete logarithm problem in the cases where these apply.

1 Introduction

The security of elliptic curve cryptography rests on the assumption that the *elliptic curve discrete logarithm problem* is hard.

Problem 1 (Elliptic Curve Discrete Logarithm Problem (ECDLP)). Let E be an elliptic curve over a finite field K. Suppose there are points $P, Q \in E(K)$ given such that P is of prime order and $Q \in \langle P \rangle$. Determine k such that $Q = [k]P$.

Throughout this paper we require P to have prime order in our hard problems. Much of what we do can be adapted for non-prime order at the cost of added complication, but the prime order case is the relevant one for runtime [1].

In this article, we explore several related hard problems with a view to expanding the theoretical foundations of the security of ECDLP as a hard problem. Our research is inspired by work of Rachel Shipsey in her thesis [2], relating the ECDLP to elliptic divisibility sequences (EDS) (see also Shipsey and Swart [3]). An elliptic divisibility sequence is a recurrence sequence $W(n)$ satisfying the relation.

$$W(n+m)W(n-m) = W(n+1)W(n-1)W(m)^2 - W(m+1)W(m-1)W(n)^2.$$

---

[*] The second author was supported by NSERC Award PGS D2 331379-2006, and this work was performed during an internship of the second author at Microsoft Research.

R. Avanzi, L. Keliher, and F. Sica (Eds.): SAC 2008, LNCS 5381, pp. 309–327, 2009.
© Springer-Verlag Berlin Heidelberg 2009

We relate Shipsey's work to the MOV and Frey-Rück attacks and explain their limitations from the EDS point of view. We also point to a specific avenue for attacking ECDLP by analysing the quadratic residuosity of elliptic divisibility sequences.

The study of elliptic divisibility sequences was introduced by Morgan Ward [4]. Let Ψ_n denote the n-th division polynomial of an elliptic curve E over the rationals. The sequence $W_{E,P} : \mathbb{Z} \to \mathbb{Q}$ of the form $W_{E,P}(n) = \Psi_n(P)$ for some fixed point $P \in E(\mathbb{Q})$ is an elliptic divisibility sequence, and Ward showed that almost all elliptic divisibility sequences arise in this way. This relationship is the basis of our work here.

The general theory has been developed by Swart [5], Ayad [6], Silverman [7][8], Everest, McLaren and Thomas Ward [9] and, more recently, generalised to higher rank *elliptic nets* by Stange [10][11]. For an overview of research, see [12]. Sections 2 and 3 provide brief background on elliptic divisibility sequences and elliptic nets, more information about which can be found in [10][11][13].

The hard problems for elliptic divisibility sequences we consider are:

Problem 2 (EDS Association). Let E be an elliptic curve over a finite field K. Suppose there are points $P, Q \in E(K)$ given such that $Q \in \langle P \rangle$, $Q \neq \mathcal{O}$, and $\mathrm{ord}(P) \geq 4$ is prime. Determine $W_{E,P}(k)$ for $0 < k < \mathrm{ord}(P)$ such that $Q = [k]P$.

Problem 3 (EDS Residue). Let E be an elliptic curve over a finite field K. Suppose there are points $P, Q \in E(K)$ given such that $Q \in \langle P \rangle$, $Q \neq \mathcal{O}$, and $\mathrm{ord}(P) \geq 4$ is prime. Determine the quadratic residuosity of $W_{E,P}(k)$ for $0 < k < \mathrm{ord}(P)$ such that $Q = [k]P$.

The *rank of zero-apparition* of an elliptic divisibility sequence is the least positive n such that $W(n) = 0$.

Problem 4 (Width s EDS Discrete Log). Given an elliptic divisibility sequence W whose rank of zero-apparition is prime, and given terms $W(k)$, $W(k+1)$, ..., $W(k + s - 1)$, determine k.

Problem 4 was considered by Shipsey [2, §6.3.1] and Gosper, Orman and Schroeppel [14, §3]. Problem 2 is also implicit in [2, §6.4.1] and [14, §3].

A perfectly periodic elliptic divisibility sequence is one which has a finite period n and whose first positive index k at which $W(k) = 0$ is $k = n$. If a sequence is not perfectly periodic, then it has period $n > k$. In Section 10, we prove the following theorem.

Theorem 1. *Let E be an elliptic curve over a finite field $K = \mathbb{F}_q$. If any one of the following problems is solvable in probabilistic sub-exponential time, then all of them are:*

1. *Problem 1: ECDLP*
2. *Problem 2: EDS Association for non-perfectly periodic sequences*
3. *Problem 4 ($s = 3$): Width 3 EDS Discrete Log for perfectly periodic sequences*

In addition, the previous problems are equivalent to the following one in the case that q is odd and $E(\mathbb{F}_q)$ is of odd order.

4. Problem 3: EDS Residue for non-perfectly periodic sequences

Section 4 relates Problems 4 and 2 to the ECDLP. Section 6 expands on Problem 2. Sections 7 and 8 discuss Problem 3. Section 9 remarks on Problem 4. Section 10 proves Theorem 1. The relation with the MOV and Frey-Rück attacks is discussed in Section 5.

The authors would like to thank the referees for their helpful suggestions, and Joe Silverman, Marco Streng and Christine Swart for corrections to the final version.

2   Background on Elliptic Nets

In this section we state the background definitions and results on elliptic divisibility sequences and elliptic nets that are needed for the rest of the paper. For details and examples, see [10][11][13].

Definition 1 (Stange [10, Def. 2.1][11, Def. 3.1.1]). *Let K be a field, $n > 0$ and integer. An* elliptic net *is any map $W : \mathbb{Z}^n \to K$ such that the following recurrence holds for all p, q, r, $s \in \mathbb{Z}^n$:*

$$W(p + q + s)\, W(p - q)\, W(r + s)\, W(r)$$
$$+ W(q + r + s)\, W(q - r)\, W(p + s)\, W(p)$$
$$+ W(r + p + s)\, W(r - p)\, W(q + s)\, W(q) = 0 \quad (1)$$

We refer to n as the rank *of the elliptic net. An elliptic net of rank one is called an* elliptic divisibility sequence.

One always has $W(-\mathbf{v}) = -W(\mathbf{v})$ and $W(\mathbf{0}) = 0$, and a restriction of an elliptic net to a sublattice of $\mathbb{Z}^n$ is again an elliptic net. The important fact for our purposes is that any elliptic curve E over K and points $P_1, \ldots, P_n \in E(K)$ gives rise to a unique elliptic net $W_{E,P_1,\ldots,P_n} : \mathbb{Z}^n \to K$. The principal theorem is as follows.

Theorem 2 (Stange [10, Thm. 6.1][11, Thm. 7.1.1]). *Let $n > 0$ be an integer. Let*

$$E : f(x, y) = y^2 + \alpha_1 xy + \alpha_3 y - x^3 - \alpha_2 x^2 - \alpha_4 x - \alpha_6 = 0$$

be an elliptic curve defined over a field K. Let $\mathbf{e}_i$ be the ith standard basis vector. For all $\mathbf{v} \in \mathbb{Z}^n$, there are functions $\Psi_\mathbf{v} : E^n \to K$ in the ring

$$\mathbb{Z}[\alpha_1, \alpha_2, \alpha_3, \alpha_4, \alpha_6][x_i, y_i]_{i=1}^n \left[(x_i - x_j)^{-1}\right]_{1 \le i < j \le n} \Big/ \langle f(x_i, y_i) \rangle_{i=1}^n \subset K(E),$$

such that

1. $W(\mathbf{v}) = \Psi_{\mathbf{v}}$ satisfies the recurrence (1).
2. $\Psi_{\mathbf{v}} = 1$ whenever $\mathbf{v} = \mathbf{e}_i$ for some $1 \leq i \leq n$ or $\mathbf{v} = \mathbf{e}_i + \mathbf{e}_j$ for some $1 \leq i < j \leq n$.
3. $\Psi_{\mathbf{v}}$ vanishes at $\mathbf{P} = (P_1, \ldots, P_n) \in E^n$ if and only if $\mathbf{v} \cdot \mathbf{P} = 0$ on E (and $\mathbf{v}$ is not one of the vectors specified in 2).

In the case of rank $n = 1$, the $\Psi_{\mathbf{v}}$ are the familiar *division polynomials* of an elliptic curve [15, p. 105]. Since the $\Psi_{\mathbf{v}}$ satisfy the elliptic net recurrence (1), we may make the following definition.

Definition 2 (Stange [10, Def. 6.1][11, Def. 7.2.1]). *For any elliptic curve E defined over K and non-zero points $P_1, \ldots, P_n \in E(K)$ such that no two are equal or inverses (or, if $n = 1$, P_1 is not a 2- or 3-torsion point), the map $W_{E,P_1,\ldots,P_n} : \mathbb{Z}^n \to K$ defined by*

$$W_{E,P_1,\ldots,P_n}(\mathbf{v}) = \Psi_{\mathbf{v}}(P_1, \ldots, P_n)$$

is an elliptic net called the elliptic net associated to $E, P_1, \ldots, P_n$.

Nearly all elliptic nets arise in this way (see [10][11]). For the remainder of this article, any elliptic net or elliptic divisibility sequence will be assumed to have this form.

Elliptic nets or elliptic divisibility sequences are arrays or sequences of values of K. The zeroes in this array are particularly important.

Definition 3. *The zeroes of an elliptic divisibility sequence or elliptic net appear as a sublattice of the lattice of indices. We call this sublattice the* lattice of zero-apparition. *In the case of a sequence, this sublattice is specified by a single positive integer – the smallest positive index of a vanishing term – and this number is called the* rank of zero-apparition.

The rank of zero-apparition of an elliptic divisibility sequence associated to a point P will equal the order of the point P. In the case of an array associated to points $P_1, \ldots, P_n$, the zeroes $(v_1, \ldots, v_n)$ correspond to linear combinations $\mathbf{v} \cdot \mathbf{P}$ that vanish.

Suppose $T : \mathbb{Z}^s \to \mathbb{Z}^t$ is a $\mathbb{Z}$-linear transformation. The following theorem relates the elliptic net associated to $\mathbf{P} \in E^s$ to that associated to $T(\mathbf{P}) \in E^t$.

Theorem 3 (Stange [10, Prop. 5.6][11, Thm. 6.2.3]). *Let T be any $t \times s$ integral matrix. Let $\mathbf{P} \in E^s$ and $\mathbf{v} \in \mathbb{Z}^t$. Then*

$$W_{E,\mathbf{P}}(T^{tr}(\mathbf{v})) = W_{E,T(\mathbf{P})}(\mathbf{v})$$

$$\times \prod_{i=1}^{t} W_{E,\mathbf{P}}(T^{tr}(\mathbf{e}_i))^{v_i^2 - v_i\left(\sum_{j \neq i} v_j\right)} \prod_{1 \leq i < j \leq t} W_{E,\mathbf{P}}(T^{tr}(\mathbf{e}_i + \mathbf{e}_j))^{v_i v_j} \quad (2)$$

This has several useful corollaries. For proofs see the cited references.

Theorem 4 (Ward [4, Thm. 8.1], Stange [11, Thm. 10.2.3][16]). *Suppose that $W_{E,P}(m) = 0$. Then for all $l, v \in \mathbb{Z}$, we have*

$$W_{E,P}(lm + v) = W_{E,P}(v)a^{vl}b^{l^2}$$

where

$$a = \frac{W_{E,P}(m+2)}{W_{E,P}(m+1)W_{E,P}(2)}, \qquad b = \frac{W_{E,P}(m+1)^2 W_{E,P}(2)}{W_{E,P}(m+2)}.$$

Furthermore, $a^m = b^2$. Therefore, there exists an $\alpha \in \bar{K}$, the algebraic closure of K, such that $\alpha^2 = a$ and $\alpha^m = b$, and so

$$W_{E,P}(lm + v) = W_{E,P}(v)\alpha^{(lm+v)^2 - v^2}.$$

Theorem 5 (Stange [11, Thm. 10.2.3][16]). *Suppose $\mathbf{r} = (r_1, r_2) \in \mathbb{Z}^2$ is such that $W_{E,P,Q}(\mathbf{r}) = 0$. For $l \in \mathbb{Z}$ and $\mathbf{v} = (v_1, v_2) \in \mathbb{Z}^2$ we have*

$$W_{E,P,Q}(l\mathbf{r} + \mathbf{v}) = W_{E,P,Q}(\mathbf{v})a_{\mathbf{r}}^{lv_1}b_{\mathbf{r}}^{lv_2}c_{\mathbf{r}}^{l^2}$$

where

$$a_{\mathbf{r}} = \frac{W(r_1+2, r_2)}{W(r_1+1, r_2)w(2,0)}, \quad b_{\mathbf{r}} = \frac{W(r_1, r_2+2)}{W(r_1, r_2+1)W(0,2)}, \quad c_{\mathbf{r}} = \frac{W(r_1+1, r_2+1)}{a_{\mathbf{r}}b_{\mathbf{r}}W(1,1)}.$$

3   Perfectly Periodic Sequences and Nets

Definition 4. *An elliptic divisibility sequence is called* perfectly periodic *if it is periodic with respect to its rank of zero-apparition. An elliptic net is called perfectly periodic if it is periodic with respect to its lattice of zero-apparition.*

Definition 5. *Let $f : \mathbb{Z}^n \to K^*$ be a quadratic function, and $k \in K^*$ a constant. Two elliptic nets W and W' are called* equivalent *if $W'(\mathbf{v}) = kf(\mathbf{v})W(\mathbf{v})$.*

As an example, let W be an elliptic divisibility sequence with rank of zero-apparition m. In one variable ($n = 1$), quadratic functions to K^* have the form $f(n) = \alpha^{n^2}$ for some $\alpha \in K^*$. Suppose we use α as defined by Theorem 4, i.e. $\alpha^2 = a, \alpha^m = b$, and let take $k = \alpha^{-1}$. Then $W'(n) = \alpha^{n^2 - 1}W(n)$, and this sequence is perfectly periodic. Suppose that $K = \mathbb{F}_q$ and $\gcd(q - 1, m) = 1$. In this case the conditions of Theorem 4 determine such an α uniquely, and it lies in K. Otherwise (if $\gcd(q-1, m) \neq 1$), two such α's will exist, equal up to sign. The two resulting perfectly periodic sequences will be equal at even-indexed locations and equal up to sign at odd-indexed locations.

The moral of the last paragraph is that any elliptic divisibility sequence is equivalent to a perfectly periodic one. We can give an explicit expression for such a perfectly periodic sequence.

Theorem 6. *Let K be a finite field of q elements, and E an elliptic curve defined over K. For all points $P \in E$ of order relatively prime to $q - 1$ and greater than 3, define*

$$\phi(P) = \left(\frac{W_{E,P}(q-1)}{W_{E,P}(q-1+\text{ord}(P))} \right)^{\frac{1}{\text{ord}(P)^2}}. \tag{3}$$

We will also define $\phi(\mathcal{O}) = 0$. Then for all n where $\gcd(n, m) = 1$,

$$\phi([n]P) = \phi(P)^{n^2} W_{E,P}(n). \tag{4}$$

In particular, for a point P of prime order not dividing $q - 1$ and greater than 3, the sequence $\phi([n]P)$ is a perfectly periodic elliptic divisibility sequence equivalent to $W_{E,P}(n)$.

More generally, let $\mathbf{P} \in E(K)^n$ be a collection of nonzero points, no two equal or inverses, and all elements of a single cyclic group and all having a fixed prime order greater than 3 not dividing $q - 1$. The n-array $\phi(\mathbf{v} \cdot \mathbf{P})$ (as $\mathbf{v}$ ranges over $\mathbb{Z}^n$) forms a perfectly periodic elliptic net equivalent to $W_{E,\mathbf{P}}(\mathbf{v})$. Specifically,

$$\phi(\mathbf{v} \cdot \mathbf{P}) = W_{E,\mathbf{P}}(\mathbf{v}) \prod_{i=1}^n \phi(P_i)^{v_i^2 - v_i(\sum_{j \neq i} v_j)} \prod_{1 \leq i < j \leq n} \phi(P_i + P_j)^{v_i v_j}.$$

Proof. The proof uses Theorem 3. We will demonstrate the method of proof in the rank one case before proceeding to the general case. Take $T = (l)$, so

$$W_{E,[l]P}(n) W_{E,P}(l)^{n^2} = W_{E,P}(nl).$$

By symmetry,

$$W_{E,[n]P}(l) W_{E,P}(n)^{l^2} = W_{E,P}(nl).$$

Let $m = \text{ord}(P)$. Thus, combining the above and using $l = q - 1$ and $q - 1 + m$ in turn,

$$\frac{W_{E,[n]P}(q-1) W_{E,P}(n)^{(q-1)^2}}{W_{E,P}(q-1)^{n^2}} = W_{E,[q-1]P}(n) = W_{E,[q-1+m]P}(n)$$

$$= \frac{W_{E,[n]P}(q-1+m) W_{E,P}(n)^{(q-1+m)^2}}{W_{E,P}(q-1+m)^{n^2}}$$

Rearranging,

$$\phi([n]P) = \phi(P)^{n^2} W_{E,P}(n).$$

When the order of P is prime, this holds for all n. Therefore, $\phi([n]P)$ is an elliptic divisibility sequence. By definition, $\phi([n]P)$ has period $\text{ord}(P)$ which is equal to the rank of apparition of $W_{E,P}$ and $\phi([n]P)$. So $\phi([n]P)$ is perfectly periodic.

For the rank n case, let m be the order of the cyclic group containing all the points under consideration. In Theorem 3, let $t = 1$ and $s = n$ and take $T = (v_1 \quad v_2 \quad v_3 \quad \cdots \quad v_n)$ to obtain

$$W_{E,\mathbf{P}}(l\mathbf{v}) = W_{E,\mathbf{v}\cdot\mathbf{P}}(l) W_{E,\mathbf{P}}(\mathbf{v})^{l^2}.$$

Now take $t = s = n$ in Theorem 3 , and $T = l\mathrm{Id}_n$ to obtain

$$W_{E,\mathbf{P}}(l\mathbf{v}) = W_{E,l\mathbf{P}}(\mathbf{v}) \prod_{i=1}^{n} W_{E,\mathbf{P}}(le_i)^{v_i^2 - v_i(\sum_{j \neq i} v_j)} \prod_{1 \leq i < j \leq n} W_{E,\mathbf{P}}(le_i + le_j)^{v_i v_j}.$$

Note that

$$W_{E,\mathbf{P}}(le_i) = W_{E,P_i}(l), \qquad W_{E,\mathbf{P}}(le_i + le_j) = W_{E,P_i+P_j}(l).$$

Combining the above, we have

$$W_{E,l\mathbf{P}}(\mathbf{v}) = \frac{W_{E,\mathbf{v}\cdot\mathbf{P}}(l) W_{E,\mathbf{P}}(\mathbf{v})^{l^2}}{\prod_{i=1}^{n} W_{E,P_i}(l)^{v_i^2 - v_i(\sum_{j \neq i} v_j)} \prod_{1 \leq i < j \leq n} W_{E,P_i+P_j}(l)^{v_i v_j}}.$$

Comparing this in the case of $l = q - 1$ and $l = q - 1 + m$ gives the required result, as before.

In light of this theorem, when the order of P is prime (which we shall always assume), we will use the convenient notation

$$\widetilde{W}_{E,P}(n) = \phi([n]P).$$

and call this the *perfectly periodic elliptic divisibility sequence associated to E and P*. The attractive property of a perfectly periodic sequence is formula (3): $\widetilde{W}_{E,P}(n)$ can be calculated *as a function of the point $[n]P$ on the curve* without knowledge of n.

Corollary 1. *Suppose that E is an elliptic curve over a field $K = \mathbb{F}_q$ and $P \in E(K)$ is of prime order $m \geq 4$. The period of the sequence $W_{E,P}$ is $m \operatorname{ord}_{K^*}(\phi(P))$.*

Proof. First, $\phi([n]P)$ has period exactly m. Since, if the period were $m' < m$, then $W_{E,P}(m') = 0$, a contradiction. The result now follows from equation (4).

The ratio between the period and the rank of zero-apparition, which we've demonstrated to be $\operatorname{ord}_{K^*}(\phi(P))$, is called τ by Morgan Ward [4, Thm. 11.1].

4   The Hard Problems

As we have seen, elliptic nets are closely related to the points on an elliptic curve. In this section, we will see specifically how to compute them, and how they relate, algorithmically, to the points.

The choice of segment $0 < k < \operatorname{ord}(P)$ is not crucial in Problem 2 (EDS Association): it could be restated for any segment $i \operatorname{ord}(P) < k < (i+1) \operatorname{ord}(P)$. This problem is trivial for a perfectly periodic sequence or net (since $\widetilde{W}(k) = \phi(Q)$ is computable in $\log q$ time). For the non-perfectly periodic case, the problem appears to be much harder. As for Problem 4 (EDS Discrete Log), on the other hand, for non-perfectly periodic elliptic divisibility sequences, it can be solved by computing an $\mathbb{F}_q^*$ discrete log. For this problem, it is the case of perfect periodicity that seems very difficult.

We will see that these hard problems are related according to the following diagram.

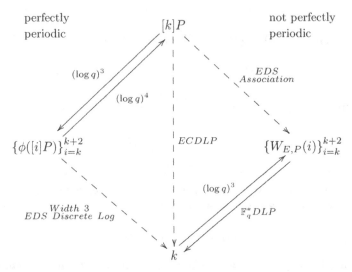

We demonstrate the complexity of solving the problems associated to the solid lines in the following series of theorems. The solid line labelled $\mathbb{F}_q^*$DLP has the complexity of a discrete logarithm problem in $\mathbb{F}_q^*$ (this is sub-exponential by index calculus). No sub-exponential algorithms are known for the dotted lines.

Since our concern is polynomial time vs. non-polynomial time, in the following we assume naive arithmetic in $\mathbb{F}_q$, i.e. we bound the time to do basic $\mathbb{F}_q$ operations by $O((\log q)^2)$ for simplicity.

Lemma 1. *Let E be an elliptic curve defined over K, and $P \in E(K)$ be a point of prime order not less than 4. The x-coordinate of $[n]P$, $x([n]P)$, can be calculated in $O((\log q)^2)$ time from the three terms $W_{E,P}(n-1)$, $W_{E,P}(n)$, and $W_{E,P}(n+1)$ or from the three terms $\widetilde{W}_{E,P}(n-1)$, $\widetilde{W}_{E,P}(n)$, and $\widetilde{W}_{E,P}(n+1)$.*

Proof. See [11, Lemma 6.2.2] for the following identity:

$$\frac{W_{E,P}(n-1)W_{E,P}(n+1)}{W_{E,P}(n)^2} = x(P) - x([n]P). \tag{5}$$

The left-hand side of (5) is invariant under equivalence, and so the same calculation applies if we put tilde's on the W's.

Theorem 7 (Shipsey [2, Thm 3.4.1]). *Let E be an elliptic curve over K, and $P \in E(K)$ a point of order not less than 4. Given a value t, the term $W_{E,P}(t)$ in the elliptic divisibility sequence associated to E, P can be calculated in $O((\log t)(\log q)^2)$ time.*

Proof. For completeness, we give a simplified version of Shipsey's algorithm here. Following Shipsey, denote by $\langle W_{E,P}(k) \rangle$ the segment or *block centred at k* of eight terms $W_{E,P}(k-3)$, $W_{E,P}(k-2)$, ..., $W_{E,P}(k+3)$, $W_{E,P}(k+4)$ of the sequence. The block centred at t can be calculated from the block centred at 1 via a double-and-add algorithm based on an addition chain for t. The calculation of the new block from the previous depends on two instances of the recurrence (one such calculation for each term of the new block):

$$W(2i-1,0) = W(i+1,0)W(i-1,0)^3 - W(i-2,0)W(i,0)^3 \ ,$$
$$W(2i,0) = \big(W(i,0)W(i+2,0)W(i-1,0)^2$$
$$-W(i,0)W(i-2,0)W(i+1,0)^2 \big) / W(2,0) \ .$$

To begin we must calculate the block centred at 1. Recalling that $W(0) = 0$, $W(1) = 1$ and $W(-n) = -W(n)$, we must calculate $W(i)$ for $i = 2, 3, 4$. Precise formulae in terms of the coordinates of P and the Weierstrass coefficients for E can be found in [15, p. 105] or for long Weierstrass equations in [17, p. 80]. This algorithm takes $O(\log t)$ steps, each of which involves a fixed number of $\mathbb{F}_q^*$ multiplications and additions, which take $O((\log q)^2)$ time at worst.

Theorem 8. *Let E be an elliptic curve over $\mathbb{F}_q$, and $P \in E(\mathbb{F}_q)$ a point of prime order not dividing $q-1$ and greater than 3. Given a point $Q = [k]P$, the term $\phi(Q) = \widetilde{W}_{E,P}(k)$ can be calculated in $O((\log q)^3)$ time without requiring knowledge of k.*

Proof. We use equation (3). Using Theorem 7 to calculate the ratio of terms inside the parentheses takes $\log(q-1+\mathrm{ord}(Q))+\log(q-1)$ steps. Since $\mathrm{ord}(Q)$ is on the order of q, this is $O((\log q)^3)$ time at worst. The other necessary operation in (3) is to find the inverse of $\mathrm{ord}(Q)^2$ modulo $q-1$, and to raise to that exponent. Both these are also $O(\log q)$ operations.

Theorem 9. *Let E be an elliptic curve over $\mathbb{F}_q$, and $P \in E(\mathbb{F}_q)$ a point of prime order not dividing $q-1$ and greater than 3. Given the $\widetilde{W}_{E,P}(k)$, $\widetilde{W}_{E,P}(k+1)$ and $\widetilde{W}_{E,P}(k+2)$, the point $Q = [k]P$ can be calculated in probabilistic $O((\log q)^4)$ time without requiring knowledge of k.*

Proof. Calculate $x([k+1]P)$ by Lemma 1. We can calculate the corresponding possible values for y in probabilistic time $O((\log q)^4)$ [18, §7.1-2]. To determine which of the two points with this x-coordinate is actually $[k+1]P$, first take one of the two candidate points, and proceed on the assumption that it is $[k+1]P$. Using the addition formula for elliptic curves, calculate $x([k+1]P+P) = x([k+2]P)$. Compare this with (5) to determine $\widetilde{W}(k+3)$. Also determine $\widetilde{W}(k+4)$ in this manner. Then, if the terms $\widetilde{W}(k), \ldots, \widetilde{W}(k+4)$ satisfy the recurrence instance

$$\widetilde{W}(k+4)\widetilde{W}(k) = \widetilde{W}(k+1)\widetilde{W}(k+3)\widetilde{W}(2)^2 - \widetilde{W}(3)\widetilde{W}(1)\widetilde{W}(k+2)^2,$$

our assumption about the point we chose is correct. If this recurrence does not hold, then the point we chose was incorrect, and the other one is the point

$[k+1]P$ we seek. For, it is impossible that both points cause the above equation to be satisfied: any sequence of four consecutive terms in an elliptic divisibility sequence determines the entire sequence uniquely. Finally, knowing $[k+1]P$, we can calculate $Q = [k]P = [k+1]P - P$.

The following theorem is implicit in the work of Shipsey; see Section 5.2 for an explanation.

Theorem 10. *Suppose P has prime order not dividing $q-1$ and greater than 3, and $\phi(P)$ is a primitive root in $\mathbb{F}_q^*$. Given $W_{E,P}(k), W_{E,P}(k+1), W_{E,P}(k+2)$, where it can be assumed that $0 < k < \mathrm{ord}(P)$, calculating k can be reduced to a single discrete logarithm in $\mathbb{F}_q^*$ in probabilistic $O((\log q)^4)$ time.*

Proof. We can deduce the x-coordinate of the point $Q = [k]P$ by Lemma 1. Compute the two corresponding y-coordinates, which takes probabilistic time $O((\log q)^4)$ [18, §7.1-2]. Choosing one of the two possible y-coordinates, we have either $Q = [k]P$ or $Q = [-k]P$. To determine which is correct, use the trick of the proof of Theorem 9. Suppose it is the former; then, from Theorem 6, we have

$$\frac{\phi([k+1]P)}{\phi([k]P)} = \phi(P)^{2k+1} \frac{W_{E,P}(k+1)}{W_{E,P}(k)}. \tag{6}$$

So k satisfies an equation of the form $A = B^{2k+1}$ where A and B are known, and B has order $q-1$ by assumption. Therefore, we are reduced to solving a discrete logarithm of the form $A = B^x$ for $0 \le x < q-1$, with the understanding that k will be one of $(x-1)/2$ or $(x+q-1)/2$. (In fact, if $q-1 < m$, there may be at most two other possible values of k to check: the above values shifted by $q-1$.)

Remark 1. Let $m = \mathrm{ord}(P)$. Suppose that $\gcd(m, q-1) = 1$. As an integer k ranges over representatives of a single coset in $\mathbb{Z}/m\mathbb{Z}$, it ranges over all possible cosets of $\mathbb{Z}/(q-1)\mathbb{Z}$. Therefore, we cannot expect to find the set of k such that $Q = [k]P$ (i.e. a coset in $\mathbb{Z}/m\mathbb{Z}$) by solving an equation of the form $A = B^k$ in $\mathbb{F}_q^*$ (i.e. solving modulo $q-1$). One solution to this problem is to attempt to solve for an *integer* k (instead of a coset) – say, for example, the smallest non-negative k with $Q = [k]P$. This is in essence what the preceeding theorem does. With this in mind, we set some terminology.

Definition 6. *Let Q be a multiple of P on an elliptic curve E. The minimal multiplier of Q with respect to P is the smallest non-negative value of k such that $Q = [k]P$.*

Note that the minimal multiplier satisfies $0 \le k < \mathrm{ord}(P)$.

5 $\mathbb{F}_q^*$ Discrete Logarithm, the Tate Pairing and MOV/Frey-Rück Attack

Theorem 10 uses terms of the elliptic divisibility sequence to give a discrete logarithm problem in $\mathbb{F}_q^*$. We demonstrate some variations on this theme, and relate these types of equations to the Tate pairing, and to an ECDLP attack given by Shipsey [2].

5.1   An $\mathbb{F}_q^*$ DLP Equation of the Form $A = B^k$ from Periodicity Properties

The $\mathbb{F}_q^*$ DLP equations we consider are consequences of Theorem 3, but many can be conveniently understood in terms of its corollary Theorem 5. The following example involves the terms $W_{E,P}(k)$ and $W_{E,P}(k+1)$, and requires knowledge of $Q = [k]P$. The following diagram is suggestive for the discussion.

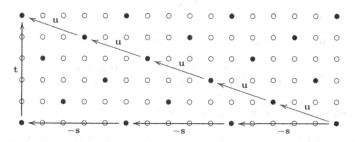

In this picture of $\mathbb{Z}^2$, $\mathbf{u} = (-3, 1)$, $\mathbf{s} = (5, 0)$ and $\mathbf{t} = (0, 5)$. Vectors $\mathbf{u}$ and $\mathbf{s}$ generate the lattice of zero-apparition Λ for some elliptic net W associated to points P and $Q = [3]P$ of order 5. The vector $\mathbf{t}$ is also in Λ. One coset of $\mathbb{Z}^2$ modulo Λ is shown as the solid discs.

Theorem 5 shows the transformation relative to translation by a vector $\mathbf{r} \in \Lambda$: it relates $W(\mathbf{v} + \mathbf{r})$ to $W(\mathbf{v})$ for each $\mathbf{v}$. This Lemma can be applied repeatedly, and different 'paths' from one point to another must agree. In the picture above, the translation property which relates $W(\mathbf{v} + (-15, 5))$ to $W(\mathbf{v})$ can be calculated by applying the transformation associated to $\mathbf{u}$ five times (the diagonal path) or by applying the transformation associated to $-\mathbf{s}$ three times followed by that associated to $\mathbf{t}$ once (the sides of the triangle).

In the general case, we have $Q = [k]P$. Then the lattice of zero-apparition Λ for $W = W_{E,P,Q}$ includes vectors $\mathbf{u} = (-k, 1)$, $\mathbf{s} = (m, 0)$ and $\mathbf{t} = (0, m)$. Suppose $\mathbf{r} = (r_1, r_2)$ is an element of Λ for $W = W_{E,P,Q}$. By Theorem 5, we have for all $l \in \mathbb{Z}$ and $\mathbf{v} \in \mathbb{Z}^2$,

$$W(l\mathbf{r} + \mathbf{v}) = W(\mathbf{v})a_{\mathbf{r}}^{lv_1} b_{\mathbf{r}}^{lv_2} c_{\mathbf{r}}^{l^2} \tag{7}$$

where

$$a_{\mathbf{r}} = \frac{W(r_1 + 2, r_2)}{W(r_1 + 1, r_2)W(2, 0)}, \quad b_{\mathbf{r}} = \frac{W(r_1, r_2 + 2)}{W(r_1, r_2 + 1)W(0, 2)}, \quad c_{\mathbf{r}} = \frac{W(r_1 + 1, r_2 + 1)}{a_{\mathbf{r}} b_{\mathbf{r}} W(1, 1)}.$$

We expect appropriate relationships between $a_{\mathbf{u}}$, $b_{\mathbf{u}}$, $c_{\mathbf{u}}$, $a_{\mathbf{s}}$, $b_{\mathbf{s}}$, etc. The $\mathbb{F}_q^*$ DLP equation we seek is one such relationship. We have

$$a_{\mathbf{s}} = \frac{W(m + 2, 0)}{W(m + 1, 0)W(2, 0)}, \quad a_{\mathbf{t}} = \frac{W(2, m)}{W(1, m)W(2, 0)}, \quad a_{\mathbf{u}} = \frac{W(2 - k, 1)}{W(1 - k, 1)W(2, 0)}.$$

For each $i \in \mathbb{Z}$, we apply (7) to obtain

$$\frac{W(-ik + 1, i - 1)W(0, -1)}{W(1, -1)W(-ik, i - 1)} = a_{\mathbf{u}}^i \tag{8}$$

Set $i = m$ in (8), and apply (7) four times:

$$a_u^m = \frac{W(-mk+1, m-1)W(0, -1)}{W(1, -1)W(-mk, m-1)}$$

$$= \left(\frac{W(-mk+1, m-1)}{W(-mk+1, -1)}\right)\left(\frac{W(-mk+1, -1)}{W(1, -1)}\right)\left(\frac{W(0, -1)}{W(-mk, -1)}\right)\left(\frac{W(-mk, -1)}{W(-mk, m-1)}\right)$$

$$= \frac{a_t^{-mk+1}b_t^{-1}c_t^1 a_s^{-k} b_s^k c_s^{k^2}}{a_t^{-mk}b_t^{-1}c_t^1 a_s^0 b_s^k c_s^{k^2}} = a_t a_s^{-k}$$

Setting $i = 1$ in (8), we obtain an expression

$$a_u = \frac{W(-k+1, 0)W(0, -1)}{W(1, -1)W(-k, 0)} = -\frac{W_{E,P}(k-1)}{W_{E,P}(k)W(1, -1)}$$

which, when substituted into the last calculation, yields

$$\left(\frac{W(m+1, 0)W(2, 0)}{W(m+2, 0)}\right)^k = \left(\frac{W_{E,P}(k-1)}{W_{E,P}(k)}\right)^m \left(-\frac{W(1, m)W(2, 0)}{W(2, m)W(1, -1)^m}\right). \quad (9)$$

5.2 An $\mathbb{F}_q^*$ DLP Equation from Shipsey's Thesis

The possibility of such an equation was observed by Rachel Shipsey in her thesis [2, (6.3)]. She uses one-dimensional periodicity properties to derive the following equation:

$$\frac{W_{E,P}((m+1)(k+1))W_{E,P}(k)}{W_{E,P}((m+1)k)W_{E,P}(k+1)} = W_{E,P}(m+1)^{2k+1} \quad (10)$$

Shipsey then argues that without knowledge of k the left hand side can be calculated up to a factor of

$$\left(\frac{W_{E,P}(k)}{W_{E,P}(k-1)}\right)^{m(m+2)}.$$

This is very much of the same spirit as equation (9), and in fact, Theorem 3 can be used to rewrite (10) in this form:

$$\frac{W_{E,P,Q}(m+1, m+1)}{W_{E,P,Q}(0, m+1)}\left(\frac{W_{E,P}(k+1)}{W_{E,P}(k)}\right)^{m(m+2)} = W_{E,P}(m+1)^{2k+1}. \quad (11)$$

By Lemma 1, knowledge of $Q, W_{E,P}(k), W_{E,P}(k-1)$ determines $W_{E,P}(k+1)$, and so this is very much equivalent to Shipsey's analysis. Note that the unknown terms in (11) are raised to the exponent $m+2$. At first blush, this may appear to lead to an ECDLP attack for $q - 1 = m + 2$ (where the unknown terms will disappear). However, this is not allowed by Remark 1. In fact, it turns out that if $q - 1 = m + 2$, then $W_{E,P}(m+1) = 1$ (this eventually follows from Theorem 3 also).

5.3 $\mathbb{F}_q^*$ DLP Equations and the Tate Pairing

Choose $m \in \mathbb{Z}^+$. Let E be an elliptic curve defined over a finite field K containing the m-th roots of unity. Suppose $P \in E(K)[m]$ and $Q \in E(K)/mE(K)$. Since P is an m-torsion point, $m(P) - m(\mathcal{O})$ is a principal divisor, say $\mathrm{div}(f_P)$. Choose another divisor D_Q defined over K such that $D_Q \sim (Q) - (\mathcal{O})$ and with support disjoint from $\mathrm{div}(f_P)$. Then, we may define the Tate pairing

$$\tau_m : E(K)[m] \times E(K)/mE(K) \to K^*/(K^*)^m$$

and Weil pairing

$$e_m : E(K)[m] \times E(K)[m] \to \mu_m$$

by

$$\tau_m(P,Q) = f_P(D_Q), \qquad e_m(P,Q) = f_P(D_Q)f_Q(D_P)^{-1}.$$

Both are non-degenerate bilinear pairings, while the Weil pairing is alternating. For details, see [19][20].

The Tate pairing and Weil pairing are used in the MOV [21] and Frey-Rück [22] attacks on the ECDLP. These use the Weil and Tate pairings, respectively, to translate an instance of the ECDLP into an $\mathbb{F}_q^*$ DLP equation, where index calculus methods may be used. The basic idea, illustrated here for the Tate pairing, is that $Q = [k]P$ implies $\tau_m(Q,S) = \tau_m(P,S)^k$ by bilinearity. If S can be chosen so that $\tau_m(P,S)$ is non-trivial, and if the Tate pairing takes values in a manageably small finite field, then index calculus methods can be used to determine k. In particular, this attack applies for curves E over $\mathbb{F}_q$ where $m = q - 1$.

In (11) and (9), all the terms may be calculated from knowledge of m, P and Q except for $W_{E,P}(k)$ and $W_{E,P}(k-1)$. However, notice that these unknown terms are raised to the power m. Therefore, in the case that $m = q - 1$, no extra information is needed and the ECDLP is reduced to an $\mathbb{F}_q^*$ DLP; this works in exactly the cases that the MOV or Frey-Rück attack applies.

These sorts of 'alternate versions' of the MOV/Frey-Rück attack do have a relation to the Tate pairing.

Theorem 11 (Stange [11, Thm. 17.2.1][13, Thm. 6]). *Let E be an elliptic curve, $m \geq 4$, and $P \in E[m]$. Let $Q, S \in E$ be such that $S \notin \{\mathcal{O}, -Q\}$. Let W be an elliptic net of rank n, associated to points $\mathbf{T} \in E(K)^n$. Let $\mathbf{s}, \mathbf{p}, \mathbf{q} \in \mathbb{Z}^n$ be such that*

$$P = \mathbf{p} \cdot \mathbf{T}, \qquad Q = \mathbf{q} \cdot \mathbf{T}, \qquad S = \mathbf{s} \cdot \mathbf{T}.$$

Let $\tau_m : E[m] \times E/mE \to K^*/(K^*)^m$ be the Tate pairing. Then

$$\tau_m(P,Q) = \frac{W(m\mathbf{p} + \mathbf{q} + \mathbf{s})W(\mathbf{s})}{W(m\mathbf{p} + \mathbf{s})W(\mathbf{q} + \mathbf{s})}.$$

Now equations (9) and (11) can be re-written as statements in terms of the Tate pairing.

Equation (9): Use Theorem 11 with $\mathbf{p} = (1,0), \mathbf{q} = (-1,0), \mathbf{s} = (2,0)$ for the left-hand side and $\mathbf{p} = (0,1), \mathbf{q} = (-1,0), \mathbf{s} = (2,0)$ for the right. This rewrites (9) as

$$\tau_m(P, -P)^k = \tau_m(Q, -P).$$

Equation (11): This is somewhat more complicated. From Theorem 4 with $m = q - 1$ and Theorem 11 with various parameters,

$$W_{E,P}(m+1)^2 \tau_m(P,P)^{-2} = \left(\frac{W_{E,P}(m+1)^2 W_{E,P}(2)}{W_{E,P}(m+2)} \right)^2 = b^2 = a^m = 1,$$

$$\tau_m(P,Q) = \frac{W_{E,P,Q}(m+1,1)W_{E,P,Q}(1,0)}{W_{E,P,Q}(m+1,0)W_{E,P,Q}(1,1)},$$

$$\tau_m(Q,P) = \frac{W_{E,P,Q}(1,m+1)W_{E,P,Q}(0,1)}{W_{E,P,Q}(0,m+1)W_{E,P,Q}(1,1)},$$

$$1 = \tau_m(P,0) = \tau_m(P,[m]Q) = \frac{W_{E,P,Q}(m+1,m+1)W_{E,P,Q}(1,1)}{W_{E,P,Q}(m+1,1)W_{E,P,Q}(1,m+1)}.$$

All of which, taken together, rewrites (11) as

$$\tau_m(P,Q)\tau_m(Q,P) = \tau_m(P,P)^{2k}.$$

Equation (4) (with $n = k$) does not, however, lend itself to this sort of re-writing in terms of pairings in the case $m = q - 1$, as the very definition of $\phi(P)$ requires the assumption that $\gcd(m, q - 1) = 1$.

6   ECDLP through EDS Association

The previous sections have demonstrated that there are a variety of ways to translate an ECDLP into an $\mathbb{F}_q^*$ DLP. The $\mathbb{F}_q^*$ DLP equation is in terms of elements of the sequence $W_{E,P}$. For example in (9), the elements are $W_{E,P}(k)$ and $W_{E,P}(k - 1)$. The problem of finding these terms (with knowledge of $Q = [k]P$ but not k) is *EDS Association*. In this example, however, it is only their quotient that is needed. Depending on the form of the $\mathbb{F}_q^*$ DLP equation, different information (certain terms or ratios of terms) suffices. We formalise the most general statement of this in the following theorem.

Proposition 1. *Fix an elliptic curve E defined over $\mathbb{F}_q$, and $P \in E(\mathbb{F}_q)$ of prime order greater than three and not dividing $q - 1$. Suppose $\phi(P)$ has order $q - 1$ in $\mathbb{F}_q^*$. With knowledge of any product*

$$\prod_{i=1}^{N} W_{E,P}(p_i(k))^{e_i}, \tag{12}$$

where the $e_i \in \mathbb{Z}$, and $p_i(x) \in \mathbb{Z}[x]$ of degree at most one, and $t(x) = \sum_{i=1}^{N} e_i p_i(x)^2$ is a non-constant polynomial in $\mathbb{Z}[x]$, the value of k can be determined in subexponential time in q.

Proof. Combine appropriate instances of equation (4) of Theorem 6 in such a way that $t(k)$ satisfies an equation in $\mathbb{F}_q^*$ of the form $A = B^{t(k)}$. That is, combine one instance for each $n = p_i(k)$ with multiplicities given by the respective e_i, and obtain an equation of the form

$$\frac{\prod_{i=1}^{N} \widetilde{W}_{E,P}(p_i(k))^{e_i}}{\prod_{i=1}^{N} W_{E,P}(p_i(k))^{e_i}} = \phi(P)^{t(k)}.$$

(Earlier in the paper we derived equation (6) in this manner. In that case, $1 = e_1 = -e_2$, $p_1(k) = k+1$, and $p_2(k) = k$, so that $t(k) = 2k+1$.)

The left hand side A includes the known product (12) as well as terms of the form $\phi([p_i(k)]P)$, while $B = \phi(P)$. The N points $[p_i(k)]P$ can each be calculated from knowledge of P and $Q = [k]P$ without knowledge of k in $O(\log q)$ curve operations. Then the various ϕ terms can be computed in time $O((\log q)^3)$ by Theorem 8. Thus we have computed A and B.

Solving the discrete logarithm $A = B^{t(k)}$ for $t(k)$ can be done in sub-exponential time by index calculus methods. Since $t(k)$ has degree at most two, solving for k from $t(k)$ requires finding square roots in $\mathbb{Z}/(q-1)\mathbb{Z}$ (see [23, §3.5.1]), which in turn depends on factoring $q-1$ which is sub-exponential using the number field sieve.

It is evident that the most costly steps are the index calculus step and the factorisation of $q-1$ if t has degree two. In many cases these algorithms have run time $r(q) = \exp(c(\log q)^{1/3}(\log \log q)^{2/3})$ [24, p.306].

7   ECDLP and Quadratic Residues

We will show that determining only one bit of information – the residuosity – about a term $W_{E,P}(k)$ may suffice to solve the ECDLP in some cases. First, we observe a hypothetical method of attack for ECDLP.

Proposition 2. *Suppose that $E(\mathbb{F}_q)$ is of odd order. Let P be a point of order relatively prime to $q-1$. Given an oracle which can determine the parity of the minimal multiplier of any non-zero point Q in $\langle P \rangle$ in time $O(T(q))$, the elliptic curve discrete logarithm for any such Q can be determined in time $O(T(q)\log q + (\log q)^2)$.*

Proof. Suppose that k is the minimal multiplier of Q with respect to P. The basic algorithm is:

1. If $Q = P$, stop.
2. Call the oracle to determine the parity of k. If k is even, find Q' such that $[2]Q' = Q$. If k is odd, find Q' such that $[2]Q' = Q - P$.
3. Set $Q = Q'$ and return to step 1.

In Step 2, since the cyclic group $\langle P \rangle$ has odd order, and the curve has no 2-torsion, there is a unique Q'. It can be found in $O(\log q)$ time (see [25] for methods). Furthermore, $Q' = [k']P$ where

$$k' = \begin{cases} k/2 & k \text{ even} \\ (k-1)/2 & k \text{ odd} \end{cases}.$$

Then k' is the minimal multiplier for Q' with respect to P. At the end of this process, the value of the original k can be deduced from the sequence of steps taken. For each even step, record a '0', and for each odd step a '1', writing from right to left, and adding a final '1': this will be the binary representation of k. The number of steps is $\log_2 k = O(\log q)$.

Proposition 3. *Fix an elliptic curve E defined over $\mathbb{F}_q$ of characteristic not equal to two, and $P \in E(\mathbb{F}_q)$ of prime order greater than three and not dividing $q - 1$. Suppose that $\phi(P)$ is a quadratic non-residue. Then, with knowledge of the quadratic residuosity of any product of the form*

$$\prod_{i=1}^{N} W_{E,P}(p_i(k))^{e_i}, \tag{13}$$

where the $e_i \in \mathbb{Z}$, and $p_i(x) \in \mathbb{Z}[x]$ of degree at most one, and $t(x) = \sum_{i=1}^{N} e_i p_i(x)^2$ is not constant as a function $\mathbb{Z}/2\mathbb{Z} \to \mathbb{Z}/2\mathbb{Z}$, the parity of k can be determined in time $O(N(\log q)^3)$.

Proof. By Theorem 6, the value $t(k)$ satisfies an equation in $\mathbb{F}_q^*$ of the form $A = B^{t(k)}$ (exactly as in the proof of Proposition 1). The quadratic residuosity of A can be calculated in time $O(N(\log q)^3)$ as in the proof of Proposition 1. Now, $B = \phi(P)$ is a quadratic non-residue. The parity of $t(k)$ can be calculated from these values in constant time (i.e. consider the question in K^* modulo $(K^*)^2$). The parity of k is determined by checking the parity of $t(0)$ and $t(1)$. This final step takes constant time.

Corollary 2. *Let E be an elliptic curve over a field of characteristic not equal to two, and suppose E has an odd number of $\mathbb{F}_q$ points. Let P have prime order greater than 3 and not dividing $q-1$, and suppose that $\phi(P)$ is a quadratic non-residue, and let k be the minimal multiplier of a multiple Q of P. Given P, Q and an oracle which can determine the quadratic residuosity of $W_{E,P}(k)$ in time $O(T(q))$, the elliptic curve discrete logarithm for any such Q can be determined in time $O((\log q)(T(q) + (\log q)^3))$.*

Proof. This follows from Proposition 3 with $N = 1, e_1 = 1, p_1(x) = x$ and Proposition 2.

A few remarks are in order.

1. If $\phi(P)$ is a quadratic residue, one solution to this obstacle is to replace the initial problem of $Q = [k]P$ with the equivalent problem of

$[n]Q = [k]([n]P)$ for any n such that $\phi([n]P)$ is a quadratic non-residue. The sequence $\widetilde{W}_{E,P}(n)$ can be calculated term-by-term until such an n is found. The existence of such an n is guaranteed when -1 is a quadratic non-residue in $\mathbb{F}_q$, in which case $\phi([m-1]P) = -\phi(P)$ suffices. Other cases are less clear.

2. The condition that the order of P is relatively prime to the even quantity $q-1$ is required in several ways. First, for the very definition of ϕ (Theorem 6). Furthermore, if the order m of the group $\langle P \rangle$ is even, in which case E has 2-torsion, then multiplication by 2 is not an automorphism, and so there is no unique 'half' of a point (this is the same difficulty that prevents this sort of parity attack on an $\mathbb{F}_q^*$ discrete log). However, if $m|(q-1)$ is odd, then k satisfies a discrete logarithm equation of the form $A = B^k$ in the group $K^*/(K^*)^m$, which has an odd number of elements. Therefore, this does not determine the parity of k.

3. Similarly, if $q-1$ is odd (i.e. $\mathbb{F}_q$ has characteristic 2), then $A = B^k$ does not carry information about the parity of k.

8   The EDS Residue Problem

In light of the preceeding section, it is natural to define the problem of EDS Residue (Problem 3). In Section 10 we will show that it is equivalent to the elliptic curve discrete logarithm in sub-exponential time. How might one determine the quadratic residuosity of $W_{E,P}(k)$? Our first observation is that knowledge of the residuosity of one term $W_{E,P}(k)$ would determine the residuosity of the next term. In this section we assume as always that P is of prime order not dividing $q-1$ and greater than 3.

Proposition 4. *Suppose Q is a known element of $\langle P \rangle$, but that its minimal multiplier k is unknown. The quadratic residuosity of $W_{E,P}(k+1)/W_{E,P}(k)$ can be calculated in $O((\log q)^3)$ time.*

Proof. From (4) with $n = k$ and $n = k+1$, we have

$$\frac{\phi(Q+P)}{\phi(Q)} = \phi(P)^{2k+1}\left(\frac{W_{E,P}(k+1)}{W_{E,P}(k)}\right).$$

The calculation of the terms $\phi(P), \phi(Q)$, and $\phi(P+Q)$ each take $O((\log q)^3)$ time.

Therefore, based on knowledge of Q but not k, the sequence

$$S(n) = \left(\frac{W_{E,P}(n)}{q}\right)\left(\frac{W_{E,P}(k)}{q}\right)$$

for $n = k, \ldots, k+N$ may be calculated in $O(N(\log q)^3)$ time. Then the sequence

$$\left(\frac{W_{E,P}(n)}{q}\right)$$

is either $S(n)$ or $-S(n)$. To determine whether it is $S(n)$ or $-S(n)$ is to determine the quadratic residuosity of $W_{E,P}(k)$.

Therefore, if some bias, or some pattern, for quadratic residues of the elliptic divisibility sequence $W_{E,P}(n)$ were known, then the correct choice of the two sequences above could be determined. However, as yet we have no evidence to suggest that the ratio of quadratic residues among the terms is not $1/2$ in general.

9   ECDLP through EDS Discrete Log in the Case of Perfect Periodicity

Problem 4 (EDS Discrete Log) is less unusual in flavour than the other problems considered here: general discrete logarithm attacks will apply. Recall the proof of Theorem 7, in which *blocks centred at* k are defined – denote this as $B(k)$. From $B(k)$, the recurrence relation can be used to calculate $B(2k)$ or $B(2k + 1)$. In fact, Shipsey goes further, and shows how two blocks $B(k), B(k')$ can be added to obtain a block $B(k + k')$ in a similarly efficient manner (see [2, p. 23]). This means that the sequence of blocks $B(n)$ is a sequence along which we can move easily by addition and $\mathbb{Z}$-multiplication. Therefore, generic algorithms such as Baby-Step-Giant-Step and Pollard's ρ can be applied to this problem.

10   Equivalence of Hard Problems

Proof (Proof of Theorem 3). (2) $\implies$ (1): Theorem 10. ; (1) $\implies$ (2): If k is known, we can assume $0 < k \leq \mathrm{ord}(P)$, and then $W_{E,P}(k)$ can be calculated in $O((\log k)(\log q)^2) = O((\log q)^3)$ time. ; (1) $\implies$ (3): Theorem 9. ; (3) $\implies$ (1): Theorem 8 allows calculation of $\phi([k]P)$, $\phi([k + 1]P)$, and $\phi([k + 2]P)$ in subexponential time. ; (4) $\implies$ (1): Corollary 2. ; (2) $\implies$ (4): Residuosity of a value in $\mathbb{F}_q^*$ can be determined in sub-exponential time (see [26] for algorithms).

References

1. Pohlig, S.C., Hellman, M.E.: An improved algorithm for computing logarithms over GF(p) and its cryptographic significance. IEEE Trans. Information Theory IT-24, 106–110 (1978)
2. Shipsey, R.: Elliptic Divibility Sequences. PhD thesis, Goldsmiths, University of London (2001)
3. Shipsey, R., Swart, C.: Elliptic divisibility sequences and the elliptic curve discrete logarithm problem (2008), http://eprint.iacr.org/2008/444
4. Ward, M.: Memoir on elliptic divisibility sequences. Amer. J. Math. 70, 31–74 (1948)
5. Swart, C.: Elliptic curves and related sequences. PhD thesis, Royal Holloway and Bedford New College, University of London (2003)
6. Ayad, M.: Périodicité (mod q) des suites elliptiques et points S-entiers sur les courbes elliptiques. Ann. Inst. Fourier (Grenoble) 43, 585–618 (1993)

7. Silverman, J.H.: Common divisors of elliptic divisibility sequences over function fields. Manuscripta Math. 114, 431–446 (2004)
8. Silverman, J.H.: p-adic properties of division polynomials and elliptic divisibility sequences. Math. Ann. 332, 443–471 (2005)
9. Everest, G., Mclaren, G., Ward, T.: Primitive divisors of elliptic divisibility sequences. J. Number Theory 118, 71–89 (2006)
10. Stange, K.E.: Elliptic nets and elliptic curves (submitted) (2007), http://arxiv.org/abs/0710.1316v2
11. Stange, K.E.: Elliptic nets and elliptic curves. PhD thesis, Brown University (May 2008)
12. Everest, G., Poorten, A.v.d., Shparlinski, I., Ward, T.: Elliptic Divisibility Sequences. In: Recurrence Sequences, pp. 163–175. American Mathematical Society, Providence (2003)
13. Stange, K.E.: The tate pairing via elliptic nets. In: Takagi, T., Okamoto, T., Okamoto, E., Okamoto, T. (eds.) Pairing 2007. LNCS, vol. 4575, pp. 329–348. Springer, Heidelberg (2007)
14. Gosper, R.W., Orman., H., Schroeppel, R.: Using somos sequences for cryptography
15. Silverman, J.H.: The arithmetic of elliptic curves. Graduate Texts in Mathematics, vol. 106. Springer, New York (1992); Corrected reprint of the, original (1986)
16. Stange, K.E.: Elliptic nets, generalised Jacobians and bi-extensions (in preparation)
17. Frey, G., Lange, T.: Background on curves and Jacobians. In: Handbook of elliptic and hyperelliptic curve cryptography. Discrete Math. Appl. (Boca Raton), pp. 45–85. Chapman & Hall/CRC, Boca Raton (2005)
18. Bach, E., Shallit, J.: Algorithmic number theory, Efficient algorithms. Foundations of Computing Series, vol. 1. MIT Press, Cambridge (1996)
19. Duquesne, S., Frey, G.: Background on pairings. In: Handbook of elliptic and hyperelliptic curve cryptography. Discrete Math. Appl. (Boca Raton), pp. 115–124. Chapman & Hall/CRC, Boca Raton (2006)
20. Galbraith, S.D.: Pairings. In: Advances in elliptic curve cryptography. London Math. Soc. Lecture Note Ser., vol. 317, pp. 183–213. Cambridge Univ. Press, Cambridge (2005)
21. Menezes, A.J., Okamoto, T., Vanstone, S.: Reducing elliptic curve logarithms to logarithms in a finite field. IEEE Trans. Inform. Theory 39, 1639–1646 (1993)
22. Frey, G., Rück, H.G.: A remark concerning m-divisibility and the discrete logarithm in the divisor class group of curves. Math. Comp. 62, 865–874 (1994)
23. Menezes, A.J., van Oorschot, P.C., Vanstone, S.A.: Handbook of applied cryptography. CRC Press Series on Discrete Mathematics and its Applications. CRC Press, Boca Raton (1997) (With a foreword by Ronald L. Rivest)
24. Crandall, R., Pomerance, C.: Prime numbers. Springer, New York (2001) (A computational perspective)
25. K. Fong, D. Hankerson, J.López., Menezes, A.: Field inversion and point halving revisited. Technical Report, CORR 2003-18, Department of Combinatorics and Optimization, University of Waterloo, Canada (2003)
26. Itoh, T., Tsujii, S.: An efficient algorithm for deciding quadratic residuosity in finite fields $GF(p^m)$. Inform. Process. Lett. 30, 111–114 (1989)

The "Coefficients H" Technique

Jacques Patarin

Université de Versailles
45 avenue des Etats-Unis, 78035 Versailles Cedex, France
jacques.patarin@prism.uvsq.fr

Abstract. The "coefficient H technique" is a tool introduced in 1991 and used to prove various pseudo-random properties from the distribution of the number of keys that sends cleartext on some ciphertext. It can also be used to find attacks on cryptographic designs. We can like this unify a lot of various pseudo-random results obtained by different authors. In this paper we will present this technique and we will give some examples of results obtained.

1 Introduction

The "coefficient H technique" was introduced in 1990 and 1991 in [11], [12]. Since then, it has been used many times (by myself , Henri Gilbert, Gilles Piret, Serge Vaudenay, etc.) to prove various results on pseudo-random functions and pseudo-random permutations. In this paper we will present in a self content way the "coefficient H technique", with different formulations when we study different cryptographic attacks (known plaintext attacks, chosen plaintext attacks, etc.). We will give proofs of some of these theorems and we will give some simple examples.

2 Notation - Definition of H

In all this paper, we will use these notations.

- KPA: Known Plaintext Attack
- CPA-1: Non-adaptive Chosen Plaintext Attack
- CPA-2: Adaptive Chosen Plaintext Attack
- CPCA-1: Non-adaptive Chosen Plaintext and Chosen Ciphertext Attack
- CPCA-2: Adaptive Chosen Plaintext and Chosen Ciphertext Attack
- $I_N = \{0,1\}^N$ (N is any integer)
- F_N will be the set of all applications from I_N to I_N
- B_N will be the set of all permutations from I_N to I_N
- ψ^k will denote the Feistel scheme of F_{2n} with k rounds with k random round functions randomly chosen in F_n (n is any integer). ψ^k is also called a random Feistel scheme or a Luby-Rackoff construction.
- $a \in_R A$ means that a is randomly chosen in A with a uniform distribution

R. Avanzi, L. Keliher, and F. Sica (Eds.): SAC 2008, LNCS 5381, pp. 328–345, 2009.

- K will denote a set of values that we will sometimes call "keys". In this paper we will consider that K is a set of k-uples of functions $(f_1, \ldots, f_k)$ of F_n. (However generally only $|K|$ will be important, not the nature of the elements of K).
- G is an application of $K \to F_N$. (Therefore, G is a way to design a function of F_N from k-uples $(f_1, \ldots, f_k)$ of functions of F_n of K).

Let m be an integer (m will be the number of queries). Let $a = (a_i)_{1 \le i \le m}$ be a sequence of pairwise distinct elements of I_N. Let $b = (b_i)_{1 \le i \le m}$ be a sequence of elements of I_N. By definition, we will denote by $H(a, b)$ or simply by H if the context of the a_i and b_i is clear, the number of $(f_1, \ldots, f_k) \in K$ such that:

$$\forall i, \ 1 \le i \le m, \ G(f_1, \ldots, f_k)(a_i) = b_i$$

Therefore, H is the number of "keys" (i.e. elements of K) that send all the a_i inputs to the exact values b_i.

3   Five Basic "coefficient H" Theorems

In this section we will formulate five theorems. These theorems are the basis of a general proof technique called the "coefficient H technique", that allows to prove security results for function generators and permutation generators (and thus applies for random and pseudo-random Feistel ciphers).

These theorems were mentioned in [12] (with proofs in french) and in [16]. Since no proof in english was easily available so far we will present in this paper, in Appendices, a proof of some of these theorems.

Theorem 1. [**Coefficient H technique, sufficient condition for security against KPA**] *Let α and β be real numbers, $\alpha > 0$ and $\beta > 0$. If:*

(1) *For random values a_i, b_i, $1 \le i \le m$ of I_N such that the a_i are pairwise distinct, with probability $\ge 1 - \beta$ we have:*

$$H \ge \frac{|K|}{2^{Nm}}(1 - \alpha)$$

Then

(2) *For every KPA with m (random) known plaintexts we have: $Adv^{KPA} \le \alpha + \beta$, where Adv^{KPA} denotes the advantage to distinguish $G(f_1, \ldots, f_k)$ when $(f_1, \ldots, f_k) \in_R K$ from a function $f \in_R F_N$*

(By "advantage" we mean here, as usual, for a distinguisher the absolute value of the difference of the two probabilities to output 1).

Theorem 2. [**Coefficient H technique, sufficient condition for security against CPA-1**] *Let α and β be real numbers, $\alpha > 0$ and $\beta > 0$. If:*

(1) *For all sequences $a = (a_i)$, $1 \le i \le m$ of m pairwise distinct elements of I_N there exists a subset $E(a)$ of I_N^m such that $|E(a)| \ge (1 - \beta) \cdot 2^{Nm}$ and such that for all sequences $b = (b_i)$, $1 \le i \le m$ of $E(a)$ we have:*

$$H \ge \frac{|K|}{2^{Nm}}(1 - \alpha)$$

Then

(2) *For every CPA-1 with m chosen plaintexts we have: $Adv^{PRF} \le \alpha + \beta$ where Adv^{PRF} denotes the advantage to distinguish $G(f_1, \ldots, f_k)$ when $(f_1, \ldots, f_k) \in_R K$ from a function $f \in_R F_N$.*

Theorem 3. [Coefficient H technique, sufficient condition for security against CPA-2] *Let α and β be real numbers, $\alpha > 0$ and $\beta > 0$. Let E be a subset of I_N^m such that $|E| \ge (1 - \beta) \cdot 2^{Nm}$.*
    If:

(1) *For all sequences a_i, $1 \le i \le m$, of pairwise distinct elements of I_N and for all sequences b_i, $1 \le i \le m$, of E we have:*

$$H \ge \frac{|K|}{2^{Nm}}(1 - \alpha)$$

Then

(2) *For every CPA-2 with m chosen plaintexts we have: $Adv^{PRF} \le \alpha + \beta$ where Adv^{PRF} denotes the probability to distinguish $G(f_1, \ldots, f_k)$ when $(f_1, \ldots, f_k) \in_R K$ from a function $f \in_R F_N$.*

Theorem 4. [Coefficient H technique, sufficient condition for security against CPCA-2] *Let α be a real number, $\alpha > 0$. If:*

(1) *For all sequences of pairwise distinct elements a_i, $1 \le i \le m$, and for all sequences of pairwise distinct elements b_i, $1 \le i \le m$, we have:*

$$H \ge \frac{|K|}{2^{Nm}}(1 - \alpha)$$

Then

(2) *For every CPCA-2 with m chosen plaintexts we have: $Adv^{PRF} \le \alpha + \frac{m(m-1)}{2 \cdot 2^N}$ where Adv^{PRF} denotes the probability to distinguish $G(f_1, \ldots, f_k)$ when $(f_1, \ldots, f_k) \in_R K$ from a function $f \in_R B_N$.*

Theorem 5. [Coefficient H technique, a more general sufficient condition for security against CPCA-2]
    Let α and β be real numbers, $\alpha > 0$ and $\beta > 0$
    If there exists a subset E of $(I_N^m)2$ such that

(1a) *For all $(a, b) \in E$, we have:*

$$H \ge \frac{|K|}{2^{Nm}}(1 - \alpha)$$

(1b) *For all CPCA-2 acting on a random permutation f of B_N, the probability that $(a,b) \in E$ is $\geq 1 - \beta$ where (a,b) denotes here the successive $b_i = f(a_i)$ or $a_i = f^{-1}(b_i)$, $1 \leq i \leq m$ that will appear.*
Then

(2) *For every CPCA-2 with m chosen plaintexts we have: $Adv^{PRF} \leq \alpha + \beta$ where Adv^{PRF} denotes the probability to distinguish $G(f_1, \ldots, f_k)$ when $(f_1, \ldots, f_k) \in_R K$ from a function $f \in_R B_N$.*

Remark. There are a lot of variants, and generalizations of these theorems. For example, in all these theorems 1, 2, 3, 4, 5, the results are also true if we change $H \geq \frac{|K|}{2^{Nm}}(1 - \alpha)$ by $H \leq \frac{|K|}{2^{Nm}}(1 + \alpha)$. However, for cryptographic uses $H \geq$ is much more practical since often it will be easier to evaluate the exceptions where H is $\ll$ average than the exceptions when H is $\gg$ average.

4   How to Use the "Coefficient H Technique"

We have used the "coefficient H technique" to obtain proofs of security (cf sections 5 and 6 below), generic attacks (cf section 7 below) and to obtain new cryptographic designs (cf section 8 below). For proofs of security, very often, the aim is to prove that a cryptographic construction A is not distinguishable from an ideal object B. For example, in the Luby-Rackoff original result of [6], A is a 3 or 4 round Feistel scheme with round functions generated from a small key k by a pseudo-random function generator, and B is a perfectly random permutation. For the proof, we introduce another ideal construction C, where all the pseudorandom functions are replaced by truly random functions (or other pseudo-random objects are replaced by truly random ones). Now the idea is that

$$Adv(A \to B) \leq Adv(A \to C) + Adv(C \to B)$$

i.e. the advantage to distinguish A from B is always smaller or equal to the advantage to distinguish A from C plus the advantage to distinguish C from B. To prove that $Adv(A \to C)$ is small is generally very easy: it comes from the hypothesis that the function generator is secure, for example. To prove that $Adv(C \to B)$ is small is sometimes more difficult. However, in A the only secret values are generally contained in a small secret cryptographic k (of 128 bits for example) while in C the secret values are much bigger since they are generally truly random secret functions. The "coefficient H" technique is very often a powerful tool to get a proof that $Adv(C \to B)$ is small (and therefore that $Adv(A \to B)$ is small, as wanted since $Adv(A \to C)$ is small). For this, we "just" have to compute some values H, as stated in Theorem 1,2,3,4,5. When the computations of these values H are easy, the proofs will be easy. (Very often these values are easy to compute when we are below a "birthday bound value", i.e. when the analysis of collisions in equations are easy since the probability to get such collisions is small). However, sometimes, the computations of the values

H are not easy. For these cases, I have developed two techniques of computations that I have called H_w and H_σ techniques.

H_w Technique

H_w stands for H "worst case" technique. The set of parameters on which we want to compute H is generally fixed from the beginning. For these computations, I sometimes use the "Theorem $P_i \oplus P_j$" (or variants of it) that I will present below. (See section 6.1 for an example of this technique).

H_σ Technique

H_σ stands for H "standard deviation" technique. The set of parameters on which we want to compute H is not fixed from the beginning, but it will automatically be fixed from the computation of the standard deviation of H. We will generally use the covariance formula to compute this standard deviation. (See section 6.2 for an example of this technique).

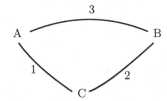

Fig. 1. Three cryptographic objects A,B,C

"Theorem $P_i \oplus P_j$"

The "Theorem $P_i \oplus P_j$" was proved in [17]. We use it sometimes to compute some difficult values H. Let us recall here what this theorem is.

Definition 1. *Let* (A) *be a set of equations* $P_i \oplus P_j = \lambda_k$, *with* $P_i, P_j, \lambda_k \in I_n$. *If by linearity from* (A) *we cannot generate an equation in only the* λ_k, *we will say that* (A) *has "no circle in P", or that the equations of* (A) *are "linearly independent in P".*

Let a be the number of equations in (A), and α be the number of variables P_i in (A). So we have parameters $\lambda_1, \lambda_2, \cdots, \lambda_a$ and $a + 1 \le \alpha \le 2a$.

Definition 2. *We will say that two indices* i *and* j *are "in the same block" if by linearity from the equations of* (A) *we can obtain* $P_i \oplus P_j = $ *an expression in* $\lambda_1, \lambda_2, \cdots, \lambda_a$.

Definition 3. *We will denote by* ξ_{max} *the maximum number of indices that are in the same block.*

Example 1. If $A = \{P_1 \oplus P_2 = \lambda_1, P_1 \oplus P_3 = \lambda_2, P_4 \oplus P_5 = \lambda_3\}$, here we have two blocks of indices $\{1, 2, 3\}$ and $\{4, 5\}$, and $\xi_{max} = 3$.

Definition 4. *For such a system* (A), *when* $\lambda_1, \lambda_2, \cdots, \lambda_a$ *are fixed, we will denote by* h_α *the number of* $P_1, P_2, \cdots, P_\alpha$ *solutions of* (A) *such that:* $\forall i, j, i \neq j \Rightarrow P_i \neq P_j$. *We will also denote* $H_\alpha = 2^{na} h_\alpha$.

Remark h_α *and* H_α *are a concise notations for* $h_\alpha(A)$ *and* $H_\alpha(A)$. *For a given value* α, h_α *and* H_α *can have different values for different systems* A.

Definition 5. *We will denote by* J_α *the number of* $P_1, P_2, \cdots, P_\alpha$ *in* I_n *such that:* $\forall i, j, i \neq j \Rightarrow P_i \neq P_j$. *So* $J_\alpha = 2^n \cdot (2^n - 1) \cdots (2^n - \alpha + 1)$.

Theorem 6 ("**Theorem** $P_i \oplus P_j$" *when* ξ_{max} *is fixed*). *Let* ξ_{max} *be a fixed integer,* $\xi_{max} \geq 2$. *Let* (A) *be a set of equations* $P_i \oplus P_j = \lambda_k$ *with no circle in* P, *with* α *variables* P_i, *such that:*

1. *We have no more than* ξ_{max} *indices in the same block.*
2. *The* $\lambda_1, \lambda_2, \cdots, \lambda_k$ *have any fixed values such that: for all* i *and* j *in the same block,* $i \neq j$, *the equation of* $P_i \oplus P_j$ *in* $\lambda_1, \lambda_2, \cdots, \lambda_\alpha$ *is* $\neq 0$ *(i.e. by linearity from* (A) *we cannot generate an equation* $P_i = P_j$ *with* $i \neq j$).

Then we have for sufficient large n: $H_\alpha \geq J_\alpha$. *(This means: for all fixed* ξ_{max}, *there exists* $n_0 \in \mathbb{N}$ *such that, for all* $n \geq n_0$, *for all system* A *that satisfies 1. and 2., we have:* $H_\alpha(A) \geq J_\alpha$).

Remark This theorem was proved in [16] if we add the condition $\alpha^3 \ll 2^{2n}$ (and also $\xi_{max}\alpha^3 \ll 2^{2n}$ since ξ_{max} is here a fixed integer).

Theorem 7 ("**Theorem** $P_i \oplus P_j$" *when* $\xi_{max}\alpha \ll 2^n$). *With the same notations, we have the same result, with the hypothesis* $\xi_{max}\alpha \ll 2^n$ *(instead of* ξ_{max} *a fixed integer).*

Remark. For cryptographic use, weaker version of this theorem will be enough. For example, instead of $H_\alpha \geq J_\alpha$ for sufficiently large n, $H_\alpha \geq J_\alpha \left(1 - f(\frac{\xi\alpha}{2^n})\right)$, where f is a function such that $f(x) \to 0$ when $x \to 0$, is enough.

Another variant of this Theorem $P_i \oplus P_j$ is:

Theorem 8 ("**Theorem** $P_i \oplus P_j$ *when* $\xi_{max} \leq O(n)$ *and* $\xi_{average} \leq 3$). *Let* $\xi_{average}$ *be the average value of* ξ, *where* ξ *is the number of variables* P_j *that are fixed from the equations* (A) *when we fix a variable* P_i. *If* $\xi_{max} \leq O(n)$ *and* $\xi_{average} \leq 3$, *then for sufficient large* n, $H_\alpha \geq J_\alpha$.

Generalizations of the "Theorem $P_i \oplus P_j$".* This theorem may have many generalizations. For example:

• Generalization 1: the theorem is still true in any group G (instead of I_n).
• Generalization 2: we have a similar property for equations with 3, 4, $\cdots$, or k variables, i.e. each equation is $P_{i_1} \oplus P_{i_2} \cdots P_{i_k} = \lambda_l$ with pairwise distinct P_i variables.

However in this paper we will only study the original "Theorem $P_i \oplus P_j$" (i.e. theorems 6 and 7) since it is this one that is needed to study random Feistel schemes.

5   First Simple Examples

5.1 ψ^2

For ψ^2 (Feistel scheme with the round functions $(f_1, f_2) \in_R F_n^2$) let $[L_i, R_i]$, $1 \leq i \leq m$ denotes the inputs, and $[S_i, T_i]$, $1 \leq i \leq m$ denotes the outputs. We have: $S_i = L_i \oplus f_1(R_i)$ and $T_i = R_i \oplus f_2(S_i)$ (*)

For random values $[L_i, R_i]$, $[S_i, T_i]$, $1 \leq i \leq m$ (such that $i \neq j \to L_i \neq L_j$ or $R_i \neq R_j$) with probability $> 1 - \frac{m^2}{2^n}$ we have that all the R_i values are pairwise distinct and all the S_i values are pairwise distinct. Moreover, if this occurs, we have exactly $H = \frac{|F_n|^2}{2^{2nm}}$ (since (*) then fix f_1 exactly on m points and f_2 exactly on m points).

So from Theorem 1 (with $\alpha = 0$ and $\beta = \frac{m^2}{2^n}$) we get:

Theorem 9. *For every KPA with m random known plaintexts, we have*

$$Adv^{KPA} \leq \frac{m^2}{2^n}$$

where Adv^{KPA} denotes the advantage to distinguish ψ^2 when $(f_1, f_2) \in_R F_n^2$ from a function $f \in_R F_{2n}$. So when $m \ll 2^{n/2}$, ψ^2 will resit all known plaintext attacks.

Remark. This result is tight, since when m^2 becomes not negligible compared with 2^n then by counting the number $\mathcal{N}$ of $(i, j)/S_i \oplus L_i = S_j \oplus L_j$ we will be able to distinguish ψ^2 from a random permutation with a known plaintext attack.

5.2   Involutive Permutations

Let assume that G is a generator of permutations that generates involutive permutations f (i.e. $f = f^{-1}$). Then we can distinguish such f from random permutations of B_N with $m = 2$ queries in CPA-2 and $m = 2$ queries in CPCA-1.

CPA-2

In CPA-2 we ask $f(a_1) = b_1$ and $f(b_1) = b_2$, and we test if $b_2 = a_1$. This gives a CPA-2 with $m = 2$ queries. It is not in contradiction with Theorem 3 since in Theorem 3, we need property (1) on **all** sequences a_i, $1 \leq i \leq m$ (and not necessary on **all** sequences b_i). Here if we have $a = (a_1, a_2)$, $b = (b_1, b_2)$ with $a_2 = b_1$ and $b_2 \neq a_1$, we will have $H = 0$. Therefore we will not be able to prove from Theorem 3 that G is secure in CPA-2 (in fact G is not secure in CPA-2) since for most (b_1, b_2) there exists (a_1, a_3) (take $a_2 = b_1$) such that $H = 0$.

CPCA-1

In CPCA-1 we ask $f(a_1) = b_1$ and $f^{-1}(a_1) = a_2$ and we test if $a_2 = b_1$. This gives a CPCA-1 distinguisher with $m = 2$ queries. We will not be able to prove

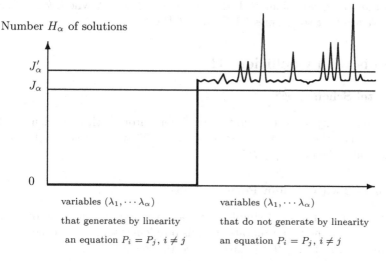

Number H_α of solutions

variables $(\lambda_1, \cdots \lambda_\alpha)$
that generates by linearity
an equation $P_i = P_j$, $i \neq j$

variables $(\lambda_1, \cdots \lambda_\alpha)$
that do not generate by linearity
an equation $P_i = P_j$, $i \neq j$

Fig. 2.

from Theorem 4 or Theorem 5 that G is secure in CPCA-1 (in fact G is not secure in CPCA-1) since in a non-adaptive chosen plaintext/ciphertext attack we can impose that $b_2 = a_1$ and if we have $a = (a_1, a_2)$, $b = (b_1, b_2)$ with $b_2 = a_1$ and $a_2 \neq b_1$ we will have $H = 0$.

$$\text{KPA} \text{ --- } \text{CPA-1} \underset{\text{CPCA-1}}{\overset{\text{CPA-2}}{\lessgtr}} \gtrdot \text{CPCA-2}$$

Fig. 3. Hierarchy of the attacks in secret key cryptography

5.3   Secret Key Security Hierarchy

In Figure 1, we have the well known hierarchy of attacks in secret key cryptography (cf [2], [4], [5]). With coefficients H technique we can easily prove on small examples this hierarchy, i.e. for example that there are some scheme secure in CPA-2 and not in CPCA-1, that some schemes are secure in CPA-1 and not in KPA etc. For example, we can easily prove that for a random involutive permutation of B_N we will have KPA and CPA-1 security in $O(\sqrt{2^N})$. Therefore the example of Section 6.2 shows that CPA-1 < CPA-2 and that CPA-1 < CPCA-1.

With f such that $f(0) = 0$ we will have that KPA < CPA-1.
With ψ^2 we will have KPA < CPA-1.
With ψ^3 we will have CPA-2 < CPCA-2 and CPCA-1 < CPCA-2.
With a random permutation such that $f^3 = \text{Id}$ we see that sometimes CPA-2 > CPCA-1

With a random permutation, such that $f^{-1}(x) = f(x) \oplus k$ where k is a secret constant we see that sometimes CPCA-1 > CPA-2.

6   Proofs with Coefficient H

6.1   Feistel Schemes ψ^k

I have proved many security results on ψ^k generators with coefficient H. For example, in [17], the security of ψ^5 when $m \ll 2^n$ was proved (with the H_w technique and "Theorem $P_i \oplus P_j$").

6.2   Xor of Two Random Permutations

Xoring two permutations is a simple very way to construct pseudorandom functions from pseudorandom permutations (this problem is sometimes called "Luby-Rackoff backwards"). In [19] we have proved this result:

Theorem 10. *For every CPA-2 on a function G of F_n with m chosen plaintexts, we have*

$$Adv^{PRF} \leq O(\frac{m}{2^n})$$

where Adv^{PRF} denotes the advantage to distinguish $f \oplus g$ with $f, g \in_R B_n$ from $h \in_R F_n$.

How to Get Theorem 10 from Theorem 3

A sufficient condition is to prove that for "most" (most since β must be small) sequences of values $b_i, 1 \leq i \leq m$, we have: the number H of $(f,g) \in B_n^2$ such that $\forall i, 1 \leq i \leq m, f \oplus g(a_i) = b_i$ satisfies: $H \geq \frac{|B_n^2|}{2^{nm}}(1 - \alpha)$ for a small value α (more precisely $\alpha \ll O(\frac{m}{2^n})$). One way to do this is to evaluate $E(H)$ and $\sigma(H)$, i.e. the mean value and the standard deviation of H when the b_i values are randomly chosen in I_n^m. (We call this technique, the "H_σ technique").

We can see that the result wanted to prove Theorem 10 exactly says that $\sigma(H) \ll E(H)$ when $m \ll 2^n$. To prove this, we can use the "covariance formula"

$$V(\sum_i N_i) = \sum_i (V(N_i)) + \sum_{i \neq j}[E(N_iN_j) - E(N_i)E(N_j)]$$

By definition, let λ_m be the number of sequences of values of I_n^3, $(f_i, g_i, h_i), 1 \leq i \leq m$ such that:

1. The m values f_i are pairwise distinct.
2. The m values g_i are pairwise distinct.
3. The m values h_i are pairwise distinct.
4. The m values $f_i \oplus g_i \oplus h_i$ are pairwise distinct.

After a change of variables we get finally that the property wanted in Theorem 10 means that

$$\lambda_m = \frac{(2^n(2^n-1)\dots(2^n-m+1))^4}{2^{nm}}\left(1+O(\frac{m}{2^n})\right)$$

(This is what was proved in [19])

I have also conjectured this property:

$$\forall f \in F_n, \text{ if } \bigoplus_{x \in I_n} f(x) = 0, \text{ then } \exists (g,h) \in B_n2, \text{ such that } f = g \oplus h.$$

Just one day after paper [19] was put on eprint, J.F. Dillon pointed to us that in fact this was proved in 1952 in [3]. We thank him a lot for this information. (This property was proved again independently in 1979 in [24]).

A New Conjecture

However I conjecture a stronger property. Conjecture:

$$\forall f \in F_n, \text{ if } \bigoplus_{x \in I_n} f(x) = 0, \text{ then the number } H \text{ of } (g,h) \in B_n2,$$

$$\text{such that } f = g \oplus h \text{ satisfies } H \geq \frac{|B_n|^2}{2^{n2^n}}.$$

Variant: I also conjecture that this property is true in any group, not only with Xor.

Remark: In this paper, I have proved weaker results involving m equations with $m \ll O(2^n)$ instead of all the 2^n equations. These weaker results were sufficient for the cryptographic security wanted.

6.3   Benes Schemes

In [18] the security of Benes schemes when $m \ll 2^n$ was finally obtained (after the beginning of some proof ideas in [1]).

7   Attacks with Coefficient H

By using the coefficient values we were able to find many generic attacks. We give here some examples.

7.1   For Feistel Schemes ψ^k

From [15] we have the results of Table 1.

Table 1. Minimum number λ of computations needed to distinguish a generator Ψ^k (with one or many such permutations available) from random permutations with an even signature of $I_n \to I_n$. For simplicity we denote α for $O(\alpha)$. $\leq$ means best known attack.

	KPA	CPA-1	CPA-2	CPCA-1	CPCA-2
Ψ	1	1	1	1	1
$\Psi 2$	$2^{n/2}$	2	2	2	2
$\Psi 3$	$2^{n/2}$	$2^{n/2}$	$2^{n/2}$	$2^{n/2}$	3
$\Psi 4$	2^n	$2^{n/2}$	$2^{n/2}$	$2^{n/2}$	$2^{n/2}$
$\Psi 5$	$\leq 2^{3n/2}$	2^n	2^n	2^n	2^n
$\Psi 6$	$\leq 2^{2n}$	$\leq 2^{2n}$	$\leq 2^{2n}$	$\leq 2^{2n}$	$\leq 2^{2n}$
$\Psi 7$	$\leq 2^{3n}$	$\leq 2^{3n}$	$\leq 2^{3n}$	$\leq 2^{3n}$	$\leq 2^{3n}$
$\Psi 8$	$\leq 2^{4n}$	$\leq 2^{4n}$	$\leq 2^{4n}$	$\leq 2^{4n}$	$\leq 2^{4n}$
$\Psi^k, k \geq 6$ *	$\leq 2^{(k-4)n}$	$\leq 2^{(k-4)n}$	$\leq 2^{(k-4)n}$	$\leq 2^{(k-4)n}$	$\leq 2^{(k-4)n}$

* If $k \geq 7$ these attacks analyze about $2^{(k-6)n}$ permutations of the generator and if $k \leq 6$ only one permutation is needed.

7.2 For Feistel Schemes ψ'^k with k Random Permutations for the Rounds Functions (Instead of Round Functions)

From [26] we have the results of Table 2.

Table 2. Maximum number of computations needed to get an attack on a k-round Feistel network with internal *permutations* (+) is shown when the values are larger than the corresponding values with internal functions.

number k of rounds	KPA	CPA-1	CPA-2	CPCA-1	CPCA-2
1	1	1	1	1	1
2	$2^{n/2}$	2	2	2	2
3	2^n(+)	$2^{n/2}$	$2^{n/2}$	$2^{n/2}$	3
4	2^n	$2^{n/2}$	$2^{n/2}$	$2^{n/2}$	$2^{n/2}$
5	$2^{3n/2}$	2^n	2^n	2^n	2^n
6	2^{3n}(+)	2^{3n}(+)	2^{3n}(+)	2^{3n}(+)	2^{3n}(+)
7	2^{3n}	2^{3n}	2^{3n}	2^{3n}	2^{3n}
8	2^{4n}	2^{4n}	2^{4n}	2^{4n}	2^{4n}
9	2^{6n}(+)	2^{6n}(+)	2^{6n}(+)	2^{6n}(+)	2^{6n}(+)
10	2^{6n}	2^{6n}	2^{6n}	2^{6n}	2^{6n}
11	2^{7n}	2^{7n}	2^{7n}	2^{7n}	2^{7n}
12	2^{9n}(+)	2^{9n}(+)	2^{9n}(+)	2^{9n}(+)	2^{9n}(+)
$k \geq 6$, $k = 0 \bmod 3$	$2^{(k-3)n}$(+)	$2^{(k-3)n}$(+)	$2^{(k-3)n}$(+)	$2^{(k-3)n}$(+)	$2^{(k-3)n}$(+)
$k \geq 6$, $k = 1$ or $2 \bmod 3$	$2^{(k-4)n}$	$2^{(k-4)n}$	$2^{(k-4)n}$	$2^{(k-4)n}$	$2^{(k-4)n}$

7.3 For Unbalanced Feistel Schemes with Contracting Functions

From [21] we have the results of Table 3.

Table 3. Results on G_k^d for any $k \ge 4$. For more than $2k$ rounds more that one permutation is needed or more than $2^{(2k-4)n}$ computations are needed in the best known attacks to distinguish from a random permutation with an even signature.

	KPA	CPA-1[a]
$G_k^d, 1 \le d \le k-1$	1	1
G_k^k	$2^{\frac{n(k-1)}{2}}$	2
G_k^{k+1}	$2^{\frac{n(k-1)}{2}}$	$2^{\frac{n}{2}}$
G_k^{k+2}	$2^{\frac{k}{2}n}$	$2^{\frac{3}{2}n}$
G_k^{k+3}	$2^{(\frac{k+1}{2})n}$	$2^{\frac{5}{2}n}$
$G_k^{k+i}, 1 \le i < k$	$2^{(\frac{k+i-2}{2})n}$	$2^{(\frac{2i-1}{2})n}$
G_k^{2k}	$2^{(2k-4)n}$	$2^{(2k-4)n}$
$G_k^d, d \ge 2k$	$2^{(d+(k-2)\lfloor \frac{d}{k}\rfloor -2k)n}$	$2^{(d+(k-2)\lfloor \frac{d}{k}\rfloor -2k)n}$

[a] Here we do not show CPA-2, CPCA-1 and CPCA-2 since for G_k^d, no better attacks are found compared with CPA-1.

7.4 For Unbalanced Feistel Schemes with Expanding Functions

From [22] we have the results of Table 4

Table 4. Best known attacks on F_k^d for $k \ge 3$

	KPA	CPA-1
$F1_k$	1	1
$F2_k$	$2^{\frac{n}{2}}$	2
$F3_k$	2^n	2
$F_k^d, 2 \le d \le k$	$2^{\frac{d-1}{2}n}$	2
F_k^{k+1}	$2^{\frac{k}{2}n}$	$2^{\frac{n}{2}}$
F_k^{k+2}	$2^{\frac{k+1}{2}n}$	2^n
F_k^{k+3}	$2^{\frac{2k+3}{4}n}$	2^{2n} or $2^{\frac{k+2}{3}n}$
$F_k^d, k+2 \le d \le 2k$	$2^{\frac{d+k}{4}n}$	$2^{(d-k-1)n}$ or $2^{\frac{d-1}{3}n}$
F_k^{2k}	$2^{\frac{3k}{4}n}$	$2^{\frac{2k-1}{3}n}$
$\vdots$	$\vdots$	$\vdots$
F_k^{3k-1}	$2^{(k-\frac{1}{8})n}$	$2^{(k-\frac{1}{2})n}$
F_k^{3k}	2^{kn}	2^{kn}
$F_k^d, 3k \le d \le k2$	$2^{(d-2k)n}$	$2^{(d-2k)n}$
F_k^{k2}	$2^{(k2-2k)n}$	$2^{(k2-2k)n}$
F_k^{k2+1}	$2^{(2k2-3k-2)n}$	$2^{(2k2-3k-2)n}$
$F_k^d, d \ge k2+1$	$2^{(\lfloor 2d(1-\frac{1}{k})\rfloor -k-3)n}$	$2^{(\lfloor 2d(1-\frac{1}{k})\rfloor -k-3)n}$

8   New Designs

8.1   Russian Doll Design

See [23] in this volume.

8.2   Design from Random Unbalanced Feistel Schemes

This design comes directly from Table 3.

8.3   Hash Function Design

From 9.1 and 9.2 we are analyzing a Hash function design (by Xoring two independent pseudorandom permutations, or by Xoring the input and the output of a pseudorandom permutation).

9   Conclusion

With the "coefficient H technique" we were able to prove many security results and to get many generic attacks. Moreover, it was a source of inspiration for the design of new schemes.

References

1. Aiello, W., Venkatesan, R.: Foiling birthday attacks in length-doubling transformations. In: Maurer, U.M. (ed.) EUROCRYPT 1996. LNCS, vol. 1070, pp. 307–320. Springer, Heidelberg (1996)
2. Bellare, M., Desai, A., Jokipii, E., Rogaway, P.: A Concrete Security Treatment of Symmetric Encryption: Analysis of the DES Modes of Operation. A Concrete Security Treatment of Symmetric Encryption and appeared in the Proceedings of 38th Annual Symposium of Computer Science, IEEE (1997)
3. Hall Jr., M.: A Combinatorial Problem on Abelian Groups. Proceedings of the Americal Mathematical Society 3(4), 584–587 (1952)
4. Katz, J., Yung, M.: Characterization of Security Notions for Probabilistic. In: Private-Key Encription – STOC 2000 (2000)
5. Katz, J., Yung, M.: Unforgeable Encryption and Chosen-Ciphertext-Secure Modes of Operation. In: Fast Software Encryption 2000 (2000)
6. Luby, M., Rackoff, C.: How to Construct Pseudorandom Permutations from Pseudorandom Functions. SIAM J. Comput. 17(2), 373–386 (1988)
7. Maurer, U.M.: A simplified and generalized treatment of luby-rackoff pseudorandom permutation generators. In: Rueppel, R.A. (ed.) EUROCRYPT 1992. LNCS, vol. 658, pp. 239–255. Springer, Heidelberg (1993)
8. Maurer, U.M.: Indistinguishability of random systems. In: Knudsen, L.R. (ed.) EUROCRYPT 2002. LNCS, vol. 2332, pp. 100–132. Springer, Heidelberg (2002)
9. Maurer, U., Pietrzak, K.: The Security of Many-Round Luby-Rackoff Pseudo-Random Permutations. In: Biham, E. (ed.) EUROCRYPT 2003. LNCS, vol. 2656, pp. 544–561. Springer, Heidelberg (2003)
10. Naor, M., Reingold, O.: On the Construction of Pseudorandom Permutations: Luby-Rackoff Revisited. J. Cryptology 12(1), 29–66 (1999)
11. Patarin, J.: Pseudorandom Permutations based on the DES Scheme. In: Charpin, P., Cohen, G. (eds.) EUROCODE 1990. LNCS, vol. 514, pp. 193–204. Springer, Heidelberg (1991)
12. Patarin, J.: Etude de Générateurs de Permutations Basés sur les Schémas du DES. Ph. Thesis. Inria, Domaine de Voluceau, France (1991)

13. Patarin, J.: New results on pseudorandom permutation generators based on the DES scheme. In: Feigenbaum, J. (ed.) CRYPTO 1991. LNCS, vol. 576, pp. 301–312. Springer, Heidelberg (1992)

14. Patarin, J.: How to construct pseudorandom and super pseudorandom permutations from one single pseudorandom function. In: Rueppel, R.A. (ed.) EUROCRYPT 1992. LNCS, vol. 658, pp. 256–266. Springer, Heidelberg (1993)

15. Patarin, J.: Generic attacks on feistel schemes. In: Boyd, C. (ed.) ASIACRYPT 2001. LNCS, vol. 2248, pp. 222–238. Springer, Heidelberg (2001)

16. Patarin, J.: Luby–rackoff: 7 rounds are enough for formula_image security. In: Boneh, D. (ed.) CRYPTO 2003. LNCS, vol. 2729, pp. 513–529. Springer, Heidelberg (2003)

17. Patarin, J.: On linear systems of equations with distinct variables and small block size. In: Won, D.H., Kim, S. (eds.) ICISC 2005. LNCS, vol. 3935, pp. 299–321. Springer, Heidelberg (2006)

18. Patarin, J.: A proof of security in $O(2^n)$ for the benes scheme. In: Vaudenay, S. (ed.) AFRICACRYPT 2008. LNCS, vol. 5023, pp. 209–220. Springer, Heidelberg (2008)

19. Patarin, J.: A proof of security in $O(2^n)$ for the xor of two random permutations. In: Safavi-Naini, R. (ed.) ICITS 2008. LNCS, vol. 5155, pp. 232–248. Springer, Heidelberg (2008)

20. Patarin, J.: Generic Attacks for the Xor of k Random Permutations (eprint) (2008)

21. Patarin, J., Nachef, V., Berbain, C.: Generic attacks on unbalanced feistel schemes with contracting functions. In: Lai, X., Chen, K. (eds.) ASIACRYPT 2006. LNCS, vol. 4284, pp. 396–411. Springer, Heidelberg (2006)

22. Patarin, J., Nachef, V., Berbain, C.: Generic attacks on unbalanced feistel schemes with expanding functions. In: Kurosawa, K. (ed.) ASIACRYPT 2007. LNCS, vol. 4833, pp. 325–341. Springer, Heidelberg (2007)

23. Patarin, J., Seurin, Y.: Building Secure Block Ciphers on Generic Attacks Assumptions. In: SAC 2008 (2008)

24. Salzborn, F., Szekeres, G.: A Problem in Combinatorial Group Theory. Ars Combinatoria 7, 3–5 (1979)

25. Schneier, B., Kelsey, J.: Unbalanced Feistel Networks and Block Cipher Design. In: Gollmann, D. (ed.) FSE 1996. LNCS, vol. 1039, pp. 121–144. Springer, Heidelberg (1996)

26. Treger, J., Patarin, J.: Generic Attacks On Feistel Schemes with Internal Permutations (paper in preparation)

A   Proof of Theorem 1

Let ϕ be an algorithm (with no limitations in the number of computations) that takes the (a_i, b_i), $1 \le i \le m$ in input and outputs 0 or 1. let P_1 be the probability that ϕ outputs 1 when $\forall i$, $1 \le i \le m$ $b_i = G(f_1, \ldots, f_k)(a_i)$ when $(f_1, \ldots, f_k) \in_R K$. Let P_1^* be the probability that ϕ outputs 1 when $b_i = F(a_i)$ when $F \in_R F_N$. We want to prove that $|E(P_1 - P_1^*)| \alpha + \beta$. Let D be the set of all pairwise distinct a_i, $1 \le i \le m$ (so $|D| \simeq 2^{Nm}(1 - \frac{m(m-1)}{2 \cdot 2^N})$). When the a_i, $1 \le i \le m$ are fixed, let $W(a)$ be the set of all $b_1, \ldots, b_m$ such that the algorithm ϕ outputs 1 on the input (a_i, b_i), $1 \le i \le m$. When the a_i, $1 \le i \le m$ are fixed in D, then we have:

$$P_1^* = \frac{|W(a)|}{2^{Nm}} \quad (1)$$

and

$$P_1 = \frac{1}{|K|} \sum_{b \in W(a)} [Numbers\ of\ (f_1, \ldots, f_k) \in K/$$

$$\forall i,\ 1 \leq i \leq m,\ G(f_1, \ldots, f_k)(a_i) = b_i]$$

so

$$P_1 = \frac{1}{|K|} \sum_{b \in W(a)} H(a, b) \quad (2)$$

Moreover, by hypothesis we have that the number $\mathcal{N}$ of (a, b) such that

$$H(a, b) \geq \frac{|K|}{2^{Nm}}(1 - \alpha)\ \text{satisfies}: \mathcal{N} \geq |D| \cdot 2^{Nm}(1 - \beta) \quad (3)$$

When the (a_i), $1 \leq i \leq m$ are fixed, let $\mathcal{N}(a)$ be the set of all b such that:

$$H(a, b) \geq \frac{|K|}{2^{Nm}}(1 - \alpha)$$

From (3) we have:

$$\sum_{a \in D} |\mathcal{N}(a)| \geq |D| \cdot 2^{Nm}(1 - \beta) \quad (4)$$

From (2) we have:

$$P_1 \geq \frac{1}{|K|} \sum_{b \in W(a) \cap \mathcal{N}(a)} H(a, b)$$

so

$$P_1 \geq \frac{(1 - \alpha)}{2^{Nm}} |W(a) \cap \mathcal{N}(a)|$$

so

$$P_1 \geq \frac{(1 - \alpha)}{2^{Nm}} (|W(a)| - |\mathcal{N}'(a)|) \quad (5)$$

where $\mathcal{N}'(a)$ is the set of all b such that $b \notin \mathcal{N}(a)$. $|\mathcal{N}'(a)| = 2^{Nm} - |\mathcal{N}(a)|$, so

$$\sum_{a \in D} |\mathcal{N}'(a)| = |D| 2^{Nm} - \sum_{a \in D} |\mathcal{N}(a)|$$

so from (4) we have:

$$\sum_{a \in D} |\mathcal{N}'(a)| \leq \beta \cdot |D| \cdot 2^{Nm}, \text{ so } E(|\mathcal{N}'(a)|) \leq \beta \cdot 2^{Nm} \quad (6)$$

(where the expectation is computed when the (a_i), $1 \leq i \leq m$ are randomly chosen in D). From (5) and (1) we have:

$$P_1 \geq (1 - \alpha)(P_1^* - \frac{|\mathcal{N}'(a)|}{2^{Nm}})$$

$$P_1 \geq (1 - \alpha)P_1^* - \frac{|\mathcal{N}'(a)|}{2^{Nm}}$$

so from (6) we get:
$$E(P_1) \geq (1 - \alpha)E(P_1^*) - \beta$$

so
$$E(P_1) \geq E(P_1^*) - \alpha - \beta \quad (7)$$

Now if we consider the algorithm ϕ' that outputs 1 if and only if ϕ outputs 0, we have $P_1' = 1 - P_1$ and $P_1'^* = 1 - P_1^*$ and from (7) we get: $E(P_1') \geq E(P_1'^*) - \alpha - \beta$ (because (7) is true for all algorithm ϕ, so it is true for ϕ'). So

$$E(1 - P_1) \geq E(1 - P_1^*) - \alpha - \beta$$

so
$$E(P_1) - E(P_1^*) \leq \alpha + \beta \quad (8)$$

From (7) and (8) we get $|E(P_1 - P_1^*)| \leq \alpha + \beta$ as claimed.

B Proof of Theorem 3

(I follow here a proof, in French, of this Theorem in my PhD Thesis, 1991, Page 27).

Let ϕ be a (deterministic) algorithm which is used to test a function f of F_n. (ϕ can test any function f from $I_N \rightarrow I_N$). ϕ can use f at most m times, that is to say that ϕ can ask for the values of some $f(C_i)$, $C_i \in I_N$, $1 \leq i \leq m$. (The value C_1 is chosen by ϕ, then ϕ receive $f(C_1)$, then ϕ can choose any $C_2 \neq C_1$, then ϕ receive $f(C_2)$ etc). (Here we have adaptive chosen plaintexts). (If $i \neq j$, C_i is always different from C_j). After a finite but unbounded amount of time, ϕ gives an output of "1" or "0". This output (1 or 0) is noted $\phi(f)$.

We will denote by P_1^*, the probability that ϕ gives the output 1 when f is chosen randomly in F_n. Therefore

$$P_1^* = \frac{\text{Number of functions } f \text{ such that } \phi(f) = 1}{|F_N|}$$

where $|F_N| = 2^{N \cdot 2^N}$.

We will denote by P_1, the probability that ϕ gives the output 1 when $(f_1, \ldots, f_k) \in_R K$ and $f = G(f_1, \ldots, f_k)$. Therefore

$$P_1 = \frac{\text{Number of } (f_1, \ldots, f_k) \in K \text{ such that } \phi(G(f_1, \ldots, f_k)) = 1}{|K|}$$

We will prove:

("**Main Lemma**"): For all such algorithms ϕ,

$$|P_1 - P_1^*| \leq \alpha + \beta$$

Then Theorem 1 will be an immediate corollary of this "Main Lemma" since Adv^{PRF} is the best $|P_1 - P_1^*|$ that we can get with such ϕ algorithms.

Proof of the "Main Lemma"

Evaluation of P_1^*

Let f be a fixed function, and let $C_1, \ldots, C_m$ be the successive values that the program ϕ will ask for the values of f (when ϕ tests the function f). We will note $\sigma_1 = f(C_1), \ldots, \sigma_m = f(C_m)$. $\phi(f)$ depends **only** of the outputs $\sigma_1, \ldots, \sigma_m$. That is to say that if f' is another function of F_n such that $\forall i, 1 \leq i \leq m$, $f'(C_i) = \sigma_i$, then $\phi(f) = \phi(f')$. (Since for $i < m$, the choice of C_{i+1} depends only of $\sigma_1, \ldots, \sigma_i$. Also the algorithm ϕ cannot distinguish f from f', because ϕ will ask for f and f' exactly the same inputs, and will obtain exactly the same outputs). Conversely, let $\sigma_1, \ldots, \sigma_n$ be m elements of I_N. Let C_1 be the first value that ϕ choose to know $f(C_1)$, C_2 the value that ϕ choose when ϕ has obtained the answer σ_1 for $f(C_1), \ldots$, and C_m the mth value that ϕ presents to f, when ϕ has obtained $\sigma_1, \ldots, \sigma_{m-1}$ for $f(C_1), \ldots, f(C_{m-1})$. Let $\phi(\sigma_1, \ldots, \sigma_m)$ be the output of ϕ (0 or 1). Then

$$P_1^* = \sum_{\substack{\sigma_1, \ldots, \sigma_n \\ \phi(\sigma_1, \ldots \sigma_m) = 1}} \frac{\text{Number of functions } f \text{ such that } \forall i, 1 \leq i \leq m, \ f(C_i) = \sigma_i}{2^{N \cdot 2^N}}$$

Since the C_i are all distinct the number of functions f such that $\forall i, 1 \leq i \leq m, \ f(C_i) = \sigma_i$ is exactly $|F_n|/2^{nm}$. Therefore

$$P_1^* = \frac{\text{Number of outputs } (\sigma_1, \ldots, \sigma_m) \text{ such that } \phi(\sigma_1, \ldots \sigma_m) = 1}{2^{Nm}}$$

Let $\mathcal{N}$ be the number of outputs $\sigma_1, \ldots, \sigma_m$ such that $\phi(\sigma_1, \ldots \sigma_m) = 1$. Then $P_1^* = \frac{\mathcal{N}}{2^{Nm}}$.

Evaluation of P_1

With the same notation $\sigma_1, \ldots, \sigma_n$, and $C_1, \ldots C_m$:

$$P_1 = \frac{1}{|K|} \sum_{\substack{\sigma_1, \ldots, \sigma_n \\ \phi(\sigma_1, \ldots \sigma_m) = 1}} [\text{Number of } (f_1, \ldots, f_k) \in K \text{ such that}$$

$$\forall i, 1 \leq i \leq m, \ G(f_1, \ldots, f_k)(C_i) = \sigma_i] \quad (3)$$

Now (by definition of β) we have at most $\beta \cdot 2^{nm}$ sequences $(\sigma_1, \ldots, \sigma_m)$ such that $(\sigma_1, \ldots, \sigma_m) \notin E$. Therefore, we have at least $\mathcal{N} - \beta \cdot 2^{Nm}$ sequences $(\sigma_1, \ldots, \sigma_m)$ such that $\phi(\sigma_1, \ldots \sigma_m) = 1$ and $(\sigma_1, \ldots, \sigma_m) \in E$ (4). Therefore, from (1), (3) and (4), we have

$$P_1 \geq \frac{(\mathcal{N} - \beta \cdot 2^{Nm}) \cdot \frac{|K|}{2^{Nm}}(1 - \alpha)}{|K|}$$

Therefore

$$P_1 \geq \left(\frac{\mathcal{N}}{2^{Nm}} - \beta\right)(1 - \alpha)$$

$$P_1 \geq (P_1^* - \beta)(1 - \alpha)$$

Thus $P_1 \geq P_1^* - \alpha - \beta$ (5), as claimed.

We now have to prove the inequality in the other side. For this, let P_0^* be the probability that $\phi(f) = 0$ when $f \in_R F_N$. $P_0^{'*} = 1 - P_1^*$. Similarly, let P_0 be the probability that $\phi(f) = 0$ when $(f_1, \ldots, f_k) \in_R K$ and $f = G(f_1, \ldots, f_k)$. $P_0 = 1 - P_1$. We will have $P_0 \geq P_0^* - \alpha - \beta$ (since the outputs 0 and 1 have symmetrical hypothesis. Or, alternatively since we can always consider an algorithm ϕ' such that $\phi'(f) = 0 \Leftrightarrow \phi(f) = 1$ and apply (5) to this algorithm ϕ').

Therefore, $1 - P_1 \geq 1 - P_1^* - \alpha - \beta$, i.e. $P_1^* \geq P_1 - \alpha - \beta$ (6). Finally, from (5) and (6), we have: $|P_1 - P_1^*| \leq \alpha + \beta$, as claimed.

Distinguishing Multiplications from Squaring Operations

Frederic Amiel[1], Benoit Feix[2], Michael Tunstall[3], Claire Whelan[4],
and William P. Marnane[5]

[1] AMESYS,
1030, Avenue Guillibert de la Lauzière,
13794 Aix-en-Provence, Cedex 3, France
f.amiel@amesys.fr
[2] Inside Contactless
41 Parc Club du Golf, 13856 Aix-en-Provence, Cedex 3, France
bfeix@insidefr.com
[3] Department of Computer Science, University of Bristol,
Merchant Venturers Building, Woodland Road,
Bristol BS8 1UB, United Kingdom
tunstall@cs.bris.ac.uk
[4] TDS (Time Data Security) Ltd.,
2060 Castle Drive, Citywest Business Campus,
Naas Road, Dublin 24, Ireland
claire.whelan@tds.ie
[5] Department of Electrical and Electronic Engineering,
University College Cork, Cork, Ireland
liam@eleceng.ucc.ie

Abstract. In this paper we present a new approach to attacking a modular exponentiation and scalar multiplication based by distinguishing multiplications from squaring operations using the instantaneous power consumption. Previous approaches have been able to distinguish these operations based on information of the specific implementation of the embedded algorithm or the relationship between specific plaintexts. The proposed attack exploits the expected Hamming weight of the result of the computed operations. We extrapolate our observations and assess the consequences for elliptic curve cryptosystems when unified formulæ for point addition are used.

Keywords: Side channel attacks, differential power analysis, modular multiplication and exponentiation, RSA, square and multiply algorithm.

1 Introduction

Side channel attacks on RSA [23] target the algorithm for modular exponentiation, the computation of which is dependent on the private key. It has been shown in the literature that an attacker can derive a private key by observing the power consumption during the computation of a naïvely implemented modular exponentiation [17]. This attack targeted implementations of the square and

R. Avanzi, L. Keliher, and F. Sica (Eds.): SAC 2008, LNCS 5381, pp. 346–360, 2009.
© Springer-Verlag Berlin Heidelberg 2009

multiply algorithm, which has been shown to be vulnerable to this technique, referred to as Simple Power Analysis (SPA). This vulnerability was present because the power consumption during the computation of a squaring operation was different to that of a multiplication, and could, therefore, be distinguished by simply monitoring the power consumption trace of the target device. This attack can allow an attacker to simply read the private key from a power consumption trace.

One of the first countermeasures proposed was a square and multiply *always* algorithm [11], which consists of a squaring operation followed by a (possibly fake) multiplication. While this algorithm achieves the effect of ensuring regular behaviour regardless of the value of the bits of the exponent, it has a large impact on efficiency. A more efficient approach, known as side channel atomicity, was proposed in [10]. While this approach does make the operations computed behave identically in terms of the instantaneous power consumption, other information being processed, such as the operand value being operated on, may leak information and provide an attacker with the necessary insight to recover the private key.

In this paper, we describe an attack that can be applied to algorithms implemented using side channel atomicity [8] without knowledge of the plaintext used. This is possible because the statistically expected Hamming weight of the result of a multiplication and a squaring operation has an exploitable difference, which is visible in the instantaneous power consumption. This highlights the importance of randomising the exponent used to calculate a modular exponentiation. A similar attack was previously proposed in [1] but requires that the architecture of a hardware implementation is known. The attack is also somewhat similar to the attack described in [27]. However, our attack is based on the distribution of the Hamming weights of the values being manipulated by a device, rather than a thorough analysis of the structure of hardware implementations of multipliers [27,29].

In some previously proposed attacks, similar power consumption traces during squaring (or doubling) operations in two separate acquisitions have been exploited by choosing or knowing the plaintexts being manipulated [14,19,30]. However, these attacks can be prevented by blinding the plaintext, and these attacks are not possible when classical padding schemes are used. The advantage of the attack described in this paper is that an attacker does not need any plaintext information. Indeed, we assume that an attacker does not have access to this information.

The implications of the proposed attack are explored further, and we analyse how attacks based on the statistically expected difference in Hamming weight of a multiplication and a squaring operation can be applied to implementations of the elliptic curve point scalar multiplication algorithm central to many elliptic curve schemes.

This paper is organised as follows. Section 2 describes why the Hamming weight is of interest in side channel analysis. Section 3 details the difference in expected Hamming weight between the results of a multiplication and squaring operation. Section 4 gives practical results using different long integer modular multiplications on a classical ARM7 microprocessor to validate the theoretical analysis given. New attacks based on this difference analysis are presented on public key algorithms in Section 5. In Section 6 we analyse the countermeasures which can be used in implementations of the algorithms discussed. We conclude our research in Section 7.

Notation: The base of a value is determined by a trailing subscript, which is applied to the whole word preceding the subscript. For example, FE_{16} is 254 expressed in base 16, $d = (d_{\ell-1}, d_{\ell-2}, \ldots, d_0)_2$ gives a binary expression for d.

2   The Hamming Weight

It has been demonstrated that in microprocessors the instantaneous power consumption is typically proportional to the Hamming weight of data being manipulated at a given point in time [8]. This difference in Hamming weight was first exploited in [17] to attack block ciphers. In this attack, an attacker acquires M power consumption traces (w_i for $i \in \{1, 2, \ldots, M\}$) during the computation of a block cipher, and chooses one bit b of an intermediate state generated during the computation of a block cipher. For a given hypothesis for a secret key value (or portion of the key) K this bit is predicted and used to determine whether a corresponding power consumption trace is a member of one of two possible sets. The first set S_0 will contain all the traces where b is equal to zero, and the second set S_1 will contain all the remaining traces, i.e. where the output bit b is equal to one.

A differential trace Δ is calculated by finding the average of each set and then subtracting the resulting values from each other, where all operations on waveforms are conducted in a pointwise fashion, i.e. this calculation is conducted on the first point of each acquisition to produce the first point of the differential trace, the second point of each acquisition to produce the second point of the differential trace, etc.

$$\Delta = \frac{\sum_{w_i \in S_0} w_i}{|S_0|} - \frac{\sum_{w_i \in S_1} w_i}{|S_1|}$$

A differential trace is produced for each value that K can take. In DES the first subkey will be treated in groups of six bits, so 64 (i.e. 2^6) differential traces will be generated to test all the combinations of six bits. The differential trace with the highest peak will validate a hypothesis for K.

In this paper we propose a novel attack based on a similar difference in Hamming weight. However, in the proposed attack it is not necessary to predict the value of a bit b, as the difference in Hamming weight is produced by the statistically expected Hamming weight of the result of the computed operations. A similar attack was previously proposed in [1] but requires that the architecture of a hardware implementation is known.

Another commonly used model to describe the power consumption is the Hamming distance model [8], where the power consumption is proportional to the Hamming weight of data being manipulated at a given point in time XORed with some previous state. An analysis of how one would perform the proposed attack in this case is beyond the scope of this paper.

In smart card implementations of RSA it has traditionally been necessary to use a cryptographic coprocessor, which would typically be modelled using the Hamming distance model [8]. However, it has been practically demonstrated in [2] that the Hamming weight model applies to many public key implementations using arithmetic coprocessors. Some modern smart card chips are using 32-bit architectures [3,20], which allow for efficient implementations of RSA without requiring a cryptographic coprocessor. In these cases the Hamming weight model is likely to apply.

3   Defining the Difference in Hamming Weight

In this section we will describe the difference in Hamming weight of a multiplication and squaring operation for random inputs, to describe why the expected difference in Hamming weight between a multiplication and squaring operation occurs.

If we consider the classical binary method of long integer multiplication, the least significant bit will be set to one, if and only if both least significant bits in the multiplicands are equal to one. The probability of the least significant bit of the output being one is, therefore, equal to $1/4$. In the case of a squaring operation the least significant bit will be equal to one if the least significant bit of the input is equal to one. For a random input this will occur with probability $1/2$.

The next least significant bit has a higher chance of being equal to one if we consider a multiplication with random inputs. However, if we conduct a squaring operation this bit will always be equal to zero. This is because there are only two bits that could affect this bit in the output. The two values that could affect this bit are 10_2 and 11_2. In the case of 10_2 only the least significant bit in the output is set to one and nothing affects the second bit, in the case of 11_2 both bits will affect the second most significant bit. The bits will therefore cancel and produce a carry. I.e. the output of every squaring operation will be equal to 0 or 1 mod 4.

This reasoning can be continued with increased complexity for more significant 'bits and will be valid for all bit lengths, until more bits than half the total number of bits being considered are included. After this point the least significant bit ceases to directly affect each bit, and will only have an effect via the carry.

Defining the exact extent of this difference for n-bit operands is a non-trivial problem. A method for defining the probability density function of the product of uniformly distributed random variables is defined in [15]. This method defines a means of computing the probability density function of the result of the product of two random values that are distributed over a continuous uniform distribution. Where, for two random values uniformly distributed in the interval $[0, \ell]$, the product can take every real value in $[0, \ell^2]$. Random values generated in a microprocessor will, by necessity, be distributed on a discrete uniform distribution. If we consider two discrete random values uniformly distributed in the interval $[0, \ell]$, the product cannot take every integer value in $[0, \ell^2]$. This is because no integer value in $(\ell, \ell^2]$ that is coprime with respect to the integer values in $[0, \ell]$ can be made from the product of two discrete random values distributed between $[0, \ell]$. The most efficient method of defining the probability distribution, and computing the expected Hamming weight of the result, is to simply count all the possible outcomes.

We will consider the multiplication and squaring of random values of bit length n, with no modular reduction. This is because we are interested in the distribution of the single-precision operations required to compute multi-precision operations. We will therefore assume that the values multiplied together will have an equal bit length. If we consider that the values multiplied together have a bit length of n, then the input values are, therefore, uniformly distributed over the integer values in the interval $[0, 2^n - 1]$.

The difference in the distributions can be demonstrated by evaluating the expected output of a multiplication and a squaring operation by calculating the mean Hamming weight of all the possible results, i.e. the expected Hamming weight of the result of squaring an n-bit value, X, is calculated as

$$\mathrm{E}(X^2) = \sum_{i=0}^{2^n-1} H(i^2) \cdot \Pr[X = i] = \frac{1}{2^n} \sum_{i=0}^{2^n-1} H(i^2),$$

and the Hamming weight of the result of multiplying two n-bit values, X and Y, is calculated as

$$\mathrm{E}(X \cdot Y) = \sum_{i=0}^{2^n-1} \sum_{j=0}^{2^n-1} H(i \cdot j) \cdot \Pr[X = i \wedge Y = j] = \frac{1}{2^{2n}} \sum_{i=0}^{2^n-1} \sum_{j=0}^{2^n-1} H(i \cdot j),$$

where H is a function that computes the Hamming weight in both cases.

This can be readily computed for bit lengths of less than, or equal to, 16. For bit lengths greater than 16 it starts to become time consuming to compute the expected Hamming weight of the output of a multiplication. Figure 1 shows the expected difference in Hamming weight for bit lengths between one and 16, and the difference appears to tend to slightly less than one as the bit length increases.

If we consider multiplication and squaring operations with 16-bit inputs, the reason for the difference in the expected result can be demonstrated if we consider the probability of each bit being equal to one. For random, uniformly distributed, 16-bit inputs the probability of each of the 32 bits in the output being equal one for a multiplication and squaring operation can be derived if all the possible inputs are considered. A plot of the probabilities for each bit for the multiplication and squaring operation is given in Figure 2. Further details on this expected difference for 32-bit variables are given in the Appendix A.

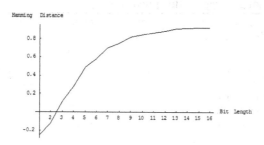

Fig. 1. The expected difference in Hamming weight between the output of a multiplication and a squaring operation, for bit lengths 1 to 16

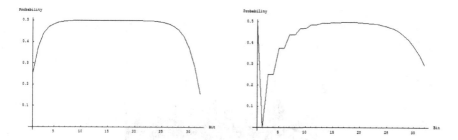

Fig. 2. The probability that each bit of the result of a multiplication (left) and a squaring operation (right) is equal to one with random 16-bit inputs

4   Demonstrating the Difference in Practice

Certain multiplication algorithms were implemented on a standard 32-bit microprocessor. The results of manipulating the power traces acquired while these multiplication algorithms were being computed are described in this section.

Long Integer Multiplication. A 128-bit multiplication using the long integer multiplication algorithm was implemented on a microprocessor and 3000 acquisitions[1] were taken for multiplications and squaring operations with random, uniformly distributed inputs. The implementation was based on the description given in [18], and is given in Algorithm 1.

The difference between the two average traces is shown in Figure 3. There are four peaks in the trace that correspond to the four squaring operations conducted by the chip to compute the square of the input, i.e. for $X = (x_3, x_2, x_1, x_0)_b$, where b is 2^{32}, there will be four occurrences in the 16 multiplications where $i = j$ when $x_i \cdot x_j$ is computed. If averaged traces corresponding to the same operation are subtracted from each other no significant peaks are produced.

---

[1] Similar results are possible with 500 traces. However, the results are not as clear.

Algorithm 1. Long Integer Multiplication

Input: $X = (x_{z-1}, \ldots, x_1, x_0)_b$, $Y = (y_{z-1}, \ldots, y_1, y_0)_b$
Output: $W = (w_{2z-1}, \ldots, w_1, w_0)_b = X \cdot Y$

$W \leftarrow 0$
for $i = 0$ **to** $z - 1$ **do**
   $c \leftarrow 0$
   for $j = 0$ **to** $z - 1$ **do**
      $(uv)_b \leftarrow w_{i+j} + x_j \cdot y_i + c$
      $w_{i+j} \leftarrow v \ ; \ c \leftarrow u$
   end
   $w_{2z-1} \leftarrow u$
end

return W

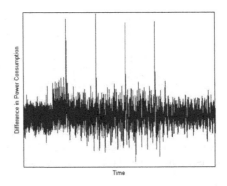

Fig. 3. The difference between two averaged power consumptions for long integer multiplication

Montgomery Multiplication. One of the most common methods of calculating modular multiplication is using Montgomery multiplication [21]. This is because of its efficiency, especially as it can be parallelised in hardware and does not require any time-consuming word-by-word divisions.

Montgomery multiplication [21] does not return the simple product of X and Y modulo M. The algorithm actually returns $XYR^{-1} \bmod M$, where $R^{-1} \bmod M$ is introduced by the algorithm ($R = b^z$), which imposes certain restrictions on its use. The conditional subtraction has been shown to be unnecessary, and undesirable in a secure implementation, and was not included in our implementation [25,26].

A description of Montgomery multiplication is given in Algorithm 2. Here b is the size of the basic data unit, usually a machine word, and z is the number of words in the representation of M, X and Y.

As previously, a 128-bit multiplication algorithm was implemented and 3000 acquisitions were taken for multiplications and squaring operations with random,

---

Algorithm 2. Montgomery Multiplication

Input: $X = (x_{z-1}, \ldots, x_1, x_0)_b$, $Y = (y_{z-1}, \ldots, y_1, y_0)_b$,
$\quad M = (m_{z-1}, \ldots, m_1, m_0)_b$, $R = b^z$ with $gcd(M, b) = 1$, and $M' = -M^{-1}$
$\quad \bmod b$
Output: $A = (a_{z-1}, \ldots, a_1, a_0)_b = X \cdot Y \cdot R^{-1} \bmod M$

$A \leftarrow 0$
for $i = 0$ **to** $z - 1$ **do**
$\quad u_i \leftarrow (a_0 + x_i \cdot y_0)M' \bmod b$
$\quad A \leftarrow (A + x_i \cdot Y + u_i \cdot M)/b$
end

if $A \geq M$ **then** $A \leftarrow A - M$

return A

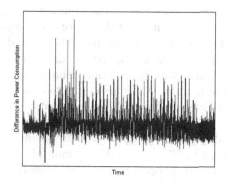

Fig. 4. The difference between two averaged power consumptions for Montgomery multiplication

uniformly distributed inputs. The difference in the average trace produced by each set of acquisitions is shown in Figure 4.

The first peak will be produced by the calculation of $a_0 + x_0 \cdot y_0 \bmod b$, as a_0 is set to zero so the difference in the distribution will be visible even where this can be calculated with one instruction (e.g. on ARM microprocessors).

In the implementation analysed the processor computed $A \leftarrow A + u_0 \cdot M$ followed by $A \leftarrow A + x_i \cdot Y$. The group of peaks following the first peak are caused by the repeated manipulation of u_0 when it is multiplied by M. The peaks are dependent on the value of M, and therefore M', that is being used and will vary from one analysis to another.

This is followed by a large peak that corresponds to the computation of $x_0 \cdot y_0$, which is subsequently combined with $u_0 \cdot M$ by adding the result to A. The next group of three peaks are created by the manipulation of A when it is combined with $u_1 \cdot M$.

This is followed by three groups of small peaks that correspond to the multiplication of three squaring operations conducted by the chip to compute the square of the input, i.e. three instances where $i = j$ when $x_i \cdot y_j$ is computed.

5   Exploiting the Difference in Tamper Resistant Cryptographic Primitives

Once an exponentiation algorithm has been chosen, for instance Barrett or Montgomery exponentiation [5,21], a common countermeasure to protect embedded implementations from Simple Power Analysis, consists in using side channel atomicity. This was introduced in [10], where they deem an algorithm to be secure if it can be broken down into indistinguishable blocks. In this section we describe how these schemes can be attacked by observing the difference between a multiplication and a squaring operation.

5.1   Recovering the Exponent in Atomic Exponentiations

The simplest exponentiation algorithm is the square and multiply algorithm, that functions by scanning the bits of an exponent from left to right. An accumulator is initially set to one and for each bit of the exponent scanned the accumulator is operated upon. For each bit the accumulator is squared, and when a bit is equal to one the accumulator is multiplied by the value being raised to the power of the exponent. The square and multiply atomic exponentiation algorithm simply means that squaring operations are computed using the same algorithm as multiplications and the side channel becomes identical [10].

If a series of power consumption traces are taken, the points corresponding to each operation (multiplication or squaring) can be identified using a method similar to that described in [27] for identifying multiplications with a constant value. The average power consumption trace of each operation can be compared to the operation preceding, or following, it by performing a pointwise subtraction. If this corresponds to subtracting the power consumption trace of a squaring from that of a multiplication peaks will be visible (as shown in Figure 3), in the case where the opposite occurs the same peaks will occur but will be negative. It is interesting to note that an attacker does not need to have any knowledge of the values being manipulated.

An attacker would therefore be able to determine a k-bit exponent by making $\frac{3}{2}k - 1$ comparisons, i.e. comparing each operation with one neighbouring operation. This can be decreased by a factor of two, where an attacker can be sure that each comparison gives noise free information, by only including each operation in a comparison once.

If we consider the (M, M^3) algorithm, as described in [10], analysing the power consumption traces is sufficient to decrease the security of the algorithm. However, we cannot recover the entire private exponent d. The (M, M^3) algorithm functions in a similar manner to the square and multiply algorithm, but there are three possible cases when parsing the bits of the private exponent from left to right. When $d_i = 0$ a squaring operation is performed. When $d_i d_{i-1} = 10_2$ the device computes a squaring operation, a multiplication with M and then a squaring operation. The third case occurs when $d_i d_{i-1} = 11_2$, where the device computes two squaring operations followed by a multiplication with M^3.

In the remainder of this section we will denote a multiplication by M and a squaring operation by S. Any sequence of operations MSM is particular because it indicates that the last two operations correspond to the secret bits $d_i d_{i-1} = 10_2$ and that $d_{i+2} d_{i+1} = 11_2$. Indeed, this sequence can only be part of a longer sequence $SSMSM$. We can also identify any bits of an exponent set to zero when there are more than two consecutive squaring operations.

Through simulations of this attack we were able to determine that an attacker can retrieve, on average, 37% of the bits of a private exponent by exploiting the sequence MSM, and a further 17% of the bits by identifying repeated squaring operations. Thus, an attacker would be able to retrieve 54% of the bits of a private exponent using the attack method proposed above.

In case where the public exponent is small (for instance 3 or $2^{16}+1$), half of the most significant bits of d are intrinsically leaked as showed in [7]. Thus, combined with the side channel leakage, up to 3/4 of the bits of a private exponent could be considered to be recoverable by an attacker. However, there are currently no factorisation techniques in the literature that can benefit from such partial information, although an interesting approach has been published in [13], where the authors assume that the exponent is modified by a small random value. How the proposed attack can be applied to an implementation where this occurs is discussed in Section 6.2.

5.2   Recovering the Scalar in ECC Using Unified Addition Formulæ

In the context of Elliptic Curve Cryptosystems (ECC), the ability to distinguish a multiplication operation from a squaring operation can also facilitate the extraction of secret information. The calculation of the point scalar multiplication of rP, where r is a secret scalar value, P is a point on the prescribed elliptic curve, and the operation of rP is known as point scalar multiplication is central to a number of ECC schemes, such as EC-DH [6]. One of the most side-channel naïve methods to calculate rP is the double and add method, which involves accumulatively doubling and adding the point P, the sequence of which is determined by the binary representation of r [4]. This method is inherently vulnerable to Simple Power Analysis and other side-channel attacks.

A countermeasure, known as unified addition formulæ, to make the double operation indistinguishable from the addition operation was proposed in [6,9]. This method defined formulæ for the calculation for point addition and point doubling, which is equivalent for both operations. Specifically, the slope for each operation is equivalent. For example, the slope calculated during the addition of the points $P = (x_1, y_1)$, $Q = (x_2, y_2)$ is

$$\lambda = \frac{x_1^2 + x_1 x_2 + x_2^2 + a_2 x_1 + a_2 x_2 + a_4 - a_1 y_1}{y_1 + y_2 + a_1 x_2 + a_3},$$

regardless of whether P is equal, or not equal, to Q. Hence, no discernible difference between the addition and doubling of a point is present in the formula. In light of the work described in this paper, a difference between these operations can be identified. The calculation of $x_1 \cdot x_2$ in the calculation of λ will allow an attacker to determine whether an addition or a doubling operation is being performed, since when a double is performed $x_1 \cdot x_2 = x_1^2$, and will be vulnerable to the attack process described in Section 5.1.

Similarly, this potential exploit can be witnessed when the elliptic curve points are represented and operated on as projective coordinates, which will be the case in most practical implementations. Unified formula for point addition and multiplication using projective coordinates was also given by [9] and further examined by [24]. In this case the addition of the points $P = (X_1, Y_1, Z_1)$, $Q = (X_2, Y_2, Z_2)$, with $x_i = X_i/Z_i$ and $y_i = Y_i/Z_i$ is

$$X_3 = 2FW \qquad Y_3 = R(G - 2W) \qquad Z_3 = 2F^3$$

where $U_1 = X_1 Z_2$, $U_2 = X_2 Z_1$, $S_1 = Y_1 Z_2$, $S_2 = Y_2 Z_1$, $Z = Z_1 Z_2$, $T = U_1 + U_2$, $M = S_1 + S_2$, $F = ZM$, $L = MF$, $G = TL$, $R = T^2 - U_1 U_2 + AZ^2$ and $W = R^2 - G$. Notice that, when a point doubling operation is being performed, the computation of Z and $U_1 U_2$ in R will be squaring operations and, hence, our attacks can also be applied to such implementations.

6   Countermeasures

As presented in the previous sections, both side channel atomicity and unified point addition formulæ are potentially vulnerable to attack according to the expected difference highlighted in Section 3. Some of the countermeasures that could be used to prevent this attack are discussed in this section.

6.1   Blinding

The most common countermeasure used to protect RSA against DPA consists in modifying plaintext with a random value, either using an additive method $m_b = m + r_1 n \mod r_2 n$, where r_1 and r_2 are random values, or in a multiplicative way $m_b = r_1^e \cdot m$ where e is the public exponent. With such a countermeasure, classical DPA [17], and related attacks (such as the attacks presented in [19] or [2]), can no longer be applied.

However, plaintext blinding is not sufficient to protect against the attack described in this paper. It is, therefore, necessary to change the order of the multiplication and squaring operations between different exponentiations. The most common solution consists of computing $d_b = d + r_1 \phi(N)$, where ϕ is Euler's Totient function and r_1 is a small random value [16]. An equivalent solution can be used to protect the double and add algorithm [11].

6.2   The Big Mac Attack

In [27], Walter presented the Big Mac attack, which demonstrates a powerful attack on devices with a high level of side channel leakage, i.e. devices where only a few power consumption traces are required to successfully conduct a power analysis. Furthermore, in [28] it is explained that using longer keys in asymmetric cryptosystems improves the probability of a Big Mac attack succeeding. This idea can be extended here to obtain a kind of Big Mac power attack which, would enable the attack described in Section 5.1 to be conducted on one power consumption trace. In such a case, the blinding of d would not provide adequate protection to defend against the attack described in this paper. The attack could also be applied to other schemes, such as the Diffie-Hellman key exchange [12] and the DSA [22].

An attack on a single power consumption trace would consist of identifying the points in a power consumption trace that correspond to the computation of $x_j \cdot y_i$ when $i = j$ (e.g. in Algorithm 1), and extract these points where each point its then treated as a separate trace. These small traces can then be used in place of a trace representing the entire operation in exactly the same manner described in Section 5.1. The points to be used can be identified by analysing an unprotected algorithm, as described in [2].

The success of this attack will depend on the length of the key and the word size of the processor, i.e. long keys and small word size will provide an accurate average and raise the probability of achieving a successful attack [28]. This demonstrates that the blinding of the private key d may not be adequate to prevent the attack presented in this paper.

7   Conclusion

This paper shows that the statistically expected difference in operations computed by a microprocessor can be used to distinguish between a multiplication and a squaring operation. All that is required is that the plaintexts used contain enough variation that the computations adhere to the distributions defined in Section 3. This is an improvement over previously published results [14,19,27,29,30], as the described attack requires no knowledge of the plaintext being manipulated or of the architecture of the multiplier. Moreover, the proposed attacks will work when classical padding schemes are used. Further work that is being conducted by the authors consists of analysing the algorithms that are potentially vulnerable to this attack, and the development of inexpensive countermeasures.

Acknowledgements

The authors would like to thanks James Curran of University College Cork for helpful discussions in the early stages of this work. The work described in this paper has been supported in part by the European Commission IST Programme under Contract IST-2002-507932 ECRYPT and EPSRC grant EP/F039638/1 "Investigation of Power Analysis Attacks". Also the support of the Informatics Commercialisation initiative of Enterprise Ireland is gratefully acknowledged.

References

1. Akishita, T., Takagi, T.: Power analysis to ECC using differential power between multiplication and squaring. In: Domingo-Ferrer, J., Posegga, J., Schreckling, D. (eds.) CARDIS 2006. LNCS, vol. 3928, pp. 151–164. Springer, Heidelberg (2006)
2. Amiel, F., Feix, B., Villegas, K.: Power analysis for secret recovering and reverse engineering of public key algorithms. In: Adams, C., Miri, A., Wiener, M. (eds.) SAC 2007. LNCS, vol. 4876, pp. 110–125. Springer, Heidelberg (2007)
3. ARM. SecurCore family,
   http://www.arm.com/products/CPUs/families/SecurCoreFamily.html
4. Avanzi, R.-M., Cohen, H., Doche, C., Frey, G., Lange, T., Nguyen, K., Verkauteren, F.: Handbook of Elliptic and Hyperelliptic Curve Cryptography. Taylor & Francis Ltd, Abington (2008)
5. Barrett, P.: Implementing the rivest shamir and adleman public key encryption algorithm on a standard digital signal processor. In: Odlyzko, A.M. (ed.) CRYPTO 1986. LNCS, vol. 263, pp. 311–323. Springer, Heidelberg (1987)
6. Blake, I., Seroussi, G., Smart, N.: Advances in Elliptic Curve Cryptography. Lecture Note Series, vol. 317. Cambridge University Press, London Mathematical Society (2005)
7. Boneh, D., Durfee, G., Frankel, Y.: An attack on RSA given a small fraction of the private key bits. In: Ohta, K., Pei, D. (eds.) ASIACRYPT 1998. LNCS, vol. 1514, pp. 25–34. Springer, Heidelberg (1998)
8. Brier, É., Clavier, C., Olivier, F.: Correlation power analysis with a leakage model. In: Joye, M., Quisquater, J.-J. (eds.) CHES 2004. LNCS, vol. 3156, pp. 16–29. Springer, Heidelberg (2004)
9. Brier, É., Joye, M.: Weierstraß elliptic curves and side-channel attacks. In: Naccache, D., Paillier, P. (eds.) PKC 2002. LNCS, vol. 2274, pp. 335–345. Springer, Heidelberg (2002)
10. Chevallier-Mames, B., Ciet, M., Joye, M.: Low-cost solutions for preventing simple side-channel analysis: Side-channel atomicity. IEEE Transactions on Computers 53(6), 760–768 (2004)
11. Coron, J.-S.: Resistance against differential power analysis for elliptic curve cryptosystems. In: Koç, Ç.K., Paar, C. (eds.) CHES 1999. LNCS, vol. 1717, pp. 292–302. Springer, Heidelberg (1999)
12. Diffie, W., Hellman, M.E.: New directions in cryptography. IEEE Transactions on Information Theory 22(6), 644–654 (1976)
13. Fouque, P.-A., Kunz-Jacques, S., Martinet, G., Muller, F., Valette, F.: Power attack on small RSA public exponent. In: Goubin, L., Matsui, M. (eds.) CHES 2006. LNCS, vol. 4249, pp. 339–353. Springer, Heidelberg (2006)
14. Fouque, P.-A., Valette, F.: The doubling attack – why upwards is better than downwards. In: Walter, C.D., Koç, Ç.K., Paar, C. (eds.) CHES 2003. LNCS, vol. 2779, pp. 269–280. Springer, Heidelberg (2003)
15. Glen, A.G., Leemis, L.M., Drew, J.H.: Computing the distribution of the product of two continuous random variables. Computaional Satatistics and Data Analysis 44(3), 451–464 (2004)
16. Kocher, P.C.: Timing attacks on implementations of diffie-hellman, RSA, DSS, and other systems. In: Koblitz, N. (ed.) CRYPTO 1996. LNCS, vol. 1109, pp. 104–113. Springer, Heidelberg (1996)
17. Kocher, P.C., Jaffe, J., Jun, B.: Differential power analysis. In: Wiener, M. (ed.) CRYPTO 1999. LNCS, vol. 1666, pp. 388–397. Springer, Heidelberg (1999)

18. Menezes, A., van Oorschot, P., Vanstone, S.: Handbook of Applied Cryptography. CRC Press, Boca Raton (1997)
19. Messerges, T.S., Dabbish, E.A., Sloan, R.H.: Power analysis attacks of modular exponentiation in smartcards. In: Koç, Ç.K., Paar, C. (eds.) CHES 1999. LNCS, vol. 1717, pp. 144–157. Springer, Heidelberg (1999)
20. MIPS-Technologies. SmartMIPS ASE, http://www.mips.com/content/Products/
21. Montgomery, P.: Modular multiplication without trial division. Mathematics of Computation 44, 519–521 (1985)
22. National Institute of Standards and Technology. Digital signature standard (DSS), FIPS-186-2 (2000)
23. Rivest, R., Shamir, A., Adleman, L.M.: Method for obtaining digital signatures and public-key cryptosystems. Communications of the ACM 21(2), 120–126 (1978)
24. Stebila, D., Thériault, N.: Unified point addition formulæ and side-channel attacks. In: Goubin, L., Matsui, M. (eds.) CHES 2006. LNCS, vol. 4249, pp. 354–368. Springer, Heidelberg (2006)
25. Walter, C.D.: Montgomery exponentiation needs no final subtractions. Electronic Letters 35(21), 1831–1832 (1999)
26. Walter, C.D.: Montgomery's multiplication technique: How to make it smaller and faster. In: Koç, Ç.K., Paar, C. (eds.) CHES 1999. LNCS, vol. 1717, pp. 80–93. Springer, Heidelberg (1999)
27. Walter, C.D.: Sliding windows succumbs to big mac attack. In: Koç, Ç.K., Naccache, D., Paar, C. (eds.) CHES 2001. LNCS, vol. 2162, pp. 286–299. Springer, Heidelberg (2001)
28. Walter, C.D.: Longer keys may facilitate side channel attacks. In: Matsui, M., Zuccherato, R.J. (eds.) SAC 2003. LNCS, vol. 3006, pp. 42–57. Springer, Heidelberg (2004)
29. Walter, C.D., Samyde, D.: Data dependent power use in multipliers. In: Montuschi, P., Shwarz, E. (eds.) 17th Symposium on Computer Arithmetic (ARITH), pp. 4–12. IEEE, Los Alamitos (2005)
30. Yen, S.-M., Lien, W.-C., Moon, S.-J., Ha, J.C.: Power analysis by exploiting chosen message and internal collisions – vulnerability of checking mechanism for RSA-decryption. In: Dawson, E., Vaudenay, S. (eds.) Mycrypt 2005. LNCS, vol. 3715, pp. 183–195. Springer, Heidelberg (2005)

A   Appendix

In Section 3 we discussed the expected Hamming weight of multiplication and squaring operations for random, uniformly distributed, 16-bit inputs. Given the complexity of evaluating all of the possible inputs to a multiplication, it is not possible to evaluate the expected Hamming weight and the corresponding distribution of the individual bits for larger bit length. Given that the implementations described in Section 4 are on a 32-bit chip, it would be helpful to attempt to describe the corresponding distribution.

To characterise the distribution of the individual bits in the result of a 32-bit squaring operation all the possible input values were evaluated and the result is plotted on the right hand side of Figure 5. It is not possible to evaluate a 32-bit multiplication in the same way as there are 2^{64} possible inputs to a

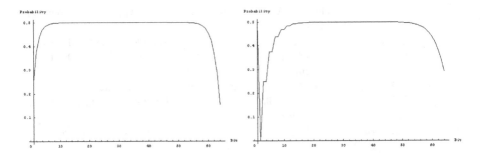

Fig. 5. The distribution of the individual bits of the result of a multiplication (left) and a squaring operation (right) with random 32-bit inputs

32-bit multiplication. An approximation to the distribution was generated by evaluating the product of 2^{32} pairs uniformly distributed 32-bit random values.

The form of the difference is similar to that shown of 16-bit operations in Section 3, but with a larger region where the distribution of the bits are identical.

Given the very regular nature of multiplication algorithms, it would seem reasonable to assume that the same difference will occur for all bit lengths. However, it is not possible to demonstrate this because of the complexity of evaluating all the possible inputs.

Subquadratic Polynomial Multiplication over $GF(2^m)$ Using Trinomial Bases and Chinese Remaindering

Éric Schost[1] and Arash Hariri[2]

[1] ORCCA, Computer Science Department, The University of Western Ontario,
London, Ontario, Canada
eschost@uwo.ca

[2] Department of Electrical and Computer Engineering, The University of Western
Ontario, London, Ontario, Canada
hariri@ieee.org

Abstract. Following the previous work by Bajard-Didier-Kornerup, McLaughlin, Mihailescu and Bajard-Imbert-Jullien, we present an algorithm for modular polynomial multiplication that implements the Montgomery algorithm in a residue basis; here, as in Bajard *et al.*'s work, the moduli are trinomials over $\mathbb{F}_2$. Previous work used a second residue basis to perform the final division. In this paper, we show how to keep the same residue basis, inspired by l'Hospital rule. Additionally, applying a divide-and-conquer approach to the Chinese remaindering, we obtain improved estimates on the number of additions for some useful degree ranges.

Keywords: Montgomery multiplication, Chinese remainder theorem, finite fields, subquadratic area complexity.

1 Introduction

Modular multiplication of polynomials is a cornerstone for many higher-level applications, from finite field arithmetic (for non-prime fields) to implementation of cryptographic protocols. In all that follows, we focus on the practically important case of polynomials over $\mathbb{F}_2$.

Given a polynomial R of degree m, the Montgomery multiplication algorithm [1] shows how to reduce a multiplication modulo a polynomial V of degree at most m to a multiplication modulo R and a division by R, assuming that R and V are coprime. This "multiplication" is slightly twisted, though, since on input A and B, it returns AB/R modulo V.

To implement this algorithm, we need to specify R. An obvious choice is $R = x^m$ [2]. Then, the computations are similar to (but distinct from) those in the Cook-Sieveking-Kung algorithm [3, Chapter 9]. Using fast polynomial multiplication [4,5], this yields an algorithm that uses $O(m \log(m) \log \log(m))$ additions and multiplications. However, the rather large constant hidden in the

R. Avanzi, L. Keliher, and F. Sica (Eds.): SAC 2008, LNCS 5381, pp. 361–372, 2009.

big-O estimate makes it desirable to devise multiplication schemes with a possibly higher asymptotic cost, but whose performance is better for moderate values of m, say $m \leq 1000$.

To fulfill this goal, we will take another approach to the Montgomery multiplication, that follows the ideas introduced in [6] and [7] (focusing on integer multiplication), and [8] and [9] (for polynomials): we take $R = r_1 \cdots r_n$, with pairwise coprime r_i. To make this approach useful, the computations modulo r_i should be easy: in [9], which focuses on large prime base fields, the r_i's are linear. Here, following [8], the r_i's will be trinomials.

Main Result. Throughout this paper, a family of n pairwise coprime trinomials $\mathscr{R} = (r_1, \ldots, r_n)$ in $\mathbb{F}_2[x]$ is fixed; their product is denoted R. We suppose that all r_i's have the same degree d. This assumption makes it possible for us to give explicit complexity bounds; however, the algorithm still works with trinomials of different degrees. We also assume that all r_i's are squarefree. This can for instance be obtained by taking d odd: in this case, the derivative of $x^d + x^e + 1$ is either x^{d-1} or $x^{d-1} + x^{e-1}$, depending on the parity of e; in both cases, it has no common factor with $x^d + x^e + 1$.

Let $m = nd$ and let V be in $\mathbb{F}_2[x]$, with $\gcd(R, V) = 1$ and $\deg(V) \leq m$; V does not have to be irreducible and can have degree less than m. Computations modulo V will be done through the Montgomery algorithm, applied in the residue basis $\mathscr{R}$; since $m \geq \deg(V)$, a polynomial of degree less than $\deg(V)$ is uniquely determined by its residues modulo $\mathscr{R}$.

Formally, our computational model is the *boolean circuit*, using multiplication (AND) and addition (XOR) gates. The *area complexity* is the number of gates we use; we distinguish between the number of multiplications and additions. The *time complexity* is the length of the longest path in the circuit, i.e., the critical path. As is customary, we write time complexities in the form $\alpha T_A + \beta T_X$, to indicate that all paths in the directed graph underlying the circuit have at most α multiplication gates and β additions gates. Time complexity estimates will depend on a function $T_{\mathrm{rem}}(d)$ defined as follows: $T_{\mathrm{rem}}(d)$ is such that for all $i \leq n$, one can compute the remainder of a polynomial of degree at most $2d - 2$ by r_i using $2d - 2$ additions, in time $T_{\mathrm{rem}}(d)$; we describe this function further in Section 2. Since our main focus is more on the total number of gates than on time complexity, we only give big-O estimates for the latter, except in very simple cases.

Theorem 1. *One can perform modular multiplication in the residue basis $\mathscr{R} = (r_1, \ldots, r_n)$ using $7nd^2$ multiplications and*

$$7nd^2 + 8n^2d - 2nd\log_2(n) + 6nd - 2n^2 - 10n$$

additions. The time complexity is $O(T_A + \log_2(d)T_X + nT_{\mathrm{rem}}(d))$.

When d is such that one can take $n \simeq d$, our algorithm uses $O(m^{1.5})$ operations. In the worst case, $T_{\mathrm{rem}}(d)$ is in $O(dT_X)$, so the time complexity is $O(T_A + ndT_X)$. If we assume that all trinomials have the form $x^d + x^e + 1$, with $e < d/2$, then $T_{\mathrm{rem}}(d)$ is in $O(T_X)$, and the time complexity is $O(T_A + \log_2(d)T_X + nT_X)$.

Previous Work. There exist several other approaches to modular multiplication; as said above, it is possible to reach a quasi-linear number of operations. Several other families of algorithms are also known, either for low weight moduli [10,11] or for arbitrary ones, such as [12], which shares some features with the family of algorithms we present now.

Our work follows previous results of Bajard *et al.* [8], who use a basis of trinomials as well (with different constraints than ours) and Newton interpolation techniques. Here, we use the Chinese remaindering with a classical divide-and-conquer approach. Mihailescu [9] uses moduli of degree one, whose roots are either roots of unity or consecutive integers; this is not immediately possible here, since we work over $\mathbb{F}_2$.

In both previous papers, a difficulty arises, since the final exact division cannot be performed in the residue basis. The same solutions are used: shifting to another residue basis to do the division. We present an alternative solution, inspired by l'Hospital rule. This enables us to work with the same set of moduli (and thus, to reach higher m for a given moduli degree d), at the cost of a slight increase in the number of operations.

Notation. We write "large" degree polynomials (of degree typically close to $m = nd$) with upper case letters, and "low" degree ones (typically, residues of degree less than d) with lower case letters. Vectors of residues are written in bold face. The equality $A = B \bmod C$ means that A and B are congruent modulo C; the stronger equality $A = B \operatorname{rem} C$ means that A is the remainder of the division of B by C, so that $\deg(A) < \deg(C)$. The notation $A = B \operatorname{div} C$ means that A is the quotient in the Euclidean division of B by C.

Outline. Section 2 consists of preliminaries. In Section 3, we consider Chinese remaindering using trinomials. In Section 4, we present our new algorithm and illustrate its performance in Section 5.

2   Preliminaries

This section reviews basic material on operations such as polynomial multiplication or reduction. Most of these results are known; the only new element here is a straightforward estimate on the cost of multiplication by several trinomials.

Polynomial Multiplication. Let a and b be in $\mathbb{F}_2[x]$, of degree less than d. Then, the product ab can be computed using $(d-1)^2$ additions and d^2 multiplications; the time complexity is $T_A + \lceil \log_2(d) \rceil T_X$.

Reduction by Trinomials. For $i \leq n$ and $a \in \mathbb{F}_2[x]$ of degree at most $2d-2$, $a \operatorname{rem} r_i$ can be computed using $2d-2$ additions [13]. As said before, we write $T_{\mathrm{rem}}(d)$ for the time complexity of this operation; the following estimates for $T_{\mathrm{rem}}(d)$ are available:

- for arbitrary trinomials r_i, we can let $T_{\text{rem}}(d) = (2d - 2)T_X$;
- if all trinomials r_i are of the form $x^d + x^{e_i} + 1$, with $e_i < d/2$, then we can take $T_{\text{rem}}(d) = 2T_X$, see [14].

Multiplication by Trinomials. We also need to estimate the cost of multiplication of a polynomial by one or several of the trinomials $r_1, \ldots, r_n$. Our result gives a reasonable operation count; however, we are not able to obtain logarithmic time bounds. Such bounds would reduce the overall time complexity of our main algorithm as well.

Proposition 1. *Let P be in $\mathbb{F}_2[x]$ of degree less than s, let $\ell \le n$ and let $a_1, \ldots, a_\ell$ be in $\{1, \ldots, n\}$. Then one can compute the product $r_{a_1} \cdots r_{a_\ell} P$ using $2(s - d)\ell + d\ell^2$ additions in time $2\ell T_X$.*

Proof. Let $P_0 = P$ and $P_i = r_{a_i} P_{i-1}$ for $i = 1, \ldots, \ell$, so that the polynomial we want to compute is P_n. Remark that P_i has degree less than $s + di$ for all i. Given P_{i-1}, one can compute P_i using $2(s + d(i - 1)) - d$ additions. Hence, the total number of additions is at most $2(s - d)\ell + d\ell^2$. Since multiplication by a single trinomial can be done in time $2T_X$, the overall time complexity is $2\ell T_X$. $\qquad\square$

3   Chinese Remaindering for Trinomials

We continue with algorithms to perform Chinese remaindering modulo the family of pairwise coprime trinomials $r_1, \ldots, r_n$, and for the inverse operation, multiple reduction.

Given residues $\mathbf{a} = (a_1, \ldots, a_n)$, with $\deg(a_i) < d$, the Chinese remainder theorem shows that there exists a unique polynomial A of degree less than $m = nd$ with $a_i = A \text{ rem } r_i$ for all i. Quasi-linear algorithms of area complexity $O(m \log(m)^2 \log\log(m))$ are known for computing A from its residues a_i, and conversely [3, Chapter 10]. However, the constant hidden in the big-O is rather large (especially for reduction, which uses fast Euclidean division).

These algorithms rely on divide-and-conquer techniques. In what follows, we reuse this idea to devise a Chinese remainder algorithm adapted to trinomials, which performs well for moderate values of m. We also give a (substantially simpler) multiple reduction algorithm with a similar cost.

Linear Combination and Chinese Remaindering. Let $\mathbf{a} = (a_1, \ldots, a_n)$ be in $\mathbb{F}_2[x]^n$, with $\deg(a_i) < d$ for all i. We consider here the question of computing the coefficients of the linear combination

$$A = \sum_{i \le n} a_i S_i, \quad \text{with} \quad S_i = r_1 \cdots r_{i-1} r_{i+1} \cdots r_n;$$

in what follows, we will write $A = \text{LinComb}(\mathbf{a}, \mathscr{R})$. Note that this does not quite solve the Chinese remaindering question, since $A \text{ rem } r_i = a_i S_i \text{ rem } r_i$: thus, one should divide a_i by S_i modulo r_i prior to the combination. However, in the cases where we apply this algorithm, we will be able to perform this preliminary step jointly with some other operation, so that the main task is indeed the linear combination.

Proposition 2. *Given* $\mathbf{a} = (a_1, \ldots, a_n)$, *one can compute* $\mathrm{LinComb}(\mathbf{a}, \mathscr{R})$ *using* $3n^2d - nd\log_2(n)$ *additions, in time* $O(nT_X)$.

Proof. The proof adapts that of [3, Theorem 10.21] to moduli that are trinomials. If $n = 1$, we have nothing to do. Otherwise, let $n' = \lfloor n/2 \rfloor$ and $n'' = n - n'$. Define next

$$B = \sum_{1 \leq i \leq n'} a_i r_1 \cdots r_{i-1} r_{i+1} \cdots r_{n'}, \quad C = \sum_{n'+1 \leq i \leq n} a_i r_{n'+1} \cdots r_{i-1} r_{i+1} \cdots r_n,$$

so that we have

$$A = B r_{n'+1} \cdots r_n + C r_1 \cdots r_{n'}.$$

This leads to a divide-and-conquer algorithm. Assuming that B and C have been computed recursively, A is obtained through multiplications by trinomials, followed by a polynomial addition.

The first step requires to multiply B and C by several trinomials, so it is handled by Proposition 1. Since B has degree less than $n'd$ and we multiply it by n'' trinomials, we obtain a number of additions of $2(n'd - d)n'' + dn''^2$, with a time complexity of $2n''T_X$. Similarly, C has degree less than $n''d$ and we multiply it by n' trinomials, so we get $2(n''d - d)n' + dn'^2$ additions, and a time complexity of $2n'T_X$.

The final polynomial addition takes an extra nd scalar additions, which are done in parallel. After simplifying, we get that the total number of additions needed to reconstruct A from B and C is at most $3n^2d/2 - nd$ for n even, and $3n^2d/2 - nd - d/2$ for n odd. Hence, the number $N(n)$ of additions satisfies the relation:

$$N(n) \leq N(n') + N(n'') + 3n^2d/2 - nd.$$

Solving the recurrence gives $N(n) \leq 3n^2d - nd\log_2(n)$. Since $2n' \leq 2n'' \leq n+1$, the time complexity $D(n)$ satisfies

$$D(n) \leq \max(D(n'), D(n'')) + (n+2)T_X,$$

which yields our claim.   □

Reduction. For modular reduction, a more direct approach turns out to work well. In cases where we need the modular reduction in Section 4, the input polynomial A will be even; hence, we present an adapted reduction algorithm, starting with a lemma.

Lemma 1. *Given* A *of degree at most* $s(d-1)$, *one can compute all* $a_i = A \operatorname{rem} r_i$ *using* $n(s-1)(2d-2)$ *additions, in time* $O(sT_{\mathrm{rem}}(d))$.

Proof. We prove that for any given $i \leq n$, $a_i = A_i \operatorname{rem} r$ can be computed using $(s-1)(2d-2)$ additions, in time $(s-1)T_{\mathrm{rem}}(d)$. Doing so in parallel for all r_i proves our proposition.

If $s = 1$, we have nothing to do. Else, we write $A = A_0 + x^{(d-1)(s-2)}A_1$, with $\deg(A_0) < (d-1)(s-2)$ and $\deg(A_1) \leq 2(d-1)$. Let $b_i = A_1$ rem r_i and $B_i = A_0 + x^{(d-1)(s-2)}b_i$, so that $A = B_i$ mod r_i and $\deg(B_i) \leq (d-1)(s-1)$. By what was said before, one can compute b_i using $2d-2$ additions, with a time complexity of $T_{\mathrm{rem}}(d)$. Continuing inductively by reducing B_i modulo r_i, the final number of additions is $(s-1)(2d-2)$, and the time is $(s-1)T_{\mathrm{rem}}(d)$.   □

Corollary 1. *Let $A \in \mathbb{F}_2[x]$ be even and of degree less than nd. Then, one can compute all $a_i = A$ rem r_i using $n(n+3)(d-1)$ additions, in time $O(nT_{\mathrm{rem}}(d))$.*

Proof. Let us write $A = B^2$, with $\deg(B) < nd/2$. Our assumptions that all r_i are coprime imply that $n \leq d$, so that the latter degree is upper-bounded by $(\lceil n/2 \rceil + 1)(d-1)$. In view of the previous lemma, we can thus compute all $b_i = B_i$ rem r_i using $n\lceil n/2 \rceil(2d-2) \leq 2(n+1)(d-1)$ additions, in time $O(nT_{\mathrm{rem}}(d))$.

Then, we obtain a_i as b_i^2 rem r_i. The cost of reducing b_i^2 modulo r_i is at most $2d-2$, in time $T_{\mathrm{rem}}(d)$. Hence, the total cost is at most $(n+1)(d-1)+2d-2 = (n+3)(d-1)$ additions for reduction by a single trinomial; the time complexity is $O(nT_{\mathrm{rem}}(d))$.   □

4   The Multiplication Algorithm

We conclude with presenting our Montgomery-like multiplication algorithm in the residue basis. We start by recalling the Montgomery original construction, then the prior extension to residue basis computations from [9,8], and finally give our new version.

The Montgomery Algorithm. As in the introduction, let R and V be of respective degrees m and m', with $\gcd(R, V) = 1$ and $m' \leq m$. Given the inverse W of V modulo R and A, B of degrees less than m', the Montgomery algorithm computes the quantities Z, H, T, Q of Figure 1.

Input:

  – A, B, V, W, R

Output:

  – $Q = AB/R \bmod V$

1. $Z \quad = \quad AB$
2. $H \quad = \quad ZW$ rem R
3. $T \quad = \quad Z - HV$
4. $Q \quad = \quad T$ div R.

Fig. 1. Montgomery multiplication

Observe that T is 0 modulo R, so that the division yielding Q is exact. Obviously, $Q = AB/R \bmod V$. Besides, since $\deg(R) = m$ and $\deg(T) \leq m+m'-1$, we have $\deg(Q) \leq m'-1$, so $Q = AB/R$ rem V.

The Montgomery Multiplication with Polynomial Residues. In both [9] and [8], the idea of computing modulo a highly composite R is raised. We recall this process here, for our case $R = r_1 \cdots r_n$, with r_i trinomials. Additions and multiplications modulo R are done component-wise modulo $\mathscr{R} = (r_1, \ldots, r_n)$. However, the final step of the algorithm cannot be performed in the residue basis, since it becomes a division by zero.

The workaround in [9,8] consists in shifting from the set of moduli $\mathscr{R}$ to another set $\widetilde{\mathscr{R}} = (r_{n+1}, \ldots, r_{2n})$ modulo which R can be inverted. This shifting process, also called *base extension*, is thus the composite of a Chinese remaindering operation (or Newton interpolation) at $r_1, \ldots, r_n$, followed by a multiple reduction at $r_{n+1}, \ldots, r_{2n}$.

To minimize the overhead, it turns out to be better to take as input the residues $\mathbf{a}, \mathbf{b}$ of A and B modulo $\mathscr{R}$, as well as their residues $\widetilde{\mathbf{a}}, \widetilde{\mathbf{b}}$ modulo $\widetilde{\mathscr{R}}$; similarly, we output the residues $\mathbf{q}, \widetilde{\mathbf{q}}$ of Q modulo both sets. Thus, the algorithm starts as before, performing the computations modulo $\mathscr{R}$. Before the division by R, though, it shifts from the basis $\mathscr{R}$ to $\widetilde{\mathscr{R}}$, divides by R in this basis, and eventually shifts back to $\mathscr{R}$. As input, it also takes the residues $\mathbf{w}$ of W modulo $\mathscr{R}$, and the residues $\widetilde{\mathbf{s}}$ of $S = 1/R$ and $\widetilde{\mathbf{v}}$ of V modulo $\widetilde{\mathscr{R}}$. The details of this algorithm are in Figure 2 (with notation adapted to our setting).

Input:

- $\mathbf{a} = (a_1, \ldots, a_n)$ and $\widetilde{\mathbf{a}} = (a_{n+1}, \ldots, a_{2n})$
- $\mathbf{b} = (b_1, \ldots, b_n)$ and $\widetilde{\mathbf{b}} = (b_{n+1}, \ldots, b_{2n})$
- $\mathbf{w} = (w_1, \ldots, w_n)$
- $\widetilde{\mathbf{s}} = (s_{n+1}, \ldots, s_{2n})$
- $\widetilde{\mathbf{v}} = (v_{n+1}, \ldots, v_{2n})$
- $\mathscr{R} = (r_1, \ldots, r_n)$ and $\widetilde{\mathscr{R}} = (r_{n+1}, \ldots, r_{2n})$

Output:

- $\mathbf{q} = (q_1, \ldots, q_n)$ and $\widetilde{\mathbf{q}} = (q_{n+1}, \ldots, q_{2n})$

1.	$(z_i)_{i \leq n}$	$= (a_i b_i \ \text{rem}\ r_i)_{i \leq n}$
1'.	$(z_{n+i})_{i \leq n}$	$= (a_{n+i} b_{n+i} \ \text{rem}\ r_{n+i})_{i \leq n}$
2.	$(h_i)_{i \leq n}$	$= (z_i w_i \ \text{rem}\ r_i)_{i \leq n}$
2'.	$(h_{n+i})_{i \leq n}$	$= \text{shift}(h_1, \ldots, h_n)$
3.	$(t_{n+i})_{i \leq n}$	$= (z_{n+i} - h_{n+i} v_{n+i} \ \text{rem}\ r_{n+i})_{i \leq n}$
4.	$(q_{n+i})_{i \leq n}$	$= (t_{n+i} s_{n+i} \ \text{rem}\ r_{n+i})_{i \leq n}$
4'.	$(q_i)_{i \leq n}$	$= \text{shift}^{-1}(q_{n+1}, \ldots, q_{2n})$

Fig. 2. Residue Montgomery multiplication as in [8,9]

Our Algorithm. Our approach rests on the following remark: when divisions by zero occur, one can still obtain a meaningful result by dividing derivatives. With the notation of Figure 1, from the equality $T = RQ$, we obtain by differentiation

$$T' = R'Q + RQ'.$$

The polynomial R is squarefree, because all r_i are, and are pairwise coprime. Hence, we can deduce the relation

$$Q = \frac{T'}{R'} \bmod R. \tag{1}$$

In contrast to the algorithm of the previous paragraph, our algorithm does not require a second set of moduli: we work with $\mathscr{R} = (r_1, \ldots, r_n)$ all along. Still, as before, we will handle more data as input and output than the mere residues of A and B modulo $\mathscr{R}$. If A is in $\mathbb{F}_2[x]$, we still write its residue representation modulo $\mathscr{R}$ as $\mathbf{a} = (a_1, \ldots, a_n)$. Besides, we denote by $\mathbf{a}^\star$ the residue representation of its derivative, i.e., A':

$$\mathbf{a}^\star = (a_1^\star, \ldots, a_n^\star), \quad \text{with} \quad a_i^\star = A' \bmod r_i.$$

Note that $a_i^\star$ is *not* the derivative of a_i.

The previous algorithm uses a function *shift* to extend the modular information from the moduli $\mathscr{R}$ to $\widetilde{\mathscr{R}}$. In a similar manner, we use a function *diff* that takes as input the residues $\mathbf{a}$ of a polynomial A of degree less than nd, and outputs the residues $\mathbf{a}^\star$ of its derivative: this is done by computing A through the Chinese remaindering, differentiating it, and reducing the result modulo all r_i.

Now, the input of the multiplication algorithm consists of the residues $\mathbf{a}, \mathbf{b}$ of A and B modulo $\mathscr{R}$, and of the residues $\mathbf{a}^\star, \mathbf{b}^\star$ of the derivatives A' and B'; the output consists of the residues $\mathbf{q}$ and $\mathbf{q}^\star$ of Q and its derivative Q'. The computation follows the same steps as before. Since $\mathbf{a}, \mathbf{a}^\star, \mathbf{b}, \mathbf{b}^\star$ are known, we can compute the residues $\mathbf{z}$ and $\mathbf{z}^\star$, using the relations $z_i = a_i b_i \bmod r_i$ and $z_i^\star = a_i b_i^\star + a_i^\star b_i \bmod r_i$.

Next, we deduce the residue representation $\mathbf{h}$ of $H = ZW \bmod R$. However, since we take remainders modulo R, the derivative of H cannot be computed term-wise, so we use the function *diff* to obtain $\mathbf{h}^\star$ (which is valid, since $\deg(H) < nd$).

In view of (1), we see that only $\mathbf{t}^\star$ is required to obtain the quotient Q. Since $T = Z - HV$, we deduce that $t_i^\star = T' \bmod r_i$ is given by

$$z_i^\star - h_i v_i^\star - h_i^\star v_i \bmod r_i.$$

Let U be the inverse of R' modulo R and let $\mathbf{u}^\star$ be the residue vector of U modulo $\mathscr{R}$. Equation (1) then implies that $q_i^\star = Q \bmod r_i$ equals $t_i^\star u_i^\star \bmod r_i$. Knowing $\mathbf{q}$, we deduce $\mathbf{q}^\star$ by applying the function *diff* (which is valid, since $\deg(Q) < nd$). Remark that $\mathbf{q}^\star$ is not needed if we perform a single multiplication. However, since Q may be reused for further multiplications, we compute $\mathbf{q}^\star$ for consistency. The details of the algorithm are given in Figure 3.

Input:

- $\mathbf{a} = (a_1, \ldots, a_n)$ and $\mathbf{a}^\star = (a_1^\star, \ldots, a_n^\star)$
- $\mathbf{b} = (b_1, \ldots, b_n)$ and $\mathbf{b}^\star = (b_1^\star, \ldots, b_n^\star)$
- $\mathbf{w} = (w_1, \ldots, w_n)$
- $\mathbf{u}^\star = (u_1^\star, \ldots, u_n^\star)$
- $\mathbf{v} = (v_1, \ldots, v_n)$ and $\mathbf{v}^\star = (v_1^\star, \ldots, v_n^\star)$
- $\mathscr{R} = (r_1, \ldots, r_n)$

Output:

- $\mathbf{q} = (q_1, \ldots, q_n)$ and $\mathbf{q}^\star = (q_1^\star, \ldots, q_n^\star)$

$$
\begin{array}{llll}
1.   & (z_i)_{i \le n}   & = & (a_i b_i \text{ rem } r_i)_{i \le n} \\
1'. & (z_i^\star)_{i \le n} & = & (a_i b_i^\star + a_i^\star b_i \text{ rem } r_i)_{i \le n} \\
2.   & (h_i)_{i \le n}   & = & (z_i w_i \text{ rem } r_i)_{i \le n} \\
2'. & (h_i^\star)_{i \le n} & = & \text{diff}(h_1, \ldots, h_n) \\
3.   & (t_i^\star)_{i \le n} & = & (z_i^\star - h_i v_i^\star - h_i^\star v_i \text{ rem } r_i)_{i \le n} \\
4.   & (q_i)_{i \le n}   & = & (t_i^\star u_i^\star \text{ rem } r_i)_{i \le n} \\
4'. & (q_i^\star)_{i \le n} & = & \text{diff}(q_1, \ldots, q_n)
\end{array}
$$

Fig. 3. Our version of the residue Montgomery multiplication

Optimization and Cost Analysis. We finally prove the complexity statement announced in Theorem 1, starting with a discussion of the function *diff*.

This function consists of a Chinese remaindering, followed by differentiation, followed by a multiple reduction. As mentioned in Section 3, the Chinese remaindering requires as a first step the modular multiplication of the residue vector by the vector $(x_i = S_i^{-1} \text{ rem } r_i)_{i \le n}$, with $S_i = r_1 \cdots r_{i-1} r_{i+1} \cdots r_n$. We apply the function diff twice. In both cases, this product can be absorbed in other modular multiplications (requiring us to slightly modify the precomputed polynomials we take as input).

- At step 2', we apply *diff* to the vector $(h_1, \ldots, h_n)$ obtained at step 2. Hence, we can modify step 2, replacing the product $z_i w_i \text{ rem } r_i$ by the product $z_i(w_i x_i \text{ rem } r_i) \text{ rem } r_i$, so that the vector $(w_i x_i \text{ rem } r_i)_{i \le n}$ is needed as input. However, this modifies h_i; since h_i is reused at step 3, we have to compensate for this extra x_i factor: this is done by replacing the product $h_i v_i^\star \text{ rem } r_i$ by $h_i(v_i^\star S_i \text{ rem } r_i) \text{ rem } r_i$, so that we take the latter vector $(v_i^\star S_i \text{ rem } r_i)_{i \le n}$ as input.
- At step 4', we apply *diff* to the vector $(q_1, \ldots, q_n)$ obtained at step 4. Then, we modify step 4, replacing the product $t_i^\star u_i^\star \text{ rem } r_i$ by $t_i^\star(u_i^\star x_i \text{ rem } r_i) \text{ rem } r_i$, so $(u_i^\star x_i \text{ rem } r_i)_{i \le n}$ is used as an extra input.

Hence, the cost of Chinese remaindering reduces to that of the LinComb function given in Proposition 2. Differentiation is free, and gives an even polynomial;

the cost of multiple reduction is given in Corollary 1. Hence, the total cost of *diff* is at most $4n^2d - nd(\log_2(n) - 3) - n^2 - 3n$ additions. The time complexity is in $O(nT_{\mathrm{rem}}(d))$.

We complete the cost analysis of the whole algorithm. The algorithm performs seven vector multiplications in size n, with polynomials of degree less than d: this is done using $7n(d-1)^2$ additions and $7nd^2$ multiplications. There are two calls to *diff*, using $8n^2d - 2nd(\log_2(n) - 3) - 2n^2 - 6n$ additions.

The extra operations are vector additions and remainders in size n. It turns out to be better to postpone the reduction at step 1' to step 3, after all additions are done. Then, we have three size-n additions to perform, on polynomials of degree up to $2d - 2$; hence, they require $3n(2d - 1)$ scalar additions. The four remainders use $4n(2d - 2)$ additions. Summing all previous contributions gives the estimate on the number of operations in Theorem 1.

The time complexity analysis requires no extra complication, except to note that the addition at step 1' can be done in parallel with one at step 3. The total time is then seen to be in $O(T_A + \log_2(d)T_X + nT_{\mathrm{rem}}(d))$.

5 Examples

Table 1 illustrates the number of additions performed by our algorithm for a few values of d. The second column gives the maximal list of trinomials one can use, under the form of a set $S = \{i_1, \ldots, i_n\}$ of integers between 1 and $d - 1$: the corresponding trinomials are $x^d + x^{i_1} + 1, \ldots, x^d + x^{i_n} + 1$. As can be seen, the squarefreeness assumption on our trinomials forces us to discard at least half of the available ones for d even.

Table 1. Numerical examples

d	indices	$n_{\max}$	$n_{\max}\,d$	additions
13	$\{1, 2, 3, 4, 6, 7, 10, 12\}$	8	104	15912
14	$\{1, 3, 5, 9, 11\}$	5	70	9564
15	$\{1, \ldots, 14\} - \{10, 12\}$	12	180	35561
16	$\{1, 3, 5, 7, 9, 13, 15\}$	7	112	18691
17	$\{3, 4, 5, 6, 9, 11, 12, 14, 15\}$	9	153	28919
18	$\{1, 3, 5, 7, 9, 11, 13, 15, 17\}$	9	162	31768
19	$\{3, 4, 5, 6, 7, 9, 10, 12, 13, 15, 16\}$	11	209	45644
20	$\{1, 3, 5, 9, 11, 15, 17\}$	7	140	27325
21	$\{1, \ldots, 20\} - \{15\}$	19	399	117393
22	$\{1, 3, 7, 9, 11, 13, 15, 19, 21\}$	9	198	44428
23	$\{2, 3, 5, 8, 9, 11, 12, 14, 15, 18, 20, 21, 22\}$	13	299	78348
24	$\{5, 7, 11, 13, 15, 17, 19, 21, 23\}$	9	216	51514
25	$\{1, 3, 4, 7, 9, 10, 13, 15, 16, 18, 19, 21, 22, 24\}$	14	350	99352
26	$\{3, 5, 7, 9, 11, 15, 17, 21, 23\}$	9	234	59104
27	$\{1, \ldots, 26\} - \{2, 4, 9, 11, 16, 18\}$	20	540	186032

Our goal was to obtain a low operation count for multiplication modulo the modulus V. We are successful in this, since our results improve on some of the best ones previously known to us. For instance, Bajard et al. [8] have 49920 additions for $m = 192$, 139400 additions for $m = 360$ and 213716 additions for $m = 486$. We obtain 44336 additions for $(n = 8, d = 24, m = nd = 192)$, 108285 additions for $(n = 18, d = 21, m = nd = 378)$ and 159872 additions for $(n = 18, d = 27, m = nd = 486)$.

6 Conclusion

The results given here easily extend to slightly more general situations: e.g., using trinomials of different degrees would enable one to extend and refine the range of accessible degrees. Harder questions concern our time complexity: as of now, our Chinese remaindering or multiple remaindering algorithms have rather bad time complexity, due to their sequential nature. It would be most interesting to obtain a similar operation count with a logarithmic time.

Acknowledgments. We would like to thank the referees for their helpful comments, as well as Paul Zimmermann for pointing out McLaughlin's work. The first author acknowledges the support of the Canada Research Chairs Program and of NSERC.

References

1. Montgomery, P.L.: Modular multiplication without trial division. Mathematics of Computation 44, 519–521 (1985)
2. Koç, C.K., Acar, T.: Montgomery multiplication in GF(2^k). Designs, Codes and Cryptography 14, 57–69 (1998)
3. von zur Gathen, J., Gerhard, J.: Modern computer algebra. Cambridge University Press, Cambridge (1999)
4. Schönhage, A.: Schnelle Multiplikation von Polynomen über Körpern der Charakteristik 2. Acta Informatica 7, 395–398 (1977)
5. Cantor, D.G.: On arithmetical algorithms over finite fields. J. Combin. Theory Ser. A 50, 285–300 (1989)
6. Bajard, J.C., Didier, L.S., Kornerup, P.: An RNS Montgomery modular multiplication algorithm. IEEE Transactions on Computers 47, 766–776 (1998)
7. McLaughlin Jr., P.: New frameworks for Montgomery's modular multiplication method. Mathematics of Computation 73, 899–906 (2004)
8. Bajard, J.C., Imbert, L., Jullien, G.A.: Parallel Montgomery multiplication in GF(2^k) using trinomial residue arithmetic. In: 17th IEEE Symposium on Computer Arithmetic, pp. 164–171. IEEE, Los Alamitos (2005)
9. Mihailescu, P.: Fast convolutions meet Montgomery. Mathematics of Computation 77, 1199–1221 (2008)
10. Sunar, B.: A generalized method for constructing subquadratic complexity GF(2^k) multipliers. IEEE Transactions on Computers 53, 1097–1105 (2004)
11. Fan, H., Hasan, M.: A new approach to subquadratic space complexity parallel multipliers for extended binary fields. IEEE Transactions on Computers 56, 224–233 (2007)

12. Giorgi, P., Nègre, C., Plantard, T.: Subquadratic binary field multiplier in double polynomial system. In: SECRYPT 2007 (2007)
13. Wu, H.: Low complexity bit-parallel finite field arithmetic using polynomial basis. In: Koç, Ç.K., Paar, C. (eds.) CHES 1999. LNCS, vol. 1717, pp. 280–291. Springer, Heidelberg (1999)
14. Ernst, M., Jung, M., Madlener, F., Huss, S., Blümel, R.: A reconfigurable system on chip implementation for elliptic curve cryptography over $GF(2^n)$. In: Kaliski Jr., B.S., Koç, Ç.K., Paar, C. (eds.) CHES 2002. LNCS, vol. 2523, pp. 381–399. Springer, Heidelberg (2003)

Bounds on Fixed Input/Output Length Post-processing Functions for Biased Physical Random Number Generators

Kyohei Suzuki and Tetsu Iwata

Dept. of Computational Science and Engineering,
Nagoya University
Furo-cho, Chikusa-ku, Nagoya, 464-8603, Japan
kyouhe_s@nuee.nagoya-u.ac.jp, iwata@cse.nagoya-u.ac.jp

Abstract. Post-processing functions are used to reduce the imperfectness of physical random number generators. At FSE '07, Dichtl considered the case where the physical random number generator outputs independent bits that have a constant bias, and the post-processing function has fixed input and output lengths. In this paper, we first present a number of bounds on $\deg(n, m)$, which is a measure of the reduction of biases with n-bit input and m-bit output post-processing functions. We next show the exact values of $\deg(n, m)$ for a large class of (n, m) such that $1 \leq m \leq n \leq 16$, by using the bounds on $\deg(n, m)$ and a computer simulation. We finally discuss how we have derived these numerical values.

Keywords: physical random number generator, bias, post-processing, entropy extractor.

1 Introduction

Background. Cryptographic schemes are designed assuming that unbiased and independent bits are available. However, when we implement them in practice, the physical sources of randomness to which we have access are not perfect, and may contain biases and correlations. For example, we may use system clocks, keyboard or mouse movements, radioactive sources, or quantum mechanical sources (see [9,13] and [20, Chap. 17] for other examples), but they usually do not produce perfect random bits. Many cryptographic schemes rely on sequences of unbiased bits. It is therefore important to be able to extract unbiased bits from an imperfect physical source, and a natural approach to the problem is to apply a *post-processing function* (also called an entropy extractor, or a corrector), a function that transforms a weak random source into an almost perfect random source. This classical problem was extensively studied in the past [1,2,5,6,7,8,12,16,17,18,19,21,22,23,24,25,26],

At FSE '07, Dichtl studied the particular source of randomness, where the output bits of physical source are independent and have a constant (but unknown) bias [10]. That is, if $x_1, x_2, \ldots$ are the output bits of the physical source,

R. Avanzi, L. Keliher, and F. Sica (Eds.): SAC 2008, LNCS 5381, pp. 373–386, 2009.
© Springer-Verlag Berlin Heidelberg 2009

then $\Pr(x_i = 1) = 1/2 + \epsilon$ holds for some ϵ. This setting may be of practical interest as some of the above sources of randomness, such as radioactive sources and quantum mechanical sources, may output data that are independent but biased. Also, post-processing functions studies in [10] have fixed input and output lengths, and this may be important in a real system as they have a fixed input/output ratio and latency, while, for example, a well known von Neumann's method [26] does not have this property.

Dichtl's result [10]. Dichtl proposed five post-processing functions, called XOR, H, H2, H3 and Solution S. These functions take 16-bit input and produce 8-bit output. For the first four functions, the 16-bit input x is divided into two 8-bit sequences a and b as $x = (a, b)$, and the output y is given by

$$
\begin{cases}
\text{XOR} : y = a \oplus b \\
\text{H} : y = a \oplus (a \ll 1) \oplus b \\
\text{H2} : y = a \oplus (a \ll 1) \oplus (a \ll 2) \oplus b \\
\text{H3} : y = a \oplus (a \ll 1) \oplus (a \ll 2) \oplus (a \ll 4) \oplus b
\end{cases}
$$

where $a \ll i$ is the i-bit cyclic left shift of a. Since our input bits have a constant bias, its probability is a polynomial in ϵ, i.e., if x is n bits and its Hamming weight is w, then $\Pr(x) = (1/2 - \epsilon)^{n-w}(1/2 + \epsilon)^w$. Now for post-processing function F, the output probability, $\Pr(y)$, is the sum of input probabilities of x such that $y = F(x)$, which is also a polynomial in ϵ whose degree is at most n. Dichtl proposed to measure the effectiveness of reduction of bias by the "lowest degree of ϵ with non-zero coefficient." Dichtl shows that, for XOR, the coefficient of ϵ in $\Pr(y)$ is zero for any y. Similarly, ϵ and ϵ^2 are zero for H, $\epsilon, \epsilon^2, \epsilon^3$ are zero for H2, and the coefficients of $\epsilon, \ldots, \epsilon^4$ are all zero for H3, thus the lowest degree with non-zero coefficient is 2, 3, 4, and 5 for XOR, H, H2, and H3, respectively. Since the lowest degree of raw input x is 1, they all reduce the bias compared to the raw input, and H3 reduces the bias the most effective way among these four functions.

Solution S is a special type of post-processing functions, where for any input x, x and its complement have the same output value. Solution S is derived by solving the system of linear equations, and the above property reduces the search space since for any y and odd i, the coefficients of ϵ^i in $\Pr(y)$ is zero. In particular, Dichtl shows that, for any y, the coefficients of $\epsilon, \ldots, \epsilon^5$ in $\Pr(y)$ are all zero, thereby reducing more bias than the previous four functions.

We note that post-processing functions in [10] are deterministic, while foundational works [16,21,22,23,24,25] assume a small amount of true randomness, and the works on deterministic extractors [1,2,6,18] directly capture the min-entropy, which is known to be an appropriate notion for random number generation to evaluate the randomness quantity of a binary sequence [3]. As in [10], in this paper, we consider the deterministic functions and use the lowest degree of the polynomial for evaluating the reduction of biases.

Lacharme's result [14]. Lacharme shows that the problem is closely related to the coding theory, i.e., if there exists an $[n, m, d]$ linear code, then there exists an

n-bit input and m-bit output post-processing function such that the coefficients of $\epsilon, \epsilon^2, \ldots, \epsilon^{d-1}$ are all zero. This is a natural generalization of [10], as H, H2 and H3 respectively correspond to generator matrices of [16,8,3], [16,8,4] and [16,8,5] linear codes. Also, a table of linear codes [11] can be used to construct linear post-processing functions. Then Lacharme proposes to use a resilient function as the post-processing function, and shows that, for an (n, m, t)-resilient function, the coefficients of $\epsilon, \epsilon^2, \ldots, \epsilon^t$ are all zero. Finally, Lacharme studies the relation between the bias and the min-entropy.

Our contributions. We first re-formalize the problem explicitly separating the general post-processing functions and the "Solution S type" post-processing functions. For any n-bit input and m-bit output post-processing function F, we let $\mathrm{mindeg}(F)$ be the minimum degree of ϵ with non-zero coefficient in the output probability, where the minimum is taken over all the output value. We then define $\deg(n, m)$ and $\deg^s(n, m)$ to be the maximum of $\mathrm{mindeg}(F)$, where the maximum is taken over all n-bit input and m-bit output post-processing functions for $\deg(n, m)$, and over all "Solution S type" post-processing functions for $\deg^s(n, m)$. In our terminology, Dichtl shows $\deg^s(16, 8) = 6$ and derives the concrete truth table of F achieving $\mathrm{mindeg}(F) = 6$, and Lacharme shows that, if there exists an $[n, m, d]$ linear code, then $\deg(n, m) \geq d$, and if there exists a (n, m, t)-resilient function, then $\deg(n, m) \geq t + 1$.

We then present a number of bounds on $\deg(n, m)$ and $\deg^s(n, m)$. Our bounds are elementary ones and we see that proving these bounds are important in understanding the basic properties of post-processing functions. Indeed, it turns out that they are actually useful in deriving the exact values of $\deg(n, m)$ and $\deg^s(n, m)$. By using a computer simulation and the bounds we have derived, we next present the exact values of $\deg(n, m)$ for $1 \leq m \leq n \leq 16$, and $\deg^s(n, m)$ for $1 \leq m < n \leq 16$. Out of 136 values of (n, m) for $\deg(n, m)$, we have determined 123 values, and out of 120 values for $\deg^s(n, m)$, 115 values are determined. While the exact values for the remaining (n, m) are open, we derive both the upper and lower bounds. We finally discuss in detail how we have derived these numerical values. Our results can be seen as the generalization of [10] from $n = 16$ and $m = 8$ to $1 \leq m < n \leq 16$, and proving the optimality and non-optimality of the results in [14] for $1 \leq m \leq n \leq 16$.

2   Preliminaries

For a positive integer n, $\{0, 1\}^n$ is the set of all n-bit strings. For any set S, $\#S$ is the cardinality of S. An n-bit input and m-bit output post-processing function is a vector output Boolean function $F : \{0, 1\}^n \mapsto \{0, 1\}^m$. Let T_F be its truth table, i.e., $T_F = (F(00 \cdots 00), F(00 \cdots 01), \ldots, F(11 \cdots 11))^t$, which is the transposed vector of $(F(00 \cdots 00), F(00 \cdots 01), \ldots, F(11 \cdots 11))$. We say that F is balanced if each $y \in \{0, 1\}^m$ appears 2^{n-m} times in T_F. A balanced n-bit input and m-bit output post-processing function is denoted (n, m)-PP, and let (n, m)-$\mathcal{PP}$ be the set of all (n, m)-PPs. As in [10], we only consider (n, m)-PPs.

Let $x \in \{0,1\}^n$ be the input of an (n,m)-PP and $y \in \{0,1\}^m$ be its output. Throughout this paper, we assume that each bit of x has a constant (but unknown) bias ϵ, i.e., if $x = (x_1, x_2, \ldots, x_n) \in \{0,1\}^n$ is the input, then $\Pr(x_i = 1) = 1/2 + \epsilon$ for $1 \leq i \leq n$. The Hamming weight of $x = (x_1, x_2, \ldots, x_n)$ is denoted $w(x)$, which is $\#\{i \mid x_i = 1\}$. The probability of input, $\Pr(x)$, depends only on $w(x)$ and is given by

$$\Pr(x) = \left(\frac{1}{2} - \epsilon\right)^{n - w(x)} \left(\frac{1}{2} + \epsilon\right)^{w(x)}. \tag{1}$$

Therefore, $\Pr(x)$ is a polynomial in ϵ. Since $0 \leq w(x) \leq n$, there are $(n+1)$ possibilities for the value of $\Pr(x)$. If $w(x) = w$, the corresponding probability is denoted p_w, and hence $p_w = (1/2 - \epsilon)^{n-w} (1/2 + \epsilon)^w$.

For any $y \in \{0,1\}^m$, $F^{-1}(y)$ is the *preimage* of y, and is defined as the set of x such that $y = F(x)$, i.e., $F^{-1}(y) = \{x \mid y = F(x)\}$. The probability of output, $\Pr(y)$, is the sum of probabilities of 2^{n-m} n-bit inputs belonging to its preimage. That is,

$$\Pr(y) = \sum_{x \in F^{-1}(y)} \Pr(x). \tag{2}$$

Since $\Pr(x)$ is a polynomial in ϵ given by (1), $\Pr(y)$ is also a polynomial in ϵ whose degree is at most n. Therefore, $\Pr(y) = a_0 + a_1\epsilon + a_2\epsilon^2 + \cdots + a_n\epsilon^n$.

Now we define $\mathrm{mindeg}(\Pr(y))$ as follows.

Definition 1. *For all $y \in \{0,1\}^m$, define*

$$\mathrm{mindeg}(\Pr(y)) \stackrel{\mathrm{def}}{=} \min\{k \mid 1 \leq k \leq n, a_k \neq 0\}.$$

For given y, $\mathrm{mindeg}(\Pr(y))$ is the minimum degree other than the constant term.

Next, we define $\mathrm{mindeg}(T_F)$ as follows.

Definition 2. *For all $F \in (n,m)$-PP, define*

$$\mathrm{mindeg}(T_F) \stackrel{\mathrm{def}}{=} \min\{\mathrm{mindeg}(\Pr(y)) \mid y \in \{0,1\}^m\}.$$

For given $F \in (n,m)$-PP, $\mathrm{mindeg}(T_F)$ is the minimum of $\mathrm{mindeg}(\Pr(y))$, where y runs all the possible values.

Then, we define $\deg(n,m)$ as follows.

Definition 3. *For all $n \geq m \geq 1$, define*

$$\deg(n,m) \stackrel{\mathrm{def}}{=} \max\{\mathrm{mindeg}(T_F) \mid F \in (n,m)\text{-PP}\}.$$

For given (n,m), $\deg(n,m)$ is the maximum value of $\mathrm{mindeg}(T_F)$, where the maximum is taken over all the possible $F \in (n,m)$-PP. It is easy to see that $\#(n,m)$-PP $= (2^n)!/\{(2^{n-m})!\}^{2^m}$, i.e., $F \in (n,m)$-PP satisfying $\mathrm{mindeg}(T_F) = \deg(n,m)$ reduces the bias most effective way among this number of possible (n,m)-PPs.

Solution S is a special type of an (n, m)-PP, where any input x and its complement, $\bar{x}$, have the same output [10]. This implies that x and $\bar{x}$ belong to the same preimage of some y. Therefore, for any $y \in \{0,1\}^m$ and for all odd i, the coefficient of ϵ^i in $\Pr(y)$ is zero since the coefficient of ϵ^i in $(\Pr(x) + \Pr(\bar{x}))$ is zero. An (n, m)-PP is said to be an (n, m)-SPP (Solution S type PP) if for any x, x and $\bar{x}$ have the same output, and let (n, m)-$\mathcal{SPP}$ be the set of all (n, m)-SPPs.

Now we define $\deg^s(n, m)$ as follows.

Definition 4. *For all $n > m \geq 1$, define*

$$\deg^s(n, m) \overset{\text{def}}{=} \max\{\text{mindeg}(T_F) \mid F \in (n, m)\text{-}\mathcal{SPP}\}.$$

Note that $\#(n, m)$-$\mathcal{SPP} = (2^{n-1})!/\{(2^{n-m-1})!\}^{2^m}$, and the maximum is taken over all these (n, m)-SPPs. Also, for any $n > m \geq 1$, we have $\deg(n, m) \geq \deg^s(n, m)$ since (n, m)-$\mathcal{SPP} \subset (n, m)$-$\mathcal{PP}$. We do not consider the case $n = m$ since any $F \in (n, n)$-$\mathcal{PP}$ is a permutation over $\{0, 1\}^n$, and thus two distinct inputs cannot have the same output. With the similar reasoning, we do not consider the case $n = 1$.

3 Bounds on $\deg(n, m)$ and $\deg^s(n, m)$

In this section, we present bounds on $\deg(n, m)$ and $\deg^s(n, m)$ with their proofs.

3.1 Bounds on $\deg(n, m)$

We show six bounds on $\deg(n, m)$.

Theorem 1. *For all $n \geq 1$, $\deg(n, n) = 1$.*

Proof. Any (n, n)-PP is a permutation over $\{0, 1\}^n$. Therefore, for all $0 \leq w \leq n$, there always exists some $y \in \{0, 1\}^n$ such that $\Pr(y) = p_w$. Now since $\text{mindeg}(p_w) \neq 0$ for $0 \leq w \leq n$, $\text{mindeg}(\Pr(y)) \neq 0$ for any $y \in \{0, 1\}^n$. This implies $\text{mindeg}(T_F) \geq 1$ for any $F \in (n, n)$-$\mathcal{PP}$.

On the other hand, $\text{mindeg}(p_w) = 1$ holds for some $0 \leq w \leq n$, cf., $w = 0$. Thus, there always exists some $y \in \{0, 1\}^n$ such that $\text{mindeg}(\Pr(y)) = 1$. This implies $\text{mindeg}(T_F) \leq 1$ for any $F \in (n, n)$-$\mathcal{PP}$, and hence $\deg(n, n) = 1$. □

Theorem 2. *For all $n \geq 2$, $\deg(n, n - 1) = 2$.*

Proof. Constructing an $(n, n-1)$-PP corresponds to dividing 2^n n-bit inputs into 2^{n-1} preimages, where each preimage consists of two inputs. Our proof proceeds in two steps. First, we derive a necessary and sufficient condition that, for each $y \in \{0, 1\}^{n-1}$, the coefficient of ϵ in $\Pr(y)$ is zero. Then we show that when the condition is satisfied, the coefficient of ϵ^2 in $\Pr(y)$ is non-zero for some y.

Now $\Pr(x)$ in (1) can be written as

$$\Pr(x) = \left\{ \sum_{i=0}^{n-w} \binom{n-w}{i} \left(\frac{1}{2}\right)^{n-w-i} (-\epsilon)^i \right\} \left\{ \sum_{j=0}^{w} \binom{w}{j} \left(\frac{1}{2}\right)^{w-j} \epsilon^j \right\},$$

where $w(x) = w$. Therefore, the coefficient of ϵ in $\Pr(x)$ is

$$\sum_{i+j=1} \binom{n-w}{i}(-1)^i \binom{w}{j}\left(\frac{1}{2}\right)^{n-i-j} = \frac{2w-n}{2^{n-1}}.$$

Fix any $y \in \{0,1\}^{n-1}$, and suppose that its preimage consists of x_1 and x_2, where $w(x_1) = w_1$ and $w(x_2) = w_2$. Since $\Pr(y) = \Pr(x_1) + \Pr(x_2)$, the coefficient of ϵ in $\Pr(y)$ is zero if and only if

$$\frac{2w_1 - n}{2^{n-1}} + \frac{2w_2 - n}{2^{n-1}} = 0,$$

which is equivalent to $w_2 = n - w_1$. Therefore, the necessary and sufficient condition is to form a preimage with two inputs of weight w and $n - w$. Now, since $\#\{x \mid w(x) = w\} = \#\{x \mid w(x) = n - w\}$ holds for any $0 \le w \le n$, it is possible to satisfy the above condition to construct $F \in (n, n-1)$-$\mathcal{PP}$ satisfying $\mathrm{mindeg}(T_F) \ge 2$.

Next, consider some $F \in (n, n-1)$-$\mathcal{PP}$ satisfying $\mathrm{mindeg}(T_F) \ge 2$. We show that the coefficient of ϵ^2 in $\Pr(y)$ is non-zero for some y. If $w(x) = w$, then the coefficient of ϵ^2 in $\Pr(x)$ is

$$\sum_{i+j=2} \binom{n-w}{i}(-1)^i \binom{w}{j}\left(\frac{1}{2}\right)^{n-i-j} = \frac{(2w-n)^2 - n}{2^{n-1}}.$$

Similarly, if $w(x') = n - w$, we see that ϵ^2 in $\Pr(x')$ has the same coefficient. Therefore, if the preimage is formed with two inputs of weight w and $n - w$, the coefficient of ϵ^2 in $\Pr(y)$ is

$$\frac{(2w-n)^2 - n}{2^{n-1}} + \frac{(2w-n)^2 - n}{2^{n-1}} = \frac{(2w-n)^2 - n}{2^{n-2}}.$$

So we need $(2w - n)^2 - n = 0$ to eliminate ϵ^2, which is equivalent to $w = (n \pm \sqrt{n})/2$. Now it is clear that we always have some w such that $0 \le w \le n$ and $w \ne (n \pm \sqrt{n})/2$ (since $n \ge 2$, we have at least three choices of w, and the right hand side takes at most two values). This implies the coefficient of ϵ^2 in $\Pr(y)$ is non-zero for some y. Therefore, for any $F \in (n, n-1)$-$\mathcal{PP}$, $\mathrm{mindeg}(T_F) \le 2$, and hence $\deg(n, n-1) = 2$. □

Theorem 3. *For all $n > m \ge 1$, $\deg(n, m) \ge \deg(n, m+1)$.*

Proof. Suppose we have $F \in (n, m+1)$-$\mathcal{PP}$, where $\deg(n, m+1) = \mathrm{mindeg}(T_F)$. We construct $F' \in (n, m)$-$\mathcal{PP}$ such that $\mathrm{mindeg}(T_{F'}) \ge \mathrm{mindeg}(T_F)$.

Since the output length of F is $(m + 1)$ bits, F has 2^{m+1} preimages, where each preimage has 2^{n-m-1} inputs. Now we divide the 2^{m+1} preimages into 2^m pairs of preimages, and regard the pair of preimages as a new preimage. We then have 2^m new preimages each of which consists of 2^{n-m} inputs, and let F' be the resulting (n, m)-PP.

Let $y_0 \in \{0,1\}^m$ be some output of F' and $\Pr(y_0)$ be its probability. By definition, $\Pr(y_0) = \Pr(y_1) + \Pr(y_2)$ for some outputs $y_1 \in \{0,1\}^{m+1}$ and $y_2 \in \{0,1\}^{m+1}$ of F. Without loss of generality, assume that $\mathrm{mindeg}(\Pr(y_1)) \leq \mathrm{mindeg}(\Pr(y_2))$. Then we have

$$\mathrm{mindeg}(\Pr(y_0)) = \mathrm{mindeg}(\Pr(y_1) + \Pr(y_2)) \geq \mathrm{mindeg}(\Pr(y_1)).$$

Therefore, for any $y' \in \{0,1\}^m$ of F', we have some $y \in \{0,1\}^{m+1}$ of F satisfying $\mathrm{mindeg}(\Pr(y')) \geq \mathrm{mindeg}(\Pr(y))$. This implies $\min\{\mathrm{mindeg}(\Pr(y')) \mid y' \in \{0,1\}^m\} \geq \min\{\mathrm{mindeg}(\Pr(y)) \mid y \in \{0,1\}^{m+1}\}$, and the result follows. $\qquad\square$

Theorem 4. *For all $n \geq m \geq 1$, $\deg(n,m) \leq \deg(n+1,m)$.*

Proof. Suppose that we have $F \in (n,m)\text{-}\mathcal{PP}$ such that $\mathrm{mindeg}(T_F) = \deg(n,m)$. We construct $F' \in (n+1,m)\text{-}\mathcal{PP}$ satisfying $\mathrm{mindeg}(T_F) = \mathrm{mindeg}(T_{F'})$.

For an input $x' \in \{0,1\}^{n+1}$, the output of F' is $F'(x') = F(x)$, where x is the least significant n bits of x', i.e., F' simply ignores the most significant bit of x'.

Now, for any $x \in \{0,1\}^n$, we have $\Pr(0\|x) = (1/2 - \epsilon)\Pr(x)$ and $\Pr(1\|x) = (1/2 + \epsilon)\Pr(x)$. Let $\Pr(y)$ be the probability that the output of F is y. Then the probability that F' outputs y is

$$\sum_{x \in F^{-1}(y)} \left(\frac{1}{2} - \epsilon\right)\Pr(x) + \sum_{x \in F^{-1}(y)} \left(\frac{1}{2} + \epsilon\right)\Pr(x) = \Pr(y).$$

Therefore we have $\mathrm{mindeg}(T_{F'}) = \mathrm{mindeg}(T_F)$. $\qquad\square$

Theorem 5. *For all $n \geq m \geq 1$ and $k \geq 1$, $\deg(n,m) \leq \deg(kn,km)$.*

Proof. Suppose that we have $F \in (n,m)\text{-}\mathcal{PP}$ such that $\mathrm{mindeg}(T_F) = \deg(n,m)$. We construct $F' \in (kn,km)\text{-}\mathcal{PP}$ satisfying $\mathrm{mindeg}(T_F) \leq \mathrm{mindeg}(T_{F'})$.

For an input $x = (x_1, x_2, \ldots, x_k) \in (\{0,1\}^n)^k$ of F', the output is $y = (y_1, y_2, \ldots, y_k) \in (\{0,1\}^m)^k$, where $y_i = F(x_i)$ for $1 \leq i \leq k$.

Since $\Pr(y) = \Pr(y_1)\Pr(y_2) \cdots \Pr(y_k)$, we have

$$\mathrm{mindeg}(\Pr(y)) = \mathrm{mindeg}(\Pr(y_1)\Pr(y_2) \cdots \Pr(y_k)).$$

Now, $\mathrm{mindeg}(\Pr(y_1)\Pr(y_2) \cdots \Pr(y_k)) \geq \mathrm{mindeg}(\Pr(y_i))$ holds for any $1 \leq i \leq k$. Also, from the definition of $\mathrm{mindeg}(T_F)$, we have $\mathrm{mindeg}(\Pr(y_i)) \geq \mathrm{mindeg}(T_F)$ for any $1 \leq i \leq k$. Therefore, $\min\{\mathrm{mindeg}(\Pr(y)) \mid y \in \{0,1\}^{km}\} \geq \mathrm{mindeg}(T_F)$, and the result follows. $\qquad\square$

Theorem 6. *For all $n \geq 1$, $\deg(n,1) = n$.*

We use the following Piling-up Lemma [15] to prove Theorem 6.

Lemma 1 (Piling-up Lemma). *Let $n \geq 1$ and $x_1, x_2, \ldots, x_n$ be independent random variables such that $\Pr(x_i = 1) = 1/2 + \epsilon_i$. Then*

$$\Pr(x_1 \oplus x_2 \oplus \cdots \oplus x_n = 1) = \frac{1}{2} + (-2)^{n-1}\prod_{1 \leq i \leq n} \epsilon_i.$$

Now Theorem 6 is proved directly from Lemma 1.

Proof (of Theorem 6). Consider $F(x_1, x_2, \ldots, x_n) = x_1 \oplus x_2 \oplus \cdots \oplus x_n$. We see that $F \in (n, 1)\text{-}\mathcal{PP}$. Lemma 1 shows that the coefficients of $\epsilon, \epsilon^2, \ldots, \epsilon^{n-1}$ are all zero, and therefore, $\deg(n, 1) \geq n$. On the other hand, by definition, we have $\deg(n, 1) \leq n$. □

3.2 Bounds on $\deg^s(n, m)$

Similarly to $\deg(n, m)$, we show six bounds on $\deg^s(n, m)$.

Theorem 7. *For all $n \geq 2$, $\deg^s(n, n-1) = 2$.*

Proof. This follows since $F \in (n, n-1)\text{-}\mathcal{PP}$ satisfying $\deg(T_F) \geq 2$ in the proof of Theorem 2 also satisfies $F \in (n, n-1)\text{-}\mathcal{SPP}$. □

Theorem 8. *For all $n-1 > m \geq 1$, $\deg^s(n, m) \geq \deg^s(n, m+1)$.*

Proof. Similarly to the proof of Theorem 3, we can construct $F' \in (n, m)\text{-}\mathcal{SPP}$ such that $\text{mindeg}(T_{F'}) \geq \text{mindeg}(T_F)$ from any $F \in (n, m+1)\text{-}\mathcal{SPP}$. □

Theorem 9. *For all $n+1 > m \geq 1$, $\deg^s(n, m) \leq \deg^s(n+1, m)$.*

Proof. A proof is similar to the proof of Theorem 4. For any $F \in (n, m)\text{-}\mathcal{SPP}$, there exists $F' \in (n+1, m)\text{-}\mathcal{SPP}$ satisfying $\text{mindeg}(T_F) \leq \text{mindeg}(T_{F'})$. □

Theorem 10. *For all $n > m \geq 1$ and $k \geq 1$, $\deg^s(n, m) \leq \deg^s(kn, km)$.*

Proof. Similarly to the proof of Theorem 5, for any $F \in (n, m)\text{-}\mathcal{SPP}$, there exists $F' \in (kn, km)\text{-}\mathcal{SPP}$ satisfying $\text{mindeg}(T_F) \leq \text{mindeg}(T_{F'})$. □

Theorem 11. *For all even $n \geq 2$, $\deg^s(n, 1) = n$.*

Proof. We see that $F(x_1, x_2, \ldots, x_n) = x_1 \oplus x_2 \oplus \cdots \oplus x_n \in (n, 1)\text{-}\mathcal{SPP}$, and the rest of the proof is the same as Theorem 6. □

Theorem 12. *For all odd $n \geq 3$, $\deg^s(n, 1) = n - 1$.*

Proof. Since n is odd, $n-1$ is even and thus we have $\deg^s(n-1, 1) \geq n-1$ from Theorem 11. From Theorem 9, we have $\deg^s(n-1, 1) \leq \deg^s(n, 1)$, and therefore, $n-1 \leq \deg^s(n, 1)$. On the other hand, we always have $\deg^s(n, 1) \leq n$, but since $\deg^s(n, 1)$ cannot be odd from the definition of $\deg^s(n, 1)$, we have $\deg^s(n, 1) \leq n - 1$. □

4   Simulation Results

4.1   Values of $\deg(n, m)$ and $\deg^s(n, m)$

We first present our simulation results in Table 1 and Table 2. Table 1 shows the values of $\deg(n, m)$ for $1 \leq m \leq n \leq 16$, and Table 2 shows $\deg^s(n, m)$ for $1 \leq m < n \leq 16$. Table 3 is our environment for this simulation.

- In Table 1 and Table 2, if the entry is a^1, then the value is derived by our computer simulation discussed in the following sections.
- In Table 1, a^2 means that the upper bound is derived by our simulation (i.e., $\deg(n, m) \leq a$) and we apply Theorem 3 to derive the lower bound (i.e., $\deg(n, m) \geq a$).
- a^3 means that the upper bound is derived by our simulation and we apply Theorem 4 to derive the lower bound.
- a^4 means that the upper bound is derived by our simulation and the lower bound is taken from Lacharme's results [14].
- If the entry is a^1, b^1, then this means $\deg(n, m) = a$ or b for Table 1 and $\deg^s(n, m) = a$ or b for Table 2, where both values are derived by our simulation. Similarly, if the entry is a^1, b^5, this means that the upper bound a is derived by our simulation and the lower bound b is taken from Lacharme's results [14]. The exact value for these entries remains as an open question.
- In Table 2, a^6 means that the upper bound is derived by our simulation and we apply Theorem 8 to derive the lower bound.
- a^7 means that the upper bound is derived by our simulation and we apply Theorem 9 to derive the lower bound.
- $\deg^s(16, 8) = 6$, which is denoted 6^8 in Table 2, is the value from [10].

The entry with underline shows that the bound is strictly better than the one given by the t-resilient functions in [14]. For example, $\deg(10, 2) = 7$, but it is known that $(10, 2, 6)$-resilient function does not exist [4, Theorem 2, 3], and hence $F \in (10, 2)$-PP such that $\text{mindeg}(T_F) = 7$ cannot be a resilient function. For all the entries with underline, we have used the bound on t from [4].

Table 1 and Table 2 may be used to determine the values of n and m (and hence the input/output ratio) given the maximum bias that can be accepted for the application.

4.2 How to Derive $\deg(n, m)$

In this section, we discuss how we have derived numerical values of $\deg(n, m)$ in Table 1. We divide 2^n n-bit inputs into 2^m preimages, where each preimage has 2^{n-m} n-bit inputs. Consider some preimage, and let q_w be the number of x such that $w(x) = w$ in that preimage. Therefore, we require that

$$\sum_{w=0}^{n} q_w = 2^{n-m}. \tag{3}$$

If $w(x) = w$, then the coefficient of ϵ^l in $\Pr(x)$ is

$$\sum_{i+j=l} \binom{n-w}{i} (-1)^i \binom{w}{j} \left(\frac{1}{2}\right)^{n-i-j}.$$

Now consider the output probability of this preimage. The necessary and sufficient condition that the coefficients of $\epsilon, \epsilon^2, \ldots, \epsilon^e$ are all zero is;

$$\text{for } 1 \leq l \leq e, \sum_{w=0}^{n} \left\{ \sum_{i+j=l} \binom{n-w}{i} (-1)^i \binom{w}{j} \left(\frac{1}{2}\right)^{n-i-j} q_w \right\} = 0. \tag{4}$$

Table 1. The values of $\deg(n,m)$ for $1 \leq m \leq n \leq 16$

$n\backslash m$	1	2	3	4	5	6	7	8	9	10	11	12	13	14	15	16
1	1^1															
2	2^1	1^1														
3	3^1	2^1	1^1													
4	4^1	2^1	2^1	1^1												
5	5^1	3^1	2^1	2^1	1^1											
6	6^1	4^1	3^1	2^1	2^1	1^1										
7	7^1	4^1	4^1	3^1	2^1	2^1	1^1									
8	8^1	5^1	4^1	4^1	2^1	2^1	2^1	1^1								
9	9^1	6^1	$\underline{5^1}$	4^1	3^1	2^1	2^1	2^1	1^1							
10	10^1	7^1	6^1	5^1	4^1	3^1	2^1	2^1	2^1	1^1						
11	11^1	8^1	6^2	$\underline{6^1}$	4^2	4^1	3^1	2^1	2^1	2^1	1^1					
12	12^1	8^1	$\underline{7^1},6^1$	6^3	$5^1,4^1$	4^1	4^1	3^1	2^1	2^1	2^1	1^1				
13	13^1	$\underline{10^1}$	8^1	$\underline{7^1},6^1$	6^1	$5^1,4^1$	4^2	4^1	3^1	2^1	2^1	2^1	1^1			
14	14^1	$\underline{10^1}$	$9^1,8^1$	$\underline{8^1},7^5$	6^3	$6^1,5^5$	$5^1,4^1$	4^2	4^1	3^4	2^1	2^1	2^1	1^1		
15	15^1	$\underline{11^1},10^1$	10^1	8^4	7^4	6^2	6^1	$5^1,4^1$	4^2	4^1	3^4	2^1	2^1	2^1	1^1	
16	16^1	$\underline{12^1}$	$\underline{10^3}$	$\underline{9^1},8^5$	8^4	$7^1,6^1$	$7^1,6^1$	6^1	4^2	4^2	4^1	2^1	2^1	2^1	2^1	1^1

Table 2. The values of $\deg^s(n,m)$ for $1 \leq m < n \leq 16$

$n\backslash m$	1	2	3	4	5	6	7	8	9	10	11	12	13	14	15
2	2^1														
3	2^1	2^1													
4	4^1	2^1	2^1												
5	4^1	2^1	2^1	2^1											
6	6^1	4^1	2^1	2^1	2^1										
7	6^1	4^1	4^1	2^1	2^1	2^1									
8	8^1	4^1	4^1	4^1	2^1	2^1	2^1								
9	8^1	6^1	4^1	4^1	2^1	2^1	2^1	2^1							
10	10^1	6^1	6^1	4^1	4^1	2^1	2^1	2^1	2^1						
11	10^1	8^1	6^1	6^1	4^1	4^1	2^1	2^1	2^1	2^1					
12	12^1	8^1	6^7	6^7	4^1	4^1	4^1	2^1	2^1	2^1	2^1				
13	12^1	10^1	8^1	6^6	6^1	4^6	4^6	4^1	2^1	2^1	2^1	2^1			
14	14^1	10^1	8^7	$8^1,6^1$	6^7	$6^1,4^1$	4^6	4^6	4^1	2^1	2^1	2^1	2^1		
15	14^1	10^1	10^1	$8^1,6^1$	6^6	6^6	6^1	4^6	4^6	4^1	2^1	2^1	2^1	2^1	
16	16^1	12^1	10^7	$8^1,6^1$	$8^1,6^1$	6^6	6^6	6^8	4^6	4^6	4^1	2^1	2^1	2^1	2^1

Table 3. Environment for the simulation

Machine	Dell OPTIPLEX GX620
CPU	Pentium(R) 4 CPU 3.40GHz
OS	Microsoft Windows XP Professional SP2
Memory	4GB
Software	Wolfram Mathematica 6.0.1.0

The first step is to derive all the possible values of $\{q_0, q_1, \ldots, q_n\}$ that satisfy both (3) and (4). Suppose that we have d solutions, $\{q_0^{(1)}, q_1^{(1)}, \ldots, q_n^{(1)}\}$, $\{q_0^{(2)}, q_1^{(2)}, \ldots, q_n^{(2)}\}, \ldots, \{q_0^{(d)}, q_1^{(d)}, \ldots, q_n^{(d)}\}$. Let $Q_k = \{q_0^{(k)}, q_1^{(k)}, \ldots, q_n^{(k)}\}$ for $1 \leq k \leq d$. If we construct a preimage with $q_t^{(k)}$ inputs of weight t, then the resulting F is balanced (from (3)) and $\epsilon, \epsilon^2, \ldots, \epsilon^e$ are all eliminated (from (4)).

Now we have to construct 2^m preimages with the constraint that we have exactly $\binom{n}{w}$ inputs $x \in \{0,1\}^n$ such that $w(x) = w$. Therefore, the next step is to solve the following linear system;

$$\begin{bmatrix} z_1 & z_2 & \cdots & z_d \end{bmatrix} \begin{bmatrix} Q_1 \\ Q_2 \\ \vdots \\ Q_d \end{bmatrix} = \begin{bmatrix} \binom{n}{0} & \binom{n}{1} & \cdots & \binom{n}{n} \end{bmatrix}. \tag{5}$$

If there exists some $(z_1, z_2, \ldots, z_d)$ satisfying (5), this means it is possible to fulfill the above mentioned constraint, and thus we conclude $\deg(n, m) \geq e + 1$. Otherwise $\deg(n, m) \leq e$.

See Appendix for an example to derive $\deg(4, 2)$.

4.3 How to Derive $\deg^s(n, m)$

In (n, m)-SPPs, any input x and its complement, $\bar{x}$, have the same output. Thus we consider x and $\bar{x}$ as the pair $(x, \bar{x})$. Let $w(x, \bar{x}) = \min(w(x), w(\bar{x}))$, i.e., $w(x, \bar{x})$ is the minimum value of $w(x)$ and $w(\bar{x})$. Now, in $F \in (n, m)$-$\mathcal{SPP}$, we have to divide 2^{n-1} input pairs into 2^m preimages, where each preimage has 2^{n-m-1} input pairs. Consider some preimage, and let $q_{w'}$ be the number of pairs $(x, \bar{x})$ such that $w(x, \bar{x}) = w'$ in that preimage. Then, we need

$$\sum_{w'=0}^{\lfloor n/2 \rfloor} q_{w'} = 2^{n-m-1}, \tag{6}$$

where $\lfloor n/2 \rfloor$ is maximum integer at most $n/2$.

If l is odd, then the coefficient of ϵ^l in $\Pr(x) + \Pr(\bar{x})$ is zero. Otherwise the coefficient is

$$2 \sum_{i+j=l} \binom{n-w'}{i} (-1)^i \binom{w'}{j} \left(\frac{1}{2}\right)^{n-i-j},$$

where $w(x, \bar{x}) = w'$. Consider the output probability of this preimage, and the coefficients of $\epsilon, \epsilon^2, \ldots, \epsilon^e$ are all zero iff;

$$\text{for even } 1 \leq l \leq e, \quad \sum_{w'=0}^{\lfloor n/2 \rfloor} \left\{ 2 \sum_{i+j=l} \binom{n-w'}{i} (-1)^i \binom{w'}{j} \left(\frac{1}{2}\right)^{n-i-j} q_{w'} \right\} = 0. \tag{7}$$

Similarly to $\deg(n, m)$, we first solve a linear system of (6) and (7). Suppose that we have d' solutions, $Q_k = \{q_0^{(k)}, q_1^{(k)}, \ldots, q_{\lfloor n/2 \rfloor}^{(k)}\}$ for $1 \leq k \leq d'$, and consider the following linear system;

$$
\begin{bmatrix} z_1 & z_2 & \cdots & z_{d'} \end{bmatrix}
\begin{bmatrix} Q_1 \\ Q_2 \\ \vdots \\ Q_{d'} \end{bmatrix}
= \begin{bmatrix} \binom{n}{0} & \binom{n}{1} & \cdots & \binom{n}{\lfloor n/2 \rfloor} \end{bmatrix}.
\tag{8}
$$

If some $(z_1, z_2, \ldots, z_{d'})$ satisfies (8), then $\deg^s(n, m) \geq e + 2$. Otherwise we conclude that $\deg^s(n, m) \leq e$.

5 Summary of Results

In this paper, we have generalized the work in [10] in various ways. We first re-defined $\deg(n, m)$ and $\deg^s(n, m)$, and then presented twelve bounds on them. We believe that these bounds are important in understanding the basic properties of post-processing functions, and some of them are useful in deriving the exact values of $\deg(n, m)$ and $\deg^s(n, m)$. We derived the tables of $\deg(n, m)$ for $1 \leq m \leq n \leq 16$, and $\deg^s(n, m)$ for $1 \leq m < n \leq 16$, and discussed how we have derived these numerical values. Several values of $\deg(n, m)$ and $\deg^s(n, m)$ are left as open questions.

Our results suggest that, for some n and m, the resilient function is not the optimal solution as a post-processing function, and it would be interesting to see systematic constructions of optimal functions.

Acknowledgements

The authors would like to thank anonymous reviewers of SAC 2008 for their extensive and useful comments that significantly improved this paper. Especially, several open problems posed in the earlier version were solved by the comments. We also would like to thank Markus Dichtl and Takeshi Koshiba for useful feedbacks.

References

1. Barak, B., Impagliazzo, R., Wigderson, A.: Extracting randomness using few independent sources. SIAM J. Comput. 36(4), 1095–1118 (2006)
2. Barak, B., Kindler, G., Shaltiel, R., Sudakov, B., Wigderson, A.: Simulating independence: New constructions of condensers, Ramsey graphs, dispersers, and extractors. In: 37th STOC, pp. 1–10 (2005)
3. Barker, E., Kelsey, J.: Recommendation for random number generation using deterministic random bit generators (revised). NIST Special Publication 800-90 (2007), http://csrc.nist.gov/publications/PubsSPs.html

4. Bierbrauer, J., Gopalakrishnan, K., Stinson, D.R.: Bounds for resilient functions and orthogonal arrays. In: Desmedt, Y.G. (ed.) CRYPTO 1994. LNCS, vol. 839, pp. 247–256. Springer, Heidelberg (1994)
5. Blum, M.: Independent unbiased coin flips from a correlated biased source: A finite Markov chain. Combinatorica 6(2), 97–108 (1986)
6. Bourgain, J.: More on the sum-product phenomenon in prime fields and its applications. International Journal of Number Theory 1, 1–32 (2005)
7. Chor, B., Friedman, J., Goldreich, O., Håstad, J., Rudich, S., Smolensky, R.: The bit extraction problem or t-resilient functions. In: 26th FOCS, pp. 396–407 (1985)
8. Chor, B., Goldreich, O.: Unbiased bits from sources of weak randomness and probabilistic communication complexity. SIAM J. Comput. 17(2), 230–261 (1988)
9. Davis, D., Ihaka, R., Fenstermacher, P.: Cryptographic randomness from air turbulence in disk drives. In: Desmedt, Y.G. (ed.) CRYPTO 1994. LNCS, vol. 839, pp. 114–120. Springer, Heidelberg (1994)
10. Dichtl, M.: Bad and good ways of post-processing biased physical random numbers. In: Biryukov, A. (ed.) FSE 2007. LNCS, vol. 4593, pp. 137–152. Springer, Heidelberg (2007)
11. Grassl, M.: Code tables: Bounds on the parameters of various types of codes (2008), http://www.codetables.de/
12. Juels, A., Jakobsson, M., Shriver, E., Hillyer, B.K.: How to turn loaded dice into fair coins. IEEE Trans. Inform. Theory 46(3), 911–921 (2000)
13. Lacy, J.B., Mitchell, D.P., Schell, W.M.: Cryptolib: Cryptography in software. In: Proc. 4th USENIX Symposium (1993)
14. Lacharme, P.: Post-processing functions for a biased physical random number generator. In: Nyberg, K. (ed.) FSE 2008. LNCS, vol. 5086, pp. 334–342. Springer, Heidelberg (2008)
15. Matsui, M.: Linear cryptanalysis method for DES cipher. In: Helleseth, T. (ed.) EUROCRYPT 1993. LNCS, vol. 765, pp. 386–397. Springer, Heidelberg (1994)
16. Nisan, N., Ta-Shma, A.: Extracting randomness: A survey and new constructions. JCSS 58(1), 148–173 (1999)
17. Peres, Y.: Iterating von Neumann's procedure for extracting random bits. The Annals of Statistics 20(3), 590–597 (1992)
18. Raz, R.: Extractors with weak random seeds. In: 37th STOC, pp. 11–20 (2005)
19. Santha, M., Vazirani, U.V.: Generating quasi-random sequences from semi-random sources. JCSS 33, 75–87 (1986)
20. Schneier, B.: Applied cryptography. John Wiley & Sons, Inc., Chichester (1996)
21. Shaltiel, R.: Recent developments in explicit constructions of extractors. Bulletin of the European Association for Theoretical Computer Science (EATCS), 77 (2002)
22. Shaltiel, R., Umans, C.: Simple extractors for all min-entropies and a new pseudorandom generator. JACM 52(2), 172–216 (2005)
23. Ta-Shma, A.: On extracting randomness from weak random sources. In: STOC 1996, pp. 276–285 (1996)
24. Ta-Shma, A., Umans, C., Zuckerman, D.: Loss-less condensers, unbalanced expanders, and extractors. Combinatorica 27(2), 213–240 (2007)
25. Ta-Shma, A., Zuckerman, D., Safra, S.: Extractors from Reed-Muller codes. JCSS 72(5), 786–812 (2006)
26. von Neumann, J.: Various techniques used in connection with random digits. Applied Mathematics Series, U.S. National Bureau of Standards, vol. 12, pp. 36–38 (1951)

A   Deriving deg(4, 2)

We show a small example to derive $\deg(4, 2)$.

We divide sixteen 4-bit inputs into four preimages, where each preimage has four 4-bit inputs. Consider some preimage, and let q_w be the number of x such that $w(x) = w$ in that preimage. Therefore, we need

$$q_0 + q_1 + q_2 + q_3 + q_4 = 4. \tag{9}$$

Now the coefficient of ϵ in $\Pr(y)$ is zero iff

$$-\frac{1}{2}q_0 - \frac{1}{4}q_1 + \frac{1}{4}q_3 + \frac{1}{2}q_4 = 0, \tag{10}$$

which corresponds to $l = 1$ in (4). Similarly, the coefficient of ϵ^2 is zero iff

$$\frac{3}{2}q_0 - \frac{1}{2}q_2 + \frac{3}{2}q_4 = 0. \tag{11}$$

We have seven solutions that satisfy both (9) and (10), and consider the following linear system;

$$\begin{bmatrix} z_1 & z_2 & z_3 & z_4 & z_5 & z_6 & z_7 \end{bmatrix} \begin{bmatrix} 0 & 0 & 4 & 0 & 0 \\ 0 & 1 & 2 & 1 & 0 \\ 0 & 2 & 0 & 2 & 0 \\ 0 & 2 & 1 & 0 & 1 \\ 1 & 0 & 1 & 2 & 0 \\ 1 & 0 & 2 & 0 & 1 \\ 1 & 1 & 0 & 1 & 1 \end{bmatrix} = \begin{bmatrix} 1 & 4 & 6 & 4 & 1 \end{bmatrix}, \tag{12}$$

where the matrix in (12) corresponds to the seven solutions. Now since

$$(z_1, z_2, \ldots, z_7) = (1, 1, 1, 0, 0, 0, 1)$$

satisfies (12), we conclude that $\deg(4, 2) \geq 2$.

On the other hand we only have one solution satisfying (9), (10) and (11), which is $(q_0, \ldots, q_4) = (0, 2, 0, 2, 0)$. Now we consider the following linear system;

$$\begin{bmatrix} z_1 \end{bmatrix} \begin{bmatrix} 0 & 2 & 0 & 2 & 0 \end{bmatrix} = \begin{bmatrix} 1 & 4 & 6 & 4 & 1 \end{bmatrix}.$$

Since there is no solution for this system, we have $\deg(4, 2) \leq 2$, and therefore, $\deg(4, 2) = 2$.

HECC Goes Embedded: An Area-Efficient Implementation of HECC

Junfeng Fan, Lejla Batina, and Ingrid Verbauwhede

Katholieke Universiteit Leuven, ESAT/SCD-COSIC and IBBT
Kasteelpark Arenberg 10
B-3001 Leuven-Heverlee, Belgium
{junfeng.fan,lejla.batina,ingrid.verbauwhede}@esat.kuleuven.be

Abstract. In this paper we describe a high performance, area-efficient implementation of Hyperelliptic Curve Cryptosystems over $GF(2^m)$. A compact Arithmetic Logic Unit (ALU) is proposed to perform multiplication and inversion. With this ALU, we show that divisor multiplication using affine coordinates can be efficiently supported. Besides, the required throughput of memory or Register File (RF) is reduced so that area of memory/RF is reduced. We choose hyperelliptic curves using the parameters $h(x) = x$ and $f(x) = x^5 + f_3 x^3 + x^2 + f_0$. The performance of this coprocessor is substantially better than all previously reported FPGA-based implementations. The coprocessor for HECC over $GF(2^{83})$ uses 2316 slices and 2016 bits of Block RAM on Xilinx Virtex-II FPGA, and finishes one scalar multiplication in 311 μs.

Keywords: Hyperelliptic Curve Cryptosystems, Modular multiplication, Modular inversion, FPGA.

1 Introduction

Public-Key Cryptography (PKC) [10], introduced in the mid 70's by Diffie and Hellman, ensures a secure communication over an insecure network without prior key agreement. PKC is widely used for digital signatures, key agreement and data encryption. The best-known and most commonly used public-key cryptosystems are RSA [26] and Elliptic Curve Cryptography (ECC) [23,19], but recently HyperElliptic Curve Cryptography (HECC) [20] is catching up. The main benefit for curve-based cryptography *e.g.* ECC and HECC is that they offer equivalent security as RSA for much smaller parameter sizes. The advantages result in smaller data-paths, less memory and lower power consumption.

Implementing HECC on a resource-constrained platform has been a challenge in both area and performance. Over the past few years, HECC have been implemented in both software [25,27] and hardware [4,7,11,15]. However, the implementations so far failed in reaching the performance of ECC implementations with comparable hardware cost. Table 1 compares the computational complexity of point/divisor operations in ECC and HECC as in [2]. Here I, M and S denote modular inversion, multiplication and squaring, respectively. Note that Table 1 is not exhaustive, and a comprehensive description of different coordinates as

R. Avanzi, L. Keliher, and F. Sica (Eds.): SAC 2008, LNCS 5381, pp. 387–400, 2009.
© Springer-Verlag Berlin Heidelberg 2009

Table 1. Modular Operations Required by Point/Divisor Operations in $GF(2^m)$ [2]

		PA/DA	PD/DD	Coordinates Conversion
ECC	Affine	I+2M+S	I+2M+S	-
	Projective	15M+3S	7M+4S	I+2M
HECC	Affine	I+22M+3S	I+20M+6S	-
	Projective	49M+4S	38M+7S	I+4M

well as their computational complexity can be found in [2]. In addition, state of the art regarding various types of coordinates for all types of curve-based cryptosystem can be found in [9]. For example, ECC over $GF(2^{163})$ and HECC over $GF(2^{83})$ are supposed to offer equivalent security as 1024-bit RSA [2]. Using projective coordinates, one EC Point Addition (PA) requires 15 multiplications and 3 squarings in $GF(2^{163})$, while one HEC Divisor Addition (DA) requires 49 multiplications and 4 squarings in $GF(2^{83})$, which is much more complex even with parameters of half bit-lengths. In order to speed up HECC implementations, parallel multipliers [4,7] or inverters [15] were used. As a result, an ALU becomes large in the area. In order to efficiently feed data to parallel multipliers and inverters, a high-throughput Register File (RF) with an additional control logic *i.e.* a MUX array connected to ALU is required. This adds even more area to implementations.

In this paper, we describe a compact HECC coprocessor on an FPGA platform. The coprocessor utilizes a unified multiplier/inverter, which supports both multiplication and inversion. This architecture brings three main advantages. First, the fast inverter makes affine coordinates very efficient. Second, as the multiplier and inverter share partial data-path, it is much smaller in area compared to previous implementations. Third, using only one multiplier/inverter, the required throughput of Memory or RF is comparably low. Therefore we can reduce the area of the memory. Note that the architecture proposed here for FPGA design can also lead to an area-efficient design in ASICs. The coprocessor was synthesized with Xilinx ISE8.1i. On Virtex-II FPGA (XC2V4000), this coprocessor finishes one scalar multiplication of HECC over $GF(2^{83})$ in 311 μs using 2316 slices and 2016 bits memory. To the best of our knowledge, this implementation is faster than all proposed FPGA-based implementations of HECC, while the area is much smaller than that of the fastest reported implementation [15].

The rest of the paper is organized as follows. Section 2 gives a brief introduction on the previous work. Section 3 describes the mathematical background of HECC and field arithmetic. Section 4 describes the architecture of the proposed HECC coprocessor. In Sect. 5 we show the implementation results. We conclude the paper and give some future work in Sect. 6.

2   Previous Work

In 2001, Wollinger described the first hardware architecture for HECC implementations using Cantor's algorithm [6] in his thesis [32]. However, the architecture was only outlined. The first complete hardware implementation of HECC was presented in [4]. It is also based on Cantor's algorithm, but with improvement on the calculation of Greatest Common Divisor (GCD). This implementation, using 16600 slices on Xilinx Virtex II FPGA, supports a genus-2 HEC over GF(2^{113}). One scalar multiplication takes 20.2 ms on this coprocessor running at 45MHz. This work was further improved in [7].

In 2002, Lange generalized the explicit formulae for HECC over finite fields with arbitrary characteristic [21]. This was first implemented on 32-bit embedded processors (ARM7TDMI and PowerPC) in [25]. The inversion in this algorithm was performed with Extended Euclidean Algorithm (EEA). The first hardware implementation of HECC using explicit formulae was described in [12]. Further improvement by using mixed coordinates and simplified curves were proposed in [11]. In [11] the coprocessor, running at 45.3MHz, deploys 25272 slices on Xilinx Virtex II FPGA. With this implementation 2.03 ms is required to perform one scalar multiplication of HECC over GF(2^{113}). There are some ASIC implementations of HECC using projective coordinates. For example, Sakiyama proposed a HECC coprocessor [28] using 0.13-μm CMOS technology. The coprocessor runs at 500 MHz, and can perform one scalar multiplication of HECC over GF(2^{83}) in 63 μs.

The first hardware implementations of HECC using affine version of explicit formulae were described in [31], which described so far the fastest FPGA-based HECC coprocessor. This coprocessor uses three modular multipliers and two modular inverters. It uses 7785 slices on Xilinx Virtex II FPGA(XC2V4000), and can reach a clock frequency of 56.7MHz. One scalar multiplication of HECC over GF(2^{81}) takes 415 μs.

3   Mathematical Background

3.1   Hyperelliptic Curve Cryptography

Hyperelliptic curves are a special class of algebraic curves; they can be viewed as generalization of elliptic curves. Namely, a hyperelliptic curve of genus $g = 1$ is an elliptic curve, while in general, hyperelliptic curves can be of any genus $g \geq 1$.

Let $\overline{GF}(2^m)$ be an algebraic closure of the field GF(2^m). Here we consider a hyperelliptic curve C of genus $g = 2$ over GF(2^m), which is given with an equation of the form:

$$C : y^2 + h(x)y = f(x) \quad in \quad GF(2^m)[x, y], \qquad (1)$$

where $h(x) \in$ GF(2^m)[x] is a polynomial of degree at most g ($deg(h) \leq g$) and $f(x)$ is a monic polynomial of degree $2g + 1$ ($deg(f) = 2g + 1$). Also, there are no solutions $(x, y) \in \overline{GF}(2^m) \times \overline{GF}(2^m)$ which simultaneously satisfy the equation (1) and the equations: $2v + h(u) = 0, h'(u)v - f'(u) = 0$. These points are called singular points. For the genus 2, in the general case the following equation is used $y^2 + (h_2x^2 + h_1x + h_0)y = x^5 + f_4x^4 + f_3x^3 + f_2x^2 + f_1x + f_0$.

A divisor D is a formal sum of points on the hyperelliptic curve C *i.e.* $D = \sum m_P P$ and its degree is $deg D = \sum m_P$. Let Div denotes the group of all divisors on C and Div_0 the subgroup of Div of all divisors with degree zero. The Jacobian J of the curve C is defined as quotient group $J = Div_0/P$. Here P is the set of all principal divisors, where a divisor D is called principal if $D = div(f)$, for some element f of the function field of C $(div(f) = \sum_{P \in C} ord_P(f)P)$. The discrete logarithm problem in the Jacobian is the basis of security for HECC. In practice, the Mumford representation according to which each divisor is represented as a pair of polynomials $[u, v]$ is usually used. Here, u is monic of degree 2, $deg v < deg u$ and $u|f - hv - v^2$ (so-called reduced divisors). For implementations of HECC, we need to implement the multiplication of elements of the Jacobian *i.e.* divisors with some scalar.

The main operation in any hyperelliptic curve based primitive is scalar multiplication, *i.e.* mD where m is an integer and D is a reduced divisor in the Jacobian of some hyperelliptic curve C. The first algorithm for arithmetic in the Jacobian is due to Cantor [6]. However, until "explicit formulae" were invented, the HECC was not considered a suitable alternative to EC based cryptosystems. For geni 2 and 3, there was some substantial work on the formulae and algorithms for computing the group law on the Jacobian have been optimized. Algorithms for the group operation for the case of genus 2 hyperelliptic curves, which we used are due to Lange [22].

The main operation in any curve-based primitive (ECC or HECC) is the scalar multiplication. Looking at the arithmetic for both ECC/HECC the only difference between ECC and HECC is in the group operations. On this level both ciphers consist of different sequences of operations. Those for HECC are more complex when compared with the ECC point operation, but they use shorter operands. The divisor scalar multiplication is achieved by repeated divisor addition and doubling. Many techniques that help to speed up ECC scalar multiplication are also applicable to HECC. For example, using Non-Adjacent Form (NAF) for scalar representation or window method can also improve HECC performance.

3.2   Field Arithmetic

An element α in $GF(2^m)$ can be represented as a polynomial $A(x) = \sum_{i=0}^{m-1} a_i x^i$, here $a_i \in GF(2)$. Addition of two elements in $GF(2^m)$ is performed as polynomial addition in $GF(2)$

$$\sum_{i=0}^{m-1} a_i x^i + \sum_{i=0}^{m-1} b_i x^i = \sum_{i=0}^{m-1} (a_i \oplus b_i) x^i,$$

where $\oplus$ is XOR operation.

Multiplication. In the literature there are various algorithms and architectures [3,30] proposed for modular multiplication in $GF(2^m)$. The bit-serial algorithms can be classified into two categories, the Most Significant Bit (MSB) first algorithms and the Least Significant Bit (LSB) first algorithms. It is important to point out that LSB-first bit-serial multiplier has shorter critical path than MSB-first bit-serial multipliers [3]. In this paper, we use the LSB-first algorithm.

Algorithm 1. LSB-first bit-serial modular multiplication in $GF(2^m)$ [3]

Input: $A(x) = \sum_{i=0}^{m-1} a_i x^i$, $B(x) = \sum_{i=0}^{m-1} b_i x^i$, irreducible binary polynomial $P(x)$ with $\deg(P(x)) = m$.
Output: $A(x)B(x) \bmod P(x)$.
  1: $C(x) \leftarrow 0$, $A'(x) \leftarrow A(x)$;
  2: **for** $i = 0$ to $m - 1$ **do**
  3:     $C(x) \leftarrow C(x) + b_i A'(x)$;
  4:     $A'(x) \leftarrow x A'(x) \bmod P(x)$;
  5: **end for**
Return: $C(x)$.

Inversion. A multiplicative inverse of $A(x)$ is a polynomial $A^{-1}(x)$ in $GF(2)$ such that $A^{-1}(x)A(x) \equiv 1 \bmod P(x)$. Compared with the other modular operations, modular inversion is considered as a computationally expensive operation. The most commonly used methods to perform the modular inversion are based on Fermat's little theorem [1], Extended Euclidean Algorithm [18] and Gaussian elimination [16]. EEA is widely used to perform inversion in practice.

The schoolbook EEA-based inversion algorithm in $GF(2^m)$ is commonly considered inefficient due to the long polynomial division in each iteration. This problem was partially solved by replacing degree comparison with a counter [5].

In [34], Yan $et\ al.$ proposed a modified inversion algorithm based on the EEA. Algorithm 2 shows this inversion algorithm. Here we use $S^i(x)$ to denote the value of $S(x)$ after i^{th} iteration, and d_0^{i-1} the LSB of d^{i-1}. The complement of C_1 is represented as $\bar{C}_1$. Unlike many other EEA variants [14,5,18], this algorithm has no modular operations, thus a short critical path delay can be easily achieved. Besides, with a fixed number of iterations, it is more secure against side-channel analysis.

4   HECC Coprocessor Architecture

In this section we describe a compact coprocessor architecture for HECC over $GF(2^m)$. Two main approaches are used to reduce the area: using compact ALU and reducing memory area. First, we propose a unified digit-serial modular multiplier/inverter, which enables a small ALU. Second, we investigate the characteristics of the ALU, and reduce area of memory block as well as its interconnecting network.

4.1   Modular Multiplier

As shown in Algorithm 1, the main operation in LSB-first multiplication is $(bA(x) + C(x))$, which can be performed by a row of AND gates and XOR gates

Algorithm 2. EEA-Based Inversion Algorithm [34]

Input: irreducible binary polynomial $P(x)$ with $\deg(P(x)) = m$, polynomial $A(x)$ with $\deg(A(x)) < m$.

Output: $A^{-1}(x) \bmod P(x)$.

1: $R^0(x) \leftarrow P(x)$, $S^0(x) \leftarrow xA(x)$, $H^0(x) \leftarrow 0$, $J^0(x) \leftarrow x^m$, $d^0 \leftarrow 2$, $sign^0 \leftarrow 1$;

2: **for** $i = 1$ to $2m - 1$ **do**

3:     $C_1 \leftarrow s_m^i$, $C_2 \leftarrow C_1 \wedge sign^{i-1}$;

$$sign^i \leftarrow \begin{cases} \bar{C_1} & \text{if } sign^{i-1} = 1; \\ d_0^{i-1} & \text{if } sign^{i-1} = 0; \end{cases}$$

$$S^i(x) \leftarrow \begin{cases} x(R^{i-1}(x) + S^{i-1}(x)) & \text{if } C_1 = 1; \\ xS^{i-1}(x) & \text{if } C_1 = 0; \end{cases}$$

$$J^i(x) \leftarrow \begin{cases} H^{i-1}(x) + J^{i-1}(x) & \text{if } C_1 = 1; \\ J^{i-1}(x) & \text{if } C_1 = 0; \end{cases}$$

$$R^i(x) \leftarrow \begin{cases} S^{i-1}(x) & \text{if } C_2 = 1; \\ R^{i-1}(x) & \text{if } C_2 = 0; \end{cases}$$

$$H^i(x) \leftarrow \begin{cases} J^{i-1}(x)/x & \text{if } C_2 = 1; \\ H^{i-1}(x)/x & \text{if } C_2 = 0; \end{cases}$$

$$d^i \leftarrow \begin{cases} 2d^{i-1} & \text{if } sign^i = 1; \\ d^{i-1}/2 & \text{if } sign^i = 0; \end{cases}$$

4: **end for**

Return: $H^{2m-1}(x)$.

---

shown in Figure 1(a). Figure 1(b) shows the architecture of a LSB-first bit-serial multiplier. Two $(m + 1)$-bit registers are used to hold the parameter $P(x)$, $A(x)$ and two m-bit registers to hold $B(x)$ and the partial product $C(x)$. Note that $B(x)$ is shifted to right by one bit in each clock cycle. Here $(a_m P(x) + A(x))$ and $(b_0 A(x) + C(x))$ is performed on the left and right side, respectively. If low Hamming weight irreducible polynomials are used, the AND-XOR cell on the left side can be simplified. For example, using $P(x) = x^{83} + x^7 + x^4 + x^2 + 1$, only 4 AND gates and 4 XOR gates are required to perform $(a_m P(x) + A(x))$.

It is clear that the critical path delay is $T_{\text{AND}} + T_{\text{XOR}}$, where T_{AND} and T_{XOR} denote the delay of a 2-input AND and XOR gate, respectively. One multiplication in $\text{GF}(2^m)$ takes m clock cycles on this bit-serial multiplier.

4.2   Unified Modular Inverter and Multiplier

We propose a unified architecture which can perform both multiplication and inversion. In [8], Daly *et al.* have proposed a unified ALU for $\text{GF}(p)$. It can perform addition, subtraction, multiplication and inversion. Compared with this ALU, our unified inverter/multiplier in $\text{GF}(2^m)$ has a shorter critical path delay, and can be implemented in a digit-serial manner to achieve a higher throughput. Figure 2 shows the data-path of our proposed bit-serial inverter and multiplier.

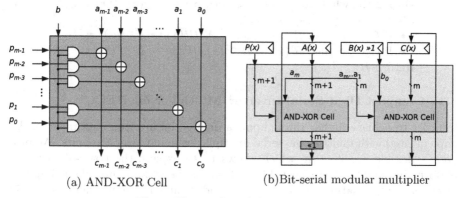

(a) AND-XOR Cell   (b)Bit-serial modular multiplier

Fig. 1. Bit-serial modular multiplier

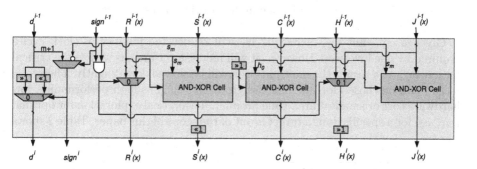

Fig. 2. Bit-serial modular multiplication/inversion unit

It realizes both Algorithm 1 and Algorithm 2. The multiplier and the inverter share one AND-XOR cell and three registers. The critical path delay is $2T_{\mathrm{MUX}}$. Here T_{MUX} denotes the delay of a 2-input multiplexer. This multiplier/inverter finishes one inversion operation in $\mathrm{GF}(2^m)$ in $(2m-1)$ clock cycles.

This data-path supports the following operations:

1. Modular Multiplication

   – Initialization $(i = 0)$, $R(x) \leftarrow P(x)$, $S(x) \leftarrow xA(x)$, $H(x) \leftarrow B(x)$, $C(x) \leftarrow 0$, $d \leftarrow 0$, $sign \leftarrow 0$;
   – During the whole loop $(0 < i < m+1)$, $d^i = 0$, $sign^i = 0$, thus, $R^i(x) = R^{i-1}(x) = P(x)$, $H^i(x) \leftarrow H^{i-1}(x)/x$, $A^i(x) \leftarrow x(A^{i-1}(x) + a_m P(x))$, and $C^i(x) \leftarrow h_0 A^{i-1}(x)/x + C^{i-1}(x)$;
   – Return $C^m(x)$.

2. Modular Inversion

   – Initialization $(i = 0)$, $R(x) \leftarrow P(x)$, $S(x) \leftarrow xA(x)$, $H(x) \leftarrow 0$, $J(x) \leftarrow x^m$, $d \leftarrow 2$, $sign \leftarrow 1$;

- During the whole loop $(0 < i < 2m)$, $S^i(x) \leftarrow x(S^{i-1}(x) + s_m R^{i-1}(x))$,
  $J^i(x) \leftarrow J^{i-1}(x) + s_m H^{i-1}(x)$,
  - If $C_2 = 1$, then $R^i(x) \leftarrow S^{i-1}(x)$, $H^i(x) \leftarrow J^{i-1}(x)/x$;
  - If $C_2 = 0$, then $R^i(x) \leftarrow R^{i-1}(x)$, $H^i(x) \leftarrow H^{i-1}(x)/x$;
- Return $H^{2m-1}(x)$.

4.3   Compact Digit-Serial Inverter/Multiplier for HECC

In order to achieve higher throughput, a digit-serial inverter/multiplier can be implemented with multiple bit-serial multiplication and inversion units. We propose a flexible architecture which allows us to explore the trade-off between performance and hardware cost. Figure 3 shows the architecture where 3 unified inversion multiplication units ($w_1 = 3$) and 4 bit-serial multipliers ($w_2 = 7$) are used. Here w_1 and w_2 denote the equivalent digit-size of this digit-serial inverter and multiplier, respectively. When choosing $m = 83$, one inversion takes $\lceil \frac{2m-1}{w_1} \rceil = 55$ clock cycles, while one multiplication takes $\lceil \frac{m}{w_2} \rceil = 14$ clock cycles.

Given a constant w_2, increasing w_1 will reduce the number of clock cycles required by one inversion. However, it will increase the area as well as the critical path delay. As a result, the multiplication will be slowed down slightly. Therefore, w_1/w_2 can be chosen for different design targets such as high performance, low hardware cost or smallest area-time product. Theoretical exploration for optimal (w_1, w_2) for a specific design target is out of the scope of this paper. Table 2 shows

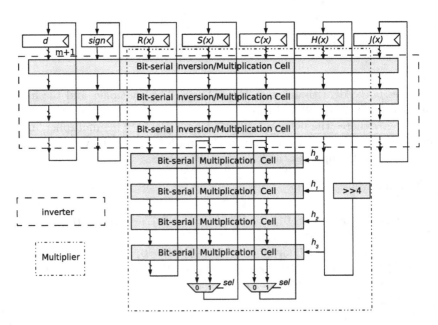

Fig. 3. Digit-serial modular multiplication inversion unit ($w_1 = 3, w_2 = 7$)

Table 2. Performance comparison of multiplication and inversion unit in $GF(2^m)$

Ref. Design	Configuration	Area [Slices]	Freq. [MHz]	Finite Field	Mul. Perf. $[ns]/[\#cycle]$	Inv. Perf. $[ns]/[\#cycle]$
	$w_1 = 1, w_2 = 14$	977	127	$GF(2^{83})$	47.1 / 6	1296 / 165
	$w_1 = 2, w_2 = 14$	1117	126	$GF(2^{83})$	47.3 / 6	654 / 83
Fig. 3	$w_1 = 3, w_2 = 14$	1500	125	$GF(2^{83})$	47.9 / 6	439 / 55
	$w_1 = 4, w_2 = 14$	1718	113	$GF(2^{83})$	52.7 / 6	372 / 42
	$w_1 = 5, w_2 = 14$	1987	104	$GF(2^{83})$	57.4 / 6	315 / 33
	$w = 8$	342	108.7	$GF(2^{81})$	101 / 11	-
Mult. [31]	$w = 16$	554	87.5	$GF(2^{81})$	69 / 6	-
	$w = 27$	882	71.0	$GF(2^{81})$	42 / 3	-
Inv. [31]	MAIA	663	87.8	$GF(2^{81})$	-	1014 / 89

the performance and area of the proposed ALU with different configurations. Here Xilinx Virtex II (XC2V4000) FPGA is used. In this HECC implementation we choose $w_1 = 3$ and $w_2 = 14$ as the best performance/area trade-off for this architecture. With this configuration, one multiplication and one inversion in $GF(2^{83})$ take 47.9 and 439 ns, respectively.

4.4 Memory/RF Analysis

Besides ALU, memory/RF is another main component that decides the overall area and performance of a coprocessor. The size, throughput and delay of memory/RF must be chosen according to the requirement of the ALU. We analyze different design strategies of HECC coprocessor here.

Both memory and RF have their own advantages and disadvantages. While registers are larger than memory of the same capacity, memory usually has one clock delay in read operation. This delay may cause performance degradation when multiple data-path work in a pipelining mode, see [31]. Thus, HECC coprocessors using multiple data-path [7,33,31] require an efficient register file to feed data to parallel multipliers and inverters. The register file and its interconnecting network make a big part of the whole area.

The area of memory/RF is dependent on the size and throughput [24,29]. Higher throughput results in a more complex decoder and a larger interconnecting network, which cause the area increase. Thus, reducing the memory/RF throughput reduces the area. Table 3 shows the required memory/RF throughput of different ALUs. Note that here we use $GF(2^{83})$ for all the ALUs, D denotes the delay of multiplication. For example, when using three multipliers, the ALU reads 6 operands from memory/RF and writes 3 data back. In [33,31], 3 clock cycles are required for one multiplication. If each operand is 84-bit, then the ALU needs to read 168 bits in each clock cycle. The proposed multiplier/inverter shown in Figure 3 requires 56-bit read and 14-bit write in each clock cycles. The required memory throughput is much smaller than that in [33] and [31].

Table 3. Comparison of memory throughput required by different ALUs

Ref. Design	Configuration	Read [Bits]	Write [Bits]	Total [Bits]
[33]	3 Mult. $(D = 3)$	168	84	252
[31]	2 Mult. $(D = 3)$	112	56	168
Fig.4	Unified M/I. $(D = 6)$	56	14	80

4.5   Coprocessor Architecture

The HECC Coprocessor is shown in Figure 4. It contains an Instruction ROM, a main controller and a unified modular multiplier/inverter. The Instruction ROM contains the field operation sequences of divisor addition and doubling. As only a single data-path is used, the coprocessor does not require high-throughput register files. Instead, a data RAM is used to keep the curve parameters, base divisor and intermediate data. On FPGAs, Block RAMs are used.

The coprocessor supports four instructions, namely,

```
Add Ra,Rb,Rc        // Ra=Rb+Rc
Mul Ra,Rb,Rc        // Ra=Rb*Rc
Mac Ra,Rb,Rc,Rd,Re  // Ra=Rb*Rc+Rd+Re
Inv Ra,Rb           // Ra=Rb^{-1}
```

Here one `Add` instruction takes two cycles. As $w_1 = 3$, one `Inv` instruction takes 55 clock cycles. One `Mul` instruction takes 6 clock cycles. One `Mac` instruction consists of one `Mul` and two `Add` instructions. However, it takes also 6 clock cycles. This is because fetching and adding data `Rd` and `Re` are performed during the multiplication. Two `Add` and one `Mul` instructions cause 6 operand fetches and 3 result stores, while one `Mac` instruction requires only 4 operand fetches and one result store. Therefore, the use of `Mac` instruction reduces the number of memory access and speeds up the scalar multiplication.

In this implementation, we choose hyperelliptic curves with the following parameters: $h(x) = x$ and $f(x) = x^5 + f_3x^3 + x^2 + f_0$. One DA operation consists of 36 instructions, which include 11 `Add`, 24 `Mac` and 1 `Inv` instructions. One DD operation consists of 14 instructions, which includes 2 `Add`, 11 `Mac` and 1 `Inv` instructions.

Note that the architecture of the coprocessor can be slightly modified so that it can be integrated into a SoC where memory is shared. The required throughput of memory needs to be further reduced. In the InsRom `Mac` instruction needs to be replaced by a `Mul` and two `Add` instructions, thus only two instead of four operands need to be loaded for each instruction. In this case, the required throughput of memory is $\frac{2*84}{6} = 28$ bits, the amount that a 32-bit dual-port SRAM is able to offer. However, the add instruction requires 6 instead of 2 clock cycles, which slightly degrades the performance of the coprocessor.

5   Implementation Results

In order to check the area and performance of the proposed coprocessor, we implemented the architecture from Figure 4 on a Xilinx Virtex-II (XC2V4000) FPGA. The coprocessor is described with Gezel [13] language and synthesized with Xilinx ISE8.1. It uses 2316 slices and 6 Block RAMs. A clock frequency of 125 MHz can be reached. Table 4 compares the area and performance with previous FPGA-based implementations of HECC in GF(2^m).

The proposed HECC coprocessor in [7] uses Cantor's method to perform divisor addition and doubling. It has two modular multipliers, one inverter, one GCD module and several other logics. Register file is connected to the datapath with MUX arrays. When supporting HECC in $GF(2^{83})$, it uses 22000 slices on Xilinx Virtex-II FPGA and can finish one scalar multiplication in 10 ms.

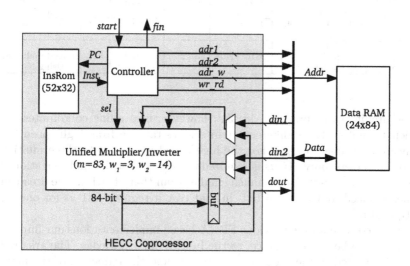

Fig. 4. Block diagram of the proposed HECC coprocessor

The proposed HECC coprocessor in [11] uses the mixed coordinates of explicit formulae proposed in [21]. The ALU contains three modules, namely divisor addition module, divisor doubling module and coordinates conversion module. Each of them has four field multipliers, while only the coordinates conversion module has a inverter. It supports Right-to-Left binary expansion method, which scans the key from LSB to MSB, and can perform divisor addition and doubling in parallel. It also supports NAF method. Here we list the performance of scalar multiplication using NAF method as it is slightly faster than the binary method.

The HECC coprocessor proposed in [17] uses projective coordinates, and a superscalar architecture is used to support parallel field operations. Several digit-serial ($w = 12$) multipliers are used. Our coprocessor, using one unified multiplier/inverter, is faster than the coprocessor in [17] that uses three multipliers.

Table 4. Performance comparison of FPGA-based HECC implementations in $GF(2^m)$

Ref. Design	FPGA	Freq. [MHz]	Area [Slices]	RAM [bits]	Finite Field	Irreducible Polynomial	Perf. [μs]	Comments
Clancy [7]	Xilinx Virtex-II	N/A	23000	0	$GF(2^{83})$	Arbitrary	10000	Two mult. One inv. Using NAF
Elias et al. [11]	Xilinx Virtex-II (XC2V8000)	45.3	25271	0	$GF(2^{113})$	Fixed	2030	12 mult. One inv. Using NAF
Sakiyama et al. [17]	Xilinx Virtex-II Pro (XC2VP30)	100	6586	8064	$GF(2^{83})$	Arbitrary	420	Three mult. Using NAF
			4749	5376	$GF(2^{83})$	Arbitrary	549	Two mult. Using NAF
			2446	2688	$GF(2^{83})$	Arbitrary	989	One mult. Using NAF
Wollinger [31]	Xilinx Virtex-II (XC2V4000)	56.7	7785	0	$GF(2^{81})$	Fixed	415	Three mult. Two inv.
		47.0	5604	0	$GF(2^{81})$	Fixed	724	Two mult. One inv.
		54.0	3955	1536	$GF(2^{81})$	Fixed	831	Two mult. One inv.
This work	Xilinx Virtex-II (XC2V4000)	125	2316	2016	$GF(2^{83})$	Fixed	311	Unified mult./inv. Using NAF

The architectures proposed in [31], however, uses affine coordinates of the explicit formulae. Three different architectures ranging from high speed to low hardware cost are proposed. For the high speed version, with three multipliers and two inverters, only 415 μs is required to finish one scalar multiplication. The area of the coprocessor is also much smaller than that of [7,11]. The area can be further reduced to 3955 slices but, in that case it requires 831 μs for one scalar multiplication.

Compared with all the previous FPGA-based implementations our implementation has the best performance, to the best of our knowledge. The area reduction is attributed to the use of compact ALU and the reduction of the memory throughput. The ALU in [31] contains two multipliers and one inverter, which in total use 2427 slices. The ALU used in this paper requires only 1500 slices. The performance gain is mainly due to the efficient inverter. When running at 56.7 MHz, the inverter in [31] requires 1570 ns in average for one inversion in $GF(2^{81})$, while the proposed ALU finishes one inversion in $GF(2^{83})$ in 439 ns. Though we use only one multiplier, which is also slower than the one in [31], the overall performance of divisor addition/doubling is better.

6   Conclusions

We describe a compact architecture for HECC over binary extension field. This architecture uses a unified modular multiplier/inverter, and reduces the throughput of the memory. Thus, the area of the coprocessor is largely reduced. On a Xilinx Virtex II (XC2V4000) FPGA, the proposed coprocessor takes 311 μs to finish one scalar multiplication in HECC over $GF(2^{83})$.

The proposed implementation can be further speeded up by exploring instruction level parallelism. Besides, if more space is available in the data memory, precomputation can be used to drastically improve the performance.

Acknowledgments

This work was supported in part by the IAP Programme P6/26 BCRYPT of the Belgian State (Belgian Science Policy), by FWO projects G.0475.05, and G.0300.07, by the European Comission through the IST Programme under Contract IST-2002-507932 ECRYPT NoE, and by the K.U. Leuven-BOF.

References

1. Asano, Y., Itoh, T., Tsujii, S.: Generalised fast algorithm for computing multiplicative inverses in $GF(2^m)$. Electronics Letters 25(10), 664–665 (1989)
2. Avanzi, R.M., Cohen, H., Doche, C., Frey, G., Lange, T., Nguyen, K., Vercauteren, F.: Handbook of Elliptic and Hyperelliptic Curve Cryptography. CRC Press, Boca Raton (2005)
3. Beth, T., Gollman, D.: Algorithm engineering for public key algorithms. IEEE Journal on Selected Areas in Communications 7(4), 458–466 (1989)
4. Boston, N., Clancy, T., Liow, Y., Webster, J.: Genus two hyperelliptic curve coprocessor. In: Kaliski Jr., B.S., Koç, Ç.K., Paar, C. (eds.) CHES 2002. LNCS, vol. 2523, pp. 400–414. Springer, Heidelberg (2003)
5. Brent, R.P., Kung, H.T.: Systolic VLSI Arrays for Polynomial GCD Computation. IEEE Trans. Computers 33(8), 731–736 (1984)
6. Cantor, D.G.: Computing in the Jacobian of a Hyperelliptic curve. Mathematics of Computation 48, 95–101 (1987)
7. Clancy, T.: FPGA-based Hyperelliptic Curve Cryptosystems. invited paper presented at AMS Central Section Meeting (April 2003)
8. Daly, A., Marnane, W., Kerins, T., Popovici, E.: An FPGA implementation of a GF(p) ALU for encryption processors. Elsevier Journal on Microprocessors and Microsystems (Special issue on FPGAs: Applications and Designs) 28(5-6), 253–260 (2004)
9. Explicit-Formulas Database, http://www.hyperelliptic.org/EFD
10. Diffie, W., Hellman, M.E.: New directions in cryptography. IEEE Transactions on Information Theory 22, 644–654 (1976)
11. Elias, G., Miri, A., Yeap, T.H.: On efficient implementation of FPGA-based hyperelliptic curve cryptosystems. Computers and Electrical Engineering 33(5-6), 349–366 (2007)
12. Yeap, T.H., Elias, G., Miri, A.: High-Performance, FPGA-Based Hyperelliptic Curve Cryptosystems. In: The Proceeding of the 22nd Biennial Symposium on Communications (May 2004)
13. GEZEL, http://rijndael.ece.vt.edu/gezel2/
14. Guo, J.-H., Wang, C.-L.: A novel digit-serial systolic array for modular multiplication. In: ISCAS 1998. Proceedings of the 1998 IEEE International Symposium on Circuits and Systems, ISCAS 1998, 31 May-3 Jun 1998, vol. 2,2, pp. 177–180 (1998)

15. Kim, H.W., Wollinger, T., Choi, Y., Chung, K.-I., Paar, C.: Hyperelliptic curve co-processors on a FPGA. In: Lim, C.H., Yung, M. (eds.) WISA 2004. LNCS, vol. 3325, pp. 360–374. Springer, Heidelberg (2005)
16. Hasan, M.A., Bhargava, V.K.: Bit-serial systolic divider and multiplier for finite fields GF(2^m). IEEE Transactions on Computers 41(8), 972–980 (1992)
17. Preneel, B., Sakiyama, K., Batina, L., Verbauwhede, I.: Superscalar coprocessor for high-speed curve-based cryptography. In: Goubin, L., Matsui, M. (eds.) CHES 2006. LNCS, vol. 4249, pp. 415–429. Springer, Heidelberg (2006)
18. Knuth, D.E.: The Art of Computer Programming, vol. 2. Addison-Wesley, Reading (1981)
19. Koblitz, N.: Elliptic Curve Cryptosystem. Math. Comp. 48, 203–209 (1987)
20. Koblitz, N.: Hyperelliptic Cryptosystems. Journal of Cryptology 1(3), 129–150 (1989)
21. Lange, T.: Inversion-free arithmetic on genus 2 hyperelliptic curves. Cryptology ePrint ARchive (2002)
22. Lange, T.: Formulae for Arithmetic on Genus 2 Hyperelliptic Curves. Applicable Algebra in Engineering, Communication and Computing 15(5), 295–328 (2005)
23. Miller, V.S.: Use of elliptic curves in cryptography. In: Williams, H.C. (ed.) CRYPTO 1985. LNCS, vol. 218, pp. 417–426. Springer, Heidelberg (1986)
24. Mulder, J.M., Quach, N.T., Flynn, M.J.: An area model for on-chip memories and its application. IEEE Journal of Solid-State Circuits 26(2), 98–106 (1991)
25. Pelzl, J.: Hyperelliptic Cryptosystems on Embedded Microprocessors. Master's thesis, Ruhr-Universitat Bochum (September 2002)
26. Rivest, R.L., Shamir, A., Adleman, L.: A Method for Obtaining Digital Signatures and Public-Key Cryptosystems. Communications of the ACM 21(2), 120–126 (1978)
27. Sakai, Y., Sakurai, K.: Design of hyperelliptic cryptosystems in small characteristic and a software implementation over F_{2^n}. In: Ohta, K., Pei, D. (eds.) ASIACRYPT 1998. LNCS, vol. 1514, pp. 80–94. Springer, Heidelberg (1998)
28. Sakiyama, K.: Secure Design Methodology and Implementation for Embedded Public-key Cryptosystems. PhD thesis, Katholieke Universiteit Leuven, Belgium (2007)
29. Shiue, W.-T.: Memory synthesis for low power ASIC design. In: ASIC 2002: Proceedings of 2002 IEEE Asia-Pacific Conference, pp. 335–342 (2002)
30. Song, L., Parhi, K.K.: Low-energy digit-serial/parallel finite field multipliers. J. VLSI Signal Process. Syst. 19(2), 149–166 (1998)
31. Wollinger, T.: Software and Hardware Implementation of Hyperelliptic Curve Cryptosystems. PhD thesis, Ruhr-University Bochum, Germany (2004)
32. Wollinger, T.: Computer Architectures for Cryptosystems Based on Hyperelliptic Curves. Master's thesis, Worcester Polytechnic Institute, Worcester, Massachusetts (May 2001)
33. Wollinger, T., Bertoni, G., Breveglieri, L., Paar, C.: Performance of HECC Coprocessors Using Inversionfree Formulae. In: International Workshop on Information Security and Hiding, Singapore (ISH 2005), May 2005, pp. 1004–1012 (2005)
34. Yan, Z., Sarwate, D.V., Liu, Z.: High-speed systolic architectures for finite field inversion. Integration, VLSI Journal 38(3), 383–398 (2005)

ECC Is Ready for RFID – A Proof in Silicon

Daniel Hein[1], Johannes Wolkerstorfer[2], and Norbert Felber[1]

[1] Swiss Federal Institute of Technology Zürich, IIS
Gloriastrasse 35, 8092 Zürich, Switzerland
Daniel.Hein@gmx.at, felber@iis.ee.ethz.ch
[2] Graz University of Technology, IAIK
Inffeldgasse 16a, 8010 Graz, Austria
Johannes.Wolkerstorfer@iaik.tugraz.at

Abstract. This paper presents the silicon chip ECCon[1], an Elliptic Curve Cryptography processor for application in Radio-Frequency Identification. The circuit is fabricated on a 180 nm CMOS technology. ECCon features small silicon size (15K GE) and has low power consumption (8.57 μW). It computes 163-bit ECC point-multiplications in 296k cycles and has an ISO 18000-3 RFID interface. ECCon's very low and nearly constant power consumption makes it the first ECC chip that can be powered passively. This major breakthrough is possible because of a radical change in hardware architecture. The ECCon datapath operates on 16-bit words, which is similar to ECC instruction-set extensions. A number of innovations on the algorithmic and on the architectural level substantially increased the efficiency of 163-bit ECC. ECCon is the first demonstration that the proof of origin via electronic signatures can be realized on RFID tags in 180 nm CMOS and below.

Keywords: Radio-Frequency Identification (RFID), Elliptic curve cryptography (ECC), Anti-Counterfeiting, Modular Multiplication.

1 Introduction

Counterfeiting is an increasing problem in the industry. Estimates assume more than 200 bn. US$ of fake products in 2005 [Org07]. In total, counterfeited products might amount up to one tenth of the total industry production. Besides legal measures, technical approaches are required to fight product piracy. This is in particular desirable for goods that affect the health of humans directly like pharmaceuticals or spare parts in aviation.

RFID (Radio-Frequency Identification) technology, which labels products with tiny chips that are powered over an air interface, is able to proof the authenticity of goods. In its simplest form, RFID tags assign a unique number to every item. This ID can be used to track goods from production to the end consumer. Central databases and online applications can verify the e-pedigree of products. The concept of e-pedigree suffers from scalability issues and assumes that unique IDs

---

[1] A portmanteaux of ECC and economical.

R. Avanzi, L. Keliher, and F. Sica (Eds.): SAC 2008, LNCS 5381, pp. 401–413, 2009.
© Springer-Verlag Berlin Heidelberg 2009

cannot be copied. The unprotected wireless communication of RFID systems in HF or UHF frequencies allows access to unique IDs without necessitating a line of sight. Also, illegitimate readers can obtain IDs from distances of 1 m (HF) to 15 m (UHF). Thus, the concept of preventing counterfeiting by using RFID technology requires cryptography to guarantee the uniqueness of a tag. A major step to bring cryptography to RFID is the landmark work of Feldhofer et al.: They showed that challenge-response authentication is useful [FDW04] and that symmetric ciphers – in particular the AES (Advanced Encryption Standard) – can be realized on RFID tags [FWR05]. AES can fulfill the hard requirements of small silicon area and minute power consumption.

The use of symmetric cryptography necessitates sophisticated key management. Its main drawback is its limitation to closed systems, where all verifiers have to be trusted and where all potential parties are known in advance. It is not realistic for open-loop systems, like global logistics and supply chain management, where this is not the case.

Asymmetric crypto, where tags possess a private key and verifiers can obtain an authentic public key, can overcome these limitations. Asymmetric crypto allows challenge-response authentication without access to an online database. Verifiers only need the public key of the device claiming a certain identity.

Although asymmetric crypto has very nice properties regarding its applicability in worldwide open-loop systems, it imposes much more effort for realization in silicon. Requirements for silicon area are roughly five times higher; power consumption also quadruples at least. Computation times are several hundred times longer than for symmetric crypto. Therefore, most asymmetric approaches for use in RFID are based on elliptic curve cryptography, which exhibits the lowest hardware requirements among standardized and secure asymmetric algorithms.

In this work, we scrutinize the requirements of passively powered RFID systems and introduce a new hardware architecture for computing ECC operations with the smallest footprint and requiring the smallest power consumption. The architecture processes ECC operations on a word length much smaller than the actual element size of 163 bits. This allows a very compact datapath, which shows excellent power characteristics: On the one hand, the proposed architecture avoids large shares of clocking power and on the other hand, it allows very flat and low power profiles which are desired for passively powered systems.

2   State of the Art

Modular multiplication is the dominating finite-field operation for computing an ECC point multiplication $k \cdot P$. Our work centers on the efficient implementation of ECC over $\mathbb{F}_{2^{163}}$ (Curve B-163 in [Nat00]). Hence, most considerations will be done for this field. Computing the point multiplication $k \cdot P$ over $\mathbb{F}_{2^{163}}$ involves 163 point doublings and additions when using Montgomery's point ladder [Mon87]. This is a widely used algorithm[Wol05, KP06, BMS+06, FW07, BBD+08] – we use it too. It is fast and has good properties to prevent side-channel analysis. In total, roughly 1000 finite-field multiplications have to be computed.

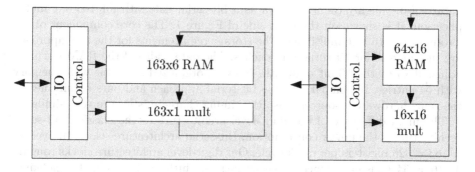

Fig. 1. Classical (bit-serial) vs. proposed architecture

The art of building ECC hardware centers on the efficient implementation of finite-field multiplication in hardware. Most ECC publications for low-resource requirements use bit-serial multipliers [Wol05, SBM$^+$06, LV07, FW07]. Bit-serial multipliers compute $c = a \cdot b$ by scheduling the operand a at full word size and multiplying it with b bit-by-bit: $c = \sum_{i=0}^{162} a \cdot b_i 2^i$.

The classical architecture when using bit-serial multipliers for ECC processors is shown on the left of Figure 1. A datapath unit *163×1 mult* computes the partial products $a \cdot b_i$ and accumulates them. Usually, modular reduction is interleaved to avoid accumulators of double size. Intermediate 163-bit values are stored in a *RAM* unit, which has to hold at least six 163-bit words during an EC point multiplication.

[SBM$^+$06] investigated the impact of the digit size on power consumption and area usage. Generally, bit-serial multipliers are more power efficient, whereas the digit-serial type is more energy efficient. Lee et al. [LV07] present an optimized architecture based on the so-called Modular Arithmetic Logic Unit [BMS$^+$06]. It uses a common-Z coordinate system for representing EC points to minimize memory requirements. Storage of intermediate values is usually the main contributor to area (roughly 66%). Area has a direct impact on the production cost of an integrated circuit.

Fürbaß et al. [FW07] departs from the pure digit-serial multiplier approach and analyzes the ramifications of using an inversion unit in conjunction with affine coordinates for operation over $\mathbb{F}_p$. Inversion is usually avoided because its computation is either much slower than multiplication or it consumes considerable silicon area. Affine coordinates reduce the memory requirements and decrease the number of cycles required to perform a point multiplication at the expense of a more complex datapath and therefore higher power consumption.

3   Architecture

The architecture presented in this article schedules both operands of the multiplication $a \cdot b$ in words of 16 bits – called digits. It thus differs totally from conventional approaches found in literature. A digit size of 16 bits leads to a datapath of roughly the same area as a bit-serial datapath for 163×1 multiplication. It is shown on the right side of Figure 1. The core component of the datapath is a 16×16 multiplier. The storage requirements for the new approach are the same as for bit-serial approaches. The data width of the RAM circuit is adjusted to fit the 16-bit datapath. RAM is organized as 16-bit wide memory with 64 entries. In comparison, the bit-serial approach and ours require nearly the same hardware resources. Also the timing characteristics are very similar.

The main difference of our digit-level architecture to the classical bit-serial architecture is the power consumption. Bit-serial architectures clock on average $2 \cdot 163 = 326$ registers per clock cycle. Our digit-level architecture clocks roughly $4 \cdot 16 = 64$ registers. Hence, it needs just one fifth of the power for clocking. The power consumption of the combinational logic in our datapath is higher. The longer combinational path causes higher signal activity. Nevertheless, the approach promises a much better power characteristic because it uses a much larger share of the power budget for computation than for clocking.

The second design choice of great impact is the restriction of ECCon to support only a single 163-bit curve over the binary extension field $\mathbb{F}_{2^{163}}$. This allows numerous optimizations, which minimize the required area, improve the running time of ECC operations and decrease the power consumption.

The selection of a binary extension field simplifies the arithmetic unit as addition is equivalent to an XOR operation. The 163-bit elements limit the size of the required memory while providing reasonable security. Support for only one finite field allows efficient modular multiplication. A tailored interleaved reduction algorithm for $\mathbb{F}_{2^{163}}$ is presented in §3.4.

3.1   Word Level Operations in $\mathbb{F}_{2^{163}}$

The small word width of the datapath necessitates splitting up finite-field operations into operations on 16-bit digits. Research with respect to this has been undertaken in the context of $\mathbb{F}_p$ [Gro02] and $\mathbb{F}_{2^m}$ [GK03] instruction-set extensions for general-purpose processors. Großschädl et al. propose a Multiply ACumulate (MAC) architecture for a word-level instruction-set extension. The ECCon processor uses the same approach: Main component of the datapath is a 16×16 MAC unit.

The point multiplication requires addition and multiplication in $\mathbb{F}_{2^{163}}$. Both necessitate to split up the 163-bit operands into 11 digits of 16 bits each. In every cycle, one digit can be fetched from memory. As soon, as two digits are available to the Arithmetic Logic Unit (ALU), either an addition, a multiplication or a MAC operation is performed. The result is then stored in an accumulator. The lower 16 bits of the accumulator can be written to memory. Thus, digit for digit a 163-bit operation is performed.

Addition is a good example for this concept. First, a digit of the input a is loaded into the ALU. Then the corresponding digit of b is added (XORed). While the next digit is fetched, the result digit is stored. This is repeated 11 times to execute one 163-bit addition. One addition takes 24 clock cycles (see tab. 1).

Table 1. Performance results

Operations	Cycles	Operations	Cycles
Addition	24	Squaring	49
Multiplication	251	Point multiplication	296,299

The two prevailing integer multiplication algorithms are operand-scanning-form and product-scanning-form multiplication (Comba multiplication). Both depend on an inner multiplication operation $a \cdot b \oplus c$, which is the reason why a MAC architecture was chosen. The ECCon processor employs the Comba algorithm. It computes the result one digit at the time which minimizes memory write cycles, at the expense of requiring an additional 16-bit adder. Furthermore, smart ordering of operand loading allows to decrease the number of memory read operations to a minimum.

A multiplication of two 163-bit elements in $\mathbb{F}_{2^{163}}$ would produce a 325-bit result. Modular reduction $a \cdot b = c \equiv d \bmod f(z)$ reduces the product to a 163-bit representation, where $f(z)$ is the irreducible polynomial. Two possible options for reduction exist. The first, interleaved reduction, performs the reduction during multiplication. The alternative is to first compute the 325-bit product and then reduce it. This has the severe drawback that it requires 162 bits of additional storage. An efficient algorithm capable of implementing the interleaved reduction on the 16-bit ALU will be detailed in §3.4.

In conformance with the state-of-the-art implementations, ECCon employs Fermat's little theorem to realize inversion. It requires nine multiplications and 162 square operations. This is not fast enough to allow using affine coordinates, therefore a projective version of the point multiplication requiring only one inversion is applied.

3.2   Choice of Word Size

The choice of a width of 16 bits for the datapath is an outcome of assessing various cost metrics for different datapath sizes. Cost is determined by the area usage, the power consumption and the clock cycles for a point multiplication. The critical path is not considered because the circuit runs well below the maximum clock frequency.

Figure 2 depicts a comparison of the ALU architecture (see also §3.3), synthesized for different bit-widths. The four graphs represent different variants of area, cycle and power products $(A(rea) \cdot C(ycles) \cdot P(ower))$. The values are computed from normalized results to allow a comparison over the different metrics. Bit-widths below 8 and above 31 were not considered because they are either far too slow (> 1 mio. cycles) or too large circuits (> 5500 gate equivalents).

On a first glance, no obvious bit-width presents itself. Considering the $A \cdot C \cdot P$ and $A \cdot C \cdot P^2$ products, starting with the transition between 15 and 16 bits, the values start to increase steeply. The $A \cdot C^2 \cdot P$ and $A \cdot C^2 \cdot P^2$ on the other hand seem to have a local minimum point at 15 bit. As the actual area, cycle and power values for a 15-bit datapath lie well within the aspired limits, this

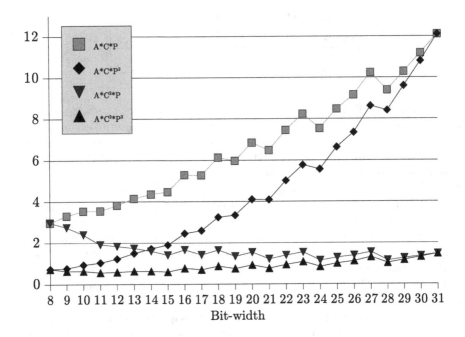

Fig. 2. Datapath bit-width comparison

bit-width was deemed optimal. Due to the fact that the RFID protocol is byte oriented and the difficulties in converting 15-bit digits to this interface, a 16-bit datapath was chosen for implementation.

3.3   MAC Unit for Interleaved Reduction

Figure 3 illustrates the top level architecture of the ALU of the ECCon processor. **B**, **RC**, $\mathbf{ACC_L}$ and $\mathbf{ACC_H}$ denominate registers. $\mathbf{ACC_L}$ and $\mathbf{ACC_H}$ compose the accumulator **ACC** which also supports a shift right-by-8 and shift right-by-16 operation. **MC**, the multiplication carry register, has the capability to select between two different inputs.

The multiplier unit $\otimes$ computes either the product of the input I times the factor register **B** or $\mathbf{ACC_L}$ times the reduction polynomial $r(z)$ (cf. 3.4). The multiplier unit uses a 16×16 polynomial array multiplier consisting of 256 AND gates and a $\mathbb{F}_2$ tree adder built from 240 XOR gates.

The select-and-add unit **SAA** contains two 16-bit $\mathbb{F}_2$ adders, implemented by two 16-bit XOR gate arrays. The input select of the higher adder (Output **DH**) allows to choose between $\mathbf{ACC_H}$ and **MC** for its first input and MulH and I for its second input. Both inputs are maskable. This is implemented with AND gates. The first input of the lower adder (Output **DL**) can be chosen from **RC**, MulL and I while the second input may be selected from $\mathbf{ACC_H}$, **MC** and $\mathbf{ACC_L}$. Again, both inputs to the adder are maskable.

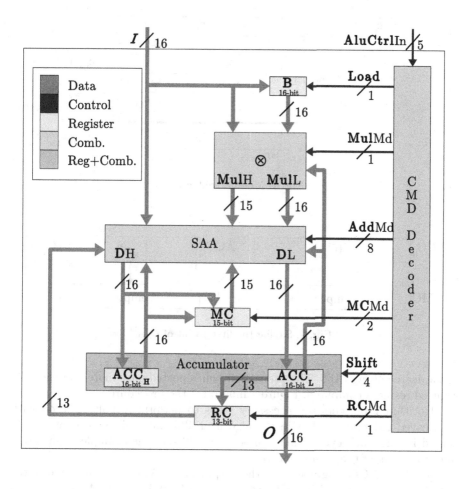

Fig. 3. ALU top level architecture

3.4 Multiplication with Interleaved Reduction

The Comba multiplication algorithm is utilized to perform the multiplication. To execute the modular reduction, the following idea is applied. We split the irreducible polynomial $f(z) = z^{163} + z^7 + z^6 + z^3 + 1$ as $z^{163} + r(z)$ and call $r(z) = z^7 + z^6 + z^3 + 1$ the reduction polynomial. The product polynomial $c(z) = a(z) \cdot b(z) | \deg\{c(z)\} \leq 2m - 2$ is thus congruent to

$$c(z) = c_{2m-2}z^{2m-2} + \cdots + c_m z^m + c_{m-1}z^{m-1} + \cdots + c_1 z + c_0$$
$$\equiv (c_{2m-2}z^{2m-2} + \cdots + c_m)r(z) + c_{m-1}z^{m-1} + \cdots + c_1 z + c_0 \ (mod \ f(z)).$$

So, by multiplying $r(z)$ and $c_H = (c_{324}z^{161} + \cdots + c_{163})$, and adding the result to $c_L = c_{162}z^{162} + \cdots + c_1 z + c_0$, a new temporary output is derived which only

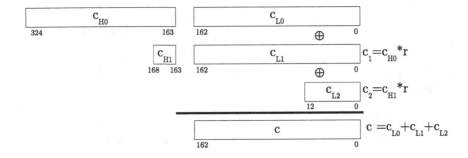

Fig. 4. Modular reduction

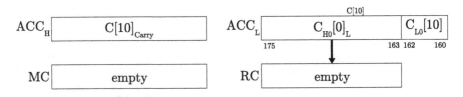

Fig. 5. Saving the first 13-bit of C_{H^0}

has a degree of $m = 168$. The process is then repeated once more and the fully reduced result is computed. Figure 4 illustrates the procedure.

Algorithm 1 presents the necessary steps to perform a multiplication with interleaved reduction. A line in the algorithm illustrates all the operations executed in one clock cycle. Line number five will serve as example to facilitate understanding of the nomenclature.

$T[i-1] \leftarrow \mathbf{ACC_L}$ signifies that the output of the ALU is to be stored in the $(i-1)^{th}$ digit of memory element T. $\mathbf{ACC} \leftarrow i - \mathcal{INDEX}(i,j) \cdot \mathbf{B} \oplus \mathbf{ACC_H}$ denotes that the result of the MAC operation is stored in the accumulator. In this case the ALU input is $B[i - \mathcal{INDEX}(i,j)]$. The function $\mathcal{INDEX}(i,j)$ computes the optimal loading sequence for the input operands, thus minimizing memory read operations. $\mathcal{INDEX}(i,j)$ is implemented by a look-up table.

For a multiplication $c = a \cdot b$, $a, b \in \mathbb{F}_{2^{163}}$, the 11th partial product computed with the product scanning form contains the polynomial $c_{175}z^{175} + \ldots + c_{160}z^{160}$. In accordance with the nomenclature used in Figure 4 the polynomial $c_{162}z^{162} + \ldots + c_{160}z^{160}$ is the last digit of C_L^0 ($C_L^0[10]$), while $c_{175}z^{175} + \ldots + c_{163}z^{163}$ is the first part of C_{H^0}.

The lower three bits representing $C_L^0[10]$ are stored in the last digit of the memory element that will contain the final product. The lower part of C_{H^0} is saved in temporary variable **RC**. This is illustrated in Figure 5.

The next partial product ($C[11]$) that is ascertained will again be split at the 3-bit boundary. This time the higher 13 bits ($c_{191}z^{191} + \ldots + c_{176}z^{176}$) are written

---

Algorithm 1. Comba multiplication with interleaved reduction

Input: $a, b \in \mathbb{F}_{2^m}$
Output: $c = a \cdot b \bmod f, c \in \mathbb{F}_{2^m}$

1 $\mathbf{B} \leftarrow A[0]$;
2 $\mathbf{ACC} \leftarrow B[0] \cdot \mathbf{B} \oplus \mathbf{ACC}$;
3 $i \leftarrow 0, j \leftarrow 0$;
4 **for** $i \leftarrow 1$ **to** $t - 1$ **do**
5 $\quad$ $T[i-1] \leftarrow \mathbf{ACC_L}; \mathbf{ACC} \leftarrow B[i - \mathcal{INDEX}(i,j)] \cdot \mathbf{B} \oplus \mathbf{ACC_H}$;
6 $\quad$ **for** $j \leftarrow 1$ **to** i **do**
7 $\quad\quad$ $\mathbf{B} \leftarrow A[\mathcal{INDEX}(i,j)]$;
8 $\quad\quad$ $\mathbf{ACC} \leftarrow B[i - \mathcal{INDEX}(i,j)] \cdot \mathbf{B} \oplus \mathbf{ACC}$;

9 $T[t-1] \leftarrow \mathbf{ACC_L}; \mathbf{ACC} \leftarrow \mathbf{ACC} \gg 16; \mathbf{MC} \leftarrow 0; \mathbf{RC} \leftarrow \mathbf{ACC_L} \gg S_R$;
10 $i \leftarrow 0, j \leftarrow 0$;
11 $\mathbf{B} \leftarrow A[\mathcal{INDEX}(i,j)]$;
12 **for** $i \leftarrow 0$ **to** $t - 2$ **do**
13 $\quad$ **if** $i \neq 0$ **then**
14 $\quad\quad$ $C[i-1] \leftarrow \mathbf{ACC_L}; \mathbf{ACC} \leftarrow B[(t+i) - \mathcal{INDEX}(i,j)] \cdot \mathbf{B} \oplus \mathbf{MC}; \mathbf{MC} \leftarrow \mathbf{ACC_H}$;
15 $\quad$ **else**
16 $\quad\quad$ $\mathbf{ACC} \leftarrow B[(t+1) - \mathcal{INDEX}(i,j)] \cdot \mathbf{B} \oplus \mathbf{ACC}$;
17 $\quad$ **for** $j \leftarrow 1$ **to** $(t-2) - i$ **do**
18 $\quad\quad$ $\mathbf{B} \leftarrow A[\mathcal{INDEX}(i,j)]$;
19 $\quad\quad$ $\mathbf{ACC} \leftarrow B[(t+1) - \mathcal{INDEX}(i,j)] \cdot \mathbf{B} \oplus \mathbf{ACC}$;
20 $\quad$ $\mathbf{ACC_H} \leftarrow T[i] \oplus \mathbf{MC}; \mathbf{ACC_L} \leftarrow \{(\mathbf{ACC_L} \ll S_L) | \mathbf{RC}\}; \mathbf{MC} \leftarrow \mathbf{ACC_H}; \mathbf{RC} \leftarrow \mathbf{ACC_L} \gg S_R$;
21 $\quad$ $\mathbf{ACC} \leftarrow \mathbf{ACC_L} \cdot r(z) \oplus \mathbf{ACC_H}$;

22 $C[t-2] \leftarrow \mathbf{ACC_L}; \mathbf{ACC_H} \leftarrow 0; \mathbf{ACC_L} \leftarrow \mathbf{MC}; \mathbf{MC} \leftarrow \mathbf{ACC_H}$;
23 $\mathbf{ACC_H} \leftarrow 0 \oplus \mathbf{MC}; \mathbf{ACC_L} \leftarrow \{(\mathbf{ACC_L} \ll S_L) | \mathbf{RC}\}; \mathbf{MC} \leftarrow \mathbf{ACC_H}; \mathbf{RC} \leftarrow \mathbf{ACC_L} \gg S_R$;
24 $\mathbf{ACC} \leftarrow \mathbf{ACC_L} \cdot r(z) \oplus \mathbf{ACC_H}$;
25 $\mathbf{ACC_H} \leftarrow \mathbf{MC}; \mathbf{ACC_L} \leftarrow T[i] \oplus \mathbf{ACC_L}; \mathbf{MC} \leftarrow \mathbf{ACC_H}; \mathbf{RC} \leftarrow \mathbf{ACC} \gg S_R$;
26 $C[i] \leftarrow \mathbf{ACC_L}; \mathbf{ACC_H} \leftarrow C[0]; \mathbf{ACC_L} \leftarrow \mathbf{RC}$;
27 $\mathbf{ACC} \leftarrow \mathbf{ACC_L} \cdot r(z) \oplus \mathbf{ACC_H}$;
28 $C[0] \leftarrow \mathbf{ACC_L}; \mathbf{ACC} \leftarrow \mathbf{ACC} \gg 16$;

---

to a second temporary space **MC**. The lower three bits are then combined with the 13 bits carried over from the last round which are currently stored in **RC**. Together they constitute the 16-bit digit $c_{178}z^{178} + \ldots + c_{163}z^{163}$, the first digit of C_H^0. This digit is restored in the accumulator (cf. Figure 6).

In the next step it is multiplied with $r(z)$. The lower 16 bit of the result in $\mathbf{ACC_L}$ are added to $C_L^0[0]$. The carry of the multiplication in $\mathbf{ACC_H}$ is than swapped with the content of **MC**. Thus, **MC** alternately stores the carry of the normal multiplication and the one of the reduction multiplication. Figure 7 corresponds to this state. Then the next partial product ($C[12]$) is computed, and the process is repeated. Thus, the interleaved multiplication and reduction

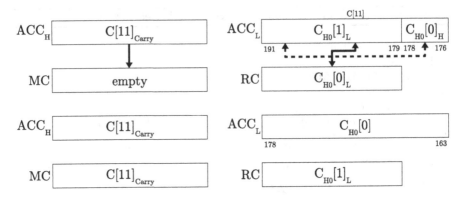

Fig. 6. Restoring C_{H^0} and saving the multiplication carry

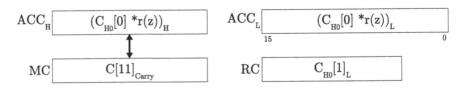

Fig. 7. Result of the first reduction multiplication

operation is performed step by step as the partial products of C_{H^0} and after that C_{H^1} become available.

4   Results and Comparison

The ECCon processor was fabricated using the UMC L180 GII 1P/6M 1.8V/3.3V CMOS technology. The ECC core consists of a 7×163-bit memory, the ALU introduced in §3 and a control unit that is capable of performing an EC point multiplication. It employs a fully registered, two phase handshake enabled 8-bit bus interface to connect to an ISO 18000-3-1 [ISO04] compatible RFID front-end.

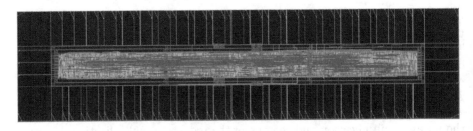

Fig. 8. The tape-out ready layout of the ECCon processor

Table 2. Synthesis results of the ECC core

Technology	Area $[\mu m^2]$	Area [GE]	Max. Frequency [MHz]	Power @ 106 KHz $[\mu W]$	Voltage [V]
UMC L180	128,098	13,250	46	8.57	1.8V
AMS C35	727,090	13,320	20	54.7	2.5V

Table 3. Area and cycle comparison

	Area [GE]	Runtime [Cycles]	Field	Bits	Technology
[BBD+08]	12,876	80K	$F_{2^{163}}$	163	INF 220nm
ECCon w/o k	11,904	296K	$F_{2^{163}}$	163	UMC 180nm
[LV07]	13,182	314K	$F_{2^{163}}$	163	TSMC 180nm
[KP06]	15,094	430K	$F_{2^{163}}$	163	AMI 350nm
[Wol05]	23,000	426K	$F_{2^{191}}$	191	AMS 350nm
[FW07]	23,656	502K	$F_{p_{192}}$	192	AMS 350nm
[OScE04]	30,333	545K	$F_{P_{(2^{167}+1)/3}}$	165	TSMC 130nm

The integrated RFID front-end is purely digital in nature and lacks an analog air interface.

The chip has a core area of 219,897 μm^2 with a utilization of approximately 70%. The minimum core area amounts to 151,126 μm^2 or 15,630 GE. An overhead of 654 GE is incurred by circuits for production testing, which reduces the core area required for the RFID front-end and the ECC processor to 14,976 GE. The circuit achieves a maximum frequency of 46 MHz and the ECC core has a power consumption of 8.57 μW at 106 kHz. This frequency was chosen because it is $\frac{1}{128}$ of the 13.56 MHz carrier signal used by the ISO-18000-3-1 RFID standard.

Table 2 presents the synthesis results of the ECC core without the RFID front-end and the test hardware. The power consumption value of the UMC 180 nm variant was obtained by measuring the fabricated IC.

ECCon was also synthesized, placed and routed in the AMS c35b4 0.35 μm CMOS technology. A power simulation with Synopsys© NanoSim was performed on the placed & routed layout. The simulated mean current, the determinative factor for an RFID application is 21.88 μA at 2.5V. During a whole EC point multiplication it varies only by 10%. This is important because high fluctuations disrupt the communication channel.

Table 3 tries to compare the ECCon processor to different architectures. The area for the ECCon processor in this table includes only a 6 × 163 bits memory. This was done to allow a fair comparison, because most related work does not include storage for the key k. [LV07], [KP06] and [OScE04] are capable to perform the point multiplication, whereas [Wol05] and [FW07] implement all operations of the ECDSA standard, except hashing and random number generation. Bock et al. ([BBD+08]) implement an ISO-18000-3-1 compatible RFID tag and a Diffie-Hellman key exchange based authentication protocol employing a digit-serial multiplier.

Table 4. Power consumption comparison

	Power [μW]	I_{mean} [μA]	f [kHz]	Tech.	Obtained by
ECCon	54.7	21.88	106	AMS350	NanoSim power simulation
[BBD+08]	96.4	n.a.	847.5	INF 220	PowerTheater power simulation
ECCon	67.23	37.35	847.5	UMC180	measured
[FW07]	141	42.73	106	AMS350	NanoSim power simulation
[BMS+06]	20 - 30	n.a.	500	CMOS130	estimated, MALU only
[OScE04]	990	n.a.	20000	TSMC130	plain synthesis results assumed

Table 4 compares the power consumption of different ECC implementations. It is important to grade the quality of these values. In this table measurement and power simulation are the most accurate, followed by synthesis results. Estimations are the most inaccurate. The power consumption of [BBD+08] is similar to that of ECCon due to their efficient latch based memory unit. They dissipate 88% of the total power in the ALU, whereas the ECCon ALU requires only 48%. Their ALU is also approximately three and a halve times larger.

5   Conclusions

This article presented a fully functional implementation of ECC over $\mathbb{F}_{2^{163}}$ on a 180 nm CMOS process. The so-called ECCon processor has a footprint of 15k GE, which is the smallest reported ECC solution so far, considering that this number also includes overhead for place-and-route, a hardwired state machine for controlling the roughly 300,000 clock cycles of one ECC operation, and a digital interface for communicating via the ISO-18000-3-1 standard. The power consumption obtained by measurement is 8.57 μW.

The ECCon processor is one of the first complete hardware solutions that can compute ECC point multiplications while fulfilling the harsh requirements of passively powered RFID tags. It will help to realize a secure Internet of things, where goods have an electronic identity that can be proved.

Acknowledgements

The authors would like to thank the anonymous reviewers for their valuable comments on an earlier version of this paper. For their help in fabricating the ECCon processor the authors also thank the staff of the Microelectronics Design Center at the Department of Information Technology and Electrical Engineering of the Swiss Federal Institute of Technology Zürich.

This research has been partially supported by the European Union Information Society Technologies (EU IST) FP6 project Collaboration @ Rural (IST-FP6-34921).

References

[BBD+08]   Bock, H., Braun, M., Dichtl, M., Hess, E., Heyszl, J., Kargl, W., Koroschetz, H., Meyer, B., Seuschek, H.: A milestone towards RFID products offering asymmetric authentication based on elliptic curve cryptography. In: Workshop on RFID Security, RFIDsec (2008)

[BMS+06]   Batina, L., Mentens, N., Sakiyama, K., Preneel, B., Verbauwhede, I.: Low-cost elliptic curve cryptography for wireless sensor networks. In: Proceedings of Third European Workshop on Security and Privacy in Ad hoc and Sensor Networks (2006)

[FDW04]   Feldhofer, M., Dominikus, S., Wolkerstorfer, J.: Strong authentication for RFID systems using the AES algorithm. In: Joye, M., Quisquater, J.-J. (eds.) CHES 2004. LNCS, vol. 3156, pp. 357–370. Springer, Heidelberg (2004)

[FW07]   Fürbass, F., Wolkerstorfer, J.: ECC processor with low die size for RFID applications. In: Proc. IEEE International Symposium on Circuits and Systems, ISCAS (2007)

[FWR05]   Feldhofer, M., Wolkerstorfer, J., Rijmen, V.: AES implementation on a grain of sand. IEEE Proceedings on Information Security 152, 13–20 (2005)

[GK03]   Großschädl, J., Kamendje, G.-A.: Instruction set extension for fast elliptic curve cryptography over binary finite fields GF(2^m). In: Proc. IEEE International Conference on Application-Specific Systems, Architectures, and Processors (2003)

[Gro02]   Großschädl, J.: Instruction set extension for long integer modulo arithmetic on RISC-based smart cards. In: Proc. 14th Symposium on Computer Architecture and High Performance Computing (2002)

[ISO04]   ISO/IEC/JTC 1/SC 31. ISO/IEC 18000 Information technology – Radio frequency identification for item management. Technical report, International Organization for Standardization, Geneva, Switzerland (2004)

[KP06]   Kumar, S., Paar, C.: Are standards compliant elliptic curve cryptosystems feasible on RFID? Printed handout of Workshop on RFID Security (RFIDSec) (July 2006)

[LV07]   Lee, Y.K., Verbauwhede, I.: A compact architecture for montgomery elliptic curve scalar multiplication processor. In: Kim, S., Yung, M., Lee, H.-W. (eds.) WISA 2007. LNCS, vol. 4867, pp. 115–127. Springer, Heidelberg (2008)

[Mon87]   Montgomery, P.L.: Speeding the Pollard and elliptic curve methods of factorization. Mathematics of Computation 48, 243–264 (1987)

[Nat00]   National Institute for Standards and Technology. Digital Signature Standard (DSS). Technical report, NIST (January 2000)

[Org07]   Organisation for Economic Co-operation and Development. The economic impact of counterfeiting and piracy. Technical report, OECD (2007)

[OScE04]   Öztürk, E., Sunar, B., Savaş, E.: Low-power elliptic curve cryptography using scaled modular arithmetic. In: Joye, M., Quisquater, J.-J. (eds.) CHES 2004. LNCS, vol. 3156, pp. 92–106. Springer, Heidelberg (2004)

[SBM+06]   Sakiyama, K., Batina, L., Mentens, N., Preneel, B., Verbauwhede, I.: Small-footprint ALU for public-key processors for pervasive security. In: Workshop on RFID Security (2006)

[Wol05]   Wolkerstorfer, J.: Is elliptic-curve cryptography suitable to secure RFID tags? In: Workshop on RFID and Lightweight Crypto (2005)

Cryptanalysis of a Generic Class of White-Box Implementations

Wil Michiels, Paul Gorissen, and Henk D.L. Hollmann

Philips Research Laboratories, High Tech Campus 34, Eindhoven, The Netherlands
wil.michiels@philips.com

Abstract. A white-box implementation of a block cipher is a software implementation from which it is difficult for an attacker to extract the cryptographic key. Chow et al. published white-box implementations for AES and DES. These implementations are based on ideas that can be used to derive white-box implementations for other block ciphers as well. In particular, the ideas can be used to derive a white-box implementation for any substitution linear-transformation (SLT) cipher. Although the white-box implementations of AES and DES have been cryptanalyzed, the cryptanalyses published use typical properties of AES and DES. It is therefore an open question whether an SLT cipher exists for which the techniques of Chow et al. result in a secure white-box implementation. In this paper we largely settle this question by presenting an algorithm that is able to extract the key from such an implementation under a mild condition on the diffusion matrix. The condition is, for instance, satisfied by all MDS matrices. Our result can serve as a basis to design block ciphers and to develop white-box techniques that result in secure white-box implementations.

Keywords: white-box cryptography, AES, Serpent, cryptanalysis, substitution linear-transformation network, MDS matrix.

1 Introduction

The classical 'black-box' attack model used for symmetric block ciphers assumes that an attacker can at most mount chosen text attacks on the implementation. An attacker is assumed to have no access to the execution of the implementation. In practice, this model is often not realistic. Consider, for instance, a content provider who sends encrypted data to a PC platform. Then the owner of this PC may benefit from illegally distributing the key for decrypting the data to other users. In this case, it is more realistic to consider the severe 'white-box attack model' in which an attacker is assumed to have full access to and full control over the implementation of a cryptographic algorithm.

White-box cryptography is the discipline that aims at solving the problem of how to implement a cryptographic algorithm in software, such that the key cannot be extracted by a white-box attack. A software implementation of a cryptographic algorithm that has the objective to resist a white-box attack on

R. Avanzi, L. Keliher, and F. Sica (Eds.): SAC 2008, LNCS 5381, pp. 414–428, 2009.

its key is called a white-box implementation. Chow et al. present white-box implementations for the block ciphers AES and DES [4,5]. These white-box implementations are based on ideas that naturally extend to any substitution-linear transformation cipher, as defined below.

Definition 1 (Substitution-Linear   Transformation   Cipher   (SLT cipher)). *A cipher is called an SLT cipher if it can be specified as follows. It consists of R rounds for an $R \geq 1$. A single round r is a bijective function $\mathcal{F}_{SLT}^{(r)}(x_1, x_2, \ldots, x_s)$ on $GF(2)^n$ with $x_i \in GF(2)^m$ and $n = m \cdot s$. This function consists of the following operations. It starts with xoring an n-bit round key $k^{(r)} = (k_1^{(r)}, k_2^{(r)}, \ldots, k_s^{(r)})$ to its input. That is, the value $y_i = k_i^{(r)} \oplus x_i$ is computed. Next, the round computes $z_i = S_i^{(r)}(y_i)$ for all y_i, where the (non-linear) invertible S-boxes $S_1^{(r)}, S_2^{(r)}, \ldots, S_s^{(r)}$ are part of the cipher specification and thus key-independent. These two steps realize confusion. The diffusion is realized by multiplying the outcome $z = (z_1, z_2, \ldots, z_s) \in GF(2)^n$ of the S-boxes with an $n \times n$ invertible matrix $M^{(r)}$ over $GF(2)$. This diffusion matrix is also part of the cipher specification.*

In our notation we will often omit the index r denoting the round when this value is clear from the context. White-box attacks have been published for extracting the 128-bit AES key and the 56-bit DES key from the white-box AES and DES implementations of Chow et al. [2,7,8,9,12]. The attacks use typical properties of AES and DES and do not apply to white-box implementations of other block ciphers. Hence, it remains an open question whether the white-box techniques proposed by Chow et al. can result in a secure white-box implementation for other SLT ciphers than AES, such as, for instance, Serpent [1]. In this paper we present an algorithm that, under a mild condition on the diffusion matrix, can extract the round keys from the white-box implementation of any SLT cipher. If the key scheduling algorithm is invertible, then having these round keys suffices to also derive the main key. Otherwise, we at least have a compact description of the main key. Although the time complexity of the algorithm depends on the choice of the S-boxes and the diffusion matrices, we were only able to find impractically large time complexities for unrealistic choices of these operators, e.g. linear S-boxes or diffusion matrices that are close to the identity matrix. To demonstrate the effectiveness of the proposed algorithm, we show in this paper that the algorithm can be applied successfully to AES and Serpent.

The remainder of this paper is organized as follows. In Section 2 we give a precise formulation of the result that is proved in this paper. Essential in this formulation is a specification of the information that is available to an attacker in a white-box implementation. In Sections 3-6 we present our cryptanalysis and in Section 7 we show how our ideas can be used to extract the round keys from a white-box AES and a white-box Serpent implementation. We end with a conclusion in Section 8.

2   Problem Formulation and Notation

In order to discuss the attack of a white-box implementation of an SLT cipher, we have to specify what kind of information we can obtain from such an implementation by a white-box attack. To answer this question, we briefly discuss how Chow et al. derive a white-box implementation of a block cipher.

First, they derive an implementation that, in each round of the block cipher, only performs a sequence of table lookups. The input to a lookup table is either the input to the round or it is obtained by concatenating the outputs of one or more other lookup tables. Such an implementation can be modeled by a network of lookup tables, where an arc from table T to T' means that (part of) the output of table T is used as (part of the) input to table T'.

In the design of the white-box implementation, they next obfuscate the lookup tables by encoding their inputs and outputs. Encoding the input and output of a table T with bijective functions f_{in} and f_{out}, respectively, corresponds to replacing table T by $f_{out} \circ T \circ f_{in}^{-1}$. Hence, we incorporate in T an input *decoding* and an output *encoding*. To see that such encodings realizes obfuscation, observe that encoding the input of a lookup table changes the order of its rows and that encoding the output changes the value of the rows.

The lookup tables are encoded in such a way that the functionality of the entire implementation does not change. Chow et al. show how this can be done with a combination of linear and a non-linear encodings. The non-linear encodings are applied as follows. The first tables in the network do not get an input decoding and the last ones do not get an output encoding. Furthermore, we choose the input encoding of a table, such that it matches the encoding that has been put on its input data by the tables that directly precede it in the network. To illustrate this, suppose that the entire output of a table T serves as the entire input of another table T'. We then encode the output of T by a randomly chosen encoding f and we decode the input of table T' accordingly, that is, as input decoding of table T' we employ f^{-1}. The result is that the output encoding of T and the input decoding of T' cancel out. This concludes the strategy employed by Chow et al. to add non-linear encodings. For the strategy to add additional linear encodings we refer to [4,5]. We note, however, that our cryptanalysis applies to the case that both the linear and non-linear encodings are applied.

Let $(x_1, x_2, \ldots, x_s)$ be the input to a round r of an SLT cipher, where x_i is the m-bit input to S-box S_i. Then, before applying the encodings, the network of lookup tables has the property that each word x_i is input to some lookup table T_i. For details, we refer to [4,5]. For the white-box implementation obtained after applying the (linear and non-linear) encodings, this has as a consequence that an attacker who has access to the inputs of all tables in the implementation, which holds in a white-box attack, has access to the encoded version $f_i^{(r)}(x_i)$ of each value x_i. Here $f_i^{(r)}$ is a secret bijective function that is used as input encoding for T_i. Furthermore, having access to $f_i^{(r)}(x_i)$ means that we can determine this value as well as set it to any given other value. This, for instance, implies that the attacker can choose the encoded input to a round r and next observe the effect of this for the input to a later round.

In our cryptanalysis we assume that for each value $f_i^{(r)}(x_i)$ an attacker knows the index i and the round r that are associated to this value. The index r can be derived by inspecting an execution of the white-box implementation. With respect to the value i we can in general at least limit the number of candidate values to a number that is feasible for performing an exhaustive search. We can do this by using the definition of the diffusion matrices and S-boxes and by generalizing parts of the cryptanalysis. For the sake of readability and as the difficulty of finding the value i is not considered to be the essence of the strength of a white-box implementation, we do not further discuss this problem in this paper and assume the value to be known.

This brings us at the following property, which specifies the information that is available to an attacker who tries to extract the round keys from a white-box implementation that is based on the techniques of Chow et al.

Property 1. In a white-box attack, an attacker has for each round r and for each m-bit input word x_i of round r access to the encoded version $f_i^{(r)}(x_i)$ of x_i. The attacker also knows the values of r and i that are associated with $f_i^{(r)}(x_i)$. The attacker does not know the value of the m-bit input word x_i nor the definition of the function $f_i^{(r)}$, which is an arbitrary m-bit bijective function.   □

In order to formulate the main result of this paper, we need the following definitions. We note that in this paper all matrices are over $GF(2)$.

Definition 2. *If N is an $n \times n$ matrix with $n = m \cdot s$, then we consider N to be partitioned into s vertical strips of size $n \times m$. We denote the jth strip by N_j That is,*

$$(N_j)_{x,y} = N_{x,(j-1)\cdot m+y} \ ,$$

where the rows and columns have indices in $\{1, 2, \ldots, n\}$.

Furthermore, we will consider each strip N_j to be partitioned further into s blocks $N_{i,j}$ of size $m \times m$. That is,

$$(N_{i,j})_{x,y} = N_{(i-1)m+x,(j-1)m+y},$$

so that

$$N = (N_1 \cdots N_s) = \begin{pmatrix} N_{1,1} & N_{1,2} & \ldots & N_{1,s} \\ N_{2,1} & N_{2,2} & \ldots & N_{2,s} \\ \vdots & \vdots & \vdots & \vdots \\ N_{s,1} & N_{s,2} & \ldots & N_{s,s} \end{pmatrix}.$$

We will refer to the ith row

$$N(i) = (N_{i,1} \ N_{i,2} \ \ldots \ N_{i,s})$$

of blocks of N as the ith block row of N.

Definition 3. *Let N be an $n \times n$ matrix with $n = m \cdot s$. We say that a subset $U \subseteq \{1, 2, \ldots, s\}$ is a spanning block set for block row i if the collection of all the m-bit columns from the blocks $N_{i,j}$ with $j \in U$ spans $\mathrm{GF}(2)^m$.*

If two subsets $U, V \subseteq \{1, 2, \ldots, s\}$ with $U \cap V = \emptyset$ both represent spanning block sets for block row i, then we say that block row i has disjoint *spanning block sets.*

MDS (Maximum Distance Separable) matrices are often used as diffusion matrix in block ciphers because of their good diffusion properties [6,10,11]. In an MDS diffusion matrix N each block $N_{i,j}$ defines the multiplication with a non-zero element in $\mathrm{GF}(2^m)$. Hence, each block $N_{i,j}$ is non-singular, which implies that any pair of blocks from a block row i defines a spanning block set. This means that the main result of this paper, which is stated below, covers the class of SLT ciphers in which the diffusion is realized by MDS matrices.

Main Result. *Consider an SLT cipher for which the diffusion matrices have the property that all their block rows have disjoint spanning block sets. Then, given a white box implementation for this cipher that satisfies Property 1, we present an algorithm for extracting the round key of any round r with $1 < r < R$.*

The result above does not cover the first and last round of the cipher. This has the following reason. In order not to change the functionality of the white-box implementation, the input of the first round and the output of the last round cannot be encoded. However, by omitting these external encodings the white-box implementations of the first and last round become less secure. As as solution to this problem, Chow et al. propose to add the external encodings and to either undo these encodings elsewhere in the software or to include these encodings in the definition of the block cipher that is implemented. In both cases it will not only be the goal of an attacker to derive the round keys of the first and last round, but also to determine the external encodings. To simplify the discussion we exclude the attack of these rounds in this paper. We note, however, that these rounds can also be attacked. The attack is based on the following result. By applying our cryptanalysis, an attacker can determine the output encoding of the first round and the input encoding of the last round. This gives the attacker the plain output of the first round and the plain input to the last round. Using this, the first and last round can be attacked.

We end this section with the description of some notational conventions used throughout this paper.

- By abuse of notation, if N is a matrix, then the map $x \mapsto Nx$ corresponding to a matrix multiplication by N will also denoted by N.
- If T denotes a lookup table, then, by abuse of notation, we also write T to denote the function that it defines.
- We define $\oplus_c$ as the map $\oplus_c(x) = x \oplus c$. Using this, we can write the key addition of an SLT cipher as $\oplus_{k^{(r)}}$.

– Let $g_1, g_2, \ldots, g_s$ be maps on m-bit vectors, and let $n = ms$. The map $g = (g_1, \ldots, g_s)$ defined by

$$g(x) = (g_1(x_1), g_2(x_2), \ldots, g_s(x_s))$$

for each n-bit vector $x = (x_1, x_2, \ldots, x_s) \in \mathrm{GF}(2)^n$ with $x_i \in \mathrm{GF}(2)^m$ is called the *diagonal map with components* $g_1, \ldots, g_s$. When considering a diagonal map h, we will always assume that the components are maps h_i; conversely, given maps $h_1, \ldots, h_s$, we will denote the diagonal map with components h_i by h.

Remark 1. Note that if c is a vector and N a matrix, then the addition map $\oplus_c$ is always diagonal and the matrix map N is diagonal if and only if N is a *block diagonal* matrix. Here, N is called a *block diagonal* matrix if all off-diagonal blocks $N_{i,j}$, $i \neq j$, are zero. More general, it is easily verified that an affine map $\alpha : x \mapsto a \oplus Ax$ is a diagonal map if and only if A is a block diagonal matrix. Note also that the ith component of the diagonal map $x \mapsto Nx$ associated with a block diagonal matrix N is just the diagonal block $N_{i,i}$ of N.

As an example of the above conventions, we can now write the function $\mathcal{F}_{\mathrm{SLT}}^{(r)}$ describing the rth round of an SLT cipher in Definition 1 as

$$\mathcal{F}_{\mathrm{SLT}}^{(r)} = M^{(r)} \circ S^{(r)} \circ \oplus_{k^{(r)}}, \qquad (1)$$

where $M^{(r)}$ is the diffusion matrix, $S^{(r)}$ is the S-box diagonal map with as components the S-boxes $S_i^{(r)}$, and $\oplus_{k^{(r)}}$ the round-key addition map $x \mapsto x \oplus k^{(r)}$, a diagonal map with as components the maps $x_i \mapsto x_i \oplus k_i^{(r)}$.

3 Determination of the Encodings Up to an Affine Part

According to Property 1, an attacker has access to the encoded version $\tilde{x}_i = f_i^{(r)}(x_i)$ of each input word x_i of a round r, where $f_i^{(r)}$ is an unknown bijective function. In the first step of our cryptanalysis, we will show how to determine the encodings up to an affine part.

Consider a fixed round r of the white-box implementation with $1 \leq r < R$, and a block row i of the diffusion matrix M. Let sets $U = \{u_1, u_2, \ldots, u_l\}$ and $V = \{v_1, v_2, \ldots, v_{l'}\}$ be two disjoint spanning block sets for block row i of M. Without loss of generality we may assume that $U \cup V = \{1, 2, \ldots, s\}$, i.e., $l' = s - l$. This partitions the s input words of a round input into two parts: words that are input to an S-box S_i with $i \in U$ and words that are input to an S-box S_i with $i \in V$. We write $\tilde{x}'$ as the vector containing all l input words $\tilde{x}_i$ with $i \in U$ and we write $\tilde{x}''$ as the vector containing all l' input words $\tilde{x}_i$ with $i \in V$. Then the ith output word $\tilde{z}_i$ of this round r is given by $\tilde{z}_i = h(\tilde{x}', \tilde{x}'')$, where

$$h(\tilde{x}', \tilde{x}'') = f_i^{(r+1)}(\psi_U(\tilde{x}') \oplus \psi_V(\tilde{x}'')).$$

Here $\psi_U(\tilde{x}') = \bigoplus_{j \in U} \psi_j(\tilde{x}_j)$ and $\psi_V(\tilde{x}'') = \bigoplus_{j \in V} \psi_j(\tilde{x}_j)$, with

$$\psi_j(\tilde{x}_j) = M_{i,j} \circ S_j \circ \oplus_{k_j} \circ (f_j^{(r)})^{-1}(\tilde{x}_j).$$

Note that by Property 1 an attacker has access to function h, but not, for instance, to functions ψ_j. In what follows, we denote the range of a function g by im(g). Now im(ψ_j) is the vector space spanned by the columns of matrix $M_{i,j}$. Hence, as U defines a spanning block, we have im$(\psi_U) = \mathrm{GF}(2)^m$. Similarly, we have im$(\psi_V) = \mathrm{GF}(2)^m$. In other words, both ψ_U and ψ_V are surjective on the vector space $\mathrm{GF}(2)^m$. In the full paper we prove Theorem 1 below, which bounds the time complexity for the construction of sets W_U and W_V which are mapped bijectively onto $\mathrm{GF}(2)^m$ by ψ_U and ψ_V, respectively.

Theorem 1. *In $\mathcal{O}(s + m \cdot 2^m)$ time, we can construct sets W_U and W_V, with $|W_U| = |W_V| = 2^m$, such that*
(i) for each fixed $\tilde{x}''$, the map $\tilde{x}' \mapsto h(\tilde{x}', \tilde{x}'')$ is a bijection on W_U, and
(ii) for each fixed $\tilde{x}'$, the map $\tilde{x}'' \mapsto h(\tilde{x}', \tilde{x}'')$ is a bijection on W_V.

Let h_c denote the bijective function $\tilde{x}' \mapsto h(\tilde{x}', c)$ from W_U onto $\mathrm{GF}(2)^m$. Let $c_1, c_2 \in W_V$, and put $d = \psi_V(c_1) \oplus \psi_V(c_2)$. Now if $\tilde{z}_i = h_{c_2}(\tilde{x}') = f_i^{(r+1)}(\psi_U(\tilde{x}') \oplus \psi_V(c_2))$, then $\psi_U(\tilde{x}') = \psi_V(c_2) \oplus (f_i^{(r+1)})^{-1}(\tilde{z}_i)$, and hence

$$h_{c_1} \circ h_{c_2}^{-1}(\tilde{z}_i) = f_i^{(r+1)}(\psi_U(\tilde{x}') \oplus \psi_V(c_1)) = f_i^{(r+1)} \circ \oplus_d \circ (f_i^{(r+1)})^{-1}(\tilde{z}_i).$$

Now fix $c_1 \in W_V$. Then to each $c_2 \in W_V$ there corresponds a unique $d \in \mathrm{GF}(2)^m$. Hence by letting $\tilde{x}'$ run through W_U, we can construct for each d in $\mathrm{GF}(2)^m$ a lookup table for the function $f_i^{(r+1)} \circ \oplus_d \circ (f_i^{(r+1)})^{-1}$. We can then use the following result of Billet et al. [2] to determine $f_i^{(r+1)}$ up to an affine part.

Theorem 2. *Let $f_i^{(r+1)}$ be an arbitrary bijective function on $\mathrm{GF}(2^m)$. Suppose that the set of functions $\{f_i^{(r+1)} \circ \oplus_d \circ (f_i^{(r+1)})^{-1} \mid d \in \mathrm{GF}(2^m)\}$ is given by means of lookup tables. Then we can construct in $\mathcal{O}(2^{3m})$ time a function $g_i^{(r+1)}$ for which the map $\alpha_i^{(r+1)} = g_i^{(r+1)} \circ f_i^{(r+1)}$ is affine.*

Combining Property 1 with the above theorem shows that for any input word x_i, an attacker can obtain the value of the affine map $\alpha_i^{(r+1)}(x_i) = g_i^{(r+1)} \circ f_i^{r+1}(x_i)$. So we have the following.

Property 2. In a white-box attack, an attacker has for each round r with $2 \leq r \leq R$ and for each m-bit input word x_i of round r access to the encoded version $\tilde{x}_i = \alpha_i^{(r)}(x_i)$ of x_i. Here, $\alpha_i^{(r)}(x_i) = A_i^{(r)} x \oplus a_i^{(r)}$ is an m-bit affine function. The attacker also knows the values of r and i that are associated with $f_i^{(r)}(x_i)$. $\qquad\square$

4   Transformation into Table Network

From Property 2 it follows that upon completion of the first step of the cryptanalysis, an attacker has for each round r with $1 < r < R$ access to the input-output behavior of the function

$$\mathcal{G}_{\mathrm{SLT}}^{(r)} = \alpha^{(r+1)} \circ \mathcal{F}_{\mathrm{SLT}}^{(r)} \circ (\alpha^{(r)})^{-1}. \tag{2}$$

As before, fix a round r. Then we have $\mathcal{G}_{\mathrm{SLT}} = \alpha^{(r+1)} \circ \mathcal{F}_{\mathrm{SLT}} \circ (\alpha^{(r)})^{-1}$, and by (1), we have that $\mathcal{F}_{\mathrm{SLT}} = M \circ S \circ \oplus_k$. Hence, if we write $N = A^{(r+1)}M$ and $R = S \circ \oplus_k \circ (\alpha^{(r)})^{-1}$, then we have that

$$\mathcal{G}_{\mathrm{SLT}}(\tilde{x}) = \oplus_{a^{(r+1)}} \circ N \circ R(\tilde{x}) = a^{(r+1)} \oplus \bigoplus_{j=1}^{s} N_j \circ R_j(\tilde{x}_j), \tag{3}$$

where N_j denotes the jth strip of matrix N and R_j denotes the jth component of diagonal map R.

We will first derive an implementation of $\mathcal{G}_{\mathrm{SLT}}$ involving s lookup tables only. More specifically, we will define tables $T_1, T_2, \ldots, T_s$ such that $\mathcal{G}_{\mathrm{SLT}}(\tilde{x}_1, \tilde{x}_2, \ldots, \tilde{x}_s)$ equals $\bigoplus_{i=1}^{s} T_i(\tilde{x}_i)$. To this end, we define T_j as

$$T_j(\tilde{x}_j) = \begin{cases} \mathcal{G}_{\mathrm{SLT}}((\tilde{x}_1, 0, \ldots, 0)), & \text{if } j = 1; \\ \mathcal{G}_{\mathrm{SLT}}((0, \ldots, 0, \tilde{x}_j, 0, \ldots, 0)) \oplus \mathcal{G}_{\mathrm{SLT}}(0), & \text{otherwise.} \end{cases}$$

We have $\mathcal{G}_{\mathrm{SLT}}(0) = \oplus_{a^{(r+1)}} \circ N \circ R(0)$ and, for all $j \geq 1$,

$$\mathcal{G}_{\mathrm{SLT}}((0, \ldots, 0, \tilde{x}_j, 0, \ldots, 0)) = \mathcal{G}_{\mathrm{SLT}}(0) \oplus N_j \circ R_j(0) \oplus N_j \circ R_j(\tilde{x}_j).$$

Hence,

$$T_j(\tilde{x}_j) = \begin{cases} a^{(r+1)} \oplus N_1 \circ R_1(\tilde{x}_1) \oplus \bigoplus_{i=2}^{s} N_i \circ R_i(0), & \text{if } j = 1; \\ N_j \circ R_j(0) \oplus N_j \circ R_j(\tilde{x}_j), & \text{for } j = 2, \ldots, s, \end{cases} \tag{4}$$

and hence we immediately see that $\mathcal{G}_{\mathrm{SLT}}(\tilde{x}) = \bigoplus_{i=j}^{s} T_j(\tilde{x}_j)$ as desired.

5   Transformation into SAT Cipher

In an SLT cipher, we can merge a key-addition operation into the S-box operation that succeeds it. The resulting S-box is then given by $S_i \circ \oplus_{k_i}$. Hence, an SLT cipher can be viewed as a generic SAT cipher, which is defined as follows.

Definition 4 (Generic Substitution-Affine Transformation Cipher (generic SAT cipher)). *A cipher is called a generic SAT cipher if it can be specified as follows. It consists of R rounds for an $R \geq 1$. A single round r is a bijective function $\mathcal{F}_{\mathrm{gen-SAT}}(x_1, x_2, \ldots, x_s)$ on $\mathrm{GF}(2)^n$ with $x_i \in \mathrm{GF}(2)^m$ and $n = m \cdot s$.*

A round consists of the following operations. First, the values $y_i = Q_i^{(r)}(x_i)$ are computed for all input words x_i, where the specification of an S-box $Q_i^{(r)}$ is derived from the key. Next, an invertible affine function $\epsilon^{(r)}(y) = E^{(r)} \cdot y \oplus e^{(r)}$ is applied to the outcome $y = (y_1, y_2, \ldots, y_s) \in \mathrm{GF}(2)^n$ of the S-boxes. The specification of this affine function $\epsilon^{(r)}$ is also derived from the key.

So the round function $\mathcal{F}_{\mathrm{gen-SAT}}$ of a generic SAT cipher can be written as $\mathcal{F}_{\mathrm{gen-SAT}} = \epsilon \circ Q$, where ϵ is affine and Q is a diagonal map, with both Q and ϵ fully specified by the key.

We can consider $\mathcal{G}_{\mathrm{SLT}}$ as a generic SAT cipher. Indeed, according to (3), $\mathcal{G}_{\mathrm{SLT}} = \theta \circ R$, where $\theta = \oplus_{a^{(r+1)}} \circ N$ is affine and R is a diagonal map. However, since the functions θ and R are not accessible to an attacker, this form is not suitable for cryptanalysis. In what follows, we will develop an alternative specification for $\mathcal{G}_{\mathrm{SLT}}$ as a SAT cipher $\mathcal{G}_{\mathrm{SLT}} = \epsilon \circ Q$, with an affine function ϵ and a diagonal map Q that are both accessible to an attacker. We will then use this expression to attack the round key of the original STL cipher $\mathcal{F}_{\mathrm{SLT}}$.

We begin with a simple observation. Since both the diffusion matrix M and the diagonal matrix $A^{(r+1)}$ from the affine map $\alpha^{(r+1)}$ are invertible, the matrix $N = A^{(r+1)}M = (N_1 \cdots N_s)$ is also invertible, and hence the columns of each of the $n \times m$ matrices N_j are also independent. Let U_j denote the m-dimensional vector space spanned by the columns of N_j. Next, we consider again Expression (4) for the lookup tables T_j. Putting $w_1 = a^{(r+1)} \oplus N \circ R(0) \oplus N_1 \circ R_1(0)$ and $w_j = N_j \circ R_j(0)$ for $j = 2, \ldots, s$, it follows from (4) that $\mathrm{im}(T_j) = w_j \oplus U_j$. Hence the 2^m rows of lookup table T_j together comprises all vectors from $w_j \oplus U_j$.

Now, select an arbitrary row v_j from each table T_j, and define

$$e = \bigoplus_{j=1}^{s} v_j.$$

Note that for each $\tilde{x}_j$, we have that $v_j \oplus T_j(\tilde{x}_j) \in U_j$. Indeed, since both v_j and $T_j(\tilde{x}_j)$ are rows of T_j, there are u_j and u'_j in U_j such that $v_j = w_j \oplus u_j$ and $T_j(\tilde{x}_j) = w_j \oplus u'_j$. But then $v_j \oplus T_j(\tilde{x}_j) = u_j \oplus u'_j \in U_j$, as claimed. Next, by selecting words $\tilde{x}_{j,1}, \ldots, \tilde{x}_{j,m}$ for which the vectors $e_{j,i} = v_j \oplus T_j(\tilde{x}_{j,i})$ in U_j are independent, we can construct a basis $e_{j,1}, \ldots, e_{j,m}$ for each U_j. We use these bases to define the submatrices E_j of $E = (E_1, E_2, \ldots, E_s)$ as

$$E_j = (e_{j,1} \cdots e_{j,m}).$$

Since the m columns of E_j span U_j and since $T_j(\tilde{x}_j) \in v_j \oplus U_j$ for each $\tilde{x}_j \in \mathrm{GF}(2)^m$, there is a vector $Q_j(\tilde{x}_j) \in \mathrm{GF}(2)^m$ such that

$$T_j(\tilde{x}_j) = v_j \oplus E_j Q_j(\tilde{x}_j).$$

We will consider Q_j as a map from $\mathrm{GF}(2)^m$ to $\mathrm{GF}(2^m)$. As all rows of table T_j are different, this map is a bijection. Now let $Q = (Q_1, \ldots, Q_s)$ be the diagonal map with components Q_j, and define the affine map ϵ by $\epsilon = \oplus_e \circ E$. By the above analysis, we have that

$$\epsilon \circ Q(\tilde{x}) = e \oplus E \circ Q(\tilde{x}) = \bigoplus_{j=1}^{s} v_j \oplus \bigoplus_{j=1}^{s} E_j Q_j(\tilde{x}_j) = \bigoplus_{j=1}^{s} T_j(\tilde{x}_j) = \mathcal{G}_{\text{SLT}}(\tilde{x}),$$

hence $\mathcal{G}_{\text{SLT}} = \epsilon \circ Q$ is a representation of $\mathcal{G}_{\text{SLT}}$ as a SAT cipher where both the affine map ϵ and the diagonal map Q are explicitly known and accessible to an attacker.

6   Extracting the Key

In this chapter we describe the last step of our cryptanalysis. We adopt the following strategy. First, we derive a relation between the S-boxes $S_i^{(r)}$ of the white-boxed SLT cipher and the S-boxes $Q_i^{(r)}$ of the generic SAT cipher that we constructed in Section 5. This relation will be of the form $Q_i^{(r)} = \gamma_i^{(r)} \circ S_i^{(r)} \circ \delta_i^{(r)}$ for affine functions $\gamma_i^{(r)}, \delta_i^{(r)}$. The diagonal map $\delta^{(r)} = (\delta_1^{(r)}, \delta_2^{(r)}, \ldots, \delta_s^{(r)})$ depends on both the round key $k^{(r)}$ of $\mathcal{F}_{\text{SLT}}^{(r)}$ and the affine encoding $\alpha^{(r)}$ that $\mathcal{G}_{\text{SLT}}^{(r)}$ puts on the input of round $\mathcal{F}_{\text{SLT}}^{(r)}$. The function $\gamma^{(r)} = (\gamma_1^{(r)}, \gamma_2^{(r)}, \ldots, \gamma_s^{(r)})$ depends on the encoding $\alpha^{(r+1)}$ that $\mathcal{G}_{\text{SLT}}^{(r)}$ puts on the output of $\mathcal{F}_{\text{SLT}}^{(r)}$. By comparing the functions $\gamma^{(r-1)}$ and $\delta^{(r)}$, we can recover the key $k^{(r)}$ contained in $\delta^{(r)}$.

We now make the last step of our cryptanalysis more precise. Fix a round r. From the previous step we get S-boxes Q_j and an affine function ϵ, such that $\mathcal{G}_{\text{SLT}} = \epsilon \circ Q$. By (1) and (2), we also have that

$$\mathcal{G}_{\text{SLT}} = \alpha^{(r+1)} \circ M \circ S \circ \oplus_k \circ (\alpha^{(r)})^{-1}.$$

Since the functions ϵ, $\alpha^{(r+1)}$, M, $\oplus_k$, and $\alpha^{(r)}$ are all affine, we conclude that

$$Q = \gamma \circ S \circ \delta, \tag{5}$$

for affine functions

$$\gamma = \epsilon^{-1} \circ \alpha^{(r+1)} \circ M \tag{6}$$

and

$$\delta = \oplus_k \circ (\alpha^{(r)})^{-1}. \tag{7}$$

Note that both γ and δ are diagonal maps. For δ this is true because $\alpha^{(r)}$ is a diagonal map. For γ this property follows from (5) and the observation that δ, Q, and S are all diagonal maps. Biryukov et al. [3] present an algorithm for efficiently determining the set Γ_i of all pairs (γ_i, δ_i) satisfying $Q_i = \gamma_i \circ S_i \circ \delta_i$. So we can use this algorithm to determine the set Γ consisting of all pairs (γ, δ) that satisfy (5). Note that this set Γ contains the pair (γ, δ) that satisfies all of (5), (6), and (7). To complete our cryptanalysis it now suffices to solve the following two problems.

- Which pairs $(\gamma^{(r-1)}, \delta^{(r-1)}) \in \Gamma^{(r-1)}$ and $(\gamma^{(r)}, \delta^{(r)}) \in \Gamma^{(r)}$ of affine functions satisfy (6) and (7)?
- If we have the pairs $(\gamma^{(r-1)}, \delta^{(r-1)}) \in \Gamma^{(r-1)}$ and $(\gamma^{(r)}, \delta^{(r)}) \in \Gamma^{(r)}$ of affine functions that satisfy (6) and (7), how can we derive round key $k^{(r)}$ from this?

Observe that the former problem need not be solved completely. If we can limit the number of candidate pairs to a value ℓ, then we can apply an algorithm for the second problem to all ℓ candidate solutions to obtain ℓ candidate round keys. The correct round key can next be derived by exhaustive search.

By (6) for round $r-1$ and (7) for round r, we have that $\epsilon^{(r-1)} \circ \gamma^{(r-1)} = \alpha^{(r)} \circ M^{(r-1)}$ and $\alpha^{(r)} = (\delta^{(r)})^{-1} \circ \oplus_{k^{(r)}}$, and hence $\delta^{(r)} \circ \epsilon^{(r-1)} \circ \gamma^{(r-1)} = \oplus_{k^{(r)}} \circ M^{(r-1)}$. So if we let $\gamma^{(t)} = \oplus_{c^{(t)}} \circ C^{(t)}$ and $\delta^{(t)} = \oplus_{d^{(t)}} \circ D^{(t)}$, then we obtain that

$$d^{(r)} \oplus D^{(r)} e^{(r-1)} \oplus D^{(r)} E^{(r-1)} C^{(r-1)} y \oplus D^{(r)} E^{(r-1)} c^{(r-1)} = k^{(r)} \oplus M^{(r-1)} y,$$

for all y. For this equality to hold, the constant parts as well as the linear parts are the same in both sides of the equation. This implies that

$$D^{(r)} E^{(r-1)} C^{(r-1)} = M^{(r-1)}$$

and that the round key $k^{(r)}$ is given by

$$k^{(r)} = d^{(r)} \oplus D^{(r)} e^{(r-1)} \oplus D^{(r)} E^{(r-1)} c^{(r-1)}.$$

The above analysis now leads to the algorithm described in Fig. 1 for finding $k^{(r)}$.

Known: Q, E, S, M.

- Step 1: For a particular pair $(r-1, r)$ of successive rounds, construct for each S-box S_i the set
$$\Gamma_i = \{(\gamma_i, \delta_i) \mid Q_i = \gamma_i \circ S_i \circ \delta_i \wedge \gamma_i, \delta_i \text{ affine}\}$$
and let Γ be such that $(\gamma, \delta) \in \Gamma$ if $(\gamma_i, \delta_i) \in \Gamma_i$ for all i.
- Step 2: Construct the subset $\Lambda^{(r)} \subseteq \Gamma^{(r-1)} \times \Gamma^{(r)}$ of pairs of affine functions $(\gamma^{(r-1)}, \delta^{(r-1)}) \in \Gamma^{(r-1)}$ and $(\gamma^{(r)}, \delta^{(r)}) \in \Gamma^{(r)}$ such that
$$D^{(r)} E^{(r-1)} C^{(r-1)} = M^{(r-1)}, \tag{8}$$
where matrices $C^{(r-1)}$ and $D^{(r)}$ define the linear part of $\gamma^{(r-1)}$ and $\delta^{(r)}$, respectively.
- Step 3: The round key $k^{(r)}$ of round r is contained in the set
$$K^{(r)} = \left\{ D^{(r)} E^{(r-1)} c^{(r-1)} \oplus D^{(r)} e^{(r-1)} \oplus d^{(r)} \mid (\gamma^{(r-1)}, \delta^{(r-1)}, \gamma^{(r)}, \delta^{(r)}) \in \Lambda^{(r)} \right\}.$$

Fig. 1. Basic algorithm for finding the round key of a round r

For implementing Step 1 of the algorithm, we already referred to [3]. We now describe how Step 2 can be implemented.

6.1   Solving the Linear Equivalence Problem for Matrices

Step 2 of the algorithm of Fig. 1 deals with the matrices C and D specifying the linear parts of the affine m-bit diagonal maps $\gamma = \oplus_c \circ C$ and $\delta = \oplus_d \circ D$. Note that as observed in Remark 1, C and D are block diagonal matrices.

Definition 5. *Let $X = X_1 \times X_2 \times \ldots \times X_s$, where X_i consists of $m \times m$ matrices. Then we denote by $\mathcal{D}(X)$ the collection of all block diagonal matrices with ith diagonal block contained in X_i, for all i.*

We can now formulate the problem of Step 2 as an instance of the Linear Equivalence Problem of Matrices (LEPM) defined below.

Definition 6 (Linear Equivalence Problem of Matrices (LEPM)). *A problem instance is defined by (M, E, X, Y) for invertible $n \times n$ matrices M and E and sets $X = X_1 \times X_2 \times \ldots \times X_s$ and $Y = Y_1 \times Y_2 \times \ldots \times Y_s$, where X_i and Y_j contain invertible $m \times m$ matrices and $n = m \cdot s$. Find all pairs of block-diagonal $n \times n$ matrices $(C, D) \in \mathcal{D}(X) \times \mathcal{D}(Y)$ such that $M = D \cdot E \cdot C$.*

In Fig. 2 we describe an algorithm for solving LEPM. The algorithm gradually reduces the sets X_i and Y_j, as follows. If a pair $(C, D) \in \mathcal{D}(X) \times \mathcal{D}(Y)$ satisfies $M = D \cdot E \cdot C$, then $M_{i,j} = D_i E_{i,j} C_j$ holds for all i, j. So if for some $C_j \in X_j$ there does not exist a $D_i \in Y_i$ for which $M_{i,j} = D_i E_{i,j} C_j$, then C_j can never be used as jth component in C, and so can be removed from X_j. A similar argument can be used to remove a matrix D_i from a set Y_i.

We proceed with such removal steps until no more removals are possible. If some set X_j or some set Y_i is empty, then the LEPM problem has no solution. Next, if all sets X_j and Y_i contain exactly one linear mapping, then the only candidate solution to the LEPM problem instance is the solution defined by these linear mappings; moreover, since no X_j or Y_i was further reduced, this solution must indeed be valid. So in this case, the LEPM problem is solved.

On the other hand, suppose that a set X_j or a set Y_i exists that contains more than one linear mapping. If all X_j have size one, then C is uniquely determined, and hence $D = E^{-1}C^{-1}M$ is also uniquely determined. As a consequence, all sets Y_i must also have size one. So we may assume without loss of generality that some set X_j has size bigger than one. In that case, for each matrix C_j in X_j, we rerun the algorithm with X_j replaced by the set $X'_j = \{C_j\}$. Obviously, in this way all solutions are found.

To know whether the algorithm presented is effective for attacking a white-box implementation, we have to know an upper bound on the number of solutions returned and the number of recursive invocations. The former number is related to the cardinality of the set K of candidate round keys in the algorithm of Fig. 1. The latter number determines the time complexity of the algorithm of Fig. 2. The problem is that we do not want to answer the question for one particular white-box implementation of a block cipher, but for any white-box implementation of that block cipher. Hence, we want to derive upper bounds on these numbers that only depend on the block cipher specification and not, for instance, on the encodings put on the input and output of a round $\mathcal{F}_{\text{SLT}}$ by the white-box implementation. The following theorem, which is proved in the full version of this paper, can be used to derive such bounds.

Theorem 3. *For a round r of an SLT cipher, let $I = (M, E, X, Y)$ be the problem instance of LEPM that is associated with the cryptanalysis of its white-box*

---

algorithm LEPM_solver(X, Y)

---

begin
  repeat
    for all X_j **do**
      for all $C_j \in X_j$ **do**
        if $\neg \exists_{D_i \in Y_i} M_{i,j} = D_i \cdot E_{i,j} \cdot C_j$ **then**
          $X_j := X_j \setminus \{C_j\}$;
    for all Y_i **do**
      for all $D_i \in Y_i$ **do**
        if $\neg \exists_{C_j \in X_j} M_{i,j} = D_i \cdot E_{i,j} \cdot C_j$ **then**
          $Y_i := Y_i \setminus \{D_i\}$;
  until X and Y do not change;
  if a set X_j or Y_i is empty **then**
    return $\emptyset$;
  else if $\forall_j |X_j| = 1 \wedge \forall_i |Y_i| = 1$ **then**
    return $\{(C, D)\}$ with $C_j \in X_j$ and $D_i \in Y_i$;
  else /* case $\exists_j |X_j| > 1$ */
    select smallest j with $|X_j| > 1$;
    return $\bigcup_{C_j \in X_j}$ LEPM_solver($X(X_j = \{C_j\}), Y$);
end;

Fig. 2. Algorithm for solving LEPM problem in pseudo code. In the algorithm $X(X_j = \{C_j\})$ denotes X, where X_j is replaced by $\{C_j\}$.

implementation. Furthermore, let $I' = (M, M, X', Y')$ be the problem instance in which X'_i and Y'_i are given by

$$X'_i = \{L \mid S_i^{(r-1)} = \lambda \circ S_i^{(r-1)} \circ \phi \text{ with } \lambda : x \mapsto l \oplus Lx \text{ and } \phi \text{ affine}\}$$

and

$$Y'_i = \{P \mid S_i^{(r)} = \lambda \circ S_i^{(r)} \circ \phi \text{ with } \lambda \text{ and } \phi : x \mapsto p \oplus Px \text{ affine}\}.$$

Then, applying the algorithm of Fig. 2 to I results in the same number of recursive invocations and the same number of solutions as when applying the algorithm to problem instance I'.

7   Proof of Concept

As proof of concept, we briefly discuss our cryptanalysis for attacking white-box AES and white-box Serpent. It can be verified that the diffusion matrices of both AES and Serpent satisfy the property that all their block rows have disjoint spanning blocks. Recall that this is a necessary property to perform the first step of the cryptanalysis. After applying the steps described in Sections 3-5, the cryptanalysis runs the algorithm of Fig. 1 to find a set K of candidate round keys for a given round r. The algorithm first derives for each S-box S_i the set Γ_i. For the AES S-box these sets can be shown to have a cardinality of 2040, while for the Serpent S-boxes the cardinality of these sets is either 4 or 1. Next, the algorithm solves an LEPM problem instance to find the set Λ. For AES and Serpent it can be shown that the pairs (γ_i, δ_i) from a set Γ_i satisfy the property that all affine functions γ_i have a unique linear part and that all affine functions δ_i have a unique linear part. Hence, the cardinality of set Λ is given by the number of solutions of this LEPM problem instance. Using Theorem 3 it can be proved that for any LEPM problem instance associated with a white-box implementation the algorithm does not go into recursion and that it returns only one solution. As a consequence, Λ consists of only one solution. It now follows from the third step of the algorithm of Fig. 1 that the set K of candidate round keys consists of only one solution as well. This is the round key we are looking for. The time complexity of the attack is dominated by the algorithm of Biryukov et al. [3] to determine the sets Γ_i.

8   Conclusion

Chow et al. published white-box implementations for AES and DES. As these white-box implementations have been broken, it is an interesting research direction to design a block cipher that results in a secure white-box implementation. This paper can serve as a basis for such research. In this paper we presented an algorithm for extracting the round keys from the white-box implementation of an SLT cipher in case that all block rows of the diffusion matrices of the cipher have disjoint spanning block sets. The condition on the diffusion matrices is, for instance, satisfied by all MDS matrices. Furthermore, we conjecture that our attack can be generalized to arbitrary diffusion matrices. From our result we can conclude that, unless we design new white-box techniques, SLT ciphers are less suited for white-box implementations. A weakness of SLT ciphers that is exploited by our attack is the linearity of the diffusion operator. A linear diffusion matrix is difficult to hide with non-linear encodings. Hence, a possible direction for deriving secure white-box implementations is to resort to alternative diffusion operators. Another weakness of SLT ciphers that we exploit is that, except for a key addition, all operations in the cipher are fixed (i.e., key-independent). It may help to make a larger part of the block cipher operations key-dependent.

References

1. Anderson, R.J., Biham, E., Knudsen, L.R.: Serpent: A proposal for the advanced encryption standard. In: Proceedings of the First AES Candidate Conference (1998)
2. Billet, O., Gilbert, H., Ech-Chatbi, C.: Cryptanalysis of a white box AES implementation. In: Handschuh, H., Hasan, M.A. (eds.) SAC 2004. LNCS, vol. 3357, pp. 227–240. Springer, Heidelberg (2004)

3. Biryukov, A., De Cannière, C., Braeken, A., Preneel, B.: A Toolbox for Cryptanalysis: Linear and Affine Equivalence Algorithms. In: Biham, E. (ed.) EUROCRYPT 2003. LNCS, vol. 2656, pp. 33–50. Springer, Heidelberg (2003)

4. Chow, S., Eisen, P., Johnson, H., van Oorschot, P.C.: A white-box DES implementation for DRM applications. In: Feigenbaum, J. (ed.) DRM 2002. LNCS, vol. 2696, pp. 1–15. Springer, Heidelberg (2003)

5. Chow, S., Eisen, P., Johnson, H., van Oorschot, P.C.: White-box cryptography and an AES implementation. In: Nyberg, K., Heys, H.M. (eds.) SAC 2002. LNCS, vol. 2595, pp. 250–270. Springer, Heidelberg (2003)

6. Daemen, J., Rijmen, V.: The Design of Rijndael. Springer, Heidelberg (2002)

7. Goubin, L., Masereel, J.-M., Quisquater, M.: Cryptanalysis of white box DES implementations. In: Adams, C., Miri, A., Wiener, M. (eds.) SAC 2007. LNCS, vol. 4876, pp. 278–295. Springer, Heidelberg (2007)

8. Jacob, M., Boneh, D., Felten, E.W.: Attacking an obfuscated cipher by injecting faults. In: Feigenbaum, J. (ed.) DRM 2002. LNCS, vol. 2696, pp. 16–31. Springer, Heidelberg (2003)

9. Link, H.E., Neumann, W.D.: Clarifying Obfuscation: Improving the Security of White-Box DES. In: International Symposium on Information Technology: Coding and Computing, pp. 679–684 (2005)

10. Schneier, B., Kelsey, J., Whiting, D., Wagner, D., Hall, C., Ferguson, N.: The Twofish Encryption Algorithm: A 128-Bit Block Cipher. Wiley, Chichester (1999)

11. Vaudenay, S.: On the Need for Multipermutations: Cryptanalysis of MD4 and SAFER. In: Proceedings of the 2nd International Workshop on Fast Software Encryption, pp. 286–297 (1995)

12. Wyseur, B., Michiels, W., Gorissen, P., Preneel, B.: Cryptanalysis of white-box DES implementations with arbitrary external encodings. In: Adams, C., Miri, A., Wiener, M. (eds.) SAC 2007. LNCS, vol. 4876, pp. 264–277. Springer, Heidelberg (2007)

New Linear Cryptanalytic Results of Reduced-Round of CAST-128 and CAST-256*

Meiqin Wang, Xiaoyun Wang, and Changhui Hu

Key Laboratory of Cryptologic Technology and Information Security,
Ministry of Education, Shandong University,
Jinan, 250100, China
mqwang@sdu.edu.cn, xywang@sdu.edu.cn, huchanghui@mail.sdu.edu.cn

Abstract. This paper presents a linear cryptanalysis for reduced round variants of CAST-128 and CAST-256 block ciphers. Compared with the linear relation of round function with the bias 2^{-17} by J. Nakahara et al., we found the more heavily biased linear approximations for 3 round functions and the highest one is $2^{-12.91}$. We can mount the known-plaintext attack on 6-round CAST-128 and the ciphertext-only attack on 4-round CAST-128. Moreover the known-plaintext attack on 24-round CAST-256 with key size 192 and 256 bits has been given, and the ciphertext-only attack on 21-round CAST-256 with key size 192 and 256 bits can be performed. At the same time, we also present the attack on 18-round CAST-256 with key size 128 bits.

Keywords: Linear Cryptanalysis, Block Cipher, CAST-128, CAST-256.

1 Introduction

CAST-128 is a block cipher designed by C. Adams and S. Tavares in 1996[1], and is used in a number of products notably as the default cipher in some versions of GPG and PGP[2,3]. It has been approved for Canadian government use by the Communications Security Establishment. CAST-256 is one of the fifteen candidate algorithms of the first AES Candidate Conference[4,5].

One way to reduce the size of the largest entry in the XOR table is to use injective substitution layer(S-boxes) such that the number of output bits from the S-box is sufficiently larger than the number of input bits. In this way, it is very likely that the entries in the XOR distribution table of a randomly chosen injective S-box will have only small values, making the block cipher resistant to differential cryptanalysis.

In order to resist to differential cryptanalysis, CAST-128 and CAST-256 use injective substitution S-boxes with 32-bit output and 8-bit input. Moreover, S-boxes are designed from bent functions to resist linear cryptanalysis. Therefore,

---

* Supported by 973 Program No. 2007CB807902, National Natural Science Foundation of China Key Project No. 90604036, National Outstanding Young Scientist No. 60525201.

R. Avanzi, L. Keliher, and F. Sica (Eds.): SAC 2008, LNCS 5381, pp. 429–441, 2009.

the cryptanalysis for them will be very difficult. As far as we know, the differential cryptanalysis of 9 quad-rounds CAST-256 and 5-round CAST-128 under weak-key assumption and the impossible differential cryptanalysis for 20-round CAST-256 have been given respectively in [6] and [7]. In addition, Wagner presented the boomerang attack on 16-round CAST-256[11].

Nakahara and Rasmussen presented the first concrete linear cryptanalysis on reduced-round CAST-128 and CAST-256. They can recover the subkey for 4-round CAST-128 with 2^{37} known plaintexts and $2^{72.5}$ times of 4-round CAST-128 encryption. The distinguishing attack for 12-round CAST-256 with 2^{101} known plaintexts and 2^{101} times of 12-round CAST-256 encryption has been given[8].

In this paper, we give the linear cryptanalysis for 6-round CAST-128 with $2^{53.96}$ known plaintexts and $2^{88.51}$ times of 6-round CAST-128 encryption, and give the linear cryptanalysis for 24-round CAST-256 with $2^{124.10}$ known plaintexts and $2^{156.20}$ times of 24-round CAST-256 encryption. Moreover, we present the ciphertext-only attack on 4-round CAST-128 and 21-round CAST-256.

The paper is organized as follows. Section 2 introduces the description of CAST-128 and CAST-256. In Section 3, we present how to find the more heavily biased linear approximations of three round functions in these two block ciphers. In Section 4, we give the linear cryptanalysis for reduced-round CAST-128. In Section 5, we give the linear cryptanalysis for reduced-round CAST-256. In Section 6, we conclude this paper.

2   Description of CAST-128 and CAST-256

2.1   Description of CAST-128

As a Feistel block cipher, CAST-128 uses a block size 64 bits, and the key size can vary from 40 bits to 128 bits, in 8-bit increments. For key sizes up to and including 80 bits, the number of round is 12. For key sizes greater than 80 bits, the cipher uses the full 16 rounds[1]. The overall operation of CAST-128 is similar to DES[9], which is described in Fig.1. CAST-128 splits the plaintext into left and right 32-bit halves L_0 and R_0. In the key schedule process, 16 pairs of subkeys K_{mi} and K_{ri} for the user key K are computed, with one pair of subkeys per round. A 32-bit key-dependent value K_{mi} is used as a "masking" key and a 5-bit K_{ri} is used as a "rotation" key of the ith round. Our cryptanalysis is not related to the key schedule, so we don't present it in detail. The encryption process is defined as follows,

- For $1 \leq i \leq 16$, compute L_i and R_i as follows:

$$L_i = R_{i-1}$$
$$R_i = L_{i-1} \oplus F_i(R_{i-1}, K_{mi}, K_{ri})$$

where F_i is the round function(F_i is of Type 1, Type 2, or Type 3) described later.

- The ciphertext is (R_{16}, L_{16}).

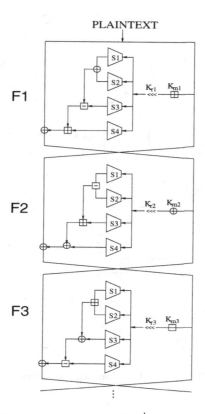

Fig. 1. CAST-128 encryption algorithm

Decryption is identical to the encryption algorithm given above, except that the subkey pairs are used in reverse order to compute (L_0, R_0) from (R_{16}, L_{16}).

Three different round functions are used in CAST-128. X is the input to the round function and I is the input to 4 S-boxes where I_a and I_d are the most significant byte and the least significant byte of I respectively($I = I_a \| I_b \| I_c \| I_d$). "+" and "−" are addition and subtraction modulo 2^{32}. "⊕" is bitwise XOR, and "<<<" is the circular left-shift operation. The round functions are defined as follows,

$$Type1 : I = ((K_{mi} + X) <<< K_{ri})$$
$$F_1 = ((S_1[I_a] \oplus S_2[I_b]) - S_3[I_c]) + S_4[I_d]$$
$$Type2 : I = ((K_{mi} \oplus X) <<< K_{ri})$$
$$F_2 = ((S_1[I_a] - S_2[I_b]) + S_3[I_c]) \oplus S_4[I_d]$$
$$Type3 : I = ((K_{mi} - X) <<< K_{ri})$$
$$F_3 = ((S_1[I_a] + S_2[I_b]) \oplus S_3[I_c]) - S_4[I_d]$$

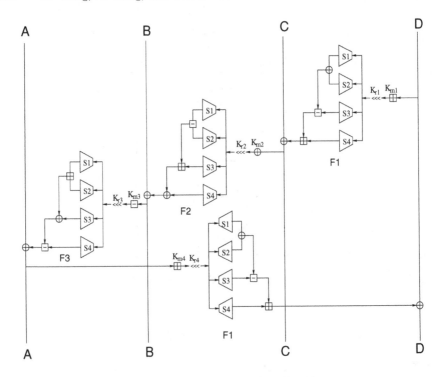

Fig. 2. CAST-256 encryption algorithm

Rounds 1, 4, 7, 10, 13, and 16 use F_1 function. Rounds 2, 5, 8, 11, and 14 use F_2 function. Rounds 3, 6, 9, 12, and 15 use F_3 function. In the above equations, S_1, S_2, S_3, and S_4 are 4 S-boxes, which input is 8-bit and output is 32-bit.

2.2   Description of CAST-256

As a candidate for the first AES conference, CAST-256 is designed based on CAST-128. The block size is 128-bit, and the key size can be 128-bit, 192-bit and 256-bit. The round number is 48 for all key size. The structure for CAST-256 is generalized Feistel Network structure in Fig. 2.

We denote 128-bit block as $\beta = (ABCD)$ where A,B,C and D are each 32 bits in length. Two types of round function, the "forward quad-round" $Q(\cdot)$ and the "reverse quad-round" $\bar{Q}(\cdot)$ are used in CAST-256.

The "forward quad-round" $\beta \longleftarrow Q_i(\beta)$ is defined as the following four rounds,

$$C = C \oplus F_1(D, K_{r1}^{(i)}, K_{m1}^{(i)})$$
$$B = B \oplus F_2(C, K_{r2}^{(i)}, K_{m2}^{(i)})$$
$$A = A \oplus F_3(B, K_{r3}^{(i)}, K_{m3}^{(i)})$$
$$D = D \oplus F_1(A, K_{r4}^{(i)}, K_{m4}^{(i)})$$

And the "reverse quad-round"$\beta \longleftarrow \bar{Q}_i(\beta)$ is defined as the following four rounds,

$$D = D \oplus F_1(A, K_{r4}^{(i)}, K_{m4}^{(i)})$$
$$A = A \oplus F_3(B, K_{r3}^{(i)}, K_{m3}^{(i)})$$
$$B = B \oplus F_2(C, K_{r2}^{(i)}, K_{m2}^{(i)})$$
$$C = C \oplus F_1(D, K_{r1}^{(i)}, K_{m1}^{(i)})$$

where $K_r^{(i)} = \{K_{r1}^{(i)}, K_{r2}^{(i)}, K_{r3}^{(i)}, K_{r4}^{(i)}\}$ is the set of rotation keys for the ith quad-round, and $K_m^{(i)} = \{K_{m1}^{(i)}, K_{m2}^{(i)}, K_{m3}^{(i)}, K_{m4}^{(i)}\}$ is the set of masking keys for the ith quad-round.

The encryption process for CAST-256 consists of 6 "forward quad-rounds" followed by 6 "reverse quad-rounds". Decryption is identical to encryption except that the sets of quad-round keys $K_r^{(i)}$ and $K_m^{(i)}$ are used in reverse order.

3   Linear Approximation for Round Functions

The S-boxes of CAST-128 have dimension 8×32 bits and are non-surjective, so their linear approximation tables are difficult to be constructed. The probability of the linear approximations for these S-boxes with the form $0 \to \Gamma$ is away from $\frac{1}{2}$ because of the non-surjective property of S-boxes, where '0' stands for a zero 8-bit mask, and 'Γ' stands for a nonzero 32-bit mask. This kind of linear approximation only represents that an exclusive-or of output bits selected by Γ is zero. Especially if there is only one non-zero bit for Γ, the probability is always equal to $\frac{1}{2} \pm \frac{1}{2^5}$. In [8], in order to obtain the linear approximation for the round function, only the linear approximation for S-boxes with the form $0 \to 1$ has been used where only the least significant output masking bit is non-zero. Then the bias for the linear approximation of the round function with the form $0 \to 1$ in Fig.3 is 2^{-17} according to the Piling-Up lemma[10] because the least significant output masking bit is not affected by the mixture operations with modular addition, modular subtraction and XOR operations. In [8], authors think the highest bias for the round function is $0 \to 1$ because the carry bits in modular addition and the borrow bits in modular subtraction of round function will reduce the bias to less than 2^{-17}, so they use the linear relations for round functions F_1, F_2 or F_3 having the following forms,

$$F_i : 00000000_X \to 00000000_X$$

$$F_i : 00000000_X \to 00000001_X$$

Based on the above line relations, 2 types of 2-round iterative linear relations for CAST-128 depicted in Fig.4(a) and Fig.4(b) respectively have been given. According to the Piling-Up lemma[10], the biases for the two 2-round iterative linear relations are all 2^{-17}[8].

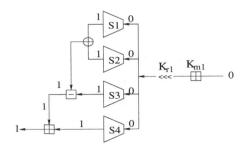

Fig. 3. Bit masks of a linear relation for round function F_1

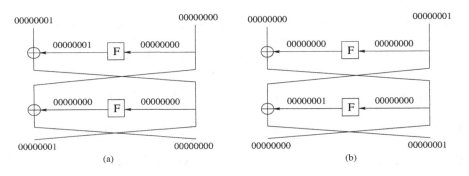

Fig. 4. 2 two-round iterative linear relations for CAST-128

However, we find an important fact that the carry-bit in the modular addition and the borrow-bit in the modular subtraction don't always decrease the bias of linear approximation, sometimes they can further increase the bias. The cryptanalysis in [8] only uses the bias for the single output bit(the least significant bit) of S-boxes. In fact, we find that the non-random properties of the consecutive output bits of S-boxes may result in the higher bias of the output bit of round function with modular addition, modular subtraction and XOR operations compared with the bias of S-boxes output. For example, two least significant bits of S-box output have 4 possible values such as '00', '01', '10' and '11'. If the distribution for the 4 values are non-random(the probabilities are not equal), the bias of the second least-significant bit of round function may be increased after the mixture operations on them. So we searched the linear approximations for the round functions F_1, F_2 and F_3 which have the form $0 \rightarrow \Gamma$ and only one non-zero bit mask of Γ, and the bias for this kind of linear approximation represents the unbalance property for each output bit of round function. The results are presented in Table1. From Table1, we identified the highest bias is not for linear approximation $0 \rightarrow 1$, but the highest biases for F_1, F_2 and F_3 are $2^{-13.71}$, $2^{-14.41}$ and $2^{-14.26}$ respectively which are corresponding to the linear approximation $0 \rightarrow 00000010_X$, $0 \rightarrow 00020000_X$, and $0 \rightarrow 00000080_X$.

Table 1. Linear approximation table for one non-zero bit mask of Γ

non-zero masking bit for Γ	$bias_{F_1} = \lvert P_r - \frac{1}{2} \rvert$	$bias_{F_2} = \lvert P_r - \frac{1}{2} \rvert$	$bias_{F_3} = \lvert P_r - \frac{1}{2} \rvert$
1	$2^{-17.00}$	$2^{-17.00}$	$2^{-17.00}$
2	$2^{-18.00}$	$2^{-17.68}$	$2^{-17.48}$
3	$2^{-18.99}$	$2^{-19.91}$	$2^{-14.48}$
4	$2^{-14.58}$	$2^{-15.00}$	$2^{-15.38}$
5	$2^{-13.98}$	$2^{-14.61}$	$2^{-15.23}$
6	$2^{-13.71}$	$2^{-16.45}$	$2^{-15.54}$
7	$2^{-16.30}$	$2^{-17.00}$	$2^{-17.81}$
8	$2^{-16.91}$	$2^{-18.79}$	$2^{-14.26}$
9	$2^{-15.24}$	$2^{-15.68}$	$2^{-18.20}$
10	$2^{-17.69}$	$2^{-18.47}$	$2^{-17.03}$
11	$2^{-17.38}$	$2^{-18.74}$	$2^{-16.60}$
12	$2^{-15.88}$	$2^{-23.68}$	$2^{-15.41}$
13	$2^{-16.08}$	$2^{-16.38}$	$2^{-16.71}$
14	$2^{-15.69}$	$2^{-14.74}$	$2^{-15.68}$
15	$2^{-17.08}$	$2^{-17.00}$	$2^{-16.80}$
16	$2^{-17.53}$	$2^{-15.19}$	$2^{-19.09}$
17	$2^{-21.54}$	$2^{-17.34}$	$2^{-16.26}$
18	$2^{-14.41}$	$2^{-14.41}$	$2^{-14.47}$
19	$2^{-15.55}$	$2^{-19.30}$	$2^{-17.43}$
20	$2^{-18.96}$	$2^{-15.88}$	$2^{-16.41}$
21	$2^{-17.66}$	$2^{-16.30}$	$2^{-20.80}$
22	$2^{-15.32}$	$2^{-16.80}$	$2^{-19.44}$
23	$2^{-17.20}$	$2^{-15.38}$	$2^{-16.17}$
24	$2^{-18.47}$	$2^{-17.93}$	$2^{-18.73}$
25	$2^{-17.23}$	$2^{-17.64}$	$2^{-15.74}$
26	$2^{-15.77}$	$2^{-16.75}$	$2^{-15.37}$
27	$2^{-14.72}$	$2^{-16.19}$	$2^{-16.44}$
28	$2^{-17.60}$	$2^{-20.46}$	$2^{-17.33}$
29	$2^{-20.12}$	$2^{-17.85}$	$2^{-17.64}$
30	$2^{-16.06}$	$2^{-15.31}$	$2^{-16.34}$
31	$2^{-16.24}$	$2^{-16.23}$	$2^{-18.09}$
32	$2^{-15.82}$	$2^{-16.03}$	$2^{-16.89}$

Additionally, the unbalance property of the single output bit of round function will result in the heavily biased linear approximation with more non-zero output masking bits. So we searched the linear approximations for 3 round functions which have the form $0 \rightarrow \Gamma$ with two and three non-zero masking bits of Γ. Further four and five non-zero masking bits of Γ for F_2 have been examined, but we have not examined four or five non-zero masking bits of Γ for F_1 and F_3 and more than five non-zero masking bits for 3 round functions because the complexity of computation is very large. Their linear relations with the highest bias we have found will be given in Table 2.

From Table 1 and Table 2, the best bias for single round function we found is $2^{-12.91}$ corresponding to the linear relation $00000000_X \rightarrow 03400000_X$ for F_2.

Table 2. Best linear approximation for more non-zero bits of Γ

Function Type	Γ	Number of non-zero bits of Γ	$bias = |P_r - \frac{1}{2}|$
F_1	$0000000C_X$	2	$2^{-14.07}$
F_2	80004000_X	2	$2^{-13.06}$
F_3	02400000_X	2	$2^{-13.71}$
F_1	02600000_X	3	$2^{-13.37}$
F_2	03400000_X	3	$2^{-12.91}$
F_3	00030020_X	3	$2^{-14.05}$
F_2	00600300_X	4	$2^{-13.64}$
F_2	32000900_X	5	$2^{-13.48}$

4   Linear Cryptanalysis for Reduced-Round CAST-128

4.1   Known-Plaintext Attack for Reduced-Round CAST-128

Based on the above linear approximations of the 3 round functions, we can obtain the 5-round linear relation in Fig 5.a. The output mask Γ in round 2 and round 4 is non-zero, but zero in round 1, 3 and 5. The input mask from the first round to the fifth round are all zero. So the probability of the linear relation in round 1, 3 and 5 are all 1. The bias of the linear relation $00000000_X \rightarrow 03400000_X$ for F_1 is $2^{-13.57}$, and the bias of the linear relation $00000000_X \rightarrow 03400000_X$ for F_2 is $2^{-12.91}$. Based on "the Piling-Up lemma", the bias for the 5-round linear approximation is $2^{-25.48}$.

The linear relation in Fig 5.a is a 5-round distinguisher from the random permutation, which can be presented as follows,

$$(P_R \oplus C_R) \cdot 03400000_X = 0$$

where P_R is the right 32-bit of the plaintext, and C_R is the right 32-bit of the ciphertext for 5-round. As a known plaintext attack, the number of known plaintext N required in linear cryptanalysis is proportional to ϵ^{-2}[10], where ϵ is the bias for the linear relation. If N is taken as $8 \cdot \epsilon^{-2}$, the attack will be successful with very high probability. So we can distinguish 5-round CAST-128 with $8 \cdot 2^{25.48 \cdot 2} = 2^{53.96}$ known plaintexts.

We can recover 37-bit subkey of 6-round using the above 5-round distinguisher in Fig 5.a. As the distinguishing attack for 5-round, the attack also requires $2^{53.96}$ known plaintexts and $2^{53.96} \cdot 2^{37} = 2^{90.96}$ one-round encryptions, which is equivalent to $2^{88.51}$ 6-round encryptions.

4.2   Ciphertext-Only Attack for Reduced-Round CAST-128

If the plaintext is ASCII encoded English text, we can attack reduced-round CAST-128 only with ciphertexts. We use the linear approximation for 3-round CAST-128 where only F_2 is active,

$$(P_R \oplus R_3) \cdot 00008000_X = 0$$

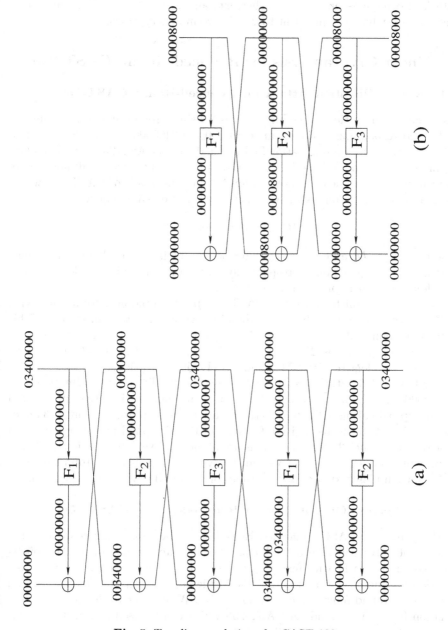

Fig. 5. Two linear relations for CAST-128

where R_3 is the right 32-bit output for round 3, and the bias for the above linear approximation is $2^{-15.19}$, so we can construct the distinguisher of 3-round CAST-128 with only $8 \cdot 2^{15.19 \cdot 2} = 2^{33.38}$ ciphertexts in Fig 5.b. Moreover we can recover 37-bit subkey of 4-round using the above 3-round distinguisher. The attack also requires only $2^{33.38}$ ciphertexts and $2^{33.38} \cdot 2^{37} = 2^{70.38}$ one-round encryptions, which is equivalent to $2^{68.38}$ 4-round encryptions.

5   Linear Cryptanalysis for Reduced-Round CAST-256

5.1   Known-Plaintext Attack for Reduced-Round CAST-256

As described in Section 3, the highest bias for single round function we found is $2^{-12.91}$ corresponding to the linear relation $0 \rightarrow 03400000_X$ for F_2. So we arrive the iterative linear approximation for one quad-round CAST-256 in Fig6.a. Only F_2 in each quad-round is active, but other 3 round functions are all non-active. We can derive the linear approximation for r quad-rounds of CAST-256 which can be used as a distinguisher, which can be represented as follows,

$$(B \oplus F) \cdot 03400000_X = 0$$

where (A, B, C, D) and (E, F, G, H) denote the plaintext block and the ciphertext block for r quad-rounds respectively. Based on "the Piling-Up lemma", the bias for the linear approximation is $2^{r-1} \cdot 2^{-12.91 \cdot r}$.

We can distinguish 21 rounds CAST-256 from a random permutation with $2^{124.1}$ known plaintexts. By the 21 rounds distinguisher, we can recover 37-bit subkey of round 22 for 24-round CAST-256 with the key size 192 or 256 bits. The time complexity is $2^{124.1} \cdot 2^{37} = 2^{161.1}$ one-round CAST-256 encryptions which is equivalent to $2^{156.2}$ 24-round CAST-256 encryptions.

For CAST-256 with key size 128 bits, we use the linear approximation $0 \rightarrow 02600000_X$ for F_1 with the bias $2^{-13.37}$ to construct the iterative quad-round linear approximation in Fig 6.b. So the iterative linear approximation for 3 quad-round CAST-256 can be derived. Only F_1 of the 4th round in each quad-round is active, but other 3 round functions are all non-active. The bias for the linear approximation is $2^{-38.11}$ and we can recover 37-bit subkey of round 16 with $2^{79.22}$ known plaintexts and $2^{111.98}$ times of 18-round CAST-256 encryption.

5.2   Ciphertext-Only Attack for Reduced-Round CAST-256

If the plaintext is ASCII encoded English text, we can attack reduced-round CAST-256 only with ciphertexts. We use the linear approximation $0 \rightarrow 00000080_X$ for round function F_3 with bias $2^{-14.26}$, so we obtain the iterative linear approximation for one quad-round CAST-256 in Fig6.c. Only F_3 in round-3 is active, but other 3 round functions are all non-active. We can derive the linear approximation for r quad-rounds of CAST-256 which can be used as a distinguisher, which can be represented as follows,

$$(A \oplus E) \cdot 00000080_X = 0$$

where (A, B, C, D) and (E, F, G, H) denote the plaintext block and the ciphertext block for r quad-rounds respectively. Based on "the Piling-Up lemma", the bias for the linear approximation is $2^{r-1} \cdot 2^{-14.26 \cdot r}$.

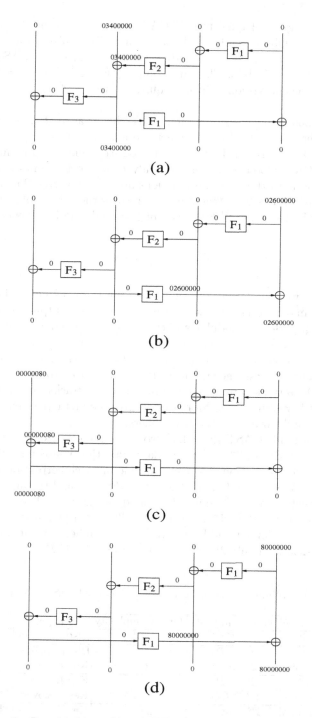

Fig. 6. One quad-round iterative linear relation for CAST-256

We can distinguish 4 quad-rounds CAST-256 from a random permutation with only $2^{111.08}$ ciphertexts. Using 4 quad-rounds distinguisher with only $2^{111.08}$ ciphertexts, we can recover the round 19 subkey for 21-round CAST-256 with the key size 192 or 256 bits. The time complexity is $2^{111.08} \cdot 2^{37} = 2^{148.08}$ one-round CAST-256 encryptions which is equivalent to $2^{143.50}$ 21-round CAST-256 encryptions.

For CAST-256 with key size 128 bits, we use the linear relation $0 \rightarrow 80000000_X$ for F_1 with the bias $2^{-15.82}$ to construct the iterative linear approximation for a quad-round CAST-256 in Fig6.d. So the iterative linear approximation for 3 quad-rounds CAST-256 can be derived. Only F_1 of the 4th round in each quad-round is active, but other 3 round functions are all non-active. The bias for the linear approximation is $2^{-45.46}$ and we can recover the subkey of round 16 with $2^{93.92}$ only-ciphertexts and $2^{126.28}$ times of 18-round CAST-256 encryption.

6   Summary

In this paper, we found that the unbalance for the consecutive bits from S-boxes output may further increase the unbalance of the output from the round function which performs modular addition, modular subtraction and XOR operations on the outputs of 4 S-boxes, This observation led us to find the heavily biased linear relation for the round functions of CAST-128 and CAST-256. After that, we present the best known linear attack on reduced-round CAST-128 and CAST-256. Our attacks are by far the best known attacks on the two ciphers without weak-key assumption. Moreover we give the first ciphertext only attack for reduced round variants of the two ciphers.

We attack 6-round CAST-128, which works for the key size more than 88 bits, with data complexity of $2^{53.96}$ known plaintexts, the time complexity of $2^{88.51}$ times of 6-round encryption. Moreover we mount a ciphertext-only attack on 4-round CAST-128 for the key size more than 68 bits, and the attack uses only $2^{33.38}$ ciphertexts and $2^{68.38}$ times of 4-round encryption. Then we present an attack on 24-round CAST-256 requiring $2^{124.10}$ known plaintexts, $2^{156.20}$ times of 24-round encryptions. In addition, we mount a ciphertext-only attack on 21-round CAST-256 with only $2^{111.08}$ ciphertexts and $2^{143.50}$ 21-round encryptions.

Table 3. Summary of linear attacks on reduced-round CAST-128

Rounds	Key Size	Data Complexity	Time Complexity	Type	Source
2	all	2^{37} KPs	2^{37}	Distinguishing	[8]
3	all	2^{37} KPs	2^{37}	Distinguishing	[8]
	>72 bits	2^{37} KPs	$2^{72.5}$	Key Recovery	[8]
4	>72 bits	2^{37} KPs	$2^{72.5}$	Key Recovery	[8]
	>68 bits	$2^{33.38}$ COs	$2^{68.38}$	Key Recovery	This Paper
6	>88 bits	$2^{53.96}$ KPs	$2^{88.51}$	Key Recovery	This Paper

KPs:Known Plaintexts, COs:Ciphertexts only

Table 4. Summary of linear attacks on reduced-round CAST-256

Rounds	Key Size	Data Complexity	Time Complexity	Type	Source
9	all	2^{69} KPs	2^{103}	Key Recovery	[8]
12	all	2^{101} KPs	2^{101}	Distinguishing	[8]
18	all	$2^{79.22}$ KPs	$2^{111.98}$	Key Recovery	This Paper
	all	$2^{93.92}$ COs	$2^{126.28}$	Key Recovery	This Paper
21	192-bit or 256-bit	$2^{111.08}$ COs	$2^{143.50}$	Key Recovery	This Paper
24	192-bit or 256-bit	$2^{124.1}$ KPs	$2^{156.20}$	Key Recovery	This Paper

^2KPs:Known Plaintexts, COs:Ciphertexts only

Table 3 and Table 4 give the comparison of our results with the previous linear attacks on CAST-128 and CAST-256.

References

1. Adams, C., Tavares, S.: The CAST-128 Encryption Algorithm. RFC 2144 (May 1997)
2. GnuPG, Gnu Privacy Guard, http://www.gnupg.org/(en)/features.html
3. PGP, Pretty Good Privacy, http://www.pgp.com/
4. Adams, C., Gilchrist, J.: The CAST-256 Encryption Algorithm. RFC 2612 (June 1999)
5. First AES Candidate Conference,
   http://csrc.nist.gov/archive/aes/round1/conf1/aes1conf.htm
6. Biham, E.: A Note on Comparing the AES Candidates, The AES Development Process, http://csrc.nist.gov/archive/aes/round1/conf2/papers/biham2.pdf
7. Seki., H., Kaneko., T.: Differential Cryptanalysis of CAST-256 Reduced to Nine Quad-rounds. Leice Transactions on Fundamentals of Electronics Communications and Computer Sciences E84A(4), 913–918 (2001)
8. Nakahara Jr., J., Rasmussen, M.: Linear Analysis of Reduced-round CAST-128 and CAST-256, SBSEG2007, pp.45–55 (2007)
9. NBS, Data Encryption Standard (DES), FIPS PUB 46, Federal Information Processing Standards Publication 46, U.S. Department of Commerce (January 1977)
10. Matsui, M.: Linear cryptanalysis method for DES cipher. In: Helleseth, T. (ed.) EUROCRYPT 1993. LNCS, vol. 765, pp. 386–397. Springer, Heidelberg (1994)
11. Wagner, D.: The boomerang attack. In: Knudsen, L.R. (ed.) FSE 1999. LNCS, vol. 1636, p. 156. Springer, Heidelberg (1999)

Improved Impossible Differential Cryptanalysis of Reduced-Round Camellia

Wenling Wu, Lei Zhang, and Wentao Zhang

State Key Laboratory of Information Security, Institute of Software,
Chinese Academy of Sciences, Beijing 100190, P.R. China
{wwl,zhanglei1015,zhangwt}@is.iscas.ac.cn

Abstract. The block cipher Camellia has now been adopted as an international standard by ISO/IEC, and it has also been selected to be Japanese CRYPTREC e-government recommended cipher and in the NESSIE block cipher portfolio. Most recently, Wu et al constructed some 8-round impossible differentials of Camellia, and presented an attack on 12-round Camellia-192/256 in [5]. Later in [6], Lu et al improved the above attack by using the same 8-round impossible differential and some new observations on the diffusion transformation of Camellia. Considering that all these previously known impossible differential attacks on Camellia have not taken the key scheduling algorithm into account, in this paper we exploit the relations between the round subkeys of Camellia, together with some novel techniques in the key recovery process to improve the impossible differential attack on Camellia up to 12-round Camellia-128 and 16-round Camellia-256. The data complexities of the two attacks are 2^{65} and 2^{89} respectively, and the time complexities of the two attacks are less than $2^{111.5}$ and $2^{222.1}$ respectively. The presented results are better than any previously published cryptanalytic results on Camellia without the FL/FL^{-1} functions and whitening layers.

Keywords: Block cipher, Camellia, Impossible differential, Cryptanalysis, Round subkey.

1 Introduction

The block cipher Camellia [1], with the same interface specification as the Advanced Encryption Standard(AES), supports 128-bit block size and 128-, 192- and 256-bit key sizes, which can usually be denoted as Camellia-128, Camellia-192 and Camellia-256 respectively. Camellia was jointly developed by NTT and Mitsubishi Electric Corporation, and it was first published at SAC 2000. Then it was submitted to some cryptographic evaluation projects such as the European NESSIE Project and the Japanese CRYPTREC Evaluation, and Camellia was selected to be CRYPTREC e-government recommended cipher in 2002 and in the NESSIE block cipher portfolio in 2003. Furthermore, it was adopted as a new international standard for 128-bit block cipher by ISO/IEC in 2005. As Camellia has become one of the most worldwide used block ciphers, in the last few years

R. Avanzi, L. Keliher, and F. Sica (Eds.): SAC 2008, LNCS 5381, pp. 442–456, 2009.

cryptanalysts had evaluated the security of Camellia against various cryptanalytic techniques, including truncated differential cryptanalysis [2,3], higher order
differential cryptanalysis [4], impossible differential cryptanalysis [3,5,6], Square
attack/Integral attack [7-11], collision attack [12,13], linear cryptanalysis [14,15]
and so on.

Impossible differential cryptanalysis [16] was first proposed by Biham et al
in 1999, and was applied to the Skipjack cipher reduced from 32 to 31 rounds.
Unlike traditional differential cryptanalysis which exploits differentials with the
highest possible probability, impossible differential cryptanalysis uses differentials which hold with probability 0, which can also be called impossible differentials. An impossible differential can usually be built in a miss-in-the-middle
manner. Recently, impossible differential cryptanalysis had received worldwide
attention, and its application to the security analysis of AES and CLEFIA both
got very good results [17-23].

The initial analysis of Camellia against impossible differential cryptanalysis
was given by M.Sugita et al [3] in 2001, they constructed a nontrivial 7-round
impossible differential for Camellia. In 2007, by exploiting some properties of the
linear diffusion function, Wu et al [5] presented some 8-round impossible differentials for Camellia, and based on it they mounted an impossible differential attack
on 12-round Camellia-192/256. Then in [6], Lu et al exploited the same 8-round
impossible differential together with the early abort technique and improved the
impossible differential cryptanalysis of Camellia. However, all of these impossible differential attacks on Camellia have not taken the key scheduling algorithm
into account. Thus in this paper, we first present some observations of the relations between round subkeys, and then by taking advantage of these relations
and some novel techniques in the key recovery process, we improve the impossible differential attack on Camellia up to 12-round Camellia-128 and 16-round
Camellia-256. As far as we know, these are the best published cryptanalytic results on Camellia without the FL/FL^{-1} functions and whitening layers, and we
summarize our results together with the previously known results on Camellia
in Table 1.

The cryptanalytic results of [6] in Table 1 come from an early version, not the
published version, so we mark them with "†". This is because there are some
mistakes in the published version. In Step 3 of the 14-round attack for camellia-
256 in [6], the authors wrote: "Finally, for every remaining pair of plaintexts we
can get the first bytes of their intermediate values just after Round 2." Byte
1,3,4,6,7,8 of K_2 should be known for calculating the first byte just after round
2. However, only byte 1,2,3,5,8 are guessed in the attack, whereas byte 4,6,7 are
unknown. There are similar mistakes in the other attacks in [6].

This paper is organized as follows. In Section 2, we give a brief description
of Camellia. In Section 3, we describe the 8-round impossible differential and
some properties of Camellia which are used in our attacks. Then in Section 4,
we present our impossible differential attacks on 12-round Camellia-128 and 16-
round Camellia-256 respectively. Finally, in Section 5 we summarize this paper.

Table 1. Summary of known cryptanalytic results on Camellia

Cipher	# of Rounds	FL/FL^{-1}	Attack Type	Data Complexity	Time Complexity	Source
Camellia-128	8	×	Truncated DC	$2^{83.6}$	$2^{55.6}$	[2]
	9	×	Collision Attack	$2^{113.6}$	2^{121}	[12]
	9	×	Square Attack	2^{66}	$2^{84.8}$	[11]
	11	×	Impossible DC	2^{120}	$2^{83.4}$	[6]†
	12	×	Impossible DC	2^{65}	$2^{111.5}$	Sec. 4.1
Camellia-192/256	11	√	Higher Order DC	2^{93}	2^{256}	[4]
	12	×	Linear Attack	2^{119}	2^{247}	[14]
	12	×	Impossible DC	2^{120}	2^{181}	[5]
	12	×	Square Attack	2^{66}	$2^{249.6}$	[11]
	13	×	Impossible DC	2^{120}	$2^{211.7}$	[6]†
	16	×	Impossible DC	2^{89}	$2^{222.1}$	Sec. 4.2

2   Description of Camellia

The overall structure of Camellia is a variant of Feistel structure, with the FL/FL^{-1} functions inserted at every 6 rounds. Before the first round and after the last round, there are pre- and post- whitening layers which employ bitwise exclusive-OR operations with 128-bit whitening subkeys respectively. In this paper, we will only consider Camellia without FL/FL^{-1} functions and whitening layers, namely the simplified variant of Camellia.

Let L_{r-1} and R_{r-1} be the left and the right halves of the r-th round input, and K_r be the r-th round subkey respectively. Then the r-th round of Camellia can be expressed as follows.

$$L_r = R_{r-1} \oplus F(L_{r-1} \oplus K_r),$$
$$R_r = L_{r-1}.$$

Here the round function of Camellia is $F = P \circ S$, and the transformations S and P are defined as follows.

$$S : (F_2^8)^8 \longrightarrow (F_2^8)^8$$
$$x_1 \mid x_2 \mid x_3 \mid x_4 \mid x_5 \mid x_6 \mid x_7 \mid x_8 \longrightarrow y_1 \mid y_2 \mid y_3 \mid y_4 \mid y_5 \mid y_6 \mid y_7 \mid y_8$$
$$y_1 = s_1(x_1), \quad y_2 = s_2(x_2), \quad y_3 = s_3(x_3), \quad y_4 = s_4(x_4),$$
$$y_5 = s_2(x_5), \quad y_6 = s_3(x_6), \quad y_7 = s_4(x_7), \quad y_8 = s_1(x_8).$$

where s_1, s_2, s_3 and s_4 are four 8×8 S-boxes.

$$P : (F_2^8)^8 \longrightarrow (F_2^8)^8$$
$$y_1 \mid y_2 \mid y_3 \mid y_4 \mid y_5 \mid y_6 \mid y_7 \mid y_8 \longrightarrow z_1 \mid z_2 \mid z_3 \mid z_4 \mid z_5 \mid z_6 \mid z_7 \mid z_8$$
$$z_1 = y_1 \oplus y_3 \oplus y_4 \oplus y_6 \oplus y_7 \oplus y_8, \quad z_5 = y_1 \oplus y_2 \oplus y_6 \oplus y_7 \oplus y_8,$$
$$z_2 = y_1 \oplus y_2 \oplus y_4 \oplus y_5 \oplus y_7 \oplus y_8, \quad z_6 = y_2 \oplus y_3 \oplus y_5 \oplus y_7 \oplus y_8,$$
$$z_3 = y_1 \oplus y_2 \oplus y_3 \oplus y_5 \oplus y_6 \oplus y_8, \quad z_7 = y_3 \oplus y_4 \oplus y_5 \oplus y_6 \oplus y_8,$$
$$z_4 = y_2 \oplus y_3 \oplus y_4 \oplus y_5 \oplus y_6 \oplus y_7, \quad z_8 = y_1 \oplus y_4 \oplus y_5 \oplus y_6 \oplus y_7.$$

Key Scheduling Algorithm of Camellia. First of all, two 128-bit variables K_L and K_R are generated from the master key K. For Camellia-128, the 128-bit key K is used as K_L, and K_R is 0. For Camellia-192, the left 128 bits of the key K is used as K_L, and concatenation of the right 64-bit of K and the complement of the right 64-bit of K is used as K_R. For Camellia-256, the left 128-bit of the key K is used as K_L and the right 128-bit of K is used as K_R. Then two 128-bit variables K_A and K_B are generated from K_L and K_R, but note that K_B is used only if the length of the master key is 192 or 256 bits. Finally, the 64-bit round subkeys K_r are generated by rotating (K_L, K_R, K_A, K_B) and then taking the left half or the right half of them. More details are shown in [1], and in the following we only present some observations which are useful for our later attacks.

For Camellia-128, the round subkeys $K_r(1 \leq r \leq 18)$ are generated by rotating (K_L, K_A), and we can get the following expressions:

$$K_1 = (K_A \lll 0)_{L\,(64)}, \qquad K_2 = (K_A \lll 0)_{R\,(64)},$$
$$K_{11} = (K_A \lll 60)_{L\,(64)}, \qquad K_{12} = (K_A \lll 60)_{R\,(64)}.$$

For Camellia-192/256, the round subkeys $K_r(1 \leq r \leq 24)$ are generated by rotating (K_L, K_R, K_A, K_B), and we can get the following expressions:

$$K_1 = (K_B \lll 0)_{L\,(64)}, \qquad K_2 = (K_B \lll 0)_{R\,(64)},$$
$$K_3 = (K_R \lll 15)_{L\,(64)}, \qquad K_4 = (K_R \lll 15)_{R\,(64)},$$
$$K_{13} = (K_R \lll 60)_{L\,(64)}, \qquad K_{14} = (K_R \lll 60)_{R\,(64)},$$
$$K_{15} = (K_B \lll 60)_{L\,(64)}, \qquad K_{16} = (K_B \lll 60)_{R\,(64)}.$$

3 Preliminaries

3.1 Notations

Camellia is a byte-oriented block cipher, in which the 128-bit intermediate variables are represented as 16 bytes and the 64-bit round subkeys are represented as 8 bytes. The subkey of the r-th round is represented as $K_r = (k_{r,1}, k_{r,2}, k_{r,3}, k_{r,4}, k_{r,5}, k_{r,6}, k_{r,7}, k_{r,8})$. Furthermore, $k_{r,1}[i \sim j](i, j = 1, 2, \ldots, 8, i \leq j)$ denotes the i-th to the j-th bits of $k_{r,1}$.

For a pair of plaintexts (L_0, R_0) and (L_0^*, R_0^*), we denote the plaintext difference as $(\Delta L_0, \Delta R_0)$, where $\Delta L_0 = L_0 \oplus L_0^*$, $\Delta R_0 = R_0 \oplus R_0^*$. $(\Delta L_r, \Delta R_r)$ denotes the output difference of the r-th round. ΔL_r and ΔR_r can be represented as 8 bytes, such as $\Delta L_r = (a, 0, 0, 0, 0, 0, 0, 0)$ and $\Delta R_r = (?, ?, ?, 0, ?, 0, 0, ?)$, where 0 denotes a zero byte difference, a denotes a nonzero byte difference and the question mark ? denotes an unknown byte difference(two bytes marked with ? may be different).

3.2 The 8-Round Impossible Differential of Camellia

In [5] Wu et al presented an impossible differential attack on 12-round Camellia-192/256, which was based on the following 8-round impossible differential.

$$(0, 0, 0, 0, 0, 0, 0, 0, a, 0, 0, 0, 0, 0, 0, 0) \nrightarrow (h, 0, 0, 0, 0, 0, 0, 0, 0, 0, 0, 0, 0, 0, 0, 0).$$

where a and h are arbitrary nonzero bytes. Refer to [5] for more details, and the 8-round impossible differential is also illustrated in Fig. 1. In this paper, we will exploit this 8-round impossible differential and improve the impossible differential cryptanalysis of Camellia up to 12-round Camellia-128 and 16-round Camellia-256.

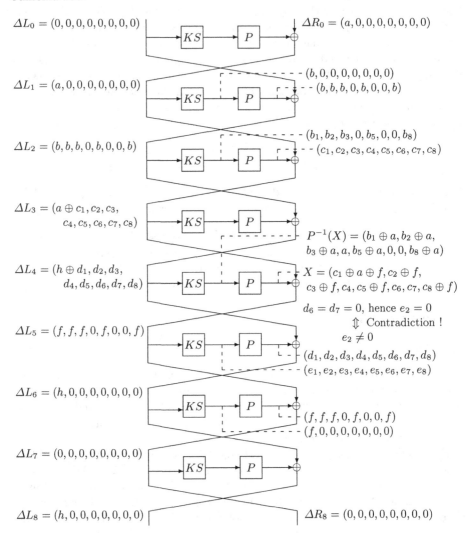

Fig. 1. 8-Round Impossible Differential of Camellia

3.3   Some Properties of Camellia

In this subsection, we exploit some properties of the key scheduling algorithm of Camellia-128 and Camellia-192/256, and present the following observations of the relations between round subkeys.

For Camellia-128, according to its key scheduling algorithm, we know that:

$$K_1 = (K_A \lll 0)_L, \quad K_{11} = (K_A \lll 60)_L,$$
$$K_2 = (K_A \lll 0)_R, \quad K_{12} = (K_A \lll 60)_R.$$

Therefore, the five bytes of K_1 and K_{12} at positions $(1, 2, 3, 5, 8)$ can be expressed as follows, respectively.

$$
\begin{aligned}
k_{1,1} &= K_A [1 \sim 8], & k_{12,1} &= K_A [125 \sim 128, 1 \sim 4], \\
k_{1,2} &= K_A [9 \sim 16], & k_{12,2} &= K_A [5 \sim 12], \\
k_{1,3} &= K_A [17 \sim 24], & k_{12,3} &= K_A [13 \sim 20], \\
k_{1,5} &= K_A [33 \sim 40], & k_{12,5} &= K_A [29 \sim 36], \\
k_{1,8} &= K_A [57 \sim 64], & k_{12,8} &= K_A [53 \sim 60].
\end{aligned}
$$

Furthermore, the first bytes of K_2 and K_{11} are expressed as follows, respectively.

$$k_{2,1} = K_A [65 \sim 72], \qquad k_{11,1} = K_A [61 \sim 68].$$

According to the above expressions, we can obtain the following property of Camellia-128:

Property 1. *For the round subkeys of Camellia-128:*

1) $(k_{1,1}, k_{1,2}, k_{1,3}, k_{1,5}, k_{1,8})$ and $(k_{12,1}, k_{12,2}, k_{12,3}, k_{12,5}, k_{12,8})$ have 28 common bits.
2) If $(k_{1,1}, k_{1,2}, k_{1,3}, k_{1,5}, k_{1,8})$ and $(k_{12,1}, k_{12,2}, k_{12,3}, k_{12,5}, k_{12,8})$ are known, there remains only 16 unknown bits of K_1, namely $(k_{1,4}[1 \sim 4], k_{1,6}, k_{1,7}[1 \sim 4])$.
3) If K_1 and $(k_{12,1}, k_{12,2}, k_{12,3}, k_{12,5}, k_{12,8})$ are known, the value of K_{12} is determined.
4) $k_{2,1}[1 \sim 4] = k_{11,1}[5 \sim 8], \quad k_{11,1}[1 \sim 4] = k_{1,8}[5 \sim 8].$

For Camellia-192/256, according to the key scheduling algorithm, we notice that the round subkeys of Rounds 1, 2, 15 and 16 are all determined by the intermediate variable K_B, and the expressions are as follows.

$$K_1 = (K_B \lll 0)_L, \quad K_{15} = (K_B \lll 60)_L,$$
$$K_2 = (K_B \lll 0)_R, \quad K_{16} = (K_B \lll 60)_R.$$

Similarly, we can obtain the following properties of Camellia-192/256.

Property 2. *For the round subkeys of Camellia-192/256:*

1) If the value of K_1 is known, then there remains only 4 unknown bits of K_{16}.
2) If the value of K_1 and K_{16} are known, there remains 60 unknown bits of K_2.
3) If the value of K_1 and K_2 are known, then the value of K_{15} is determined.

Furthermore, according to the key scheduling algorithm of Camellia-192/256, the round subkeys of Rounds 3, 4, 13 and 14 are all determined by the intermediate variable K_R, and the expressions are as follows.

$$K_3 = (K_R \lll 15)_L, \quad K_{13} = (K_R \lll 60)_L,$$
$$K_4 = (K_R \lll 15)_R, \quad K_{14} = (K_R \lll 60)_R.$$

Based on these expressions, we can get the following relations between subkey bytes.

$$k_{14,1} = k_{4,6}[6 \sim 8] \parallel k_{4,7}[1 \sim 5],$$
$$k_{14,2} = k_{4,7}[6 \sim 8] \parallel k_{4,8}[1 \sim 5],$$
$$k_{14,3} = k_{4,8}[6 \sim 8] \parallel k_{3,1}[1 \sim 5],$$
$$k_{14,5} = k_{3,2}[6 \sim 8] \parallel k_{3,3}[1 \sim 5],$$
$$k_{14,8} = k_{3,5}[6 \sim 8] \parallel k_{3,6}[1 \sim 5],$$
$$k_{13,1} = k_{3,6}[6 \sim 8] \parallel k_{3,7}[1 \sim 5].$$

Therefore, we can obtain another property of Camellia-192/256.

Property 3. *For the round subkeys of Camellia-192/256:*

1) $(k_{3,1}, k_{3,2}, k_{3,3}, k_{3,5}, k_{3,8})$ and $(k_{14,1}, k_{14,2}, k_{14,3}, k_{14,5}, k_{14,8})$ have 16 common bits.

2) If $(k_{3,1}, k_{3,2}, k_{3,3}, k_{3,5}, k_{3,8})$ and $(k_{14,1}, k_{14,2}, k_{14,3}, k_{14,5}, k_{14,8})$ are known, there remains only 19 unknown bits of K_3, namely $(k_{3,4}, k_{3,6}[6 \sim 8], k_{3,7})$.

3) If K_3 and $(k_{14,1}, k_{14,2}, k_{14,3}, k_{14,5}, k_{14,8})$ are known, the value of K_{14} is determined.

4) $k_{13,1} = k_{3,6}[6 \sim 8] \parallel k_{3,7}[1 \sim 5]$.

Finally, according to the analysis in [6], the linear diffusion function P of Camellia satisfies the following property.

Property 4. [6] *For $X, X^* \in (F_2^8)^8$, if there exists an h such that $P^{-1}(X \oplus X^* \oplus (h, 0, 0, 0, 0, 0, 0, 0))$ has the form of $(?, ?, ?, 0, ?, 0, 0, ?)$, then there is only one possible value of h.*

4   Impossible Differential Cryptanalysis of Reduced-Round Camellia

4.1   Impossible Differential Attack on 12-Round Camellia-128

We set the 8-round impossible differential at Rounds 3 to 10, and present an impossible differential attack on 12-round Camellia-128, which is illustrated in Fig. 2. The first step of the attack is data collection, and we only choose the pairs whose output differences of Round 2 satisfy the above impossible differential distinguisher. According to the round function of Camellia, we can know that the required plaintext difference must have the following form:

$$\Delta L_0 = (u, u, u, 0, u, 0, 0, u),$$
$$\Delta R_0 = P(?, ?, ?, 0, ?, 0, 0, ?) \oplus (?, 0, 0, 0, 0, 0, 0, 0).$$

Then by constructing appropriate plaintext structures, we can obtain plaintext pairs with the required difference, and this technique helps us reduce the data complexity.

The second step of the attack is data filtering. Based on certain property of ciphertext difference, we can filter out part of the wrong pairs and this may help

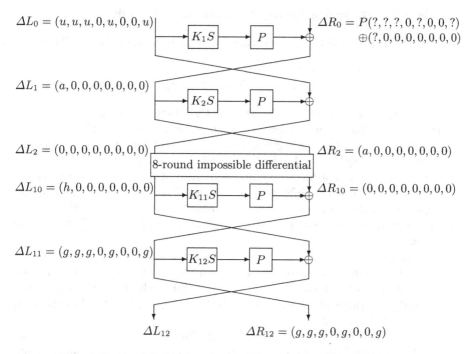

$\Delta L_0 = (u, u, u, 0, u, 0, 0, u)$
$\Delta R_0 = P(?, ?, ?, 0, ?, 0, 0, ?)$
$\oplus (?, 0, 0, 0, 0, 0, 0, 0)$

$\Delta L_1 = (a, 0, 0, 0, 0, 0, 0, 0)$

$\Delta L_2 = (0, 0, 0, 0, 0, 0, 0, 0)$
$\Delta R_2 = (a, 0, 0, 0, 0, 0, 0, 0)$

8-round impossible differential

$\Delta L_{10} = (h, 0, 0, 0, 0, 0, 0, 0)$
$\Delta R_{10} = (0, 0, 0, 0, 0, 0, 0, 0)$

$\Delta L_{11} = (g, g, g, 0, g, 0, 0, g)$

ΔL_{12}
$\Delta R_{12} = (g, g, g, 0, g, 0, 0, g)$

Fig. 2. Impossible Differential Attack on 12-Round Camellia-128

us reduce the time complexity of the following computation. Note that all the useful ciphertext pairs must satisfy the following condition:

$$\Delta L_{10} = (h, 0, 0, 0, 0, 0, 0, 0), \quad \Delta R_{10} = (0, 0, 0, 0, 0, 0, 0, 0).$$

where h denotes a nonzero byte, namely h has 255 possible values. Moreover, for every S-box of Camellia, when the input difference of S-box is nonzero, there are at most 2^7 possible output differences. Therefore, there are at most 255×2^7 possible output differences $(\Delta L_{11}, \Delta R_{11})$ after Round 11. Considering that there are 5 nonzero bytes of ΔL_{11}, namely bytes at positions (1,2,3,5,8), then there are at most $255 \times 2^7 \times (2^7)^5 \approx 2^{50}$ possible output differences $(\Delta L_{12}, \Delta R_{12})$ after Round 12. Therefore, in the data filtering step, the probability that a random pair remains after the test is about $2^{-78} = 2^{50} \times 2^{-128}$.

The third step of the attack is key recovery. According to Property 4, we can compute the output differences of S-boxes used in Round 1 and Round 12. Then by utilizing the difference distribution tables of S-boxes, we can recover 5 bytes $(k_{1,1}, k_{1,2}, k_{1,3}, k_{1,5}, k_{1,8})$ of K_1 and 5 bytes $(k_{12,1}, k_{12,2}, k_{12,3}, k_{12,5}, k_{12,8})$ of K_{12}. Furthermore, the corresponding pairs must be discarded if the relations between the subkeys contradict with Property 1-1. Lastly, we recover the correct key by discarding all the wrong subkeys using the impossible differential. In this step, we employ the divide-and-conquer technique when decrypting with the subkey guesses value and this also helps us reduce the time complexity of the attack.

In the following, we describe the attack procedure in detail.

1. Data Collection: Choose 2^m structures and each structure is as follows:

$$L_0 = (x, x, x, a_4, x, a_6, a_7, x),$$
$$R_0 = P(y_1, y_2, y_3, b_4, y_5, b_6, b_7, y_8) \oplus (y, c_2, c_3, c_4, c_5, c_6, c_7, c_8).$$

where (a_i, b_j, c_l) are fixed constants, and the 7 bytes $(x, y_1, y_2, y_3, y_5, y_8, y)$ take all possible values. Therefore, each structure contains 2^{56} plaintexts, which can generate about $2^{56} \times 2^{56}/2 = 2^{111}$ plaintext pairs. Hence 2^m structures can generate about 2^{111+m} plaintext pairs.

2. Data Filtering: According to the above analysis of the ciphertext differences, there are 2^{50} possible ciphertext differences. Therefore, after this test the expected number of remaining pairs is about $2^{111+m} \times 2^{50} \times 2^{-128} = 2^{33+m}$.

3. For each remaining pair $(L_0||R_0, L_{12}||R_{12})$ and $(L_0^*||R_0^*, L_{12}^*||R_{12}^*)$, do as follows:

   (a) Compute $P^{-1}(L_{12} \oplus L_{12}^* \oplus (h, 0, ..., 0))$ for all the 255 possible values of h. According to Property 4, we can obtain only one value of h such that it has the form $(?, ?, ?, 0, ?, 0, 0, ?)$. Similarly, we can compute the only one value of a such that $P^{-1}(R_0 \oplus R_0^* \oplus (a, 0, ..., 0))$ has the form $(?, ?, ?, 0, ?, 0, 0, ?)$.

   (b) Using the obtained input and output differences of the S-box in Round 1 and Round 12, together with the value of L_0 and R_{12}, we can calculate subkey bytes $(k_{1,1}, k_{1,2}, k_{1,3}, k_{1,5}, k_{1,8})$ and $(k_{12,1}, k_{12,2}, k_{12,3}, k_{12,5}, k_{12,8})$ by searching the difference distribution tables of S-boxes. Check if the deduced subkey bytes satisfy the 28-bit condition suggested by Property 1-1, and if this is not the case, discard the pair and return to Step 3 to try another pair. After this test, there remains about $2^{33+m} \times 2^{-28} = 2^{5+m}$ pairs.

   (c) For every guess of the 16 unknown bits $(k_{1,4}[1 \sim 4], k_{1,6}, k_{1,7}[1 \sim 4])$, do as follows. Note that according to Property 1-3, we can determine the value of K_{12} now.

       i. For every remaining pairs, encrypt the first round to get (L_1, L_1^*) using K_1, and decrypt the last round to get (R_{11}, R_{11}^*) using K_{12}.

       ii. Utilizing the difference distribution tables of S-boxes, we can calculate the value of $k_{2,1}$ using (L_1, L_1^*), a and $P^{-1}(L_0 \oplus L_0^*)$; Similarly we can calculate the value of $k_{11,1}$ using (R_{11}, R_{11}^*), h and $P^{-1}(R_{12} \oplus R_{12}^*)$.

       iii. Check if the subkey bytes satisfy the following equation suggested by Property 1-4.

$$k_{11,1} = k_{1,8}[5 \sim 8]||k_{2,1}[1 \sim 4].$$

       If there exists a plaintext pair that passes this test, then discard the 76-bit subkey guess value $(k_{1,1}, k_{1,2}, k_{1,3}, k_{1,5}, k_{1,8}, k_{12,1}, k_{12,2}, k_{12,3}, k_{12,5}, k_{12,8}, k_{1,4}[1 \sim 4], k_{1,6}, k_{1,7}[1 \sim 4], k_{2,1}, k_{11,1})$, as this is an impossible differential and the subkey guess satisfying it must be wrong.

Furthermore, the probability that a subkey guess may remain after this test is about $1 - 2^{-8}$. We choose $m = 9$, hence the number of remaining wrong subkey is about $2^{76}(1 - 2^{-8})^{2^{5+m}} \approx 2^{76} \times e^{-2^6} < 1$.

The data complexity of the attack is about $2^{56} \times 2^9 = 2^{65}$ CP, and the time complexity of the attack is estimated as follows. The time complexity of Step 1 is 2^{65} encryptions, and the memory spaces needed to store the plaintexts and ciphertexts are about 2^{66} blocks, where one block means 128 bits. In Step 2, choosing the qualified pairs requires about $2^{65} \times 2^{50} = 2^{115}$ MA (Memory Access), and the memory spaces needed to store the possible ciphertext differences are about 2^{50} blocks. In Step 3, the time complexity of Step (a) is about $2^{42} \times 2/12 < 2^{40}$ encryptions; the time complexity of Step (b) is less than 2^{39} encryptions, since the calculation of key using difference distribution table of S-box is only about one F computation; and the time complexity of Step (c) is about $2^{14} \times 2^{16} \times 2/12 < 2^{28}$ encryptions.

As a rule, one MA is equivalent to about one-round encryption of Camellia. Therefore, the total data complexity of the attack is 2^{65} CP, and the time complexity of the attack is less than $2^{111.5}$ encryptions, and the memory complexity of the attack is about 2^{66} blocks.

4.2   Impossible Differential Attack on 16-Round Camellia-256

We set the 8-round impossible differential at Rounds 5 to 12, and present an impossible differential attack on 16-round Camellia-256, which is illustrated in Fig. 3. The first step of the attack is data collection, and we also exploit the plaintext structure to reduce data complexity.

The second step of the attack is data filtering. In this step we try to filter out part of the wrong pairs whose plaintext and ciphertext differences can not satisfy the impossible differential, so as to reduce the computation workload for later analysis. According to the 8-round impossible differential, the output differences of a useful pair after Round 4 and Round 12 must be as follows, respectively.

$$\Delta L_4 = (0,0,0,0,0,0,0,0), \quad \Delta R_4 = (a,0,0,0,0,0,0,0),$$

$$\Delta L_{12} = (h,0,0,0,0,0,0,0), \quad \Delta R_{12} = (0,0,0,0,0,0,0,0).$$

where a and h are nonzero bytes. Therefore, for a useful pair, the left half of output difference after Round 1 must have the form $P(?,?,?,0,?,0,0,?) \oplus (?,0,0,0,0,0,0,0)$, and the left half of the output difference after Round 2 must have the form $(u,u,u,0,u,0,0,u)$. Similarly, the right half of the input difference before Round 16 must have the form $P(?,?,?,0,?,0,0,?) \oplus (?,0,0,0,0,0,0,0)$, and the right half of the input difference before Round 15 must have the form $(u,u,u,0,u,0,0,u)$. Hereafter, we denote the set of differences with the form $P(?,?,?,0,?,0,0,?) \oplus (?,0,0,0,0,0,0,0)$ as Π_1, and the set of differences with the form $(u,u,u,0,u,0,0,u)$ as Π_2. Obviously, there are 2^{48} elements in the set Π_1 and 2^8 elements in the set Π_2, namely $\#\{\Pi_1\} = 2^{48}$ and $\#\{\Pi_2\} = 2^8$. Therefore, the probability that a random plaintext pair is a useful pair for our analysis is about $2^{-144} = \frac{2^{48}}{2^{64}} \times \frac{2^{48}}{2^{64}} \times \frac{2^8}{2^{64}} \times \frac{2^8}{2^{64}}$.

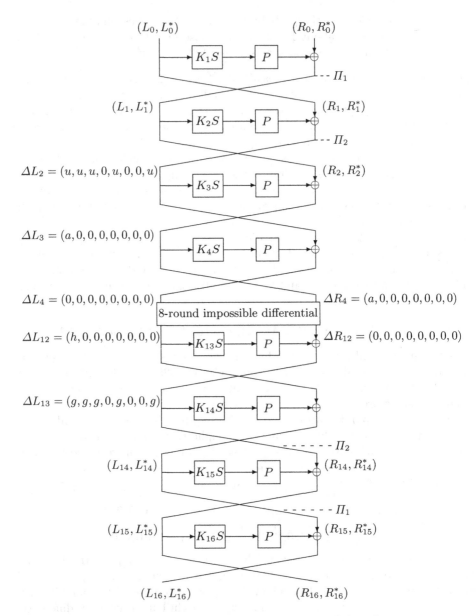

Fig. 3. Impossible Differential Attack on 16-Round Camellia-256

The third step of the attack is subkey guessing and sieving, and the divide-and-conquer technique is also used to reduce the time complexity. First of all, we need to guess part of the round subkeys to encrypt and decrypt the first and last two rounds, respectively. Then based on Property 4, we can compute the output differences of S-boxes used in Round 3 and Round 14. Utilizing the difference distribution tables of the S-boxes, we can calculate 5 bytes $(k_{3,1}, k_{3,2}, k_{3,3}, k_{3,5},$ $k_{3,8})$ of K_3 and 5 bytes $(k_{14,1}, k_{14,2}, k_{14,3}, k_{14,5}, k_{14,8})$ of K_{14}, respectively. Then we use Property 2 and Property 3 to filter out the wrong pairs. Lastly, using the remaining pairs we can discard all the wrong subkey guesses based on the impossible differential, and thereby recover the correct key.

In the following, we describe the attack procedure in detail.

1. Data Collection: Choose 2^{89} plaintexts as follows:

$$L_0 = (x_1, ..., x_8), \qquad R_0 = (y_1, ..., y_8)$$

where $x_i (1 \le i \le 8)$ and y_1 all take 64 arbitrary values chosen from F_2^8, and $y_j (1 < j \le 8)$ all take 32 arbitrary values chosen from F_2^8. This way we can get about $2^{89} \cdot 2^{89}/2 \approx 2^{177}$ plaintext pairs, and these pairs need to be stored for later analysis which requires about 2^{178} blocks memory.

2. For every guess of K_1, do the followings:
   (a) Encrypt the first round for each of the 2^{89} plaintexts, and check if the output difference $F(L_0, K_1) \oplus F(L_0^*, K_1) \oplus R_0 \oplus R_0^*$ satisfies the form of $P(?,?,?,0,?,0,0,?) \oplus (?,0,0,0,0,0,0,0)$. If this is not the case, discard the corresponding plaintext pair. After this test, there remains about $2^{177} \times \frac{2^{48}}{2^{64}} = 2^{161}$ pairs.
   (b) According to Property 2-1, K_{16} has 60 common bits with K_1, and thus there are only 2^4 possible values of K_{16}. For every possible value of K_{16}, decrypt the last round for each of the 2^{89} plaintexts. Check if the input difference $F(R_{16}, K_{16}) \oplus F(R_{16}^*, K_{16}) \oplus L_{16} \oplus L_{16}^*$ of Round 16 satisfies the form of $P(?,?,?,0,?,0,0,?) \oplus (?,0,0,0,0,0,0,0)$. If this is not the case, discard the corresponding pair. After this test, there remains about $2^{161} \times \frac{2^{48}}{2^{64}} = 2^{145}$ pairs.
   (c) For every guess of K_2, encrypt the second round for the 2^{89} plaintexts. Note that there are only 60 unknown bits of K_2 after guessing the values of K_1 and K_{16} according to Property 2-2. Check if $F(L_1, K_2) \oplus F(L_1^*, K_2) \oplus R_1 \oplus R_1^*$ has the form of $(u, u, u, 0, u, 0, 0, u)$. If this is not the case, discard the pair. After this test there remains about $2^{145} \times \frac{2^8}{2^{64}} = 2^{89}$ pairs.
   (d) Decrypt Round 15 for each of the 2^{89} pairs using K_{15} which can be deduced by K_1 and K_2. Check if the difference $F(R_{15}, K_{15}) \oplus F(R_{15}^*, K_{15}) \oplus L_{15} \oplus L_{15}^*$ satisfies the form $(u, u, u, 0, u, 0, 0, u)$. Discard the unsatisfied pairs, and after this step there remains about $2^{89} \times \frac{2^8}{2^{64}} = 2^{33}$ pairs.
3. For each of the 2^{128} possible candidates of $(K_1, K_2, K_{15}, K_{16})$, and for each of the 2^{33} remained pairs $(L_0 || R_0, L_{16} || R_{16})$ and $(L_0^* || R_0^*, L_{16}^* || R_{16}^*)$, do as follows:
   (a) Encrypt the first two rounds and decrypt the last two rounds to get $(L_2 || R_2, L_{14} || R_{14})$ and $(L_2^* || R_2^*, L_{14}^* || R_{14}^*)$.
   (b) Compute $P^{-1}(L_{14} \oplus L_{14}^* \oplus (h, 0, ..., 0))$ for all the 255 possible values of h. According to Property 4, we can obtain only one value of h such that it has the form $(?,?,?,0,?,0,0,?)$. Similarly, we can compute the

only one value of a such that $P^{-1}(R_2 \oplus R_2^* \oplus (a, 0, ..., 0))$ has the form $(?, ?, ?, 0, ?, 0, 0, ?)$.

(c) Using the obtained input and output differences of the S-box in Round 2 and Round 14, together with the value of L_2 and R_{14}, we can calculate subkey bytes $(k_{3,1}, k_{3,2}, k_{3,3}, k_{3,5}, k_{3,8})$ and $(k_{14,1}, k_{14,2}, k_{14,3}, k_{14,5}, k_{14,8})$ by searching the difference distribution tables of S-boxes. Check if the deduced subkey bytes satisfy the 16-bit condition suggested by Property 3-1, and if this is not the case, discard the pair and return to Step 3 to try another pair. After this test, there remains about $2^{33} \times 2^{-16} = 2^{17}$ pairs.

(d) For each possible value of the 19 unknown bits $(k_{3,4}, k_{3,6}[6 \sim 8], k_{3,7})$ of K_3, do as follows. Note that according to Property 3-3, we can know the value of K_{14} now.

 i. For every remaining pairs, encrypt Round 3 using K_3 and decrypt Round 14 using K_{14} to get (L_3, L_{12}) and (L_3^*, L_{12}^*).

 ii. Utilizing the difference distribution tables of S-boxes, calculate the value of $k_{4,1}$ using (L_3, L_3^*), a and $P^{-1}(L_2 \oplus L_2^*)$; calculate the value of $k_{13,1}$ using (L_{12}, L_{12}^*), h and $P^{-1}(L_{13} \oplus L_{13}^*)$.

 iii. Check if the subkey bytes satisfy the following equation suggested by Property 3-4.

$$k_{13,1} = k_{3,6}[6 \sim 8] \| k_{3,7}[1 \sim 5].$$

If there exists a plaintext pair that passes this test, then discard the $221 (= 128 + 64 + 19)$ bits subkey guess(K_B, K_3, K_{14}), as this is an impossible differential and the subkey guess satisfies it must be wrong. Furthermore, the probability that a subkey guess may remain is about $1 - 2^{-8}$, and the number of remaining wrong subkey is about $2^{221}(1 - 2^{-8})^{2^{17}} \approx 2^{221} \times e^{-2^9} < 1$.

The data complexity of the attack is about 2^{89} CP, and the time complexity of the attack can be estimated as follows. Step 2(a) has a time complexity of about $2^{64} \times 2^{89} \times 2^{-4} = 2^{149}$, and needs about $2^{48} \times 64$ bits memory spaces to store the elements in the set Π_1, and then requires about $2^{64} \times 2^{89} \times 2^{48} = 2^{201} MA$(Memory Access) for testing the qualified pairs. Step 2(b) has a time complexity of about $2^{64} \times 2^4 \times 2^{89} \times 2^{-4} = 2^{153}$, and requires about $2^{64} \times 2^4 \times 2^{89} \times 2^{48} = 2^{205} MA$(Memory Access) for testing the qualified pairs. Step 2(c) has a time complexity of about $2^{64} \times 2^{64} \times 2^{89} \times 2^{-4} = 2^{213}$, and needs about $2^{64} \times 2^{64} \times 2^{89} \times 2^8 = 2^{225} MA$ for testing the qualified pairs. Step 2(d) has a time complexity of about $2^{64} \times 2^{64} \times 2^{89} \times 2^{-4} = 2^{213}$ encryptions and $2^{225} MA$. In Step 3, the time complexity of Step 3(a) is about $2^{128} \times 2^{33} \times 2/4 = 2^{160}$; the time complexity of Step 3(b) is about $2^{128} \times 2^{33} \times 2/16 = 2^{158}$; the time complexity of Step 3(c) is less than 2^{158} encryptions, since the time to calculate subkey using difference distribution table of S-box is only about one F computation; and the time complexity of Step 3(d) is about $2^{128} \times 2^{19} \times 2^{17} \times 2/16 < 2^{151}$ encryptions.

Therefore, the total data complexity of the attack is 2^{89} CP, and the time complexity of the attack is less than $2^{222.1}$ encryptions, and the memory complexity of the attack is about 2^{178} blocks.

5 Conclusion

In [5] Wu et al constructed some 8-round impossible differentials of Camellia, and based on it they successfully attacked Camellia reduced up to 12 rounds using impossible differential cryptanalysis. Then in [6] Lu et al observed some new properties of the linear diffusion function P, and by using the same 8-round impossible differential they improved the impossible differential cryptanalysis of Camellia. However, all of these impossible differential attacks on Camellia have not taken the key scheduling algorithm into account. In this paper, we present some observations of the relations between round subkeys of Camellia, and by taking advantage of these relations and some novel techniques (such as differential cryptanalysis, divide-and-conquer etc.), we improve the impossible differential attack on Camellia up to 12-round Camellia-128 and 16-round Camellia-256. These results are better than any previously published cryptanalytic results on Camellia without the FL/FL^{-1} functions and whitening layers. Note that our method used in this paper does not apply to Camellia-192 effectively, since the relations between round subkeys of Camellia-192 are difficult to exploit in the attack process.

Acknowledgments. This work is supported by the National High-Tech Research and Development Program of China (No.2007AA01Z470), the National Natural Science Foundation of China (No.90604036), and the National Grand Fundamental Research 973 Program of China (No.2004CB318004). Moreover, the authors are very grateful to the anonymous reviewers for their comments and editorial suggestions.

References

1. Aoki, K., Ichikawa, T., Kanda, M., et al.: Specification of Camellia–a 128-bit Block Cipher. In: Selected Areas in Cryptography-SAC 2000. LNCS, vol. 2012, pp. 183–191. Springer, Heidelberg (2001)
2. Lee, S., Hong, S.H., Lee, S.-J., Lim, J.-I., Yoon, S.H.: Truncated differential cryptanalysis of camellia. In: Kim, K.-c. (ed.) ICISC 2001. LNCS, vol. 2288, pp. 32–38. Springer, Heidelberg (2002)
3. Sugita, M., Kobara, K., Imai, H.: Security of reduced version of the block cipher camellia against truncated and impossible differential cryptanalysis. In: Boyd, C. (ed.) ASIACRYPT 2001. LNCS, vol. 2248, pp. 193–207. Springer, Heidelberg (2001)
4. Hatano, Y., Sekine, H., Kaneko, T.: Higher order differential attack of (II). In: Nyberg, K., Heys, H.M. (eds.) SAC 2002. LNCS, vol. 2595, pp. 129–146. Springer, Heidelberg (2003)
5. Wu, W., Zhang, W., Feng, D.: Impossible Differential Cryptanalysis of Reduced-Round ARIA and Camellia. Journal of Computer Science and Technology 22(3), 449–456 (2007)
6. Lu, J., Kim, J.-S., Keller, N., Dunkelman, O.: Improving the efficiency of impossible differential cryptanalysis of reduced camellia and MISTY1. In: Malkin, T.G. (ed.) CT-RSA 2008. LNCS, vol. 4964, pp. 370–386. Springer, Heidelberg (2008)

7. He, Y., Qing, S.: Square attack on reduced camellia cipher. In: Qing, S., Okamoto, T., Zhou, J. (eds.) ICICS 2001. LNCS, vol. 2229, pp. 238–245. Springer, Heidelberg (2001)
8. Yeom, Y., Park, S., Kim, I.: On the Security of Camellia against the Square Attack. In: FSE 2002. LNCS, vol. 2356, pp. 89–99. Springer, Heidelberg (2002)
9. Yeom, Y., Park, S., Kim, I.: A Study of Integral Type Cryptanalysis on Camellia. In: The 2003 Symposium on Cryptography and Information Security-SCIS 2003, Hamamatsu, Japan, pp. 26–29 (2003)
10. Lei, D., Chao, L., Feng, K.: New observation on camellia. In: Preneel, B., Tavares, S. (eds.) SAC 2005. LNCS, vol. 3897, pp. 51–64. Springer, Heidelberg (2006)
11. Duo, L., Li, C., Feng, K.: Square like attack on camellia. In: Qing, S., Imai, H., Wang, G. (eds.) ICICS 2007. LNCS, vol. 4861, pp. 269–283. Springer, Heidelberg (2007)
12. Wenling, W., Dengguo, F., Hua, C.: Collision attack and pseudorandomness of reduced-round camellia. In: Handschuh, H., Hasan, M.A. (eds.) SAC 2004. LNCS, vol. 3357, pp. 252–266. Springer, Heidelberg (2004)
13. Jie, G., Zhongya, Z.: Improved collision attack on reduced round camellia. In: Pointcheval, D., Mu, Y., Chen, K. (eds.) CANS 2006. LNCS, vol. 4301, pp. 182–190. Springer, Heidelberg (2006)
14. Shirai, T.: Differential, Linear, Boomerang and Rectangle Cryptanalysis of Reduced-Round Camellia. In: Proceedings of the Third NESSIE Workshop, Munich, Germany, November 6-7 (2002), https://www.cosic.esat.kuleuven.be/nessie/
15. Wu, W., Feng, D.: Differential-Linear Cryptanalysis of Camellia. In: Progress on Cryptography, pp. 173–180. Kluwer Academic Publishers, Dordrecht (2004)
16. Biham, E., Biryukov, A., Shamir, A.: Cryptanalysis of Skipjack Reduced to 31 Rounds Using Impossible Differentials. In: EUROCRYPT 1999. LNCS, vol. 2595, pp. 12–23. Springer, Heidelberg (1999)
17. Cheon, J.H., Kim, M., Kim, K., Lee, J.-Y., Kang, S.: Improved impossible differential cryptanalysis of rijndael and crypton. In: Kim, K.-c. (ed.) ICISC 2001. LNCS, vol. 2288, pp. 39–49. Springer, Heidelberg (2002)
18. Phan, R.C.-W.: Impossible Differential Cryptanalysis of 7-round AES. Information Processing Letters 91(1), 33–38 (2004)
19. Jakimoski, G., Desmedt, Y.: Related-Key Differential Cryptanalysis of 192-bit key AES Variants. In: Matsui, M., Zuccherato, R.J. (eds.) SAC 2003. LNCS, vol. 3006, pp. 208–221. Springer, Heidelberg (2004)
20. Zhang, W., Wu, W., Zhang, L., Feng, D.: Improved related-key impossible differential attacks on reduced-round AES-192. In: Biham, E., Youssef, A.M. (eds.) SAC 2006. LNCS, vol. 4356, pp. 15–27. Springer, Heidelberg (2007)
21. Biham, E., Dunkelman, O., Keller, N.: Related-key impossible differential attacks on 8-round AES-192. In: Pointcheval, D. (ed.) CT-RSA 2006. LNCS, vol. 3860, pp. 21–33. Springer, Heidelberg (2006)
22. Zhang, W., Wu, W., Feng, D.: New results on impossible differential cryptanalysis of reduced AES. In: Nam, K.-H., Rhee, G. (eds.) ICISC 2007. LNCS, vol. 4817, pp. 239–250. Springer, Heidelberg (2007)
23. Tsunoo, Y., Tsujihara, E., Saito, T., Suzaki, T., Kubo, H.: Impossible Differential Cryptanalysis of CLEFIA. In: Fast Software Encryption-FSE 2008. Springer, Heidelberg (2008)

Author Index